Introduction to Pharmaceutical Analytical Chemistry

Introduction to Pharmaceutical Analytical Chemistry

STIG PEDERSEN-BJERGAARD

Department of Pharmacy, University of Oslo, Norway and Department of Pharmacy, University of Copenhagen, Denmark

BENTE GAMMELGAARD

Department of Pharmacy, University of Copenhagen, Denmark

TRINE GRØNHAUG HALVORSEN

Department of Pharmacy, University of Oslo, Norway

Second Edition

WILEY

This edition first published 2019
© 2019 John Wiley & Sons Ltd

Edition history:
"John Wiley & Sons Ltd. (1e, 2012)".

The right of Stig Pedersen-Bjergaard, Bente Gammelgaard and Trine Grønhaug Halvorsen to be identified as the authors of the editorial material in this work has been asserted in accordance with law.

Registered Offices
John Wiley & Sons, Inc., 111 River Street, Hoboken, NJ 07030, USA
John Wiley & Sons Ltd, The Atrium, Southern Gate, Chichester, West Sussex, PO19 8SQ, UK

Editorial Office
The Atrium, Southern Gate, Chichester, West Sussex, PO19 8SQ, UK

For details of our global editorial offices, customer services, and more information about Wiley products visit us at www.wiley.com.

Wiley also publishes its books in a variety of electronic formats and by print-on-demand. Some content that appears in standard print versions of this book may not be available in other formats.

Limit of Liability/Disclaimer of Warranty
In view of ongoing research, equipment modifications, changes in governmental regulations and the constant flow of information relating to the use of experimental reagents, equipment and devices, the reader is urged to review and evaluate the information provided in the package insert or instructions for each chemical, piece of equipment, reagent or device for, among other things, any changes in the instructions or indication of usage and for added warnings and precautions. While the publisher and authors have used their best efforts in preparing this work, they make no representations or warranties with respect to the accuracy or completeness of the contents of this work and specifically disclaim all warranties, including without limitation any implied warranties of merchantability or fitness for a particular purpose. No warranty may be created or extended by sales representatives, written sales materials or promotional statements for this work. The fact that an organization, website or product is referred to in this work as a citation and/or potential source of further information does not mean that the publisher and authors endorse the information or services the organization, website or product may provide or recommendations it may make. This work is sold with the understanding that the publisher is not engaged in rendering professional services. The advice and strategies contained herein may not be suitable for your situation. You should consult with a specialist where appropriate. Further, readers should be aware that websites listed in this work may have changed or disappeared between when this work was written and when it is read. Neither the publisher nor authors shall be liable for any loss of profit or any other commercial damages, including, but not limited to, special, incidental, consequential or other damages.

Library of Congress Cataloging-in-Publication Data

Names: Pedersen-Bjergaard, Stig, author. | Gammelgaard, Bente, author. |
    Halvorsen, Trine G. (Trine Grønhaug), 1975- author.
Title: Introduction to pharmaceutical analytical chemistry / Stig
    Pedersen-Bjergaard, Department of Pharmacy, University of Oslo, Norway and
    Department of Pharmacy, University of Copenhagen, Denmark, Bente
    Gammelgaard, Department of Pharmacy, University of Copenhagen,
    Denmark,Trine Grønhaug Halvorsen, Department of Pharmacy, University of Oslo,
    Norway.
Other titles: Introduction to pharmaceutical chemical analysis.
Description: Second edition. | Hoboken, NJ : Wiley, 2019. | Revision of:
    Introduction to pharmaceutical chemical analysis / Steen Hansen, Stig
    Pedersen-Bjergaard, Knut Rasmussen. 2012. | Includes bibliographical
    references and index. |
Identifiers: LCCN 2018051420 (print) | LCCN 2018053744 (ebook) | ISBN
    9781119362739 (Adobe PDF) | ISBN 9781119362753 (ePub) | ISBN 9781119362722
    (paperback)
Subjects: LCSH: Drugs–Analysis. | Pharmaceutical chemistry. | BISAC: SCIENCE
    / Chemistry / Analytic.
Classification: LCC RS189 (ebook) | LCC RS189 .H277 2019 (print) | DDC
    615.1/9–dc23
LC record available at https://lccn.loc.gov/2018051420

Cover design: Wiley
Cover Images: © Background © REB Images/Getty Images,
Testing image © TEK IMAGE/SCIENCE PHOTO LIBRARY/Getty Images,
Research image © TEK IMAGE/SCIENCE PHOTO LIBRARY/Getty Images,
Formula © ALFRED PASIEKA/SCIENCE PHOTO LIBRARY/Getty Images

Set in size of 10/12pt and TimesLTStd by SPi Global, Chennai, India

10 9 8 7 6 5 4 3 2 1

Contents

Preface to the Second Edition

This textbook is an extensive revision of '*Introduction to Pharmaceutical Analysis*' from 2012. We have revised the manuscript totally, and updated the content according to current practice in pharmaceutical analytical chemistry, and according to current versions of European and United States Pharmacopeia. Additionally, we have added a new chapter on chemical analysis of biopharmaceuticals, improved the illustrations throughout and provided illustrations in colour. The intention of these efforts has been to provide the reader with a textbook at the level expected in 2018.

We have changed the title to emphasize that this textbook is about analytical chemistry, and that the applications described are all related to pharmaceuticals. However, the philosophy is the same as with the first edition. The textbook is primarily for pharmacy and chemistry students (and other scientists approaching the pharmaceutical sciences) at university level, requesting basic knowledge on chemical analysis of pharmaceutical ingredients and preparations, and chemical analysis of drug substances in biological fluids. In the first part of the textbook, we teach the fundamentals of the main analytical techniques. Compared to textbooks in pure analytical chemistry, we go into less detail but we still teach to a level where the reader can understand the details in current pharmacopeia and bioanalytical methods. The second part of the textbook is unique, as we focus on identification, purity testing and assay of pharmaceutical ingredients, identification and quantitation of active ingredients in pharmaceutical preparations, and identification and quantitation of drugs in biological fluids. Such systematic discussion of pharmaceutical applications is not found in any other textbook on the market.

Originally, this textbook was written in Norwegian by Stig Pedersen-Bjergaard and Knut Rasmussen. The first Norwegian edition came in 2004 (ISBN 82-7674-844-9), and this was revised in 2010 (ISBN 978-82-450-1013-8). The manuscript was translated to English and improved by Stig Pedersen-Bjergaard, Knut Rasmussen, and Steen Honoré Hansen in 2012. Since that first English edition, the author team has changed. Knut Rasmussen has retired and Steen Honoré Hansen passed away in the autumn 2017. Knut is acknowledged for his pioneering work in the period 2004–2012, and for his highly valuable advice during preparation of the current edition. Steen is acknowledged for his work on translation and improvements in preparation of the first English edition of the book, and the valuable discussions of the content and improvements of the present edition until autumn 2017. We also thank our colleagues and students at the University of Oslo and the University of Copenhagen for inspiration, discussions, advice, proof reading, chromatograms, titration curves, and fun.

Oslo/Copenhagen, June 2018

Stig Pedersen-Bjergaard
University of Oslo
University of Copenhagen

Bente Gammelgaard
University of Copenhagen

Trine Grønhaug Halvorsen
University of Oslo

Abbreviations

AAS	Atomic absorption spectrometry
ADME	Absorption, distribution, metabolism, excretion
AES	Atomic emission spectrometry
AFID	Alkali flame ionization detector (GC)
APCI	Atmospheric pressure chemical ionization
API	Active pharmaceutical ingredient
ATR	Attenuated total reflectance (IR spectrophotometry)
BP	British pharmacopoeia
CAD	Charge aerosol detector
CE	Capillary electrophoresis
CI	Chemical ionization
CRS	Chemical reference substance
CZE	Capillary zone electrophoresis
DAD	Diode array detector
DNA	Deoxyribonucleic acid
DL	Detection limit
ECD	Electron capture detector
EDL	Electrode-less discharge lamp
EDTA	Ethylene diamine tetra-acetic acid
EI	Electron ionization
EIC	Extracted ion chromatogram
ELSD	Evaporative light scattering detector
EMA	European Medicines Agency
EOF	Electro-osmotic flow
ESI	Electrospray ionization
FDA	Food and Drug Administration (USA)
FID	Flame ionization detector
FTIR	Fourier transform infrared (spectrophotometry)
GC	Gas chromatography
GLP	Good laboratory practice
GMP	Good manufacturing practice
GPC	Gel permeation chromatography
HETP	Height equivalent to theoretical plate
HIC	Hydrophobic interaction chromatography
HILIC	Hydrophilic interaction chromatography

HPLC	High performance liquid chromatography
HR	High resolution (MS)
HVPE	High voltage paper electrophoresis
ICH	International Council for Harmonization (for Technical Requirements for Pharmaceuticals for Human Use)
IEC	Ion exchange chromatography
IEF	Isoelectric focusing
IS	Internal standard
ICP	Inductively coupled plasma (spectrometry)
IR	Infrared
JP	Japanese pharmacopoeia
LC	Liquid chromatography
LLOD	Lower limit of detection (= DL)
LLOQ	Lower limit of quantification (= QL)
LOD	Limit of detection (= DL)
LOQ	Limit of quantitation (= QL)
LLE	Liquid–liquid extraction
MA	Marketing application
MALDI	Matrix assisted laser desorption ionization (MS)
MS	Mass spectrometry
MRM	Multiple reaction monitoring
NDA	New drug application
NIR	Near infrared
NP	Normal phase
NPD	Nitrogen–phosphorous detector (GC)
ODS	Octadecylsilane
OMCL	Official Medicines Control Laboratories
OTC	Over the counter (drugs)
PAT	Process analytical technology
PCHV	Paper chromatography high voltage
PCR	Polymerase chain reaction
PDE	Permitted daily exposure
PE	Polyethylene
PEG	Polyethylene glycol
Ph. Eur.	European Pharmacopoeia
PVC	Polyvinyl chloride
PVDF	Polyvinylidine difluoride
POM	Prescription only medicine (drugs)
PP	Polypropylene
PP	Protein precipitation
PS-DVB	Polystyrene–divinyl benzene
rDNA	Recombinant DNA
R	Reagent (Ph. Eur.)
RI	Refractive index (detector)
RNA	Ribonucleic acid
RP	Reversed phase

QP	Qualified person
RV	Substance used as primary standard (suffix) in volumetric analysis
RSD	Relative standard deviation
SAX	Strong anion exchanger
SCX	Strong cation exchanger
SDS-PAGE	Sodium dodecyl sulfate–polyacrylamide gel electrophoresis
SEC	Size exclusion chromatography
SFC	Supercritical fluid chromatography
SD	Standard deviation
SIM	Selected ion monitoring
SLE	Solid–liquid extraction
SOP	Standard operating procedure
SPE	Solid phase extraction
SRM	Selected reaction monitoring
TDM	Therapeutic drug monitoring
TIC	Total ion chromatogram
TID	Thermionic detector
TLC	Thin-layer chromatography
TOF	Time-of-flight
UHPLC	Ultra high performance liquid chromatography
USP	United States Pharmacopoeia
UV-Vis	Ultraviolet–visible (spectrophotometry)
WADA	World Anti-Doping Agency
WAX	Weak anion exchanger
WCX	Weak cation exchanger

Symbols and Units

The units in the book do not strictly follow the SI units. The units are adjusted to the dimensions in analytical work.

Symbol		Unit
A	Absorbance	—
$A(1\%, 1\ \text{cm})$	Specific absorbance	—
A_S	Symmetry factor	—
a	Activity	—
α	Relative retention (separation factor)	—
$[\alpha]_D^{20}$	Specific optical rotation	(°) degrees
c	Concentration	g/L, mol/L
d	Dextrorotary (optical rotation)	° (degrees)
D	Distribution ratio (also named distribution coefficient or partition coefficient)	—
E	Potential	V
E	Electrical field (CE)	V/cm
E^0	Standard electrode potential (standard reduction potential)	V
ε	Molar absorption coefficient	$\text{cm}^{-1} \cdot \text{mol}^{-1} \cdot \text{L}$
ε^0	Relative elution strength	—
η	Viscosity	cPoise
F	Flow rate (chromatography)	mL/min
F	Fluorescence	—
h	Peak height	mm
H	Height equivalent to theoretical plate	
I	Intensity	—
I_A	Acid value	mg
I_I	Iodine value	mg
I_{OH}	Hydroxyl value	mg
I_S	Saponification value	mg

Symbol		Unit
K_a	Acid ionization constant (= acid dissociation constant, acidity constant)	M
K_b	Base ionization constant (= basicity constant)	M
K_D	Partition ratio (= distribution constant)	
K_w	Autoprotolysis equilibrium constant of water (= ion product of water)	M^2
k	Retention factor	—
λ	Wavelength	nm
L	Length	m (mm)
l	Levorotary (optical rotation)	° (degrees)
μ_{app}	Apparent mobility	$cm^2 \cdot min^{-1} \cdot V^{-1}$
μ_e	Electrophoretic mobility	$cm^2 \cdot min^{-1} \cdot V^{-1}$
μ_{eo}	Electroosmotic mobility	$cm^2 \cdot min^{-1} \cdot V^{-1}$
M	Molarity	mol/L^{-1}
M	Molar mass	g/mol
M	Molecular mass	u = Da
M_r	Relative molar mass	—
N	Number of theoretical plates	—
v	Frequency	Hz (s^{-1})
pI	Isoelectric point	—
P	Distribution ratio between 1-octanol and aqueous solution, pH 7.4	—
P'	Polarity index	—
ϕ	Quantum yield (fluorescence)	—
r	Radius	m (mm)
R_f	Retention factor (TLC)	—
R_S	Resolution (chromatography)	—
ρ	Density	g/cm^3
σ	Standard deviation	—
s	Standard deviation	
T	Temperature	K, C
T	Transmittance	—
t_R	Retention time	min
t_M	Hold-up time	min
t'_R	Adjusted retention time	min
u	Linear velocity (flow rate)	cm/s
v	Velocity	m/s
V	Volume	L, $1\,mL = 1\,cm^3$
V_M	Hold-up volume (void volume) (LC); total permeation volume (SEC)	mL

Symbol		Unit
V_O	Exclusion volume	mL
V_R	Retention volume	mL
W	Peak width	min
W_h	Peak width at half height	min
$\bar{x}$	Mean	—
z	Charge	—

Constants

Avogadro's number	N	$6.0221 \times 10^{23} \text{ mol}^{-1}$
Faraday's constant	F	$9.649 \times 10^4 \cdot \text{C(oulomb)} \cdot \text{mol}^{-1} = 96.485 \text{ kJ} \cdot \text{mol}^{-1}$
Gas constant	R	$8.314 \text{ J} \cdot \text{K}^{-1} \cdot \text{mol}^{-1}$
Speed of light in vacuum	c	$2.998 \times 10^8 \text{ m/s}$
Planck's constant	h	$6.626 \times 10^{-34} \text{ J} \cdot \text{s}$
$\ln(\log_e) = \log_{10} \times 2.303$		

Greek alphabet

Upper case	Lower case	Name	English
A	α	Alpha	a
B	β	Beta	b
Γ	γ	Gamma	g
Δ	δ	Delta	d
E	ε	Epsilon	e
Z	ζ	Zeta	z
H	η	Eta	h
Θ	θ	Theta	th
I	ι	Iota	i
K	κ	Kappa	k
Λ	λ	Lambda	l
M	μ	Mu	m
N	ν	Nu	n
Ξ	ξ	Xi	x
O	o	Omicron	o
Π	π	Pi	p
P	ρ	Rho	r
Σ	σ, ς *	Sigma	s
T	τ	Tau	t
Y	υ	Upsilon	u
Φ	ϕ	Phi	ph
X	χ	Chi	ch
Ψ	ψ	Psi	ps
Ω	ω	Omega	o

1

Introduction to Pharmaceutical Analytical Chemistry

1.1 Introduction

In daily conversation, words like pills and drugs are typically in use. However, when entering the pharmaceutical world, wording becomes very important and correct terms should be used. A *pharmaceutical preparation* contains a substance that is pharmacologically active, which is called the *active pharmaceutical ingredient (or active ingredient)*. The active pharmaceutical ingredient is often abbreviated to *API*. A large number of APIs exists and two examples are shown in Figure 1.1, namely *paracetamol* (acetaminophen), which is used against pain, and *insulin aspart*, which is used in the treatment of diabetes. Paracetamol is a *small molecule API* (or *small molecule drug*) produced by *organic synthesis* and with a molecular mass of 151 Da. Insulin aspart, on the other hand, is a two-chain peptide produced by *recombinant DNA technology*. It is a large molecule drug with a molecular mass of 5826 Da and is termed a *biopharmaceutical* due to its biological origin.

An active ingredient is not given (*administered*) to the patient as a pure substance, but is combined with *excipients* (synonymous with *inactive ingredients*) into a *dosage form* in order to be able to give an exact dose to the patient. The excipients are not

Introduction to Pharmaceutical Analytical Chemistry, Second Edition.
Stig Pedersen-Bjergaard, Bente Gammelgaard and Trine Grønhaug Halvorsen.
© 2019 John Wiley & Sons Ltd. Published 2019 by John Wiley & Sons Ltd.

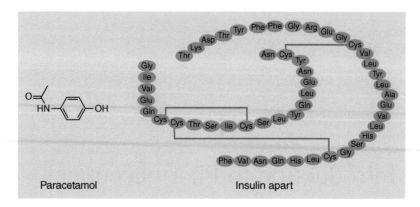

Paracetamol

Insulin apart

Figure 1.1 *Paracetamol (small molecule drug) and insulin aspart (biopharmaceutical)*

pharmacologically active. The dosage form can be a *tablet*, a *capsule*, or a *syrup* for *oral administration*, an *injection* for *parenteral administration*, or an *ointment* for *topical administration*. The excipients used in pharmaceutical preparations serve several functions, and these can be summarized as follows:

- Ensure that the dosage form has a shape and size that is easy to use for the patient.
- Ensure that the API is released and delivered to the patient in the correct amount.

Table 1.1 *Excipients in paracetamol tablets and paracetamol syrup (example)*

Content	Amount (mg)	Function
Tablet (mass 285 mg)		
Paracetamol	250	Active ingredient
Hydroxypropyl cellulose		Binder
Maize starch		Disintegrant
Talcum		Glidant
Magnesium stearate		Lubricant
Syrup (volume 1 mL)		
Paracetamol	24	Active ingredient
Sorbitol		Sweetener
Glycerol		Sweetener
Polyvinylpyrrolidone		Thickening agent
Saccharine sodium salt		Sweetener
Methylparaben		Preservative
Ethylparaben		Preservative
Propylparaben		Preservative
Sodium metabisulfite		Antioxidant
Citric acid		pH regulator
Sodium citrate		pH regulator
Strawberry aroma		Flavouring agent
Water		Solvent

- Ensure that the pharmaceutical preparation has an acceptable stability.
- Ensure that the pharmaceutical preparation does not have an unpleasant taste or odour.
- Facilitate production of the pharmaceutical preparation.

The excipients vary widely for different preparations. To exemplify this, Table 1.1 shows the excipients of tablets and syrup containing *paracetamol* as the active ingredient. Paracetamol has both an *analgesic* and an *antipyretic* effect, which means that it is used against pain and fever. Each paracetamol tablet has a total mass of 285 mg. Paracetamol constitutes 250 mg, while the remaining 35 mg is made up of excipients. The excipients include a disintegrating agent, a lubricant, a glidant, and a binder. *Binders, lubricating agents,* and *gliding agents* are added to facilitate manufacture. The *disintegrating agent* ensures rapid disintegration of the tablet in the stomach of the patient and rapid release of paracetamol.

Paracetamol syrup (liquid preparation) is a 24 mg/ mL solution of paracetamol in water. In addition, several excipients are added. *Sweetening* and *flavouring agents* are added for

---

Box 1.1   Official European Pharmacopoeia definitions

Medicinal product:
(i) Any substance or combination of substances presented as having properties for treating or preventing disease in human beings and/or animals or (ii) any substance or combination of substances that may be used in or administered to human beings and/or animals with a view either to restoring, correcting, or modifying physiological functions by exerting a pharmacological, immunological, or metabolic action, or to making a medical diagnosis.

Pharmaceutical preparation:
Pharmaceutical preparations are medicinal products generally consisting of active substances that may be combined with excipients, formulated into a dosage form suitable for the intended use, where necessary after reconstitution, presented in a suitable and appropriately labelled container.
Example: Paracetamol tablets as received from the pharmacy

Dosage form:
Physical manifestation of a product that contains the active ingredient(s) and/or excipient(s) that are intended to be delivered to the patient.
Examples: Tablets, syrups

Active pharmaceutical ingredient:
Any substance intended to be used in the manufacture of a medicinal product and that, when so used, becomes an active ingredient of the medicinal product. Such substances are intended to furnish a pharmacological activity or other direct effect in the diagnosis, cure, mitigation, treatment, or prevention of disease, or to affect the structure and function of the body.
Example: Paracetamol

Excipient:
Any constituent of a medicinal product that is not an active substance. Adjuvants, stabilizers, antimicrobial preservatives, diluents, antioxidants, for example, are excipients.
Example: Hydroxypropyl cellulose

better taste. *Antimicrobial agents* are added to prevent bacterial growth and *antioxidants* are added to reduce chemical degradation of the API. The latter aspect is particularly important with liquid pharmaceutical preparations, because chemical compounds in solution are more sensitive to oxidative degradation. In addition, agents are added to increase the viscosity of the syrup and to control pH. Controlling pH is important in order to keep the dissolved active substance stable and high viscosity makes the syrup easier to handle by the patient.

The terms pharmaceutical preparation, dosage form, API, excipient, and *medicinal product* have strict definitions; these are summarized in Box 1.1. The definitions are important to have in mind when reading this textbook.

Pharmaceutical preparations may be divided into 'over the counter drugs' (OTC drugs), which may be sold directly to the consumer in pharmacies and supermarkets without restrictions, and 'prescription only medicine' (POM) that must be prescribed by a licensed practitioner (normally a medical doctor). Paracetamol tablets are a typical OTC drug, whereas paroxetine tablets used against serious depression is a typical POM.

1.2   Pharmaceutical Analytical Chemistry

1.2.1   A Brief Definition

This textbook is about pharmaceutical analytical chemistry (or pharmaceutical analysis as a short name). Pharmaceutical analysis is the scientific discipline of analytical chemistry applied to pharmaceuticals. This textbook teaches how to *identify* and *quantify* (measure the content of) drug substances in a given sample. The sample can be a pharmaceutical ingredient, a pharmaceutical preparation, or a biological fluid such as blood and urine. The following sections briefly explain where such samples are examined by pharmaceutical analysis, to give an understanding of the importance of the subject of the current textbook.

1.2.2   Manufacture of Pharmaceuticals

Most pharmaceutical preparations are produced industrially by *pharmaceutical manufactures*, but some small-scale production also occurs in hospitals and pharmacies. Figure 1.2 outlines the work-flow for a typical industrial production of a pharmaceutical preparation. Production starts by ordering the *pharmaceutical ingredients*, namely the API and the necessary excipients. In some cases, the manufacturer produces some of these ingredients in-house, but most commonly they are produced elsewhere by different industrial suppliers. The pharmaceutical ingredients arrive in large quantities and are typically packed in cardboard drums or in large plastic containers.

Upon arrival, all the pharmaceutical ingredients are registered in the manufacturer's documentation system, tagged with internal labels, and stored in a separate area. Here the ingredients are temporarily in *quarantine*. Then, samples of the pharmaceutical ingredients are collected and analysed to ensure that they have the correct *identity* and are of high *purity*. Such testing involves pharmaceutical analysis and this is discussed in details in Chapter 18 for small molecule drugs and in Chapter 21 for biopharmaceuticals. The results from testing are compared with the *specifications* (requirements) of the manufacturer, and provided that the results comply with the specification, the pharmaceutical ingredients are labelled as

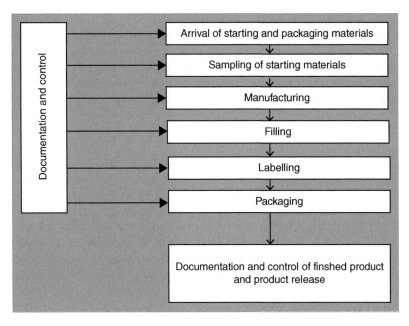

Figure 1.2   Illustration of a typical work-flow for pharmaceutical manufacturing

released material and transferred to production. Production starts with weighing or measuring the different ingredients in appropriate amounts for the subsequent production. Then, the ingredients are transferred to the manufacture. Manufacture of tablets uses several types of equipment such as machinery for *granulation, drying*, and *tablet pressing*. The manufacture of liquid preparations is carried out in large *tanks*, while the production of ointments and creams are carried in large *pots* with agitation and heating. During the manufacturing process, critical process parameters are measured to ensure the quality of the pharmaceutical preparation, and systems for this are defined as *process analytical technology (PAT)*. When the pharmaceutical preparation leaves the production site, samples are taken for the final testing. This testing again involves pharmaceutical analysis, and is intended to confirm that the API is identified and is present in the correct amount in the pharmaceutical preparation. Such final testing of pharmaceutical preparations is discussed in detail in Chapter 19. The pharmaceutical preparation is then filled in appropriate containers (*filling*), the containers are marked with labels (*labelling*), and the containers are packed in large units (*packaging*). The pharmaceutical preparation is in quarantine until the *final assessment*. Here, the results from pharmaceutical analysis need to be in compliance with the specification. The assessment also embraces many other factors, including production conditions, results of in-process testing, a review of the manufacturing (including packaging), documentation, compliance with finished product specifications, and examination of the final finished pack. After leaving the manufacturer, the pharmaceutical preparations are sent to *pharmaceutical wholesalers*, and from here the preparations are distributed to pharmacies, hospitals, or other retailers where they become available to the patients.

The industrial manufacture of pharmaceutical preparations is a complicated process involving many different steps. Typically, production is a *batch process*, which means that the products are made in limited batches. Each time a batch is produced, a new manufacturing process is started from the beginning with new ingredients. Between each manufacture of a given pharmaceutical preparation, the equipment is often used for the production of other preparations. Consequently, the production facility must be cleaned thoroughly between each batch to prevent earlier production ingredients from contaminating the next preparation (this is termed *cross-contamination*).

1.2.3 Development of New Drugs

New APIs are developed by *pharmaceutical companies*, based on *drug discovery* research and subsequent *preclinical development*. Drug discovery recalls innovation and heavy research activities. In this work, new chemical/biochemical substances are identified and tested for their pharmacological activity. A successful drug candidate is then tested in animals for its *effect* and for its *toxicity*. At this stage the *absorption, distribution, metabolism*, and *excretion* (abbreviated to *ADME*) of the drug candidate are studied and the appropriate dose is settled. Pharmaceutical analysis is involved in all parts of these processes for characterization, identification, and quantitation of the drug candidate as a pure substance, in preparations and in blood, urine, and tissue samples.

New APIs are patented to give the pharmaceutical companies an exclusive right to produce and market them for a certain number of years. Before entering the market, new APIs first have to enter *clinical trials* on humans (*phases I, II*, and *III*) to ensure efficacy and safety. Again pharmaceutical analysis is involved, and a large number of blood samples from the clinical trials are analysed to quantify the new API. Then, all data from drug discovery, preclinical development, and clinical trials I, II, and III are combined into a *Marketing Application* (*MA*, for Europe) or a *New Drug Application* (*NDA*, for the United States) and submitted to the *regulatory authorities*. In Europe, the MA is examined and approved by the *European Medicines Agency* (*EMA*), while in the US the examination and approval is performed by the *US Food and Drug Administration* (*FDA*). Active pharmaceutical ingredients with no patent protection, or with expired patents, can be produced and marketed as *generic drugs* by other pharmaceutical companies without restrictions or licences.

1.2.4 Use of Pharmaceuticals

At the start of any *medication*, it is common to treat patients with a standard *dose*, but it is well known that different patients may exhibit large variations in response to a given pharmaceutical product. In such cases it is important to adjust the dose. One example is the treatment of *hypertension*. The dose may be reduced when blood pressure is too low and the dose may be increased when blood pressure is too high. For other types of treatment, such as *depression, psychosis*, and *epilepsy*, the efficacy of the medication is more challenging to evaluate, and in those cases *therapeutic drug monitoring* (*TDM*) is advised. In TDM a blood sample is collected from the patient and analysed to ensure that the drug level is appropriate. The analysis of drugs in biological fluids is termed *bioanalysis*. In addition to TDM, and the previously mentioned ADME studies, bioanalysis is crucial in drug development programs (clinical trials) and for the detection of drugs of abuse in biological samples (blood, urine, saliva) from humans (*forensic investigations* and *doping control*). Bioanalysis is another major area of pharmaceutical analysis, which is discussed in Chapter 20.

1.3 This Textbook

From the discussions above, it appears that pharmaceutical analysis plays a major role in the life cycle of pharmaceuticals. Thus, pharmaceutical analysis is important for people working in the pharmaceutical industry, hospital laboratories, contract analytical laboratories, pharmaceutical and medical research institutions, and institutions investigating cases of drug abuse and doping in sports (forensic and doping laboratories).

The textbook is especially written for pharmacy students. In Europe, the training of pharmacists has to be in compliance with Directive 2005/36/EC of the European Parliament and of the Council. Box 1.2 summarizes some of the requirements of this directive.

Box 1.2   Part of directive 2005/36/EC on the recognition of professional qualifications in article 44: training as a pharmacist

(Subjects related to pharmaceutical analysis are given in bold.)

Training for pharmacists shall provide an assurance that the person concerned has acquired the following knowledge and skills:

(a) Adequate knowledge of medicines and the substances used in the manufacture of medicines;
(b) Adequate knowledge of pharmaceutical technology and the physical, **chemical**, biological and microbiological **testing of medicinal products**;
(c) Adequate knowledge of the metabolism and the effects of medicinal products and the action of toxic substances, and of the use of medicinal products;
(d) Adequate knowledge to evaluate scientific data concerning medicines in order to be able to supply appropriate information on the basis of this knowledge;
(e) Adequate knowledge of the legal and other requirements associated with the pursuit of pharmacy.

Course of training for pharmacists

- Plant and animal biology
- Physics
- General and inorganic chemistry
- Organic chemistry
- **Analytical chemistry**
- Pharmaceutical chemistry, including **analysis of medicinal products**
- General and applied biochemistry (medical)
- Anatomy and physiology; medical terminology
- Microbiology
- Pharmacology and pharmacotherapy
- Pharmaceutical technology
- Toxicology
- Pharmacognosy
- Legislation and, where appropriate, professional ethics

The general teaching in 'Analytical chemistry' as defined by the DIRECTIVE is covered by Chapters 3 to 17. Basically, this can be found in textbooks in analytical chemistry as well, but the content in the current textbook has been carefully selected to cover the analytical techniques and concepts most relevant for pharmaceutical analysis. The level of detail is less than in comprehensive analytical chemistry textbooks to fit the subject into the broad pharmacy curriculum. In some cases, the reader may require more technical details, but they are easily found in analytical chemistry textbooks or Internet resources based on the fundamental understanding from reading the current textbook.

The teaching in 'chemical testing of medicinal products' and 'analysis of medicinal products' as defined by the DIRECTIVE is covered by Chapters 18 to 21. These chapters focus on key pharmaceutical issues, including:

- Chemical analysis of pharmaceutical ingredients
- Chemical analysis of pharmaceutical preparations
- Chemical analysis of biopharmaceuticals
- Chemical analysis of drug substances in biological fluids

Also in this part, the level of detail has been selected to fit the subject into the broad curriculum of pharmaceutical sciences. Readers looking for more details can find these in pharmacopoeias and Internet resources.

2

Marketing Authorizations, Pharmaceutical Manufacturing, and International Pharmacopoeias

2.1 Introduction

The purchaser of food and beverages normally discovers that a product is associated with a significant quality problem if it has either an abnormal taste, or an unusual smell or appearance. This is not the case with pharmaceutical preparations, and there is normally no way patients can decide whether a tablet contains the active ingredient, whether the amount of active ingredient is correct, or whether any contaminants or degradation products are present in the preparation. Therefore, the patient is not in a position to recognize that a pharmaceutical preparation is incorrect or defective. The patient literally takes the pharmaceutical preparation entirely on trust, which extends back to those responsible for the manufacture of the preparation. It is therefore mandatory that the *pharmaceutical industry* maintains the highest standards of quality in the development, manufacture, and marketing of pharmaceutical preparations. This is regulated by *national medicinal agencies* and by a number of laws and guidelines as discussed briefly in the following.

Introduction to Pharmaceutical Analytical Chemistry, Second Edition.
Stig Pedersen-Bjergaard, Bente Gammelgaard and Trine Grønhaug Halvorsen.
© 2019 John Wiley & Sons Ltd. Published 2019 by John Wiley & Sons Ltd.

2.2 Marketing Authorization and Industrial Production

A *marketing authorization* is required for a pharmaceutical preparation in order to be released to market. Pharmaceutical manufactures and companies apply for marketing authorizations, and these are evaluated and granted by national medicinal agencies. Marketing authorization is granted based on the stated quality, safety, and efficacy of the pharmaceutical preparation. In EU, a common legislation is available for marketing authorization in more than one EU country. This is managed by the *European Medicines Agency (EMA)*. In the United States, the *Food and Drug Administration (FDA)* serves as a national medicinal agency. The marking authorization is granted for a limited period of time (often five years) and may thereafter be renewed. The marketing authorization may be withdrawn before expiration if:

- The product is no longer considered to meet the requirements for quality, safety, or efficacy.
- The product does not have the specific qualitative or quantitative composition.
- The provisions that apply to pharmaceutical manufacture have not been followed.

The main sections of applications for marketing authorization are summarized in Table 2.1. Each application contains detailed information about the chemical methods for identification and purity testing as detailed in Table 2.2. In this way, marketing authorizations are linked to pharmaceutical analysis.

As seen from Table 2.2, detailed information has to be filed about the chemical methods to be used to control the pharmaceutical ingredients before production, about the chemical methods to be used during production, and the chemical methods to be used for control of the finished pharmaceutical preparation. In addition, the application should provide results from test analysis of the different trial production batches, with documentation that the chemical methods used have been tested and found suitable for use (validation). The entire

Table 2.1 Requirements for the content of an application for marketing authorization

Administrative information (name of applicant, name of manufacturer)
Name of the drug and its composition
Chemical, pharmaceutical, and biological documentation
Toxicological and pharmacological documentation including possible interactions with other drugs
Clinical documentation
Expert reports
Proposals for advertisements
Proposed labelling preparations
Proposed leaflet
Proposals for prescription status
Documentation of the manufacturing authorization
Confirmation of fees paid to the Medicinal Agency
A copy of the marketing authorization for the drug in other EEA countries[a] (Europe only)
A copy of the advertisement and package leaflet approved in other EEA countries[a] (Europe only)

[a] *European Economic Area.*

Table 2.2 *Content requirements for application for marketing authorization (only the topics most relevant to pharmaceutical analysis are included)*

Topic	Requirements
Control of active substance (active pharmaceutical ingredient; API)	Characteristics of the active substance and purity requirements (specification)
	Detailed description of the applicant's chemical methods for the confirmation of the identity of the active substance
	Detailed description of the applicant's chemical methods for control of the purity
	Detailed information on the active substance
	Nomenclature of the active substance
	Description of the active substance
	Manufacturing method for the active substance (chemical synthesis)
	Quality control methods of the manufacturer of the active substance
	Known impurities in the active substance
	Results of the chemical control of previously produced batches of the active substance
Control of excipients	Characteristics of the excipients and purity requirements (specification)
	Detailed description of the applicant's chemical methods for confirmation of identity of the excipient
	Detailed description of the applicant's chemical methods for control of the purity of the excipient
	Information on excipients
Control of production preparations	Detailed description of the applicant's chemical methods for control of production mixtures
Control of finished product	Detailed description of the applicant's chemical methods for the confirmation of the identity and determination of the content of the active substance in the preparation
	Detailed description of the applicant's chemical methods for the confirmation of the identity of a dye (excipient)
	Detailed description of the applicant's chemical methods for the determination of the levels of antimicrobial additives and preservatives
	Documentation (validation) of the suitability of all the chemical methods
	Results of the chemical control of previously produced batches of the product

Table 2.3   *Main elements of the GMP/GLP regulations*

Paragraph	Description
1	Demands for a quality department and for quality control in all stages of production
2	Requirements for staff
3	Requirements for premises and equipment
4	Documentation requirements
5	Requirements for production
6	Requirements for quality control
7	Requirements in connection with external contract work
8	Requirements for the withdrawal of products
9	Requirements for self-inspection

control scheme based on pharmaceutical analysis for the forthcoming preparation should therefore be documented in the application for marketing approval, and this, together with the rest of the information shown in Table 2.1, forms the basis for the authorities' evaluation of the application. If a pharmaceutical manufacturer or company later wishes to make changes to the composition of a pharmaceutical preparation, it must be approved by the authorities. This is even the case if changes are to be made for the control methods of an existing product, since these are part of the marketing authorization.

Any pharmaceutical preparation can only be marketed if marketing authorization is granted to the company. A second requirement is that the company has a *manufacturing authorization* from the authorities. The word *manufacturing* refers to production, packaging, repackaging, labelling, relabelling, and release of pharmaceutical preparations, as well as the necessary controls in connection with these activities. All activities and facilities shall be in accordance with *good manufacturing practice (GMP)*, including all activities related to pharmaceutical analysis, which is described by *good laboratory practice (GLP)*. GMP and GLP are comprehensive regulatory frameworks, and the main elements are presented in Table 2.3.

As seen from the table, GMP regulates among others personnel, premises, and equipment, and set requirements to production, quality control, and documentation. Also, GMP demands a *quality department* in the pharmaceutical company and release of batches of pharmaceutical preparations is their responsibility.

The company has to follow the manufacturing operations as described in the approved marketing authorization. This also applies to all activities involving pharmaceutical analysis. Thus, the pharmaceutical ingredients as well as the pharmaceutical preparation have to be tested. This has to be done for each production batch using the analytical methods described in the approved marketing authorization. In addition, the company has to appoint at least one *qualified person (QP)* who is approved by the national medicine agency. Only qualified persons can assess the results from control of the pharmaceutical ingredients, production, and pharmaceutical preparation, and together with additional documentation release the production batch. Box 2.1 summarizes typical responsibilities in this respect.

Box 2.1   Typical batch release requirements

- The production batch and its manufacture comply with the provisions of the marketing authorization.
- The production has been carried out in accordance with GMP.
- The principal production and testing processes have been validated (validation is defined as the documented act of demonstrating that processes will consistently lead to the expected results).
- Any deviations or planned changes in production or quality control have been authorized by the responsible persons.
- All the necessary checks and tests have been performed.
- All necessary production and quality control documentation has been completed.
- The QP should in addition take into account any other factors of which he/she is aware that are relevant for the quality of the production batch.

A high level of documentation constitutes an essential part and is vital for batch release and certification by the QP. Clearly written documentation and *standard operating procedures* (SOPs) prevent errors from spoken communication and permit tracing of batch history. Box 2.2 summarizes typical documentation requirements.

Box 2.2   Typical documentation requirements

- Specifications that in detail describe the requirements that must be fulfilled prior to quality evaluation.
- Manufacturing formulae, processing, and packaging instructions.
- Procedures that give directions for performing operations such as cleaning, sampling testing, and equipment operation.
- Records providing a history of each batch or product.
- The batch documentation shall be retained for at least one year after the expiry date of the batches.

2.3   Pharmacopoeias

Standards for pharmaceutical ingredients and for pharmaceutical preparations have been given in *pharmacopoeias* for many years. It all started in the seventeenth century, and historically many countries had their own local or national *pharmacopoeia*. *Pharmacopoeia Nordica* was published in 1963 as the first pharmacopoeia to be authorized in more than one country, and this covered the Nordic countries. Currently, the *European Pharmacopoeia* (*Ph. Eur.*) is authorized in European countries, the *British Pharmacopoeia* (*BP*) in the United Kingdom, and the *United States Pharmacopoeia* (*USP*) is authorized in the United States of America.

The *standards* (also called *monographs*) in the pharmacopoeias are the official require-ments for pharmaceutical ingredients and preparations. Standards are minimum require-ments and shall ensure that medicines are of a high quality. In order to obtain marketing authorization, it is important that the manufacturer's own requirements for the pharmaceu-tical ingredients and the preparation, which are termed *specifications*, meet the standards of Ph. Eur. and USP.

Standards for a large number of pharmaceutical ingredients are found in Ph. Eur. and USP. The individual standards prescribe the requirements for identity and purity. The phar-macopoeias also provide detailed procedures for identification and control of purity for each pharmaceutical ingredient. In the application for marketing authorization, the manu-facturer refers to this information for the documentation of the pharmaceutical ingredients (Table 2.2).

Due to globalization and expansion in international trade, there is a growing need to develop global quality standards for pharmaceutical ingredients and preparations. As stan-dards are a vital instrument for registration, market surveillance, and free movement and trade of medicines among countries, harmonization among the world's three major pharma-copoeias (Ph. Eur., USP, and the Japanese Pharmacopoeia (JP)) is an important task. This harmonization process is now well under way between the three pharmacopoeias but there is still long way to go.

The *International Council for Harmonization of Technical Requirements for Pharma-ceuticals for Human Use (ICH)* is bringing together the regulatory authorities and the phar-maceutical industry to discuss scientific and technical aspects of drug registration. Since 1990, ICH has gradually evolved, to respond to the increasing global drug development. ICH's mission is to *achieve greater harmonization worldwide to ensure that safe, effective, and high quality medicines are developed and registered in the most resource-efficient man-ner*. ICH has published a number of guidelines, which have become important documents, especially for the pharmaceutical industry.

The pharmacopoeias prescribe the procedures for chemical analysis of pharmaceuti-cal ingredients and preparations. In this textbook, the discussion is restricted to Ph. Eur. and USP.

2.4 Life Time of Pharmaceutical Preparations and Ingredients

Although some pharmaceutical preparations as tablets can be stable for many years, there is a maximum life time for all pharmaceutical preparations and ingredients of five years. This is to avoid any discussion on how long pharmaceutical preparations and ingredients may be stored. However, pharmaceutical ingredients should always comply with the monographs in the pharmacopoeias or similar standards, and life time can be shorter.

Generally, the pharmaceutical preparations should at the time of production not deviate more than ±5% from the declared content of an active pharmaceutical ingredient (API). Pharmaceutical preparations are stability tested, and within the life time up to ±10% deviation from the declared content is accepted. Thus, the life time of a pharmaceutical preparation, often termed *shelf-life*, is based on stability testing. The ±10% requirement is general, but tighter limits may be applied when necessary. Box 2.3 shows an example of the ±10% requirement.

Box 2.3   Paracetamol tablets and ±10% requirement for content of API

Paracetamol tablets with a declared content of 250 mg should be within:

250 ± 25 mg, which is equivalent to:
225–275 mg

3

Fundamentals of Bases, Acids, Solubility, Polarity, Partition, and Stereochemistry

3.1 Acids, Bases, pH, and pK$_a$

Water can react with itself to form *hydronium ions* (H_3O^+) (also called hydroxonium ions) and *hydroxide ions* (OH^-):

$$2H_2O \rightleftharpoons H_3O^+ + OH^- \qquad (3.1)$$

This is termed *autoprotolysis* and water acts as an *acid* (donating a proton, H^+) as well as a *base* (accepting a proton). The *autoprotolysis constant* K_w (equilibrium constant) is

$$K_w = [H_3O^+][OH^-] = 10^{-14} \, M^2 \qquad (3.2)$$

Introduction to Pharmaceutical Analytical Chemistry, Second Edition.
Stig Pedersen-Bjergaard, Bente Gammelgaard and Trine Grønhaug Halvorsen.
© 2019 John Wiley & Sons Ltd. Published 2019 by John Wiley & Sons Ltd.

As seen from Eq. (3.2) only a very small amount of water is ionized. The concentration of hydronium ions, $[H_3O^+]$, and hydroxide ions, $[OH^-]$, in pure water is therefore 10^{-7} M. $[H_3O^+]$ is equivalent to $[H^+]$ and both terms will be used in the following.

The *pH value* of an aqueous solution is defined as the negative logarithm to the activity (a_{H^+}), and the latter is approximated to the concentration of protons $([H^+])$:

$$pH = -\log(a_{H^+}) \approx -\log[H_3O^+] = -\log[H^+] \tag{3.3}$$

Aqueous solutions with pH < 7 are *acidic*. If pH = 7, the solution is *neutral*, and if pH > 7 the solution is *alkaline*. An aqueous solution with pH 1 is strongly acidic, whereas a solution with pH 5 is slightly acidic. Similarly, at pH 13 solutions are strongly alkaline and at pH 9 solutions are slightly alkaline.

Strong acids, such as HCl, are fully dissociated and ionized in dilute aqueous solution. Hydrochloric acid (HCl) dissociates completely according to the following reaction when dissolved in water:

$$HCl + H_2O \rightarrow Cl^- + H_3O^+ \tag{3.4}$$

High concentrations of strong acids dissolved in water provide a strongly acidic solution. Thus, a 100 mM solution of HCl in water provides a pH value of 1.0 (pH = $-\log(0.1) = 1.0$).

In contrast, weak acids are not completely dissociated in aqueous solution and the ionized and unionized forms are in equilibrium. Acetic acid is an example of a weak acid and dissolved in water, the following equilibrium exists:

$$CH_3COOH + H_2O \rightleftharpoons CH_3COO^- + H_3O^+ \tag{3.5}$$

The dissociation reaction for a weak acid (HA) can be written more generally in the following way:

$$HA + H_2O \rightleftharpoons A^- + H_3O^+ \tag{3.6}$$

The corresponding equilibrium constant (K_a) is defined as follows:

$$K_a = \frac{[H_3O^+][A^-]}{[HA]} \tag{3.7}$$

The stronger an acid, the more the equilibrium in Eq. (3.6) is shifted towards the right and the higher is the K_a value. The strength of an acid can therefore be expressed by the value of K_a, which is termed the *acid ionization constant* (also termed the *acid dissociation constant*). More commonly, however, strength is expressed by the pK_a value defined as follows:

$$pK_a = -\log K_a \tag{3.8}$$

Strong acids have low pK_a values, whereas weak acids have high pK_a values.

Similar for bases, strong bases are completely dissociated in water. The strong base sodium hydroxide, NaOH, is completely dissociated to Na^+ and OH^- in water, while weak bases are not completely dissociated. Ammonia is an example of a weak base and is only partly dissociated in water:

$$NH_3 + H_2O \rightleftharpoons NH_4^+ + OH^- \tag{3.9}$$

Dissociation of a weak base (B) can generally be written as

$$B + H_2O \rightleftharpoons BH^+ + OH^- \tag{3.10}$$

The corresponding equilibrium constant (K_b) is defined as follows:

$$K_b = \frac{[OH^-][BH^+]}{[B]} \tag{3.11}$$

The stronger the base, the more equilibrium in Eq. (3.10) is shifted towards the right. The strength of bases can be expressed by pK_b, defined as follows:

$$pK_b = -\log K_b \tag{3.12}$$

The equilibrium constant, K_b is the *base ionization constant*.

For corresponding acids and bases, the following equation is valid:

$$K_a K_b = [H^+][OH^-] = K_w = 10^{-14} \, M^2 \tag{3.13}$$

and taking the logarithm to this expression

$$pK_a + pK_b = 14 \tag{3.14}$$

In most cases, the strength of basic substances is expressed by pK_a of the *corresponding acid*, which is the protonated base. Strong bases are thus associated with high pK_a values. According to Eq. (3.9), the ammonium ion, NH_4^+, is the corresponding acid to the base ammonium, NH_3. The pK_a value of NH_3 is 9.2.

3.2 Buffers

A *buffer solution* is a mixture of an acid and its corresponding (conjugated) base. Such solutions resist changes in pH when they are diluted or when acids or bases are added to the solution. Buffer solutions are extremely important for keeping pH constant and are often used in pharmaceutical analysis.

Rearranging Eq. (3.7) leads to the *Henderson–Hasselbalch* equation (*buffer equation*):

$$pH = pK_a + \log \frac{[A^-]}{[HA]} \tag{3.15}$$

Here [HA] is the concentration of the acid and [A$^-$] is the concentration of the corresponding base. The buffer equation shows that when pH = pK_a, the concentrations of the acid and its corresponding base are equal. Acetic acid and its corresponding base, acetate, is an example of a buffer system. In this case [HA] is the concentration of acetic acid and [A$^-$] is the concentration acetate. A 0.1 M solution of this buffer can be produced by dissolving 0.05 mol acetic acid and 0.05 mol sodium acetate in 1 L water. The pK_a value of acetic acid is 4.76. At this pH, the concentrations of acid and base are equal, and the ratio [A$^-$]/[HA] is therefore 1.0. The value of log (1.0) is zero and pH of the buffer is 4.76. Adding base to this solution changes the ratio [A$^-$]/[HA] as the concentration of [HA] decreases while the concentration of [A$^-$] increases correspondingly. When there is 10 times as much base as acid,

Table 3.1   Common buffers in pharmaceutical analysis

Acid	Acid/base	pK_a
Acetic acid	CH_3COOH/CH_3COO^-	4.76
Ammonia	NH_4^+ / NH_3	9.25
Carbonic acid	H_2CO_3/HCO_3^-	6.35
	HCO_3^-/CO_3^{2-}	10.33
Formic acid	$HCOOH/HCOO^-$	3.74
Phosphoric acid	$H_3PO_4/H_2PO_4^{2-}$	2.15
	$H_2PO_4^{2-}/HPO_4^-$	7.20
Tris(hydroxymethyl)aminomethane	$(HOCH_2)_3C\text{-}NH_3^+/(HOCH_2)_3C\text{-}NH_2$	8.07

the ratio becomes 10 and the pH of the solution becomes 1 pH unit higher ($pK_a + \log 10$). Similarly, if the acid concentration is increased so it becomes 100 times higher than the base concentration, the pH of the solution decreases two units ($pK_a + \log 0.01$). When choosing a buffer system, an acid with a pK_a value close to the desired pH of the solution is chosen to obtain a system with high *buffer capacity*. Common buffers are given in Table 3.1.

3.3   Acid and Base Properties of Drug Substances

Most active pharmaceutical ingredients (drug substances) are either weak bases or acids. Base properties are normally caused by the presence of one or more *amino groups*, as shown in Figure 3.1. *Amines* are bases, as the nitrogen in the amine group has an unshared pair of electrons that can accept a proton. Amines can be *primary*, *secondary*, or *tertiary*. Note that nitrogen in an *amide* group (Figure 3.1) is not basic and is therefore not protonated. Acid properties of drug substances are often linked to carboxylic acid functionalities, phenolic groups, or sulfonamides (Figure 3.1). In all these structures, a proton can be donated from the functional group, and thus provide acidity.

Figure 3.2 lists a few examples of active pharmaceutical ingredients with their corresponding pK_a values. Three of them are bases: among these, fluoxetine has the highest pK_a value, and therefore fluoxetine is the strongest base. Morphine with the lowest pK_a value is the weakest base. Of the two acids, ibuprofen has the lowest pK_a value and is therefore the strongest acid.

The base or acid strength of pharmaceutical substances largely impact analytical procedures. If the basic substance fluoxetine (B) is dissolved in water, the reaction according to Eq. (3.10) describes the situation. If acid (H_3O^+) is added to the solution, such as hydrochloric acid (HCl), the equilibrium is shifted towards the right and more molecules become protonated (BH^+). On the other hand, if a base, such as sodium hydroxide (NaOH), is added to the solution, the equilibrium is shifted more to the left, and more fluoxetine molecules will be present in the free base form (B). The solution behaves in principle as a buffer system. As pK_a of fluoxetine is 10.0, equal concentrations of free base (B) and protonated base (BH^+) are present in the solution at pH 10.0 ($[B] = [BH^+]$). If pH is decreased from 10.0 to 9.0, equilibrium is shifted and 90% of the molecules will be protonated (BH^+). If pH is lowered to 8.0, 99% of the molecules will be protonated (BH^+) according to Eq. (3.15).

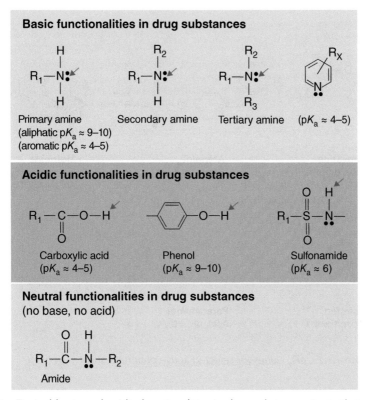

Figure 3.1 *Typical basic and acidic functionalities in drug substances (note that amides are neutral)*

This is further visualized in Figure 3.3. On the other hand, increasing pH from 10.0 to 11.0, 90% of the molecules will be on free base form (B). Basic pharmaceutical substances, such as fluoxetine, are much more soluble in water when present in the protonated state (BH^+). Thus for bases, water solubility is increasing as pH is decreased below the pK_a value.

For acids it is opposite. At $pH = pK_a$, 50% of the molecules are on acid form (HA) and 50% of the molecules are on deprotonated form (A^-). Increasing pH above the pK_a value, the solubility is increasing because more and more of the molecules become deprotonated (A^-). In solutions of pH below the pK_a value, acidic substances become less ionized and solubility in water is decreasing. Thus, for acids, water solubility is increasing as pH is increased above the pK_a value. Solubility largely impacts analytical procedures, and therefore the fundamental understanding of acid/base equilibria as discussed above is important.

3.4 Distribution Between Phases

Pharmaceutical substances can *partition* between two immiscible phases, such as between two immiscible liquids, between a liquid and a gas, or between a solid and a liquid. The first type of partition is termed *liquid–liquid partition*, while the two latter are termed *liquid–gas*

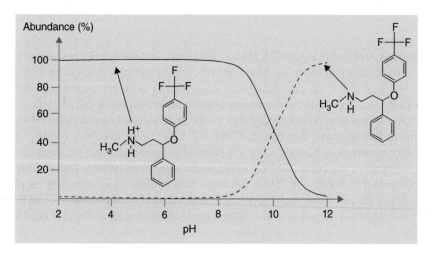

Fluoxetine
Base, pK_a = 10.0

Atenolol
Base, pK_a = 9.4

Morphine
Base, pK_a = 8.2
Acid, pK_a = 10.2

Ibuprofen
Acid, pK_a = 4.4

Paracetamol
Acid, pK_a = 9.7

Figure 3.2   pK_a *values for selected active pharmaceutical ingredients*

Figure 3.3   *Free and protonated form of fluoxetine as function of pH in aqueous solution*

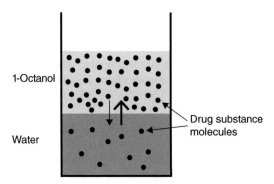

Figure 3.4 *Liquid–liquid partition of a drug substance between water and 1-octanol. Distribution is in favour of 1-octanol in this case ($K_D > 1$)*

partition and *solid–liquid partition*, respectively. A common example of an experimental liquid–liquid partition system is the two-phase system of water and 1-octanol, as illustrated in Figure 3.4. Partition plays an important role in the understanding of analytical procedures. In the following, partition is discussed with a primary focus on liquid–liquid partition.

Liquid–liquid partition of a drug substance between an aqueous phase and an organic phase can be described in the following way:

$$S_{\text{Aqueous}} \rightleftharpoons S_{\text{Organic}} \quad\quad (3.16)$$

where S stands for *solute*, which in this context is a drug substance. The solute is *distributed* between the two phases, and at equilibrium this distribution can be described by the *partition ratio* K_D (also termed *distribution constant*, see Box 3.1):

$$K_D = \frac{[S]_{\text{Organic}}}{[S]_{\text{Aqueous}}} \qu\quad (3.17)$$

Box 3.1   IUPAC definitions

Partition coefficient	Not recommended
Partition constant	The ratio of *activity* of a given species A in the extract to its activity in the other phase
Partition ratio (K_D)	The ratio of the concentration of a substance in *a single definite form*, A, in the extract to its concentration in the same form in the other phase
Distribution constant (K_D)	Synonymous with the partition ratio
Distribution ratio (D)	The ratio of *the total analytical concentration* of a solute in the extract (regardless of its chemical form) to its total analytical concentration in the other phase

$[S]_{Organic}$ is the concentration of the solute in the organic phase and $[S]_{Aqueous}$ is the concentration of solute in the aqueous phase at equilibrium. K_D is constant for a given solute under given conditions in the aqueous and organic phases. The higher the partition ratio, the more efficiently the analyte is distributed into the organic phase. Solutes with high water solubility will mainly be present in the aqueous phase, and the value for K_D is low. More hydrophobic solutes will have a higher solubility in the organic phase, and will have higher values for K_D.

For neutral solutes, Eq. (3.17) describes the situation well. However, active pharmaceutical ingredients are mainly basic or acidic compounds, and for these the situation becomes a little more complicated because they dissociate in aqueous solution. Thus, an acidic solute (*HA*) will dissociate in an aqueous solution according to Eq. (3.6) and the dissociation can be described by the acid dissociation constant K_a according to Eq. (3.7). For this solute the *distribution ratio, D*, can be defined according to the following equation:

$$D = \frac{[HA]_{Organic,total}}{[HA]_{Aqueous,total}} \tag{3.18}$$

where $[HA]_{Organic,total}$ is the total concentration of the solute in the organic phase and $[HA]_{Aqueous,total}$ is the total concentration of the solute in the aqueous phase. The key point is that only the uncharged fraction of the solute molecules can enter the organic phase, whereas the charged fraction will remain in the aqueous sample. Thus, the following equations can be written:

$$[HA]_{Organic,total} = [HA]_{Organic} \tag{3.19}$$

$$[HA]_{Aqueous,total} = [HA]_{Aqueous} + [A^-]_{Aqueous} \tag{3.20}$$

By substitution of Eqs. (3.19) and (3.20) into (3.18) the following equation is obtained:

$$D = \frac{[HA]_{Organic}}{[HA]_{Aqueous} + [A^-]} \tag{3.21}$$

By substitution of Eqs. (3.7) and (3.17) into (3.21), the following equation is derived:

$$D = K_D \frac{[H^+]}{[H^+] + K_a} \tag{3.22}$$

In a similar way, the following equation describes partition of a basic solute (*B*) that is partly dissociated in the aqueous sample:

$$D = K_D \frac{K_a}{[H^+] + K_a} \tag{3.23}$$

K_a is the dissociation constant for the corresponding acid BH^+. Equations (3.22) and (3.23) show that the following parameters affect the distribution (*D*) of basic and acidic drug substances:

- Partition ratio (K_D)
- pH in the aqueous sample ($[H^+]$)
- pK_a value of the analyte (K_a)

For a neutral solute, with no acidic or basic functional groups, the distribution is independent of pH in the aqueous phase, and is determined by the partition ratio (K_D) only. For acidic drug substances, distribution is highly dependent on pH in the aqueous phase. Thus, distribution is in favour of the organic phase at low pH and in favour of the aqueous phase at high pH. This can be derived from Eq. (3.22) and is illustrated in Figure 3.5 for distribution of ibuprofen (acidic drug substance) between 1-octanol and water as an example. At low pH, the ibuprofen molecules are mainly neutral (*HA*), solubility in the aqueous phase is limited, and the molecules easily distribute into the organic phase. At high pH, most ibuprofen molecules are negatively charged (A^-). Under such conditions, solubility in the aqueous phase is much higher and partition into the organic phase is strongly reduced.

For basic drug substances, distribution is influenced by pH in the opposite way, as shown in Figure 3.5 with fluoxetine (basic drug substance). When increasing pH in the aqueous phase, distribution into the organic phase is increasing due to the decreased ionization of the drug molecules.

Information about distribution of pharmaceutical substances between the aqueous phase and organic phase can be derived from log D values. When the distribution ratio is determined as the distribution ratio between 1-octanol and the aqueous buffer at pH 7.4 (physiological pH), it is most often termed P. Examples of log P values for selected active pharmaceutical ingredients are presented in Table 3.2.

When the distribution ratio is determined as the ratio between 1-octanol and the aqueous buffer at pH values different from 7.4, it is termed log D. The log P values provide a measure of *lipophilicity*, and drug substances with a high log P value are non-polar and lipophilic compounds. Thus, fluoxetine (log $P = 3.93$) distributes strongly into the organic phase

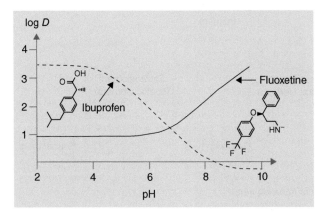

Figure 3.5   *Distribution of ibuprofen (acidic drug substance) and fluoxetine (basic drug substance) between 1-octanol and water as a function of pH*

Table 3.2   Log P values for selected active pharmaceutical ingredients

Compound	Log P
Fluoxetine	3.93
Ibuprofen	3.50
Hydrocortisone	1.76
Morphine	0.87
Paracetamol	0.48
Atenolol	0.34

in a two-phase system of 1-octanol and water, whereas paracetamol (log P = 0.48) distributes much less into the organic phase. Although log P values are only valid for two-phase systems of 1-octanol and water, they provide an indication also of the level of distribution in other organic–water two-phase systems.

3.5   Stereoisomers

Describing the stereochemistry of a drug substance is to visualize the spatial orientation of its components in space. Biological systems including the human body contain large biomolecules that are constructed from building blocks with unique stereochemistry. Biological systems are therefore able to distinguish between isomers that only differ in their spatial configuration. Such isomers may therefore also have different biological effects.

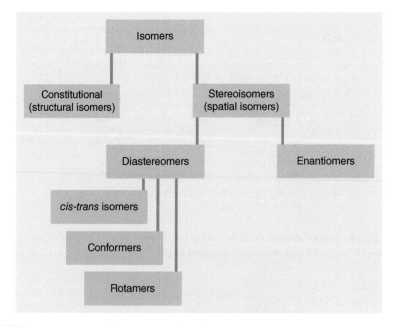

Figure 3.6   Classification of isomers

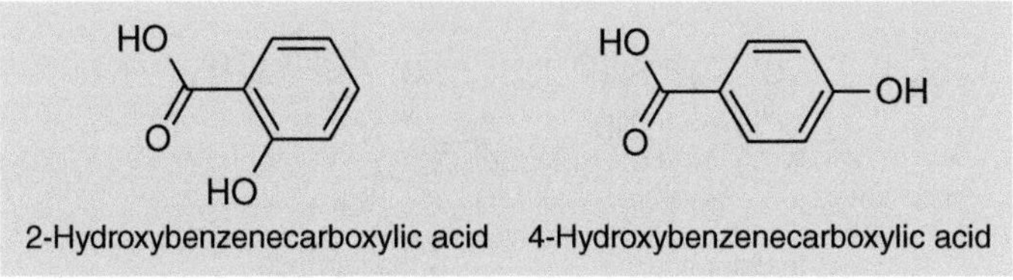

Figure 3.7 *Salicylic acid (2-hydroxybenzenecarboxylic acid) and 4-hydroxybenzene-carboxylic acid are constitutional isomers*

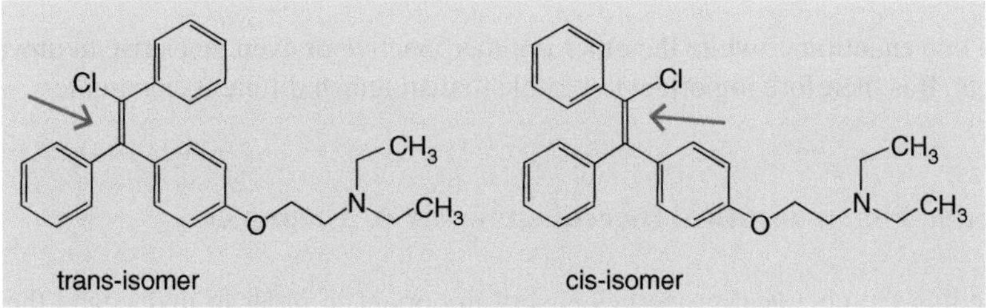

Figure 3.8 Cis–trans *isomers of clomiphene*

Organic compounds (including active pharmaceutical ingredients) can exist as different *isomers*, and this plays a vital role in pharmacy and in pharmaceutical analysis. As shown in Figure 3.6, isomers can be divided into *constitutional isomers* and *stereoisomers*. Constitutional isomers are different compounds with different chemical structures, as illustrated in Figure 3.7 using salicylic acid (2-hydroxybenzenecarboxylic acid) and a related impurity (4-hydroxybenzene-carboxylic acid) as an example.

Stereoisomers are spatial isomers, divided into diastereomers and enantiomers. *Diastereomers* have different physicochemical characteristics such as a different melting point and partition ratio. *Cis–trans* isomers belong to this group and a number of active pharmaceutical ingredients can be found in this group. Clomiphene is one example as illustrated in Figure 3.8, and as active pharmaceutical ingredient clomiphene is a mixture of both the *cis* and the *trans* isomers.

Enantiomers constitute a special group of stereoisomers. A pair of two enantiomers contains a *chiral centre* and is a mirror image of each other. The most abundant chiral centre is where a carbon atom is connected to four different ligands, but also nitrogen, phosphor, and sulfur can be chiral centres. A solution of a compound containing one or more chiral centres is able to rotate plane-polarized light either left or right. This is denoted (−) or (+), respectively. However, this is not an unambiguous way to describe the configuration of the chiral centre for a given compound, as the direction and size of the rotation is dependent on the solvent used to dissolve the compound. To give an unambiguous description of the configuration of the chiral centre, the *R/S nomenclature* should be used. This nomenclature gives the absolute configuration of the position of groups connected to the chiral atom. Two enantiomers are exemplified in Figure 3.9 for thalidomide.

Enantiomers have identical physicochemical characteristics, but can be distinguished based on their different ability to rotate plane-polarized light. When they enter a chiral environment such as the human body, they may behave differently. Many active pharmaceutical ingredients are chiral and it is often observed that the pharmacological effect is

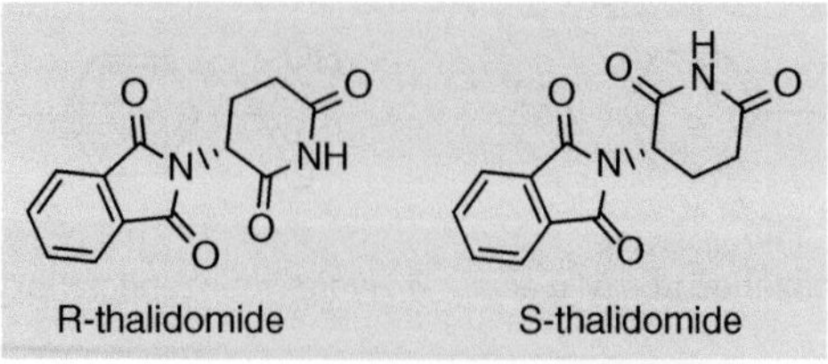

Figure 3.9 *Two enantiomers of thalidomide*

related to one enantiomer while the other is either inactive or even gives rise to unwanted side effects. It is therefore important to be able to distinguish different enantiomers.

3.6 Active Pharmaceutical Ingredients – A Few Examples

Understanding the chemical properties is very important in order to understand the analytical procedures. In the following, the chemical properties of six active pharmaceutical ingredients will be discussed based on the theory described in the previous sections.

3.6.1 Fluoxetine – A Basic and Lipophilic Drug

Fluoxetine is an antidepressant drug, and a typical small molecule drug substance with basic and lipophilic properties. The active pharmaceutical ingredient is delivered as fluoxetine hydrochloride (Figure 3.10). Fluoxetine has one chiral centre, and the active pharmaceutical ingredient is a mixture of both enantiomers. Both the European Pharmacopoeia (Ph. Eur.) and the United States Pharmacopoeia (USP) publish a monograph for fluoxetine hydrochloride. USP also publishes monographs for fluoxetine tablets, capsules, and oral solution, which are common pharmaceutical preparations with fluoxetine. Tablets and

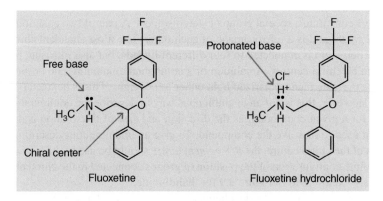

Figure 3.10 *Structure of fluoxetine and fluoxetine hydrochloride*

capsules typically contain 20 mg fluoxetine. The recommended blood plasma concentration of fluoxetine is in the range 150–500 ng/ mL.

The structures of fluoxetine as a free base (B) and fluoxetine hydrochloride (BH^+Cl^-) are illustrated in Figure 3.10. Fluoxetine is a base due to the amino group (—NH—). The protonated base (BH^+) is the corresponding weak acid of fluoxetine. Addition of HCl to the structural formula is commonly used to represent the hydrochloride form. Fluoxetine hydrochloride is a salt and when dissolved in aqueous solution the substance will dissolve and dissociate according to the following scheme:

$$BH^+Cl^- \rightarrow BH^+\,(aq) + Cl^-\,(aq) \tag{3.24}$$

The pK_a value of fluoxetine is 10.0. This is a relatively high pK_a value for basic drugs, and compared with other basic drugs fluoxetine is therefore a relatively strong base. If fluoxetine is dissolved in aqueous solution at pH 10.0, 50% of the molecules will be protonated (BH^+) and 50% will be on free-base form (B). Decreasing pH in the solution will increase the number of positively charged molecules and increase the solubility in water. This is illustrated in Figure 3.11, where the water solubility of fluoxetine is plotted as a function of pH. In acidic solution (pH 2), the solubility is about 70 g/L, which decreases to about 0.10 g/L at pH 10. The distribution of fluoxetine between water and 1-octanol is affected correspondingly, as shown in Figure 3.11. Thus, at pH 2, the log D value is 0.83, which increases to 3.60 at pH 10. Generally, log D is high due to the lipophilic nature of fluoxetine. This is caused by the relatively large hydrocarbon skeleton of the compound. Fluoxetine is also soluble in organic solvents, but solubility depends on the chemical properties of the organic solvent. According to Ph. Eur., fluoxetine is freely soluble in methanol and sparingly soluble in methylene chloride.

3.6.2   Atenolol – A More Polar Basic Drug

Atenolol is an antiadrenergic and antihypertensive drug (heart and blood pressure drug). The active pharmaceutical ingredient of atenolol is the free base, and is delivered as a mixture of two enantiomers (one chiral centre, Figure 3.12). Both Ph. Eur. and USP publish

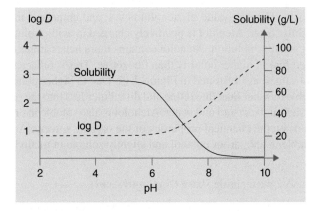

Figure 3.11   *Solubility in water and log D (1-octanol-water) of fluoxetine as function of pH*

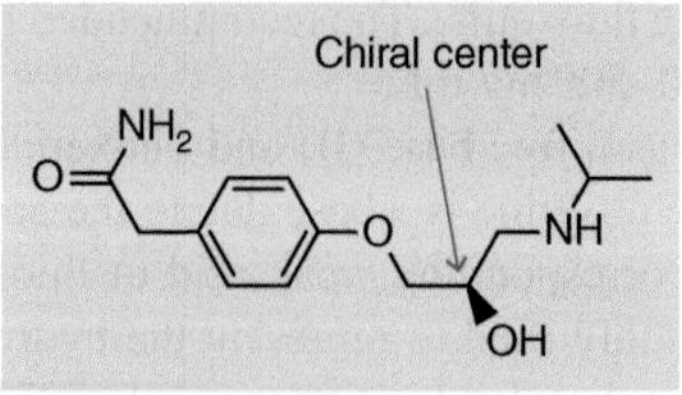

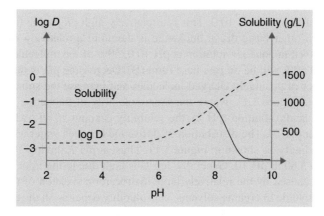

Figure 3.12 *Structure of atenolol*

Figure 3.13 *Solubility in water and log D (1-octanol-water) of atenolol as function of pH*

a monograph for atenolol. USP also publishes monographs for atenolol injection, tablets, oral solution, and oral suspension, which are common pharmaceutical preparations with atenolol. Tablets typically contain 25–100 mg atenolol per tablet. The recommended blood plasma concentration of atenolol is in the range 100–1000 ng/mL.

The structure of atenolol is presented in Figure 3.12. Atenolol contains a single basic secondary amine group (—NH—). The terminal —NH_2 group is an amide, which is neither a base nor an acid. The pK_a value of atenolol is 9.4, and compared to fluoxetine the base strength is slightly lower. Atenolol is positively charged in acidic solution (BH^+) and neutral (B) in strong alkaline solution. Atenolol contains more heteroatoms (N and O) than fluoxetine and is therefore of higher polarity than fluoxetine. This is reflected in both water solubility and log *D* values, as illustrated in Figure 3.13. Due to higher polarity, atenolol is generally more soluble in water than fluoxetine and distributes less into the organic phase in two-phase systems such as 1-octanol and water. Atenolol is also soluble in organic solvents, but solubility depends on the chemical properties of the organic solvent. According to Ph. Eur., atenolol is soluble in anhydrous ethanol and slightly soluble in methylene chloride.

3.6.3 Morphine – A Zwitterionic Drug (Base and Acid)

Morphine is an analgesic drug (natural opium alkaloid). Morphine is delivered as morphine hydrochloride or morphine sulfate as the active pharmaceutical ingredient. Both Ph. Eur.

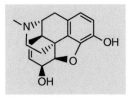

Figure 3.14   *Structure of morphine*

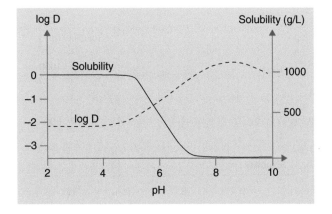

Figure 3.15   *Solubility in water and log D (1-octanol-water) of morphine as function of pH*

and USP publish a monograph for morphine. USP also publishes monographs for morphine capsules, injection, and suppositories, which are common pharmaceutical preparations with morphine. Injections typically contain 10 or 20 mg/mL morphine. The recommended blood plasma concentration of morphine is in the range 10–100 ng/mL.

The structure of morphine as a free base is illustrated in Figure 3.14. Morphine contains a tertiary amine group (–N) and is therefore basic. The pK_a value of this functionality is about 8.2. Morphine also contains a phenol group (OH) and therefore exhibit weakly acidic properties. The pK_a value of the acidic functionality is about 10.2. In acidic solution, morphine exists mainly as positively charged ions and in strongly alkaline solution the substance is dissolved in anionic form. However, at neutral and slightly alkaline conditions, morphine molecules carry both positive and negative charges and are *zwitterions*. This pattern is also reflected in the solubility and log *D* curves plotted in Figure 3.15. According to the Ph. Eur., morphine is soluble in water, slightly soluble in ethanol, and practically insoluble in toluene.

3.6.4   Ibuprofen – An Acidic Drug

Ibuprofen is an anti-inflammatory drug. Ibuprofen contains a single chiral centre (Figure 3.16) and the pharmaceutical ingredient is delivered as a mixture of the two enantiomers. The pharmaceutical ingredient is the free carboxylic acid form of ibuprofen.

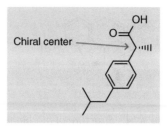

Figure 3.16   Structure of ibuprofen

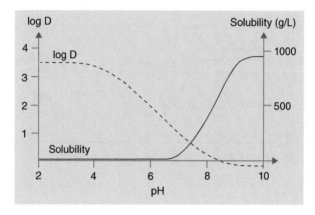

Figure 3.17   Solubility in water and log D (1-octanol-water) of ibuprofen as function of pH

Both Ph. Eur. and USP publish a monograph for ibuprofen. USP also publishes monographs for ibuprofen tablets and oral suspension, which are common pharmaceutical preparations containing ibuprofen. Tablets typically contain 200–600 mg ibuprofen per tablet. The recommended blood plasma concentration of ibuprofen is in the range 15–30 µg/mL. Thus, compared with fluoxetine and atenolol discussed above, ibuprofen is a high dose substance.

The structure of ibuprofen is illustrated in Figure 3.16. Ibuprofen contains a single carboxylic acid group (COOH) and the substance is therefore a weak acid. The pK_a value is 4.4. Thus, in strongly acidic solution, ibuprofen molecules are mainly neutral (HA) and solubility in water is very low. In neutral and alkaline solution, the ibuprofen molecules deprotonate and negative ions (A^-) are formed. Thus, the solubility of ibuprofen is substantially higher in neutral and alkaline solutions, as shown in Figure 3.17. The ibuprofen molecule mainly contains hydrocarbon, and therefore the log P value is relatively high (3.50). Thus, especially in acidic solution, ibuprofen distributes mainly into 1-octanol (high log D) and other organic phases, as illustrated in Figure 3.17. Ibuprofen is soluble in organic solvents, but its solubility depends on the chemical properties of the organic solvent. According to Ph. Eur., ibuprofen is freely soluble in acetone, methanol, and methylene chloride.

3.6.5 Paracetamol – A Weak Acid

Paracetamol (or *acetaminophen*) is an analgesic drug. Both Ph. Eur. and USP publish a monograph for paracetamol. USP also publishes monographs for paracetamol capsules, oral solution, suppositories, oral suspension, extended-release tablets, and different combination tablets with other active pharmaceutical ingredients. Tablets typically contain 500–1000 mg of paracetamol per tablet. The recommended blood plasma concentration of paracetamol is in the range 5–25 µg/mL. Thus, similar to ibuprofen, paracetamol is a high dose substance.

The structure of paracetamol is illustrated in Figure 3.18. Paracetamol contains a phenolic group (—OH) and the substance is therefore acidic. However, the pK_a value is 9.7 and the acidic properties are very weak. Thus, to dissolve paracetamol as negatively charged ions, the solution has to be strongly alkaline. In neutral and acidic solutions, paracetamol is present in the neutral form. The —NH— functionality is part of an amide group, and is therefore not basic. Paracetamol is a relatively small molecule, with many heteroatoms. Therefore, the log P value is low (log P = 0.48). Paracetamol is therefore more soluble in water than ibuprofen. The water solubility is plotted versus pH in Figure 3.19. Due to ionization, the solubility is increasing in alkaline solution. The log D value versus pH is also illustrated in Figure 3.19. Partition into 1-octanol is highest in acidic solution where the molecules are neutral. Paracetamol is soluble in organic solvents, but its solubility depends on the chemical properties of the organic solvent. According to Ph. Eur., paracetamol is freely soluble in alcohol and is very slightly soluble in methylene chloride.

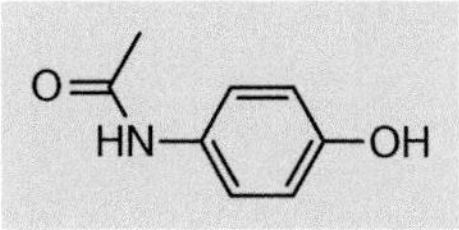

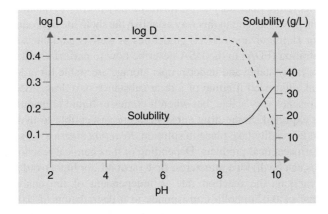

Figure 3.18   *Structure of paracetamol*

Figure 3.19   *Solubility in water and log D (1-octanol-water) of paracetamol as function of pH*

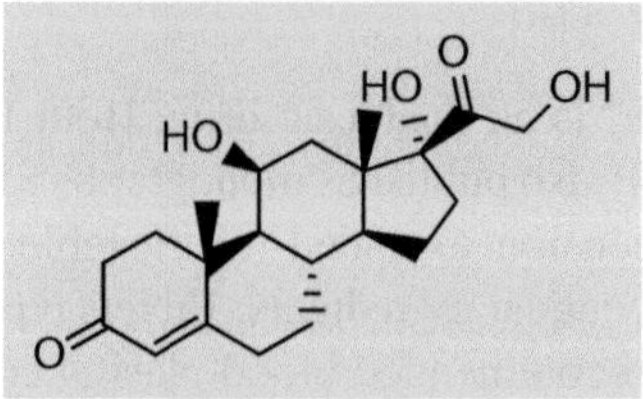

Figure 3.20 *Structure of hydrocortisone and hydrocortisone hydrogen succinate*

3.6.6 Hydrocortisone – A Neutral Drug

Hydrocortisone is a corticosteroid. Hydrocortisone is delivered as a pharmaceutical ingredient in different forms, including hydrocortisone, hydrocortisone acetate, hydrocortisone butyrate, hydrocortisone sodium phosphate, hydrocortisone valerate, and hydrocortisone hydrogen succinate. Both Ph. Eur. and USP publish monographs for the different forms of hydrocortisone. USP also publishes monographs for a large number of pharmaceutical preparations containing hydrocortisone.

The structure of hydrocortisone is illustrated in Figure 3.20. Hydrocortisone contains neither basic nor acidic functionalities. Thus, in aqueous solution all the molecules remain unchanged independently of pH and water solubility is not influenced by pH. The log *P* value for hydrocortisone is 1.76. Therefore, water solubility is strongly limited and hydrocortisone is distributed mainly into 1-octanol. This distribution is independent of pH. Hydrocortisone is sparingly soluble in ethanol and acetone, and slightly soluble in methylene chloride.

3.7 Stability of Drug Substances

Drug substances and drug products should be stable or only degrade to a small extent during *shelf-life*. It is therefore necessary to perform *stability testing* to obtain knowledge of possible degradation processes and in this way establish the shelf-life of products. The International Council for Harmonization (ICH) guidelines as well as guidelines from the Food and Drug Administration (FDA) in the USA describe how to perform such studies. Many drug substances are fairly stable and under proper storage are stable for at least five years, which is the normal authorized lifetime of a drug substance or a drug. Also a number of dry drug formulations are very stable, but when it comes to liquid preparations long time stability cannot be expected. Some drug substances are susceptible to hydrolysis and/or oxidation, which more readily take place in solution. *Reaction kinetics* is used to calculate the shelf-life of pharmaceutical products. Depending of the chemical reaction taking place the reaction kinetics can be divided into zero-, first-, second-, or higher-order reactions.

In zero-order reactions the reaction rate is independent of the analyte concentration. A rate constant of 0.02 mmol/h corresponds to a degradation of 0.02 mmol/h and 0.048 mmol/24 h. A solution containing 1 mol of a drug substance will in this case degrade to 90% of its original concentration (10 mmol of total degradation) within about 208 days. This will give the product a *shelf-life* of half a year.

In most cases, reaction kinetics are considered to be of first order or are approximated to a first-order reaction, also denoted pseudo first-order. In this case the reaction rate is dependent on the analyte concentration [A], and the unit of the rate constant, k, is time^{-1} (e.g. h^{-1} or s^{-1}):

$$-\frac{d[A]}{dt} = k[A] \tag{3.25}$$

The degradation process by time can be described as

$$\frac{dx}{dt} = k(a-x) \iff \int_0^x \frac{dx}{a-x} = \int_0^t k \, dt \tag{3.26}$$

Here x is the concentration of degraded product and a is the initial concentration of analyte. The half-life of the substance can be calculated from the following equation:

$$t_{0.5} = \frac{1}{k} \ln \frac{a}{a - \frac{1}{2}a} = \frac{1}{k} \ln 2 \tag{3.27}$$

The hydrolysis of aspirin (acetylsalicylic acid) is considered as a pseudo first-order reaction, and at pH 7.4 and 25 °C the rate constant is about 1.4×10^{-2} h^{-1}. Using Eq. (3.27) gives a half-life of 49 hours and 10% degradation will take place within 7.5 hours. It is therefore not possible to store liquid preparations of aspirin as the shelf-life would be only a few hours. Due to the relative fast degradation it is also important to consider the stability of the prepared sample when performing an analysis of aspirin tablets. In order not to bias the obtained quantitative analytical data, the extracted tablet solution should be analysed as fast as possible. Degradation to 0.5% or 1.0% will take place within 0.36 h or 0.7 h, corresponding to about 20 min or 40 min, respectively. Longer storage of the sample solution will increase the bias of the analytical data.

4

Fundamentals of Pharmaceutical Analytical Chemistry

4.1 Pharmaceutical Analytical Chemistry

Pharmaceutical analysis is intended to either *identify* or *quantify* one or more substances in a given sample of pharmaceutical interest. In pharmaceutical analysis, the substance or substances of interest are normally active pharmaceutical ingredients (APIs), excipients,

Introduction to Pharmaceutical Analytical Chemistry, Second Edition.
Stig Pedersen-Bjergaard, Bente Gammelgaard and Trine Grønhaug Halvorsen.
© 2019 John Wiley & Sons Ltd. Published 2019 by John Wiley & Sons Ltd.

contaminants, degradation products, impurities, drug substances, and drug metabolites. A substance to be identified or quantified is termed *analyte*. The samples in pharmaceutical analysis are typically pharmaceutical ingredients, pharmaceutical preparations, or biological fluids like human blood and urine. The samples consist of one or several analytes, and a sample *matrix* which is defined as the rest of the sample. *Identification* is intended to confirm the identity of the analytes. Identification can also be referred to as *qualitative analysis*. *Quantitative analysis* (also termed *quantitation* or *quantification*) is intended to measure the exact concentration or the exact amount of the analyte in a given sample. As an example, paracetamol tablets containing 250 mg of paracetamol per tablet as the active ingredient have to be controlled prior to release from production. This is accomplished by pharmaceutical analysis. Paracetamol is the analyte, whereas the rest of the tablet, consisting of different inactive substances, is the sample matrix. Identification of paracetamol in the tablets is performed to make sure that the tablets contain the correct active ingredient, whereas quantitation is performed to measure the content of paracetamol and to check that this result is within the specification of the pharmaceutical manufacturer. The latter may state that the content should be within the range of 250 ± 12.5 mg per tablet.

Procedures for pharmaceutical analysis are often complicated and consist of several steps, as illustrated in Figure 4.1. First, a sampling is performed, where the required number of samples are taken. During sampling, it is essential that samples are taken in a representative manner. Manufactured tablets have to be sampled in a systematic way during the entire time scale of production to give an average of the total production. Sampling is beyond the scope of this book and is not discussed further here. However, sampling is subject to much focus by pharmaceutical manufacturers.

Often, the samples must be stored until further analysis. The principal challenge with *sample storage* is to avoid degradation and compositional changes of the sample. If such changes occur, the final quantitation will not reflect the original composition of the sample. To protect samples against degradation, they are often stored at low temperature and protected from light, as in a refrigerator or a freezer. This is especially important for liquid samples where degradation chemistry is likely to occur. In the case with 250 mg paracetamol tablets, sample storage is not a critical issue as the tablets are stable at room temperature. Sample storage is also beyond the scope of this book and is not discussed further here.

After sampling and storage, samples are normally pretreated in some way, which is termed *sample preparation*. Sample preparation can be very simple or quite complicated,

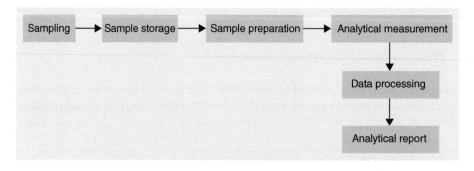

Figure 4.1 *Different steps in a typical procedure in pharmaceutical analysis*

depending on the complexity of the sample. Sample preparation serves the following objectives:

- The sample is compatible with the instrument for the analytical measurement.
- The analyte is present in sufficient concentration for the analytical measurement.
- The sample matrix that can interfere with the analytical measurement is removed.

In the case of tablets, this involves pulverization of tables and dissolution of the tablet powder and filtration of the solution to remove insoluble material.

At the end, the *analytical measurement* is performed, where the analyte or analytes are identified and quantified. In the example with tablets, chromatography is typically used to identify paracetamol and to measure the quantity of paracetamol in the tablets. After the analysis, measurements are processed (*data processing*), where the results are calculated and presented in an *analytical report*.

4.2   How to Specify Quantities, Concentrations, and Compositions of Mixtures

In most cases, the analytical measurement is performed when the analyte (drug substance) is present in solution. A *solution* is a homogeneous mixture of two or more substances. The species dissolved in a solution are termed *solutes*. The analyte is therefore an example of a solute. The liquid is termed *solvent*. When 0.100 g of paracetamol is dissolved in 500.0 mL of water, paracetamol is the solute and water is the solvent. The amount of analyte or solute in solution is normally expressed as *concentration*. Concentration means the amount of solute per volume unit of solution. In this example the concentration could be expressed in SI units (g/L) as $0.100 \, \text{g}/0.5 \, \text{L} = 0.200 \, \text{g/L}$. Often concentrations are expressed in milligram per millilitre (mg/mL) instead. Micrograms per millilitre (µg/mL), nanogram per millilitre (ng/mL), or picogram per millilitre (pg/mL) can also be used:

$$1 \, \text{g/L} = 1 \, \text{mg/mL} = 10^3 \, \text{µg/mL} = 10^6 \, \text{ng/mL} = 10^9 \, \text{pg/mL} \qquad (4.1)$$

An example of how to convert from mg/mL to µg/mL is shown in Box 4.1.

Box 4.1   Conversion from mg/mL to µg/mL

The concentration of paracetamol in a solution is 0.0125 mg/mL. This corresponds to the following concentration in µg/mL:

$$0.0125 \, \text{mg/mL} \times 10^3 \, \text{µg/mg} = 12.5 \, \text{µg/mL}$$

Often the term *molarity* is used to express the concentration. One mol is Avogadro's number ($N_A = 6.022 \times 10^{23} \, \text{mol}^{-1}$) of particles (atoms, molecules, ions). It is important to realize that mol is a number. The molarity of a certain solution is defined as follows:

$$\text{Molarity} = \text{number of mol of solute per litre of solution} \qquad (4.2)$$

The unit of molarity is mol/L = M. In Box 4.2 the molarity of paracetamol is calculated when 0.100 g of this substance is dissolved in 500.0 mL. In pharmaceutical analysis, molarity is often used to express concentration of chemical reagents in solution, such as solutions of hydrochloric acid, sodium hydroxide, and potassium hydrogen phthalate used for titration.

Box 4.2   Calculation of molarity

0.100 g of paracetamol (molar mass = 151.2 g/mol) is dissolved in water and the total volume is adjusted to 500.0 mL. The molarity of the solution with respect to paracetamol is calculated as

$$\frac{\dfrac{0.100\,g}{151.2\,g/mol}}{0.5000\,L} = 1.32 \times 10^{-3}\,M$$

When the concentrations are low, it is more convenient to use millimolar (mM), micromolar (µM), or nanomolar (nM):

$$1\,M = 10^3\,mM = 10^6\,\mu M = 10^9\,nM \tag{4.3}$$

An example of how to convert from M to µM is shown in Box 4.3.

Box 4.3   Conversion from M to µM

A solution is 1.62×10^{-5} M with respect to paracetamol. This corresponds to the following concentration in µM:

$$1.62 \times 10^{-5}\,M \times 10^6\,\mu M/M = 16.2\,\mu M$$

An example of how to convert from mg/mL to M is shown in Box 4.4.

Box 4.4   Conversion from mg/mL to M

The concentration of paracetamol in a solution is 0.0125 mg/mL. The molar mass for paracetamol is 151.2 g/mol. The concentration in M is calculated as follows:

$$0.0125\,mg/mL = 0.0125\,g/L$$

$$\frac{0.0125\,g/L}{151.2\,g/mol} = 8.27 \times 10^{-5}\,M$$

For a solute present in a given sample, the term *percentage* is often used to express the exact quantity of solute in the sample. However, the term percentage is ambiguous, and one of the following specific definitions should be used:

$$\% \text{ weight} = \%(w/w) = \frac{\text{mass of solute}}{\text{mass of sample}} \times 100\% \tag{4.4}$$

$$\% \frac{\text{weight}}{\text{volume}} = \%(w/v) = \frac{\text{mass of solute}}{\text{volume of sample}} \times 100\% \tag{4.5}$$

$$\% \text{ volume} = \%(v/v) = \frac{\text{volume of solute}}{\text{volume of sample}} \times 100\% \tag{4.6}$$

The term % weight, % (w/w), is synonymous with mass percentage, weight percent, and weight–weight percentage. This term is typically used to express the content of an API in a solid or semi-solid pharmaceutical preparation, such as a cream. Box 4.5 shows an example on how to calculate this using the term % (w/w).

Box 4.5   Calculation of % weight

One gram of cream contains 10 mg of hydrocortisone (active ingredient). This corresponds to

$$\frac{10 \text{ mg}}{1000 \text{ mg}} \times 100\% = 1.0\%(w/w)$$

The term % weight/volume, % (w/v), is typically used to express the content of an API in a liquid pharmaceutical preparation, such as eye drops. The density of the liquid is normally set to the value of water (1.0 g/mL). Thus, a 1 g/100 mL solution is 1% (w/v), regardless of whether the solvent is water (density 1.0 g/mL) or ethanol (density 0.79 g/mL). Box 4.6 shows an example on how to convert % (w/v) to the unit mg/mL.

Box 4.6   Conversion of % weight/volume to mg/mL

An eye drop preparation is labelled to contain 0.5% (w/v) chloramphenicol. This corresponds to

$$\frac{0.5 \text{ g}}{100 \text{ mL}} = \frac{500 \text{ mg}}{100 \text{ mL}} = 5 \text{ mg/mL}$$

Generally % (w/v) is related to mg/mL in the following manner:

$$1\%(w/v) = \frac{1 \text{ g}}{100 \text{ mL}} = \frac{1000 \text{ mg}}{100 \text{ mL}} = 10 \text{ mg/mL}$$

1% (w/v) = 10 mg/mL is a practical conversion factor.

The term % volume/volume, % (*v/v*), is used when different liquids are mixed, to specify the composition of the mixture. One example is the mixing of water and acetonitrile as the mobile phase for liquid chromatography.

At very low concentrations, it is convenient to use the terms *parts per million* (ppm) or *parts per billion* (ppb). These are defined as follows:

$$\text{ppm} = \frac{\text{mass of solute}}{\text{mass of sample}} \times 10^6 \tag{4.7}$$

$$\text{ppb} = \frac{\text{mass of solute}}{\text{mass of sample}} \times 10^9 \tag{4.8}$$

Often the terms are expressed in mass/volume instead; hence

$$1\,\text{ppm} = 1\,\mu g/mL = 1\,mg/L \text{ and } 1\,\text{ppb} = 1\,ng/mL = 1\,\mu g/L \tag{4.9}$$

Box 4.7 shows an example on how to calculate using ppm.

Box 4.7   Calculation of ppm

0.4001 g of $Pb(NO_3)_2$ is dissolved in dilute nitric acid and the volume is adjusted to 250.0 mL. This solution is subsequently diluted by a factor of 100 with dilute nitric acid. The concentration of lead (ppm) in the final solution is calculated as
Molar mass of $Pb(NO_3)_2$ is 331.2 g/mol.
0.4001 g of $Pb(NO_3)_2$ is equal to

$$\frac{0.4001\,g}{331.2\,g/mol} = 1.208 \times 10^{-3}\,mol\ Pb(NO)_3$$

The molar mass of Pb is 207.2 g/mol, and g of lead is calculated as follows:

$$1.208 \times 10^{-3}\,mol\ Pb = 1.208 \times 10^{-3}\,mol \times 207.2\,g/mol = 0.2503\,g\ Pb$$

In the first solution, the content of Pb is 0.2503 g of Pb in 250 g solution (assuming the density is 1.0 g/mL)
The final solution is 100 times diluted and therefore the content of Pb is

$$\frac{0.2503\,g}{250\,mL} \times \frac{1}{100} = 1.001 \times 10^{-5}\,g/mL$$

$$\frac{1.001 \times 10^{-5}\,g}{1\,mL} = \frac{1.001 \times 10^{-5}\,g}{1\,g} = \frac{10.01 \times 10^{-6}\,g}{1\,g} \times 10^6 = 10\,\text{ppm}$$

For testing of pharmaceutical ingredients and pharmaceutical preparations, concentrations are relatively high, and normally mg/mL, % (*w/v*), or % (*w/w*) are used to express concentrations. However, for the analysis of drug substances in biological samples, concentrations are normally much lower, and here it is customary to express the concentrations in ng/mL or nM.

4.3 Laboratory Equipment

4.3.1 The Analytical Balance

Quantitation in pharmaceutical analysis is based on accurate weighing of drug substances, chemicals, and samples. Such weighing is performed with *analytical balances, semi-microbalances, microbalances,* or *ultra-microbalances* to meet the requirements for high accuracy. The different types of balance are defined in Table 4.1.

Balances are fundamental instruments in the laboratory. Many laboratories have *laboratory balances* or *precision balances* in addition, but these are not intended for highly accurate weighing. A typical analytical balance is shown in Figure 4.2. Balances are equipped with a digital display with direct recording of the mass. A balance is characterized by its capacity, readability, and reproducibility. The *capacity* defines the maximum mass that can be weighed, and analytical balances have a weighing capacity up to 100 or 200 g. Semi-microbalances and microbalances have lower capacities. The *readability* is the smallest increment of mass that can be measured, and analytical balances have a readability of 0.1 mg. Reading the mass from the digital display of an analytical balance therefore gives

Table 4.1 Different types of balances used in pharmaceutical analysis

Balance name	Readability	Quantity of decimal digits (g)
Ultra-microbalance	0.1 µg	0.0000001
Microbalance	1 µg	0.000001
Semi-microbalance	0.01 mg	0.00001
Analytical balance	0.1 mg	0.0001

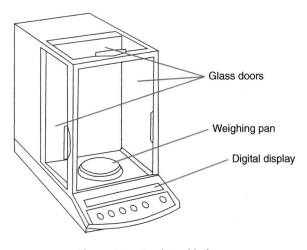

Glass doors

Weighing pan

Digital display

Figure 4.2 Analytical balance

a four-digit value such as 0.1006 g. The *repeatability* of the balance is a measure of the standard deviation when a standard mass is weighed several times.

Analytical balances are used extensively in pharmaceutical analysis. The analytical balance should be located on a heavy table, such as a marble table, to minimize vibrations and ensure stable readings. Analytical balances have adjustable feet and a bubble metre that allows the balance to be kept at level.

A normal weighing procedure with an analytical balance includes the following steps:

- Place an empty weighing vessel on the weighing pan (use a vessel of such a size that the loading capacity is not exceeded).
- Reset the reading of the balance (*tare* to 0.0000 or 0.00000 g).
- Fill the substance or sample to be weighed into the weighing vessel.
- Record the mass on the digital display.
- Empty the vessel into the sample container.
- Place the vessel on the balance again, read the balance, and subtract this mass from the mass from the first reading.
- Clean the balance after use.

It is important never to place drug substances or samples directly on the pan, as this will contaminate the balance. Therefore, the substances or the sample to be weighed should always be placed in a *weighing vessel* placed on the pan. First, place an empty vessel on the pan. Use a paper tissue or tweezers to handle the vessel, because fingerprints may change its mass. Let the balance stabilize for a few seconds and then reset the digital display to show 0.0000(0) g. Make sure that the weighing vessel is centred on the pan and that the glass doors protecting the pan are closed to protect it from drafts. Then place the substance or the sample to be weighed in the weighing vessel, make sure that the weighing vessel is still centred, close the doors, and wait for the balance to stabilize before the mass is recorded on the digital display.

Temperature plays an important role, and it is essential that the object to be weighed has the same temperature as the standard mass used for calibration. Since calibration is normally performed at room temperature, it is therefore important that the objects for weighing are kept at room temperature. A heated object must, therefore, be cooled to room temperature before it can be weighed. Static electricity from the weighing vessel or the object itself can also be a problem for exact weighing.

The balance has to be calibrated on a regular basis. Analytical balances calibrate themselves by a built-in procedure placing an internal mass below the weighing pan. This is termed *internal calibration*. In addition to internal calibration, the operator regularly has to perform *external calibration*. During external calibration, a standard mass is weighed to check that the reading is within the allowed limits (*tolerance*; limit of the error). The weighing is repeated several times to verify satisfactory reproducibility and measurement uncertainty. Balances are within different classes defined by their tolerance. The tolerances for *Class 1* and *Class 2* balances as used in the United States Pharmacopoeia (USP) are summarized in Table 4.2 (Classes 3 and 4 are not included).

For *assays* (quantitation) related to pharmaceutical ingredients and preparations, the selection of balance is highly important. Selection of balance is not based on Table 4.1, as the same type of balances (such as semi-microbalances) can vary significantly in terms

Table 4.2   Tolerance for Class 1 and Class 2 balances (according to American Society for Testing and Materials, Standard E 617)

Standard mass	Class 1 tolerance (mg)	Class 2 tolerance (mg)
100 g	0.25	0.50
10 g	0.050	0.074
1 g	0.034	0.054
100 mg	0.010	0.025
10 mg	0.010	0.014
1 mg	0.010	0.014

of weighing *accuracy*. Therefore, it is important to emphasize that readability and weighing accuracy are not the same. For selection of an appropriate balance, the following should be considered:

- The capacity of the balance must be larger than, or equal to, the largest load to be handled in the application.
- The uncertainty, when weighing the smallest sample, must be smaller than or equal to 0.1%.

If a balance meets these two criteria, it is in principle suitable for the application. The second condition is also known as the 'minimum weight condition'. For small sample masses, the *minimum sample weight* can be calculated according to the following equation:

$$\text{Minimum sample weight} = 1000 \times 2 \times \text{standard deviation} \qquad (4.10)$$

The standard deviation (repeatability) should be calculated from at least 10 replicate weighings. The balance can be used to weigh in the range from the minimum sample weight and up to the capacity. This is termed the *operating range*. Box 4.8 shows an example of how to calculate the minimum sample weight and how to establish the operating range.

Box 4.8   Calculation of minimum sample weight and operating range for a semi-microbalance

A semi-microbalance with a specified capacity of 60 g has a reported repeatability of 0.015 mg. The minimum sample weight is calculated as follows:

$$\text{Minimum sample weight} = 1000 \times 2 \times 0.015 = 30 \text{ mg}$$

This balance can therefore be used in the range from 30 mg up to 60 g (operating range).

USP has specific requirements for the repeatability and accuracy of balances. The requirements, and examples on how to check these, are given in Box 4.9.

Box 4.9   Testing repeatability and accuracy for a semi-microbalance

REPEATABILITY

Repeatability is satisfactory if two times the standard deviation (s) divided by the *desired smallest net weight* (N) does not exceed 0.10%. The 'desired smallest net weight' is the smallest mass the user plans to use on that balance (the user dictates this value). This test will determine if that 'desired smallest net weight' chosen is satisfactory or not.

As an example, a standard mass of 1.00000 g is weighed 10 times on a semi-microbalance with a specified capacity of 60 g, with the following results (g):

1.00001	1.00003	1.00002	1.00005	1.00001
1.00005	1.00002	1.00001	1.00003	1.00000

The standard deviation (repeatability) is calculated as

$$\text{Standard deviation} = 0.00002\,g = 0.02\,mg$$

The desired smallest net weight (N) is set to 50 mg and the criterion is

$$\frac{2s}{N} \times 100\% \leq 0.10\%$$

$$\frac{2s}{N} \times 100\% = \frac{2 \times 0.02}{50} \times 100\% = 0.08\% \leq 0.10\%$$

Thus, repeatability is satisfactory from an N value of 50 mg.

ACCURACY

The accuracy of a balance is satisfactory if its weighing value, when tested with a suitable weight, is within 0.10% of the test weight value. A test weight is suitable if it has a mass between 5% and 100% of the balance's capacity.

As an example, a standard mass of 100 g is weighed on a semi-microbalance with a specified capacity of 100 g, with the following result: 100.00236 g.
The criterion is

$$-0.1\% < \frac{\text{test weight} - \text{standard weight}}{\text{standard weight}} \times 100\% < 0.1\%$$

Calculation:

$$\frac{\text{Test weight} - \text{standard weight}}{\text{Standard weight}} \times 100\% = \frac{100.00236\,g - 100.00000\,g}{100.00000\,g} \times 100\%$$

$$= 0.002\% < 0.1\%$$

Thus, the accuracy is satisfactory.

Generally, Class 1 balances may be used for quantities greater than 10 mg, Class 2 balances may be used for quantities greater than 20 mg, Class 3 balances for quantities greater than 50 mg, and Class 4 balances for quantities greater than 100 mg. However, it should be emphasized that balances may be prone to larger uncertainties than specified by the manufacturer, depending on the local conditions in the laboratory of installation. This can be checked as illustrated in Boxes 4.8 and 4.9.

4.3.2   Pipettes

Pipettes are used to deliver exact volumes of liquid. Figure 4.3 shows two different types of pipettes, namely a transfer pipette and a measuring pipette. *Transfer pipettes* deliver a specified volume, such as 5.00 or 10.00 mL. *Tolerance* of transfer pipettes with different volumes are given in Table 4.3 for pipettes belonging to *Class A*. Class A pipettes are the most accurate and are recommended for assays related to pharmaceutical ingredients and preparations. If a lower accuracy is acceptable, *Class B* pipettes and measuring pipettes can be used. *Measuring pipettes* are graded and used to deliver variable volumes within a specified range, such as, for example, 8.4 mL.

Correct use of pipettes is essential in order to obtain high accuracy. Place the tip of the pipette into the solution. Use a rubber balloon to suck the liquid up into the pipette. Never use your mouth to suck up the liquid. The pipette is filled to a level higher than the *calibration mark* of the desired volume. Then remove the pipette from the solution and remove traces of solution on the outer surface of the pipette with a paper towel. Hold the pipette tip against the wall of an empty beaker (waste) and drain the liquid until the meniscus just reaches the calibration mark. Touching the beaker draws liquid from the pipette without leaving part of a drop hanging when the liquid reaches the calibration mark. The top of the liquid in the pipette will form a meniscus, as shown in Figure 4.4, and the bottom of this liquid meniscus should be exactly at the calibration mark of the desired volume. When the

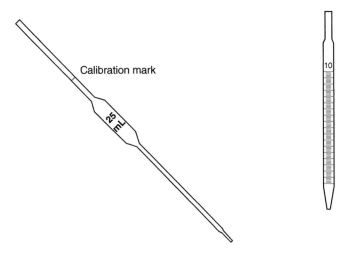

Figure 4.3   *Transfer pipette and measuring pipette*

Table 4.3   *Tolerance of Class A transfer pipettes (according to American Society for Testing and Materials)*

Designated volume (mL)	Limit of error (mL)	Relative error (%)
1	0.006	0.60
2	0.006	0.30
5	0.01	0.20
10	0.02	0.20
25	0.03	0.12
50	0.05	0.10

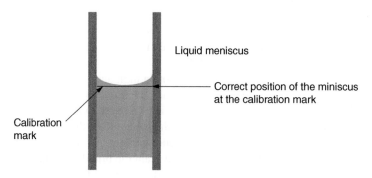

Figure 4.4   *Correct position of the meniscus at the calibration mark*

liquid level is adjusted to the mark of the desired volume, the pipette is transferred to the receiving vessel. The liquid is drained from the pipette by placing the pipette tip at the glass wall; allow the liquid to flow slowly along the wall. When seemingly all the liquid has drained from the pipette, it is still held for a few seconds against the glass wall before being removed. In the pipette tip, there will still remain a drop of liquid, but this should not be blown out of the pipette. A correction is made for the liquid remaining in the tip during calibration of the pipette. After using the pipette, it is important to rinse it well before using it again.

Another type of pipette is the *micropipette* shown in Figure 4.5. Micropipettes are typically used to deliver liquid volumes of between 1 and 5000 μL. By means of a screw at the end of the micropipette, the volume to be delivered can easily be adjusted. To use a micropipette, place a fresh tip tightly in the barrel. The tips are made of plastic and are intended for single use only. Then set the required volume with the screw on the top of the pipette. Depress the plunger to the first stop, which corresponds to the desired volume. Hold the pipette vertically and dip the tip 3–5 mm into the solution. It is important that the pipette is held vertically, because the amount that is drawn up in a micropipette depends on

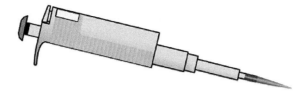

Figure 4.5 *Micropipette*

Table 4.4 *Tolerances for micropipettes (under optimal conditions)*

Pipette volume (μL)	At 10% pipette volume accuracy (%)	At 100% pipette volume accuracy (%)
2.5–25	±4.5	±0.8
10–100	±1.8	±0.6
100–1000	±1.6	±0.3

the angle between the pipette and the liquid. Then release the plunger slowly and carefully so that the pipette tip is filled with the desired volume of liquid. When the pipette tip is to be removed from the liquid, the tip should slide along the wall to remove excess liquid. This ensures that liquid located on the outer surface of the pipette tip is removed. Then transfer the liquid volume to the desired container by placing the pipette tip along the wall and push the plunger gently down to the first stop. After a second pause press the plunger to the bottom and the remaining liquid will be pressed out of the pipette tip. Unlike transfer pipettes and measuring pipettes, micropipettes should be emptied completely.

Micropipettes are very convenient, but precision and accuracy are not at the level of transfer pipettes, and performance may decline with time and use. Therefore, micropipettes require periodic maintenance (cleaning, seal replacement, and lubrication) and calibration. Performance of micropipettes can reach the level indicated in Table 4.4 under optimal conditions.

The accuracy (and precision) of pipettes can be tested by the operator. For micropipettes in particular, the volume reading can be adjusted according to accuracy testing. The latter is termed *calibration*, and micropipettes have to be calibrated frequently. Calibration is carried out by pipetting water, followed by exact weighing of the pipetted water using an analytical balance. Using a *correction factor*, which takes into account the density of water (which is temperature-dependent) and the *buoyancy*, the exact volume given by a pipette can be calculated using the following equation:

$$\text{Accurate volume} = \text{weight of water} \times \text{correction factor} \qquad (4.11)$$

The correction factor at different temperatures is given in Table 4.5. Box 4.10 exemplifies calibration of a micropipette.

Table 4.5 *Correction factor for the calibration of pipettes using water*

Temperature (°C)	Correction factor (mL/g)	Temperature (°C)	Correction factor (mL/g)
15	1.0020	16	1.0021
17	1.0023	18	1.0025
19	1.0027	20	1.0029
21	1.0031	22	1.0033
23	1.0035	24	1.0038
25	1.0040	26	1.0043
27	1.0046	28	1.0048
29	1.0051	30	1.0054

Box 4.10   Calibration of a micropipette

The volume reading of a 5-ml micropipette is to be checked. The pipette is set to 5.00 ml and water is pipetted and placed in a weighing vessel. The net mass of the pipetted water is 4.9905 g. The temperature in the laboratory is 23 °C. At this temperature, the correction factor is 1.0035 ml g^{-1} according to Table 4.5. The exact volume to be taken with the transfer pipette is calculated as follows:

$$Accurate\ volume = 4.9905\ g \times 1.0035\ mL/g = 5.01\ mL$$

Normally, the experiment above is repeated several times to establish an average value for the exact volume.

4.3.3   Volumetric Flasks

To prepare a solution with exact concentration of a given solute, the solute has to be weighed exactly and subsequently dissolved in an exact volume of solvent. For exact weighing the analytical balance is used and for dissolution in an exact volume a *volumetric flask* is used.

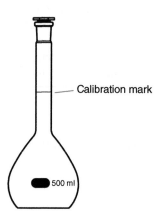

Calibration mark

500 ml

Figure 4.6   *Volumetric flask*

Table 4.6 *Tolerances of Class A volumetric flasks (according to American Society for Testing and Materials)*

Designated volume (mL)	Limit of error (mL)	Relative error (%)
10	0.02	0.20
25	0.03	0.12
50	0.05	0.10
100	0.08	0.08
250	0.12	0.05
500	0.15	0.03
1000	0.30	0.03

A volumetric flask is shown in Figure 4.6. A volumetric flask has a narrow neck with a calibration mark that shows to what exact level the solvent should be filled to get the exact and specified volume. Volumetric flasks are thus made for a particular volume at a particular temperature. This temperature is printed on the flask and is typically 20 °C.

During preparation of solutions with an exact concentration, the first step includes transfer of an exact amount of the solute into the volumetric flask. This can be either solid material weighed on an analytical balance or a liquid delivered with a pipette. It is important that the solute is transferred to the volumetric flask without losses. Then add some of the solvent and make sure that the solute is dissolved. Finally, add solvent to the calibration mark. The lower meniscus should be at level with the calibration mark, as illustrated in Figure 4.4. Volumetric flasks are available in several different qualities. The highest quality is *Class A* and Table 4.6 shows the tolerance of these with different volumes.

4.3.4 Burettes

A *burette* is a long tube of glass with a tap at the lower end. A burette is shown in Figure 4.7. Along the tube there is a graduation (millilitre scale) that makes it possible to continuously read the volume of liquid delivered from the burette. Burettes are used in titration, where a solution (*titrant*) is added gradually to a sample (*titrate*) until a given point (*endpoint*) where the titration is terminated. At this point, the operator can read off the consumption of titrant with high accuracy on the burette. Readings on the burette are made the same way as for pipettes and volumetric flasks. This means that it is the level of the lower meniscus that is used (Figure 4.4). When reading the titrant level on a burette, the eye of the operator should be at the same height as the top of the liquid. For burettes, there are also several different qualities, with the most accurate graded as *Class A*. Class A burettes should be used for assays related to pharmaceutical ingredients and preparations. Tolerances of Class A burettes are summarized in Table 4.7.

To perform a titration, the burette is first filled with the titrant. The level of titrant is read on the millilitre scale before starting. The starting level can be 0.00 mL or another exact reading like 0.67 mL. Then the tap is opened gently and the titrant flows slowly from the burette and into the titrate. Towards the end of the titration, the titrant is delivered at a reduced speed by gently opening and closing the tap. At the end of the titration, the level of the titrant is read again, and the difference between the start level and end level represents the exact volume consumed during the titration. If the titration starts at 0.67 mL and ends at

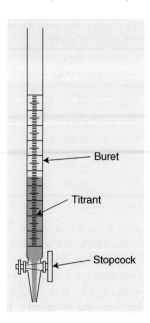

Figure 4.7 Burette

Table 4.7 Tolerances of Class A burettes (according to American Society for Testing and Materials)

Designated volume (mL)	Subdivisions (mL)	Limit of error (mL)
10	0.02	0.02
25	0.10	0.03
50	0.10	0.05

15.75 mL, the total volume of titrant is $15.75 - 0.67 = 15.08$ mL. If the titration is performed with a 25-mL burette with a tolerance of 0.03 mL, the true value may be between 15.05 and 15.11 mL. In some cases, an air bubble may be located in the tap before starting, and it is important that this is removed prior to titration. This can be done by allowing some titrant to drain out of the burette with the tap in the fully open position.

In high throughput laboratories, the manual pipettes are exchanged with software controlled automated systems, or laboratory robots that can dispense volumes automatically. Titrations are also performed by automatic titration systems.

4.4   How to Make Solutions and Dilutions

To make a solution of a solute with an exact concentration, the first step is to weigh out the correct mass of the solute on an analytical balance. Then the solute is transferred from the weighing vessel and into a volumetric flask. To transfer all the solute from the weighing vessel, it is normal to wash with the solvent by flushing the solvent directly into the volumetric flask. More solvent is added to dissolve the solute. When all solute has dissolved, the volumetric flask is filled to the calibration mark with solvent. In order not to overfill the volumetric flask, it is common towards the end to add the solvent carefully and exactly to the calibration mark. If it is undesirable to wash the weighing vessel to transfer 100% of solute into the volumetric flask, the weighing vessel can be weighed after transfer of most solute into the volumetric flask, and the difference in mass from the original recording is the exact amount of solute transferred into the volumetric flask.

Frequently, solutions with very low concentrations of a given solute have to be prepared. A 1 mg/L (1 μg/mL) solution is such an example, and theoretically this solution can be prepared by weighing 1 mg of solute and dissolve it in 1 L of solvent. However, 1 mg of solute is a very small amount, and weighing at this level is challenging and associated with uncertainty. In addition, 1 L of solvent is required, which is a large and inconvenient volume, and may be expensive. To prepare very dilute solutions, it is therefore usual to first make a more concentrated solution, termed the *stock solution*, and then *dilute* this to a lower concentration. Thus, to prepare the 1 mg/L solution, 50.00 mg of solute can be dissolved in 100.0 mL (0.1000 L) of solvent. The concentration of this stock solution is 500.0 mg/L. Then 5.00 mL of the stock solution is pipetted and diluted to a total volume of 100.0 mL, resulting in a 20 times dilution, and the concentration is now 25.0 mg/L. Finally, 2.00 mL of this solution is pipetted and diluted to a total volume of 50.0 mL, resulting in a 25 times dilution of this solution, and the concentration is now 1.00 mg/L. Thus, by preparing the solution in three steps and based on weighing 50 mg, uncertainty has been reduced and the total volume of solvent used is now 250 mL. (The stock solution was diluted 500 (20 × 25) times during the procedure. This could also be achieved by diluting 0.2 mL of the stock solution to 100.0 mL).

Calculations related to dilutions can be accomplished with the *dilution equation*:

$$C_{undilute} \times V_{undilute} = C_{dilute} \times V_{dilute} \qquad (4.12)$$

Box 4.11   Dilution of hydrochloric acid

A 1.05 M solution of hydrochloric acid is to be diluted to obtain a solution that is 1.00×10^{-2} M. A volume of 500.0 mL is requested for the diluted solution. The number of millilitres of concentrated solution to be added to a 500 mL volumetric flask is calculated as follows. The dilution equation is used where $C_{undilute} = 1.05$ M, $C_{dilute} = 1.00 \times 10^{-2}$ M, $V_{dilute} = 500.0$ mL, and where $V_{undilute}$ is unknown:

$$1.05 \text{ M} \times V_{undilute} = 1.00 \times 10^{-2} \text{M} \times 500.0 \text{ mL}$$

By rearrangement:

$$V_{undilute} = \frac{1.00 \times 10^{-2} \text{ M} \times 500.0 \text{ mL}}{1.05 \text{ M}} = 4.76 \text{ mL}$$

Thus, 4.76 mL of 1.05 M HCl should be diluted to 500.0 mL to give a final concentration of 1.00×10^{-2} M.

$C_{undilute}$ is the concentration of solute in the undiluted solution, $V_{undilute}$ is the volume of the undiluted solution, C_{dilute} is the concentration of solute in the diluted solution, and V_{dilute} is the volume of the diluted solution. $C_{undilute} \times V_{undilute}$ represents the number of mol (or mg) of solute taken from the concentrated (undiluted) solution, which equals the amount placed in the dilute solution ($C_{dilute} \times V_{dilute}$). Calculation examples related to dilution are shown in Boxes 4.11 and 4.12.

Box 4.12   Dilution of paracetamol solution

50.0 mg of paracetamol is dissolved in methanol in a volumetric flask and the final volume is adjusted to 100.0 mL. From this solution, 1.00 mL is collected by a pipette and diluted to 100.0 mL with methanol in a new volumetric flask. The concentration of paracetamol in the final solution is calculated as follows. The first solution of paracetamol will have the following concentration ($C_{undilute}$):

$$C_{undilute} = \frac{50.0 \text{ mg}}{100.0 \text{ mL}} = 0.500 \text{ mg/mL}$$

The concentration in the final solution (C_{dilute}) is calculated according to the dilution equation, where

$$C_{undilute} = 0.500 \text{ mg/mL}$$

$$V_{dilute} = 100.0 \text{ mL}$$

$$V_{undilute} = 1.00 \text{ mL}$$

Based on the dilution equation:

$$0.500 \text{ mg/mL} \times 1.00 \text{ mL} = C_{dilute} \times 100.0 \text{ mL}$$

By rearrangement:

$$C_{dilute} = \frac{0.500 \text{ mg/mL} \times 1.00 \text{ mL}}{100.0 \text{ mL}} = 5.00 \times 10^{-3} \text{ mg/mL} = 5 \text{ μg/mL}$$

4.5   Errors, Accuracy, and Precision

4.5.1   Systematic and Random Errors

In all quantitative pharmaceutical analysis, there will be some uncertainty associated with measurements because minor errors occur. This means that the analytical result is an *estimate* for the real content (*true value*) of analyte. It is of course important to reduce uncertainties to a minimum, to ensure that a given analytical result is as close to the true value as possible. Thus, care has to be taken both during the development of analytical methods and when the analytical methods are used for quantitative purposes.

Errors made during analytical procedures can be either gross errors, systematic errors, or random errors. *Gross errors* are due to major problems during the laboratory work and are recognized by the laboratory personnel. Gross errors can be such as spilt samples and incorrect use of analytical instruments. If a gross error occurs, the results are rejected and the procedure is repeated. *Systematic errors* (or determinate errors) are errors that can be detected and therefore can be corrected. Systematic errors will affect all the results in a sample sequence in the same direction, which means that all the results are either too low or too high. An example of a systematic error is if a pipette, which is intended to deliver 1.00 mL, is taking up 1.09 mL. If the pipette is not calibrated, the error may not be recognized and the pipette will give a larger volume than expected every time it is used. This will give rise to a systematic error that will affect any analysis for which the pipette is used.

Random errors (or indeterminate errors) are errors that cannot be detected and which therefore cannot be corrected. Random errors affect the results in a random fashion. In some cases, the analytical results will be too high and in other cases the results will be too low. As an example, filling of pipettes is prone to random error. In some cases, the meniscus hits the calibration mark exactly, providing an exact volume of liquid. In other cases, the meniscus is slightly below the calibration mark, resulting in a slightly lower volume, and in yet other cases the pipette is filled slightly more than marked. The error will turn out differently in the different cases, and corrections cannot be made.

4.5.2   Accuracy and Precision

Systematic and random errors affect analytical measurements, and the analytical results are basically estimates of the real content (true value). To describe the quality of analytical measurements, the terms *accuracy* and *precision* are used. In everyday life it is common to use these terms interchangeably, but in pharmaceutical analysis they have strict definitions. The accuracy expresses how well the analytical result matches the 'true value'. Accuracy can be expressed by the difference between the true value (X_t) and the analytical result (X), which is the *error*. The lower the error the better is the accuracy. Errors can be given in absolute values (*absolute error*) or relative values (*relative error*) using the following equations:

$$\text{Absolute error} = X - X_t \tag{4.13}$$

$$\text{Relative error} = \frac{X - X_t}{X_t} \times 100\% \tag{4.14}$$

Small errors in an analytical method mean that the method provides high accuracy, the analytical result is close to the true value, and this is of course preferable.

Quantitative measurements are often performed by multiple measurements (repetitions or *replicates*) to ensure that results are not affected by random errors. The precision of a method indicates how close the different measurements are. Precision is expressed by the *standard deviation* (*s*) or *relative standard deviation* (*RSD*). The terms accuracy and precision are further illustrated in Figure 4.8. Establishment of accuracy and precision of an analytical method is part of the method of validation.

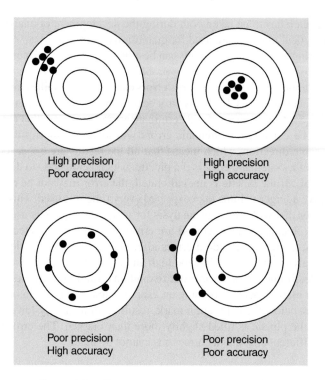

Figure 4.8   *Illustration of accuracy and precision*

4.6   Statistical Tests

4.6.1   Mean Value and Standard Deviation

Most measurements in pharmaceutical analysis follow a *Gaussian distribution*. These measurements can be characterized by a *mean value* ($\overline{x}$) and a *standard deviation* (s). These terms are defined by the following equations:

$$\overline{x} = \frac{x_1 + x_2 + x_3 + \ldots + x_n}{n} \tag{4.15}$$

$$s = \sqrt{\frac{\sum_{i=1}^{i=n}(x_i - x)^2}{n - 1}} \tag{4.16}$$

where $x_1, x_2, x_3, \ldots, x_n$ are the individual measurements and n is the number of measurements. The quantity $n - 1$ is termed the *degree of freedom*. RSD is often used instead of standard deviation and is calculated by the following equation:

$$\text{RSD} = \frac{s}{x} \times 100\% \tag{4.17}$$

To report an analytical result, which is often based on replicate measurements, both the mean value (average) and the standard deviation (or RSD) are calculated. This is illustrated in Box 4.13. In practice, these calculations are performed in a spreadsheet, such as Excel.

Box 4.13   Calculation of x, s, and RSD

Quantitation of paracetamol in an oral solution, with a specified content of 24 mg/mL of paracetamol, gave the following results when six replicate measurements were conducted:

24.3 mg/mL	23.7 mg/mL	24.7 mg/mL	23.2 mg/mL	23.9 mg/mL	24.2 mg/mL

The mean value is calculated as follows:

$$\bar{x} = \frac{(24.3 + 23.7 + 24.7 + 23.2 + 23.9 + 24.2)\ \text{mg/mL}}{6} = 24.0\ \text{mg/mL}$$

The standard deviation is calculated as follows:

$$s = \sqrt{\frac{\begin{array}{c}(24.3 - 24.0)^2 + (23.7 - 24.0)^2 + (24.7 - 24.0)^2 \\ + (23.2 - 24.0)^2 + (23.9 - 24.0)^2 + (24.2 - 24.0)^2\end{array}}{6 - 1}} = 0.52\ \text{mg/mL}$$

The RSD is calculated as follows:

$$\text{RSD} = \frac{0.52\ \text{mg/mL}}{24.0\ \text{mg/mL}} \times 100\% = 2.2\%$$

In the final report, both the mean value of $24.0\ \text{mg ml}^{-1}$ and the standard deviation of $0.5\ \text{mg ml}^{-1}$ (or RSD of 2.2%) is given to describe the analytical result.

Calculations of the mean value and standard deviation are normally performed in Excel as follows:

(1) The numbers are arranged in a column, e.g. (A1:A6).
(2) The mean value is calculated by the function: =AVERAGE(A1:A6).
(3) The standard deviation is calculated by the function: =STDEV.S(A1:A6).

	A	B	C	D	E
1	24.3				
2	23.7				
3	24.7				
4	23.2				
5	23.9				
6	24.2				
7	**24.0**		A7: =AVERAGE(A1:A6)		
8	**0.5215**		A8: =STDEV.S(A1:A8)		

4.6.2 Confidence Intervals

If a sample is analysed by an infinite number of replicate measurements, and if the measurements are not affected by any systematic errors, the calculated average value (x) will be equal to the true value (μ), and in a similar way, the standard deviation (s) will be equal to the true standard deviation (σ). Performing a large number of replicate measurements on a single sample is not feasible, and in practice only a small number (typically 3–6) of replicate measurements are performed. In the latter case, the *confidence interval* can be calculated, which is an interval that includes the true value (μ) with a given probability (assuming no systematic errors). Confidence intervals are calculated using the following equation:

$$\mu = \bar{x} \pm \frac{t_{(n-1)}s}{\sqrt{n}} \tag{4.18}$$

Here $\bar{x}$ is the mean value, s is the estimated standard deviation, n is the number of replicate measurements, and $t_{(n-1)}$ is obtained from Table 4.8. Note that the degrees of freedom are equal to $n-1$.

A confidence interval calculated at 95% confidence level implies that there is 95% probability that the true value is within the calculated confidence interval. An example of how to calculate a confidence interval is given in Box 4.14.

4.6.3 Comparison of Standard Deviations with the *F*-Test

In some cases it is important to compare the standard deviations (precisions) of two methods. Consider that a new and faster analytical method has been developed and that this method apparently gives the same result as the old method. However, it is important that the precision of the new method is not poorer than the precision of the old method. This can be examined by an *F*-test, where *variances* of the two methods are compared (the variance is the square of the standard deviation). In order to test whether the difference

Table 4.8   Values of student's t

Degree of freedom ($n - 1$)	Confidence level			
	90%	95%	99%	99.9%
1	6.314	12.706	63.657	636.619
2	2.920	4.303	9.925	31.598
3	2.353	3.182	5.841	12.924
4	2.132	2.776	4.604	8.610
5	2.015	2.571	4.032	6.869
6	1.943	2.447	3.707	5.959
7	1.895	2.365	3.500	5.408
8	1.860	2.306	3.355	5.041
9	1.833	2.262	3.250	4.781
10	1.812	2.228	3.169	4.587
15	1.753	2.131	2.947	4.073
20	1.725	2.086	2.845	3.850

Box 4.14   Calculation of confidence interval

In Box 4.13 the content of paracetamol was measured to 24.0 mg/mL with a standard deviation of 0.52 mg/mL. The number of replicate measurements was six and the degrees of freedom is therefore five. A confidence interval for the true value at 95% confidence level is calculated as follows, utilizing a value of 2.571 for t as found in Table 4.8:

$$\mu = 24.0 \pm \frac{2.571 \times 0.52 \text{ mg/mL}}{\sqrt{6}} = [24.0 \pm 0.5] \text{ mg/mL}$$

This means that, with 95% probability, the true value of the content of paracetamol is within the interval $[24.0 \pm 0.5]$ mg/mL.

Calculations of confidence intervals are normally performed in Excel as follows:

(1) Confidence interval is calculated by the function: =CONFIDENCE(alpha, standard_dev,size).
(2) Alpha is the confidence level (95% corresponds to 0.05), standard_dev is the calculated standard deviation, and size is the sample size.
(3) In this case: =CONFIDENCE(0.05,0.52,6) = 0.546.

---

between two variances is significant, the hypothesis H_0: $\sigma_1^2 = \sigma_2^2$ is tested by calculating $F_{\text{calculated}}$:

$$F_{\text{calculated}} = \frac{s_1^2}{s_2^2} \tag{4.19}$$

Notice that the larger standard deviation is allocated in the equation so that $F_{\text{calculated}} \geq 1$. The number of degrees of freedom for n measurements is $(n-1)$. $F_{\text{calculated}}$ is compared with the critical value (F_{critical}) in Table 4.9. If $F_{\text{calculated}} > F_{\text{critical}}$ the difference is significant.

Table 4.9   *Critical values of F at 95% confidence level (two-tailed)*

Degrees of freedom for s_2	Degrees of freedom for s_1									
	1	2	3	4	5	6	7	8	9	10
1	647.79	799.50	864.16	899.58	921.85	937.11	948.22	956.66	963.28	968.63
2	38.51	39.00	39.17	39.25	39.30	39.33	39.36	39.37	39.39	39.40
3	17.44	16.04	15.44	15.10	14.88	14.73	14.62	14.54	14.47	14.42
4	12.22	10.65	9.98	9.60	9.36	9.20	9.07	8.98	8.90	8.84
5	10.01	8.43	7.76	7.39	7.15	6.98	6.85	6.76	6.68	6.62
6	8.81	7.26	6.60	6.23	5.99	5.82	5.70	5.60	5.52	5.46
7	8.07	6.54	5.89	5.52	5.29	5.12	4.99	4.90	4.82	4.76
8	7.57	6.06	5.42	5.05	4.82	4.65	4.53	4.43	4.36	4.30
9	7.21	5.71	5.08	4.72	4.48	4.32	4.20	4.10	4.03	3.96
10	6.94	5.46	4.83	4.47	4.24	4.07	3.95	3.85	3.78	3.72

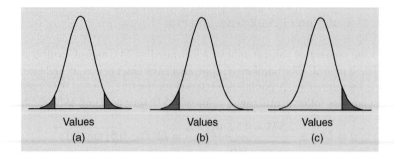

Figure 4.9   *Normal distribution of (a) two-tailed and (b) and (c) one-tailed testing*

Simple statistic tests related to pharmaceutical analyses, including the *F*-test and *t*-test discussed in this textbook, are based on the assumption that the experimental data are normally distributed (*Gaussian distribution*, Figure 4.9). The *F*-test tests if two variances (σ_1 and σ_2) differ significantly, whereas the *t*-test tests if two mean values (x_1 and x_2) differ significantly. The tests can either be one-sided (*one-tailed*) or two-sided (*two-tailed*), as illustrated in Figure 4.9. A two-tailed approach tests both the possibility that σ_1 or x_1 is higher or lower than σ_2 or x_2. The one-tailed approach only tests one of the possibilities. Most often two-tailed tests are used. An example of the use of the *F*-test for comparison of standard deviations of two analytical methods is given in Box 4.15.

4.6.4   Comparison of Means with a *t*-Test

Often the analytical results from two different samples have to be compared, to check whether or not they are different with respect to the content of analyte. If the content of paracetamol in one sample is measured to 24.02 mg ml^{-1} and in another sample measured to 24.36 mg/mL, it is tempting to believe that the contents of the two samples are different. This may or may not be true and is entirely dependent on the standard deviation of the two measurements. If the standard deviations are very small, and thus the method has high precision, the content of the analyte in the two samples is probably different. If, however, the measurements are associated with larger standard deviations, the measurements of 24.02 and 24.36 mg/mL, respectively, are not significantly different. In order to determine whether two sets of measurements are significantly different, a *t-test* has to be performed.

A *t*-test is performed by calculating the mean value for both measurement series (series 1 and 2), which are referred to as x_1 and x_2, respectively. The number of measurements in series 1 is n_1 and n_2 is the number of measurements in series 2. First, the *pooled standard deviation* (S_{pooled}) for the two measurement series is calculated with the following equation,

Box 4.15 Comparison of standard deviations of two analytical methods

Quantitation of paracetamol in a 500 mg tablet formulation was performed by ultraviolet (UV)-spectrophotometry (UV) and by liquid chromatography with UV (LC-UV) detection. The precision of the methods was compared to examine whether they were different.

The results of the two methods are shown below, where six replicate measurements of mg paracetamol per tablet was conducted with each method:

UV	LC-UV
503.54	494.74
502.06	498.13
500.27	498.84
502.23	500.99
498.72	499.50
497.12	500.75

The standard deviations of the two methods are calculated in Excel, and F is calculated using the largest standard deviation as the numerator, making $F<1$:

=AVERAGE	500.66	498.83
=STDEV.S	2.42	2.28
s^2	5.83	5.20
$F = s_1^2/s_2^2$	1.12	

$F_{calculated}$ is compared to $F_{critical}$ from Table 4.9. As both methods were applied to six samples, the number of degrees of freedom is $(6-1) = 5$ for both data sets and the value of $F_{critical}$ is 7.15 according to Table 4.9. Thus, the calculated F value is lower than the table value, leading to the conclusion that the standard deviations of the two methods are not different.

Calculation in Excel

An F-test can also be performed in Excel. This is done by activating *Data* followed by *Data analysis* and selecting F-test in the dropdown menu. However, it is important to note that the test in Excel is a one-tailed test and thus the confidence level should be changed from 0.05 to 0.025 to obtain the correct value of $F_{critical}$. This is illustrated below in the output for $(p = 0.025)$ on the left. The $F_{critical}$ here is equal to the value from Table 4.9.

F-test: Two-sample test for variance ($p = 0.05$)

	Variable 1	Variable 2
Average	500.656667	498.825
Variance	5.83330667	5.20379
Observations	6	6
Degrees of freedom	5	5
F	1.12097273	
$P(F \leq f)$ one-tailed	0.45166468	
F-critical one-tailed	5.05032906	

F-test: Two-sample test for variance ($p = 0.025$)

	Variable 1	Variable 2
Average	500.65667	498.825
Variace	5.8333067	5.20379
Observations	6	6
Degrees of freedom	5	5
F	1.1209727	
$P(F \leq f)$ one-tailed	0.4516647	
F-critical one-tailed	7.1463818	

where s_1 and s_2 are the standard deviations for the two individual series of measurements (series 1 and 2, respectively):

$$S_{pooled} = \sqrt{\frac{s_1^2(n_1 - 1) + s_2^2(n_2 - 1)}{n_1 + n_2 - 2}} \tag{4.20}$$

Then, a $t_{calculated}$ value is calculated from the following formula:

$$t_{calculated} = \frac{|x_1 - x_2|}{S_{pooled}} \sqrt{\frac{n_1 n_2}{n_1 + n_2}} \tag{4.21}$$

The value for $t_{calculated}$ is then compared with a corresponding t value from Table 4.8, for example at the 95% confidence level. The number of degrees of freedom is $n_1 + n_2 - 2$. If $t_{calculated}$ is *numerically* higher than the value found in Table 4.8, the probability is 95% that the true values for the two measurement series are different. Similarly, the results are not significantly different if $t_{calculated}$ is numerically less than or equal to the value found in Table 4.8. An example of the use of the *t*-test is shown in Box 4.16.

The examples above demonstrate the simplicity of using a spreadsheet for testing differences between standard deviations and means for two samples.

By first testing if there is a difference between variances by an *F*-test, the correct *t*-test can be applied for testing difference between means. There are three possibilities: *t-test: two-sample assuming equal variances, t-test: two-sample assuming unequal variances*, and *paired t-test*. The latter is used for comparison of two analytical methods after applying the method to the same set of samples, e.g. analysing samples from 20 different batches, so each batch is characterized by a pair of samples analysed by the two different methods. In this way, differences in the tablet contents will not influence the difference between the methods.

Box 4.16   Example of *t*-test

Using the same data as for the *F*-test in Box 4.15 comparing the precision of two analytical methods, it is now examined if the analytical results are significantly different. The standard deviations were already calculated:

	Method 1	Method 2
=AVERAGE	500.66	498.83
=STDEV.S	2.415	2.281
s^2	5.83	5.20

First, the pooled standard deviation is calculated to

$$s_{pooled} = \sqrt{\frac{(2.415^2 \times (6-1) + 2.281^2 \times (6-1)}{(6+6-2)}} = 2.349$$

Then, the $t_{calculated}$ value is calculated:

$$t_{calculated} = \frac{|500.657 - 498.825|}{2.349} \times \sqrt{\frac{6 \times 6}{6+6-2}} = 1.35$$

The number of degrees of freedom is: $n = n_1 + n_2 - 2 = 6 + 6 - 2 = 10$.

From Table 4.8, the *t* value is found to be 2.228. Since $t_{calculated}$ is below the value in the table (1.35 < 2.228), the analytical results obtained in the two series are not significantly different (95% probability).

Calculation in Excel

The hypothesis tested is that $H_0 : \mu_1^2 = \mu_2^2$; thus the expected difference between the means is 0.

The *t*-test is opened via *Data* and *Data analysis* and selecting *t-test: two-sample assuming equal variances* in the dropdown menu. It has already been demonstrated that the variances are not different by the *F*-test in Box 14.15. The result of the test is:

t-test: Two-sample assuming equal variances		
	Variable 1	Variable 2
Average	500.656667	498.825
Variance	5.833307	5.20379
Observations	6.000000	6.000000
Pooled variance	5.518548	
Hypothesized mean difference	0.000000	
Degrees of freedom	10.000000	
t-stat	1.350500	
$P(T \le t)$ one-tailed	0.103312	
t-critical one-tailed	1.812461	
$P(T \le t)$ two-tailed	0.206624	
t-critical two-tailed	2.228139	

It is noticed that the values of *t*-stat are equal to the value calculated above, and as *t*-stat < *t*-critical is two tailed, there is no difference between the means.

Table 4.10 $Q_{critical}$ at 95% confidence level

Number of measurements	Critical value 95% confidence level
3	0.970
4	0.829
5	0.710
6	0.625
7	0.568
8	0.526
9	0.493
10	0.466

4.6.5   *Q*-Test to Reject Outliers

In some cases, a single measurement in an analytical series of replicates deviates significantly from the rest of the results. Often it is tempting to remove such *outliers* from the data material. This can be done with gross errors, when a serious error occurred in the experimental work for this particular and single replicate measurement. For other outliers, a *Q*-test (also called Dixon's test) must be performed before rejection. To perform a *Q*-test, the individual measurements are arranged by an ascending numerical value; if the outlier (suspect value) is larger than all other values, it will appear as the last value:

$$x_1, x_2, x_3, \ldots, x_i, x_{suspect} \tag{4.22}$$

Here x_1 is the lowest value, x_i is the second largest value, and $x_{suspect}$ is the largest suspect value that is requested to be rejected. Alternatively, if the lowest value is the suspect value, the readings are arranged as follows:

$$x_{suspect}, x_i, \ldots, x_3, x_2, x_1 \tag{4.23}$$

Box 4.17   Example of *Q*-test

A quantitative determination of paracetamol in oral solution with a specified content of 24 mg/mL, gave the following results when six individual measurements were conducted:

21.6 mg/mL	23.1 mg/mL	23.2 mg/mL	23.3 mg/mL	23.6 mg/mL	23.7 mg/mL

The value of 21.6 mg/mL is apparently different from the other values. Before this can be rejected, a *Q*-test must be performed. The $Q_{calculated}$ value is calculated as follows:

$$Q_{calculated} = \frac{|21.6 - 23.1|}{|23.7 - 21.6|} = 0.71$$

From Table 4.10, the Q value at the 95% confidence level (six measurements) was found to be 0.625.

Since $Q_{calculated} > Q_{critical}$ from the table, the outlier can be rejected.

Here $x_{suspect}$ is the smallest suspect value that is desired to be rejected and x_i is the second smallest value. Then, a Q value is calculated ($Q_{calculated}$) according to the following equation:

$$Q_{calculated} = \frac{|\text{suspect value} - \text{nearest value}|}{|\text{largest value} - \text{lowest value}|} \tag{4.24}$$

From Table 4.10, the critical value ($Q_{critical}$) is read based on the number of measurements in the series. If $Q_{calculated} > Q_{critical}$, the outlier can be rejected. If $Q_{calculated} \leq Q_{critical}$, the outlier cannot be rejected. An example of the use of a Q-test is shown in Box 4.17.

4.7 Linear Regression Analysis

As discussed in Section 4.6, it is necessary to establish the relationship between the measured signal of the analytical instrument (y) and the concentration of analyte in the sample (x) prior to quantitative analysis. This is termed *calibration* and, during this process, several *standard solutions* are analysed. The standard solutions are solutions of the pure analyte substance (chemical reference substance) dissolved in a pure solvent and prepared with exact concentrations. The measured signals of the standard solutions are plotted as a function of the standard concentrations. Normally (and preferably), there is a linear relationship between concentration (x) and signal (y). A straight line is described using the following equation:

$$y = mx + b \tag{4.25}$$

Here m is the slope and b is the y-intercept. For n data points, representing n different standard solutions with different concentrations, generally written in the form (x_i, y_i), the slope m and the y-intercept b can be calculated by the *least squares method* according to the following equations:

$$m = \frac{n \sum (x_i y_i) - \sum x_i \sum y_i}{D} \tag{4.26}$$

$$b = \frac{\sum (x_i^2) \sum y_i - \sum (x_i y_i) \sum x_i}{D} \tag{4.27}$$

where D is calculated from the following equation:

$$D = n \sum (x_i)^2 - \left(\sum x_i \right)^2 \tag{4.28}$$

Thus, based on n data points obtained for the standard solutions, the best fit for a linear relationship can be established between the measured signal and analyte concentration by using Eqs. (4.25) to (4.28). To express how well the n data points fit the linear relationship, the *square of the correlation coefficient* (R^2) is calculated from the following equation:

$$R^2 = \frac{\left[\sum (x_i - x)(y_i - y) \right]^2}{\sum (x_i - x)^2 \sum (y_i - y)^2} \tag{4.29}$$

Here x is the mean of all the x values and y is the mean of all the y values. R^2 should be very close to 1.0 to represent a linear fit, and normally values above 0.999 can be obtained in simple aqueous solutions, while values of 0.99 are considered as acceptable for linear correlation in more complicated samples. Linear regression using the least squares method and calculation of the square of the correlation coefficient are normally and very easily performed in a spreadsheet. Box 4.18 shows an example of how to establish a linear calibration curve based on the least squares method. Linearity should always be controlled by visual inspection of the regression line.

Calibration curves should be based on at least five concentrations and the range of standard concentrations should be chosen so the expected concentrations of the samples are measured in the central part of the curve. The reason for this is that the uncertainty on the concentrations determined by the calibration curve is lowest in the central part, as illustrated in Figure 4.10. More detailed information on the regression line, including confidence intervals for the slope and y-axis intercept, can be obtained in Excel. This is shown in Box 4.19.

Box 4.18   Calculation of calibration curve based on the least squares method

Five standard solutions with exact concentrations of a new API (prepared from a pure substance) are analysed by liquid chromatography to establish a calibration curve for an assay of samples with unknown concentration of the API. The results, area (=instrumental response) versus concentration, are summarized as follows:

Standard concentration/mg/ml	Area
0.048	3349
0.096	6469
0.144	9942
0.192	12 704
0.288	18 702

The data are plotted in Excel showing the calibration equation and R^2.

The residual plot shows that the residuals (a measure of the deviation from the calculated regression line) are randomly positioned on both sides of zero. If the line was curved, the residuals would not be randomly distributed.

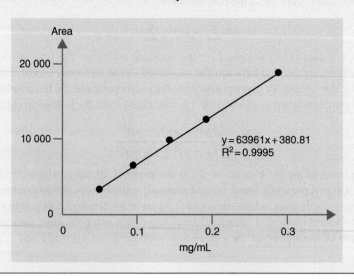

$y = 63961x + 380.81$
$R^2 = 0.9995$

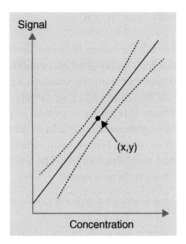

Figure 4.10   *Confidence limits for a concentration determined by an unweighted regression line*

Box 4.19   Data of analysis of a regression line

Using the same data as in Box 4.18, information on the uncertainty on the slope and intercept can be obtained. Selecting Data and Data analysis in Excel and selecting Regression in the dropdown menu, the following appear:

Regression statistics	
Multiple R	0.9998
R-square	0.9995
Adjusted R-square	0.9993
Standard error	151.9634
Observations	5

ANOVA

	df	*SS*	*MS*	*F*	*Significance F*
Regression	1	139 499 572.2	139 499 572.2	6040.809 59	4.69428E-06
Residual	3	69 278.58108	23 092.860 36		
I alt	4	139 568 850.8			

	Coefficients	*Standard error*	*t-stat*	*P value*	*Lower 95%*	*Upper 95%*
Intercept	380.8	143.51	2.65	0.07677	−75.92	837.54
X-variable 1	63 960.9	822.94	77.72	4.69E-06	61 341.91	66 579.82

It appears that the 95% confidence interval of the intercept is [−75.92 to 837.54]; thus the intercept is not significantly different from zero. The 95% confidence interval of the slope is [61 341.91–66 579.82]. This can be used to evaluate whether the slope, and thereby the sensitivity of the method, is different from another method.

4.8 How to Present an Analytical Result

As mentioned earlier, analytical results are often based on several replicate measurements. In such cases, the analytical result may be reported as a confidence interval. Alternatively, the mean value can be reported together with the standard deviation associated with the measurement. Any outliers should be removed before calculating the confidence interval, the mean, and the standard deviation. It is highly important to present the analytical result with the correct number of digits (significant figures). In cases where the precision is high, several digits should be included in the analytical result, while results from methods with low precision should be presented with fewer digits. The following setup can be used to determine the number of digits in an analytical result:

- Calculate the mean value (x) and standard deviation (s) for the set of replicate measurements.
- Enter (x) in the form $a \times 10^n$, where a is a number between 1 and 10 and n is an integer.
- Write (s) in the form $b \times 10^n$, where n has the same value as above.
- Align b under a. Draw a vertical line to the right of the first digit in b different from 0. Then remove any digits in a to the right of this line.
- The correct number of digits in the result is the number of digits that are left in a.
- If the first digit removed from a is ≥ 5, the last digit in a should be rounded upwards.

Box 4.20 Adjustment of analytical result to correct number of digits

The following results (with four digits) were obtained in Box 4.17 for a quantitative measurement of paracetamol in oral solution (mg/mL):

23.34	23.56	23.12	23.17	23.67	21.56

As mentioned in Box 4.17, the value of 21.56 was rejected based on a Q-test, and the following measurements remained (mg/mL):

23.34	23.56	23.12	23.17	23.67

For this series the mean value was calculated to be 23.3720 mg/mL. The standard deviation was 0.2395 mg/mL. Adjustment of the analytical result to the correct number of digits was accomplished as follows:

$x = 2.33$	720×10^1
$s = 0.02$	395×10^1

The result should therefore be printed as 23.4 mg/mL with a standard deviation of 0.2 mg/mL.

Figure 4.11   *Reading of a burette*

Box 4.20 shows an example of how to adjust the analytical result with the correct number of digits.

All numbers should be given only with *significant figures*, which are all the digits known with certainty plus the first uncertain digit. With a 50-mL burette, for example, with graduations every 0.1 mL, it is easy to see that the liquid level is greater than 24.5 mL but less than 24.6 mL (Figure 4.11). Thus, for all the digits 24.5 mL are certain. However, the position of the liquid between the graduations can be estimated to perhaps ±0.02 mL, and in this particular example it is estimated to 24.56 mL. The last digit is the first uncertain digit. Thus, the reading should be reported as 24.56 mL, which contains four significant figures. The reading can also be reported as 0.02456 L, but still it contains four significant figures.

The numbers 1036, 103.6, 10.36, 1.036 all contain four significant figures. The numbers 1036 and 1.036×10^3 also both contain four significant figures. The numbers 0.001036, 0.01036, 0.1036, 1.036, 1.036×10^{-3}, and 10.36×10^{-2} also contain four significant figures. Thus, to summarize:

(1) Zeros before the decimal point are not significant.
(2) Zeros between non-zero digits are significant.

Terminal zeros may or may not be significant. For example, if the volume of a beaker is expressed as 1.0 L, the presence of the otherwise unnecessary zero implies that the volume is known to a few tenths of a litre. Both 1 and 0 are significant figures.

Care is required in determining the appropriate number of significant figures in values obtained from arithmetic combinations of two or more numbers. The following rules can be used:

• Addition and subtraction: significant figures after the decimal place not higher than those of that number having the fewest significant figures after the decimal point.
• Multiplication or division: number of significant figures equal to the smallest number carried by any of the value being multiplied and/or divided.
• Taking logarithms: quote the logarithm with the mantissa having as many figures as the significant figures in the original number.

Box 4.21   Significant figures

Calculate the following:

$$\frac{24 \text{ mL} \times 4.52 \text{ M}}{100.0 \text{ mL}} = 1.0_{848} \text{ M} = 1.1 \text{ M}$$

The answer should be 1.1 M because 24 mL determines the number of significant figures (=2).

Determining the appropriate number of significant figures is exemplified in Box 4.21.

4.9   Additional Words and Terms

Earlier in this chapter, the words analyte, matrix, qualitative analysis, quantitative analysis, accuracy, precision, calibration, chemical reference substance, and standard solutions were discussed. In this final section, some additional terms are explained because they will be used in the following chapters.

In pharmaceutical analysis, the terms analysis and determination are used. The word *analysis* is used in connection with the samples to be examined, while the word *determination* refers to the substance to be measured quantitatively. Thus, blood samples, urine samples, raw materials, and tablets are analysed, while the concentrations of nortriptyline in blood, amphetamine in urine, paracetamol in raw materials, and diazepam in tablets are determined.

For a given sample, it is common to subject several small portions to analytical measurement, and different portions of the same sample are called *sample replicates*. Two replicates are often referred to as *duplicates* and three replicates as *triplicates*. The purpose of analysing several sample replicates is to ensure that the analytical result is representative for the entire sample. For each sample replicate, several individual measurements may be performed, and these measurements are termed *measurement replicates*. The purpose of taking several measurement replicates is to ensure that the final analytical result is not affected by serious random errors.

Quantitative pharmaceutical analysis is normally performed by measurement of a physical parameter (y), which in a well-known and reproducible way depends on the concentration of analyte in the sample (x). In some cases, substances other than the analyte may contribute to y, and this is termed *interference*. Interferences are very important to eliminate, as they give rise to errors in quantitative measurements. During the development of analytical methods, it is therefore crucial to ensure that the method is not affected by interferences from any other substances in the sample. This is accomplished by analysing blind samples as defined below.

A *blank sample* is a sample that does not contain the substance or substances to be determined, but all or as many as possible of the other ingredients in the sample. Blank samples undergo the same analysis procedure as the samples. Blank samples are used to control other substances in the samples in order to ensure that they do not cause interference. During development of methods for quantitation of active ingredients in pharmaceutical preparations, it is common to use a *placebo preparation* as a blind sample, which is a preparation that contains all the ingredients except the API. For the analysis of drugs in plasma samples and urine samples, it is common to use drug-free plasma and urine as the blind samples, respectively.

5

Titration

5.1 Introduction

Titration is a quantitative technique and methods based on titration provide high accuracy (99.5–100.5%) and precision (<0.5% relative standard deviation (RSD)). Titration is an official method in the European Pharmacopoeia (Ph. Eur.) and in the United States Pharmacopoeia (USP). In pharmaceutical analysis, titration is mainly used for quantitative analysis (*assay*) of active pharmaceutical ingredients (APIs) and excipients with the purpose of assessing the purity of a given substance (analyte).

Titrations are based on measurement of volumes. An accurate amount of analyte is dissolved in a specific volume of solution. This solution is termed the *titrate*. Increments of a standardized solution of a reagent are added gradually until the entire amount of analyte has reacted. The standardized solution is termed the *titrant*. The concentration and exact volume of titrant added are used to calculate the amount of analyte. More generally, a titration can be described by the following titration equation:

$$x(\text{Analyte}) + y(\text{Titrant}) \rightarrow \text{Products} \tag{5.1}$$

Here x and y are the number of mol of analyte (pharmaceutical compound) and titrant involved to complete the titration. The titration is completed when all the analyte has reacted and transformed to products. The titration has then reached the *equivalence point*.

Introduction to Pharmaceutical Analytical Chemistry, Second Edition.
Stig Pedersen-Bjergaard, Bente Gammelgaard and Trine Grønhaug Halvorsen.
© 2019 John Wiley & Sons Ltd. Published 2019 by John Wiley & Sons Ltd.

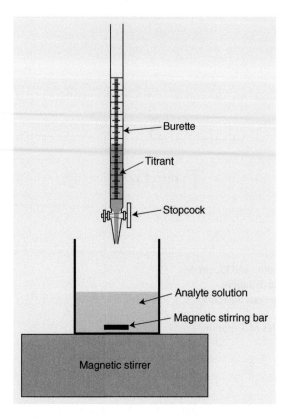

Figure 5.1   *Schematic view of the burette used for titration*

The titrant is added either manually from a *burette* (see Figure 5.1) or automatically from an automatic titration apparatus termed an *automatic burette* or *titrator*.

Detection of the equivalence point can be based on visual inspection of colour change (*Indicator detection*) or based on electrochemical measurements (*Potentiometric endpoint detection*). In the latter case, the titration is often termed *Potentiometric titration*. The latter is preferred in the pharmaceutical industry because it can be automated.

The volume of titrant consumed is read from the burette/titrator and the quantity of analyte in the sample solution can be calculated on the basis of the underlying titration equation and the exact concentration of reagent in the titrant. An example of such a calculation is shown in Box 5.1. As in all calculations related to pharmaceutical analysis, the numbers of significant figures are important. In Box 5.2 the numbers of significant figures related to the example in Box 5.1 are discussed.

Box 5.1   Titration of a pharmaceutical ingredient and calculation of purity (assay)

To establish the purity of the API ketoprofen (anti-inflammatory drug), 0.1996 g is weighed on an analytical balance and dissolved in 25 mL of ethanol 96% (*v/v*); 25 mL of water is added to the solution (note that knowledge of the exact volume of the final solution is not needed in this case as the exact mass of the analyte is known). The solution is titrated with 7.85 mL of 0.1006 M NaOH to the equivalence point. The structure of ketoprofen is illustrated below:

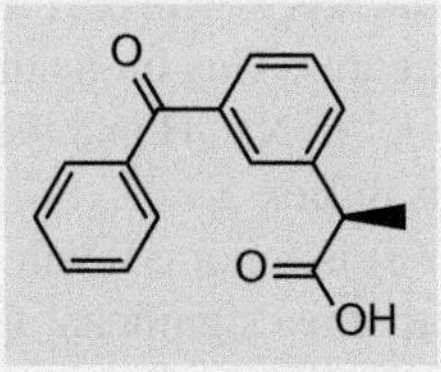

Ketoprofen is an acid and the carboxylic group (–COOH) will react with NaOH during the titration according to the follow scheme:

$$-COOH + OH^- \rightarrow -COO^- + H_2O$$

As seen from the reaction, ketoprofen and sodium hydroxide react in the molar ratio 1 : 1, and the number of mol of sodium hydroxide used during the titration is therefore equal to the number of mol of pure ketoprofen present in the 0.1996 g portion of API. The number of mol of NaOH (n_{NaOH}) reacted is calculated as follows:

$$n_{NaOH} = \frac{0.1006 \text{ mol/L} \times 7.85 \text{ mL}}{1000 \text{ mL/L}} = 7.90 \times 10^{-4} \text{mol}$$

This is equal to the number of mol of pure ketoprofen present in 0.1996 g portion of API:

$$7.90 \times 10^{-4} \text{mol} \times 254.3 \text{ g/mol} = 0.201 \text{ g}$$

The purity of the active ingredient is calculated as follows:

$$\text{Purity} = \frac{0.201 \text{ g}}{0.1996 \text{ g}} \times 100\% = 100.6\%$$

Box 5.2   Significant figures in titration calculations

Significant figures must be considered for all numbers included in a calculation. From the example in Box 5.1, the measured data are

(a) Recorded mass of active ingredient (0.1996 g) – four significant figures
(b) Molarity of the titrant (0.1006 M) – four significant figures
(c) Volume of titrant (7.85 mL) – three significant figures

During multiplication and division, the number of significant figures is equal to the smallest number carried by the values being multiplied or divided. The volume of titrant is based on reading a 10 mL burette of class A with 0.02 mL subdivisions (and tolerance). This is done with three significant figures. This limits the number of significant figures to three. Therefore, the number of mol of NaOH and the amount of ketoprofen are calculated with three significant figures in Box 5.1.

However, the final purity (100.6%) is reported with four significant figures. Theoretically, this should be reported with three significant figures as 101%, whereas purity less than 100% reported with three significant figures will be a decimal number such as 99.8%. To avoid this confusion, assay data are reported with one decimal digit. This is also reflected in the purity definitions for pharmaceutical ingredients in Ph. Eur. and USP, which are typically from 99.0% to 101.0%.

When several calculations are performed in a series, like in Box 5.1, intermediate results used in subsequent calculations should not be rounded. Thus, only the final number should be rounded to correct number of significant figures.

A number of requirements should be fulfilled for a successful titration:

- The titration reaction must be well-defined and without any side reactions.
- The reaction must be virtually complete ($\approx$100% of analyte must be converted to product).
- Other substances in the sample should not react with the titrant.
- The equivalence point should be clearly detected.
- The exact concentration of titrant must be known (standardized solution).

These requirements are all of vital importance to obtain accurate titration results. The first requirement implies that the stoichiometry of the reaction must be known in order to calculate the amount of analyte in the sample (as exemplified in Box 5.1). This also means that the analyte and the titrant must react in a well-defined manner without reactions. If the reaction does not progress to $\approx$100% or if other substances are present in the titrate, the consumption of titrant cannot be exactly correlated to the amount of analyte. Clear detection of the equivalence point is mandatory as this is the point where the titration is terminated and the consumption of titrant is read. As shown in Box 5.1, the exact volume of the titrant used to reach the equivalence point, as well as the exact concentration of titrant are needed for the calculations. Furthermore, a high reaction rate is advantageous, but not essential, as a fast reaction between the analyte and the reagent reduces the analysis time.

Reactions used in titration can be of several types, and in this chapter the following types are described:

- Acid–base reactions (*Acid–base titration*)
- Reduction–oxidation reactions (*Redox titration*)
- Complexometric reactions (*Complexometric titration*)
- Precipitation reactions (*Precipitation titration*)

The emphasis will be placed on acid–base and redox titrations, while complexometric and precipitation titrations will only be discussed briefly.

During titration, the titrant is added to the titrate until the equivalent point is reached, where virtually all analyte has been converted into products. All titrations are equilibrium reactions, but those used in pharmaceutical analysis have equilibrium completely shifted to the product side. The titration is stopped when the *endpoint* of titration is reached. The endpoint is defined as the point at which the operator or the titration apparatus terminate the titration. Ideally, the endpoint coincides with the equivalence point. If not, the result is prone to *titration error*.

To achieve high accuracy in titration, it is essential to know the exact concentration of titrant. Thus, the titrant has to be *standardized* prior to use. For standardization, a *primary standard* is needed. A primary standard is a chemical reagent with the following characteristics:

- High chemical purity (>99.95%)
- Well-defined chemical composition (including the amount of crystal water)
- High stability during storage or in contact with air and light
- High solubility in the titration solvent (water or organic solvent)
- Non-volatile
- Strong electrolyte

A variety of chemicals meet these requirements and a few examples are given in Table 5.1. In Ph. Eur., primary standards are indicated with the suffix *RV*. Titrants of primary standards can be made directly by accurate weighing of the primary standard (analytical balance) and subsequent dissolution and dilution to a specific volume (volumetric flask). The concentration of such solutions is known with high accuracy. Direct preparation of a titrant based on a primary standard is exemplified in Box 5.3.

Table 5.1   Examples of primary standards in Ph. Eur

Potassium hydrogen phthalate	Sodium carbonate	Potassium bromate	Benzoic acid
$M = 204.2$ g/mol	$M = 106.0$ g/mol	$M = 167.0$ g/mol	$M = 122.12$ g/mol

Na_2CO_3 $KBrO_3$

Box 5.3   Preparation of 0.1 M potassium hydrogen phthalate (primary standard)

20.4326 g of potassium hydrogen phthalate RV (reported purity > 99.95%) is weighed on an analytical balance and is transferred to a 1000 mL volumetric flask, dissolved in anhydrous acetic acid, and diluted with the same solvent to the mark. Using a class A volumetric flask, the volume of the final solution is set to 1000.0 mL (1.0000 L). The exact molarity ($M_{Titrant}$) is calculated as follows, taking 204.2 g/mol as the molar mass:

$$M_{Titrant} = \frac{20.4326 \text{ g}}{204.2 \text{ g/mol} \times 1.0000 \text{ L}} = 0.1001 \text{ M}$$

The numbers included in the calculation are 20.4326 g with six significant figures and 1.0000 L with five significant figures. Nevertheless, the number of significant figures for $M_{Titrant}$ is set to four. The reason for this is that the purity of the primary standard is not 100%, but >99.95%. Therefore, although the weighing is done with six figures, only four figures are significant.

  The primary standard prepared in this example can be used to standardize 0.1 M solutions of NaOH, which are often used for titration of APIs as discussed in the following.

Box 5.4   Standardization of 0.1 M hydrochloric acid with sodium carbonate as primary standard

A 0.1 M solution of hydrochloric acid is standardized to determine the exact concentration of HCl. Sodium carbonate is used as the primary standard. An amount of 0.1006 g of Na_2CO_3 (molar mass = 105.99 g/mol) is dissolved in water and diluted to 20 mL. This solution is titrated with 0.1 M HCl, and the consumption of 0.1 M HCl is 18.55 mL. The following reaction occurs:

$$2HCl + CO_3{}^{2-} \rightarrow 2Cl^- + H_2CO_3$$

The number of moles of primary standard in the solution (n_{PS}) is calculated as follows:

$$n_{PS} = \frac{0.1006 \text{ g}}{105.99 \text{ g/mol}} = 9.492 \times 10^{-4} \text{mol}$$

Since the primary standard and HCl react in the molar ratio 1 : 2, the equivalent amount of HCl is $2 \times 9.492 \times 10^{-4}$ mol = 1.898×10^{-3} mol. This amount of HCl is 18.55 mL and the molarity of the HCl solution (M_{HCl}) is calculated as follows:

$$M_{HCl} = \frac{1.898 \times \text{mol} \times 1000 \text{ mL/L}}{18.55 \text{ mL}} = 0.1023 \text{ M}$$

In most cases, however, it is not possible to obtain a primary standard that is suitable for direct titration of the pharmaceutical ingredient. In these cases, the titrant is first *standardized* against a primary standard to determine the exact concentration of the titrant. An example of standardization is shown in Box 5.4 for 0.1 M hydrochloric acid.

5.2   Potentiometric Titration and Electrodes

Potentiometric endpoint detection is an alternative to detection by colour indicators. The advantage of potentiometric endpoint detection is that the technique allows for automation (automatic burette). Titration with potentiometric detection is termed *potentiometric titration*. The principle of potentiometric titration is shown in Figure 5.2. A two-electrode system is placed in the titration solution coupled to a *voltmeter*. One electrode acts as *indicator electrode* and the potential of this electrode changes as a function of titrant concentration. The second electrode is a *reference electrode* and the potential of this does not change during titration. The voltmeter measures the potential difference between the indicator electrode and the reference electrode. Plotting the potential, E, as function of the

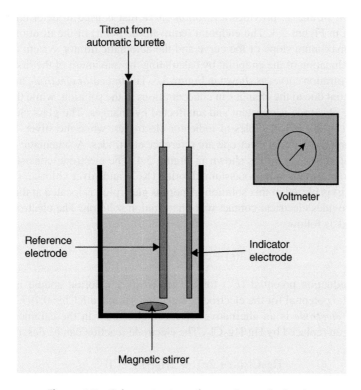

Figure 5.2   *Schematic view of potentiometric titration*

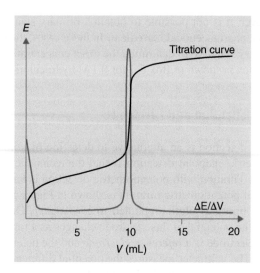

Figure 5.3   Titration curve of E versus V and the first derivative ($\Delta E/\Delta V$) versus V

added volume of titrant, V, provides a *titration curve* that is used to determine the end-point, as shown in Figure 5.3. The endpoint (equivalence point) of the titration is located at the point of maximum slope of the curve and the automatic titrator system often determines the exact location of the endpoint by calculating the maximum of the first derivative ($\Delta E/\Delta V$) of the titration curve, as shown in Figure 5.3. The *indicator electrode* measures the change in potential due to the changes in concentrations in the solution, while the potential of the reference electrode is constant and unaffected by changes. The glass electrode and the platinum electrode are examples of indicator electrodes, while the silver–silver chloride electrode and the calomel electrode are reference electrodes. A schematic view of the *silver–silver chloride electrode* is shown in Figure 5.4. The electrode consists of a tube filled with saturated solution of potassium chloride (KCl) and silver chloride (AgCl), and a silver wire (Ag) is placed in this solution. A porous glass plug is located at the bottom of the tube that provides electrical contact with the titration solution. The electrode reaction can be described as follows:

$$AgCl(s) + e^- \rightarrow Ag(s) + Cl^- \tag{5.2}$$

The standard reduction potential (E^0) for the silver/silver chloride couple is $+0.222$ V (at 25 °C) and the potential for the electrode containing saturated KCl is 0.197 V.

The *calomel electrode* is an alternative reference electrode. In the calomel electrode, Ag/AgCl has been replaced by Hg/Hg_2Cl_2. The electrode reaction can be described as follows:

$$Hg_2Cl_2(s) + 2e^- \rightarrow 2Hg(l) + 2Cl^- \tag{5.3}$$

The standard reduction potential (E^0) of the calomel electrode is 0.268 V and the potential for the electrode containing saturated KCl is 0.241 V.

The *glass electrode* is used for measurement of pH and is a very important indicator electrode used for acid–base titrations. The glass electrode belongs to the group of electrodes

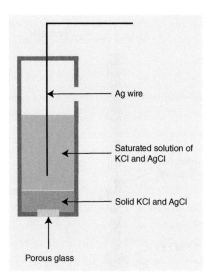

Ag wire

Saturated solution of
KCl and AgCl

Solid KCl and AgCl

Porous glass

Figure 5.4 *Schematic view of the silver–silver chloride reference electrode*

termed *ion-selective electrodes*. The glass electrode consists of a pH-sensitive glass membrane shaped like a bubble at the bottom of the electrode. Inside this bubble, there is a solution of hydrochloric acid (HCl) saturated with silver chloride (AgCl). This solution is in contact with a silver wire connected to a voltmeter, acting as a silver–silver chloride electrode. The potential of the glass electrode is measured by connection to a reference electrode, which can be either a silver–silver chloride electrode or a calomel electrode, located in the analyte solution. Often the glass electrode and the reference electrode are combined in one embodiment, termed a *glass combination electrode*. A schematic view of a glass combination electrode is shown in Figure 5.5. The glass electrode serves as an indicator electrode for H^+ ions in the titration solution. When the glass membrane is in contact with an external solution (aqueous), the inner and outer surfaces swell and become hydrated gel layers. Metal ions (Na^+ in the glass) diffuse out of the glass into solution and H^+ ions can diffuse into the membrane to replace the metal ions, and this process results in an ion-exchange equilibrium. The glass electrode responds selectively to H^+ as this is the main ion that binds to the glass membrane. The H^+ ions are not transported across the membrane, instead the charge is transported by Na^+ ions in the glass. The potential difference across the membrane is measured by the inner and outer silver–silver chloride reference electrodes. The potential difference between these is dependent on the chloride concentrations in their departments and the potential difference across the membrane. As the chloride concentrations are constant (saturated) and the hydrogen concentration is fixed on the inside of the membrane (0.1 M), the potential is only dependent on the H^+ concentration in the external solution. The embodied reference electrode is in contact with the external solution via a porous plug at the side of the electrode that functions as a salt bridge.

The potential across the membrane depends on the activity in the external solution:

$$E = \text{constant} + \frac{RT}{nF} \ln(a_{H^+}) \tag{5.4}$$

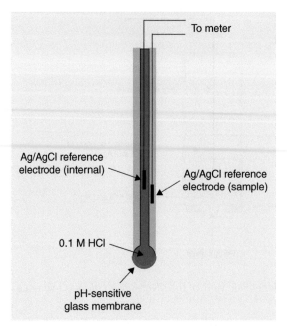

Figure 5.5 *Schematic view of the glass combination electrode with the internal reference electrode*

where R is the gas constant, T is the temperature, n is the charge of the analyte ion, and F is Faraday's constant. Inserting the values of R, T, and F and converting to log, the general expression for membrane selective electrodes at 25 °C (298 K) is

$$E = \text{constant} + \frac{0.05916}{n} \log a_{H^+} \qquad (5.5)$$

For diluted solutions, activity can be replaced by concentration.

In redox titrations, *platinum electrodes* simply consisting of a platinum wire are the most commonly used indicator electrodes in combination with a reference electrode. Other indicator electrodes used can be made of gold or silver. The function of the metal electrodes is simply to transmit electrons to and from the solution during the chemical reaction of the titration.

Compared to indicator endpoint detection, potentiometric endpoint detection has a slightly slower response time. That is, there is often a slight delay from the addition of titrant until the reading of the voltmeter has stabilized. For this reason, the titrant is added slowly when approaching the endpoint.

5.3 Aqueous Acid–Base Titrations

Most titrations of pharmaceutical ingredients are acid–base titrations. Acid–base titrations are used to assay basic or acidic pharmaceutical ingredients. In aqueous solution, strong

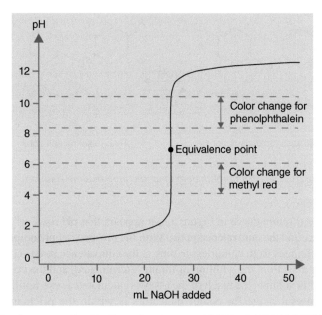

Figure 5.6 *Titration curve for titration of a strong acid (50 mL of 0.1 M HCl) with a strong base (0.2 M NaOH)*

acids and bases are entirely dissociated and the reaction for titration of a strong acid (titrate) with a strong base (titrant) is therefore

$$H_3O^+ + OH^- \rightarrow 2H_2O \tag{5.6}$$

When a strong acid is gradually titrated with a strong base, pH changes according to Figure 5.6. This curve is termed a *titration curve*. As appears from the titration curve, pH is very low in the strong acid titrate at the onset. At this point, the titrate comprises an aqueous solution of a strong acid. As titrant is added gradually, pH rises slightly due to the acid–base reaction. Just before the equivalence point, pH increases sharply, and in the equivalence point the stoichiometric reaction is complete. If the titration is continued after the equivalence point, the curve levels off at high pH values. Now, the titrate comprises an aqueous solution of a strong base, because the acid has been neutralized and the base is in excess. The equivalence point is located at the point of maximum slope of the titration curve and is the *inflection point* (point where curvature changes) of the curve.

To detect the endpoint of an acid–base titration visually, a colour indicator can be added to the titrate to perform *indicator endpoint detection*. Colour indicators of acid–base titration are acidic or basic substances, which change colour as they are transformed to their basic or acidic form, respectively. An example of an acid–base indicator is *phenolphthalein*, which is shown in Figure 5.7. Phenolphthalein has a pK_a value of 9.4. Thus, at pH 9.4, 50% of the indicator is in acidic form and 50% is in basic form. At this pH, the indicator is coloured pink. At pH 8.4 only 10% of the indicator is in basic form and the pink colour is very weak, while at pH 10.4 about 90% of the indicator is in basic form and the pink colour is very intense. In practice, the colour change of phenolphthalein is seen clearly between pH 8.4

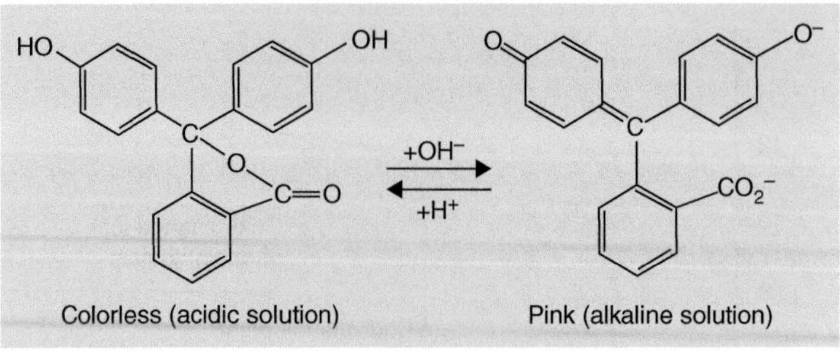

Figure 5.7   *Phenolphthalein in acidic and alkaline solution*

and 10.4. From the titration curve in Figure 5.6, it appears that pH rises very rapidly from about pH 4 to 12 around the equivalence point. With the presence of phenolphthalein in the titrate, the colour changes from colourless to pink at the equivalence point. This colour shift serves as endpoint detection, the addition of titrant is terminated, and the volume of titrant is read. Based on this volume, the result of the assay is calculated as exemplified in Box 5.1.

Generally, the useful range for colour indicators is within the pH range of $pK_a \pm 1$. A variety of acid–base indicators are used in pharmaceutical analysis and Table 5.2 shows a few examples from Ph. Eur. In the titration curve shown in Figure 5.6, pH increases rapidly at the equivalence point and alternative acid–base indicators to phenolphthalein, like methyl red with a pK_a value of 5.1 can be used. With this indicator, the colour changes from red to yellow in the equivalence point. Another alternative is methyl orange (pK_a 3.5) with a similar colour change.

For titration of a strong acid with a strong base (Figure 5.6), pH is initially very low and the shift in pH around the equivalence point corresponds to approximately 10 units (pH 2 → 12). Titration of a weak acid with a strong base is different and is illustrated in Figure 5.8. In this case, pH in the initial phase of the titration is higher because the titrate is a weak acid, and the shift in pH in the equivalence point now corresponds to approximately eight units (pH 4 → 12). In this case, phenolphthalein can still be used as a colour indicator since the colour shift falls within the range of the inflection of the titration curve. Methyl orange is not a suitable indicator as it will change colour prior to the equivalence point.

In general, either the titrate (analyte) or the titrant (reagent) has to be a strong acid or base to ensure that the reaction is virtually complete (very close to 100% of analyte must

Table 5.2   *Colour indicators for acid–base titrations (common examples)*

Indicator	pK_a	Colour change pH	Colour Acid	Base
Methyl orange	3.5	2.5–4.7	Red	Yellow
Bromophenol blue	4.0	3.0–5.0	Yellow	Purple
Methyl red	5.1	4.1–6.1	Red	Yellow
Cresol red	8.3	5.3–9.3	Yellow	Red
Phenolphthalein	9.4	8.4–10.4	Colourless	Pink

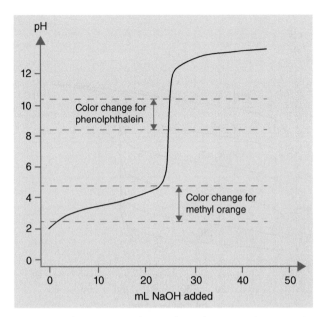

Figure 5.8 *Titration curve for titration of a weak acid (25 mL of 0.1 M solution of acetylsalicylic acid) with a strong base (0.1 M NaOH)*

be converted to product). Pharmaceutical ingredients are often weak bases or acids, and therefore strong acids or bases are used as titrants. For weak acids and bases, the pH changes in the equivalence point are less prominent, as shown in Figure 5.9.

Use of visual indicator endpoint detection may not be possible in these cases. As a rule of thumb, titration of acids with $pK_a > 6.0$ and bases with $pK_a < 8.0$ is not possible. These limits may be extended somewhat using potentiometric endpoint detection, but generally weak acids and bases must be titrated in non-aqueous solvents instead of water.

Some pharmaceutical ingredients have more than one acid or base functionality. In such cases, the stoichiometry of the titration reaction is no longer 1 : 1. If the pK_a values of the acidic or basic functionalities differ by more than four units, the titration curve has more than one equivalence point. Sodium carbonate is such an example, and this diprotic base (pK_a values of 10.33 and 6.35) can be titrated with hydrochloric acid according to the following reactions:

$$CO_3^{2-} + H^+ \rightarrow HCO_3^- \tag{5.7}$$

$$HCO_3^- + H^+ \rightarrow H_2CO_3 \tag{5.8}$$

The titration curve for titration of 0.1 M sodium carbonate with 0.1 M hydrochloric acid is shown in Figure 5.10. Phenolphthalein can be used to detect the first endpoint point, while methyl orange can be used to detect the second endpoint.

Basic pharmaceutical ingredients are titrated with strong acids like 0.1 M hydrochloric acid. This titrant is standardized against the primary standard, sodium carbonate, as shown in Box 5.4. Acidic pharmaceutical ingredients are titrated with strong base titrants and typically 0.1 M solutions of sodium hydroxide are used. Sodium hydroxide is not a primary

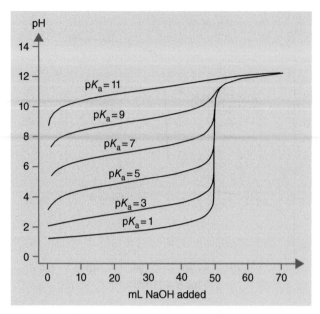

Figure 5.9 *Titration curve for titration of weak acids (50 mL of 0.1 M acid of varying pK_a value) with a strong base (0.1 M NaOH)*

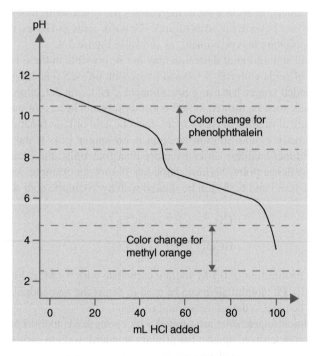

Figure 5.10 *Titration curve for titration of 50 mL of 0.1 M sodium carbonate with 0.1 M HCl*

standard and sodium hydroxide titrants must be standardized prior to use. According to Ph. Eur., 0.1 M solutions of NaOH are standardized by titration with 0.1 M solutions of hydrochloric acid. Alternatively, 0.1 M sodium hydroxide solutions can be standardized against benzoic acid, which is a primary standard.

In some cases, the titration reaction is too slow or slow side reactions may occur. Both cases may result in erroneous results. In these situations, an exact excess of titrant can be added to the analyte solution, which is allowed to react, and afterwards the unreacted titrant is quantified by titration. This procedure is called *back titration.*

Sometimes a blank titration is performed to establish whether the reagents used in the titration consume titrant. By titration of a solution *without analyte* in exactly the same way as titration of the analyte, this source of error can be determined. This procedure is termed *blank titration.* The Ph. Eur. assay for acetylsalicylic acid (ASA) is an example of a back titration that also includes a blank titration and is described in Box 5.5.

Box 5.5   Titration of acetylsalicylic acid according to Ph. Eur

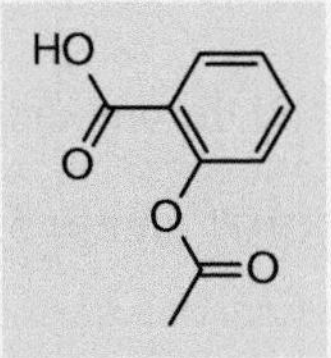

Molecular structure of ASA, $C_9H_8O_4$, $M = 180.158$ g/mol

ASA contains a carboxylic acid group, which quickly reacts with NaOH. However, ASA also comprises an ester group, which will slowly hydrolyse to acetic acid by reaction with NaOH. In a direct titration of ASA with NaOH, the titrant volume used to reach the endpoint will therefore be the result of the volume used to titrate the acid plus the volume used to titrate the unknown amount of hydrolysed ester. Therefore, the assay is a back titration according to the following procedure.

In a flask with a ground-glass stopper, dissolve 1.000 g ASA in 10 mL of ethanol (96%). Add 50.0 mL of 0.5 M sodium hydroxide. Close the flask and allow to stand for 1 hour. Using 0.2 mL of phenolphthalein solution R as indicator, titrate with 0.5 M hydrochloric acid. Carry out a blank titration. One mL of 0.5 M sodium hydroxide is equivalent to 45.05 mg of $C_9H_8O_4$.

The solution of ASA and NaOH is allowed to stand for one hour and the following reaction is completed:

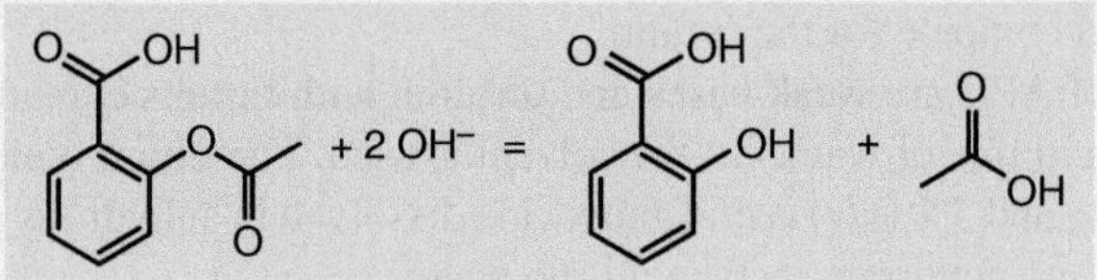

Thus, 1 mol ASA is equivalent with 2 mol NaOH, or 1 mol NaOH is equivalent with 0.5 mol ASA. Therefore 1 mL of 0.5 M NaOH corresponds to

$$0.5 \times \frac{1\,\mathrm{mL}}{1000\,\mathrm{mL/L}} \times 0.5\,\mathrm{mol/L} \times 180.158\,\mathrm{g/mol} = 0.04504\,\mathrm{g} = 45.04\,\mathrm{mg\ ASA}$$

Back titration is used to determine the unreacted NaOH and a blank titration determines how much NaOH is consumed by reagents.

As an example, 1.0010 g ASA is weighed and dissolved in 10 mL ethanol, and 50.0 mL of 0.5004 M NaOH is added together with 0.2 mL indicator. After the pre-scribed reaction time, the solution is titrated. An amount of 27.96 mL of 0.4999 M HCl was consumed. A blank titration of 10 mL of ethanol, 50.0 mL of 0.5004 M NaOH and 0.2 mL of indicator consumed 49.90 mL of 0.4999 M HCl.

The volumes consumed were 49.90 mL for the blank and 27.96 mL for the ASA solution, respectively. Thus, the amount of OH^- reacted with ASA can be calculated as follows:

$$n_{OH^-} = \frac{49.90\,\mathrm{mL} - 27.96\,\mathrm{mL}}{1000\,\mathrm{mL/L}} \times 0.4999\,\mathrm{M} = 1.097 \times 10^{-2}\,\mathrm{mol}$$

ASA and OH^- react in the ratio $1:2$, and the amount of ASA (n_{ASA}) is therefore

$$n_{ASA} = 0.5 \times 1.097 \times 10^{-2}\,\mathrm{mol} = 5.484 \times 10^{-3}\,\mathrm{mol}$$

This amount corresponds to
$5.484 \times 10^{-3}\,\mathrm{mol} \times 180.158\,\mathrm{g/mol} = 0.9880\,\mathrm{g\ ASA}$
The purity is calculated as follows:

$$\mathrm{Purity} = \frac{0.9880\,\mathrm{g}}{1.0010\,\mathrm{g}} \times 100\% = 98.7\%(w/w)$$

5.4 Titration in Non-aqueous Solvents

Acid–base titrations in aqueous solution are only successful when the analytes are soluble in aqueous solution and the acid strength or base strength is adequately high. Detection of the endpoint for weak acids and bases is challenging due to the small shifts in pH at the equivalence point (Figure 5.9), and, in general, bases with $pK_a < 8$ and acids with $pK_a > 6$ cannot be titrated accurately in aqueous solution. If even weaker acids or bases are titrated in aqueous solution, water can interfere with the titration as it can act as both a weak acid and a weak base and compete for the titrant.

A large number of APIs are weak bases and titration with titrants containing anhydrous acetic acid (acetic acid free of water) as the solvent is used. The titrant typically comprises 0.1 M perchloric acid and 3% (v/v) acetic anhydride dissolved in anhydrous acetic acid. The analyte is dissolved in anhydrous acetic acid (no water).

If an acid stronger than H_3O^+ is dissolved in water, it protonates H_2O to make H_3O^+, and if a base stronger than OH^- is dissolved in water, it deprotonates H_2O to make OH^-.

Thus, in aqueous solution, strong acids like $HClO_4$ and HCl both protonate water to H_3O^+ and thereby behave as if they had the same acid strength:

$$HClO_4 + H_2O \rightarrow H_3O^+ + ClO_4^- \tag{5.9}$$

In acetic acid solvent, which is less basic than H_2O, acetic acid functions as a base and combines with protons donated by perchloric acid to form protonated acetic acid, termed *acetate acidium ions*:

$$HClO_4 + CH_3COOH \rightleftharpoons CH_3COOH_2^+ + ClO_4^- \tag{5.10}$$

The equilibrium constant of this reaction is small (the equilibrium is shifted towards the left side of the reaction). The same reaction would be seen for HCl, but the equilibrium constant would be even smaller. Thus, $HClO_4$ is a stronger acid than HCl in acetic acid. Therefore, titration of acidic or basic pharmaceutical ingredients with weak acid/base properties is performed in a solution of perchloric acid in anhydrous acetic acid.

Since the $CH_3COOH_2^+$ ion readily donates its proton to a base, a solution of perchloric acid in anhydrous acetic acid functions as a strongly acidic solution. Among the most common acids, perchloric acid is the strongest acid in anhydrous acetic acid (perchloric acid > sulfuric acid > hydrochloric acid > nitric acid), and therefore perchloric acid is preferred.

The weak base API (B) is titrated according to the following reaction:

$$B + CH_3COOH_2^+ \rightarrow BH^+ + CH_3COOH \tag{5.11}$$

The net reaction of (5.10) and (5.11) can be summarized to

$$HClO_4 + B \rightarrow BH^+ + ClO_4^- \tag{5.12}$$

During preparation of the titrant, perchloric acid (70–72%) is added to anhydrous acetic acid. However, perchloric acid contains water, which can interfere with the titration. Water is removed prior to titration by addition of acetic anhydride to the titrant. Acetic anhydride reacts and removes water according to the following reaction:

$$(CH_3CO)_2 + H_2O \rightarrow 2CH_3COOH \tag{5.13}$$

Thus, acetic anhydride is mandatory to remove water from the titration system. Alternative solvents for the substances to be titrated are dioxane or other organic solvents that are miscible with anhydrous acetic acid.

The endpoint may be determined by addition of a colour indicator, and crystal violet, quinaldine red, and methyl red are examples of indicators used in USP. Several indicators for titration in non-aqueous solvents are also used in aqueous titration, but it should be noted that the colour change of an indicator in water is different from the change in anhydrous acetic acid, since the indicator has different pK_a values in the two solvents. Indicators for titration in non-aqueous solvents must be dissolved in a suitable organic solvent, not in water.

For titrations in non-aqueous solvents, potentiometric endpoint detection is preferred. In a water-free environment, a glass electrode is used as an indicator electrode, while the reference electrode is a silver wire coated with silver chloride. Silver–silver chloride-, calomel-, or combined glass electrodes are not used for titrations in non-aqueous solvents because these leach out small amounts of chloride ions and water, both of which can interfere with the titration.

Weak base (B) APIs are often produced as salts in the form of hydrochlorides (BH^+Cl^-) or acetates ($BH^+CH_3COO^-$). The weak base is thus protonated (BH^+) and the positive charges are balanced with *counter ions* such as chloride and acetate. The APIs therefore behave as acidic compounds. Such substances can also be titrated by perchloric acid in anhydrous acetic acid. In these cases, the counter ions are titrated, but since the stoichiometric balance is known, the titration still gives an accurate measure of the API. For acetates, phosphates, sulfates, tartrates, and maleates, the API is dissolved in anhydrous acetic acid and titrated according to the following reaction with maleate ($C_4H_3O_4^-$) as an example:

$$CH_3COOH_2^+ + C_4H_3O_4^- \rightarrow CH_3COOH + C_4H_4O_4 \tag{5.14}$$

For hydrochlorides and hydrobromides, the API is dissolved in anhydrous formic acid mixed with acetic anhydride and titrated with perchloric acid dissolved in formic acid according to the following reactions with chloride (Cl^-) as an example:

$$HCOOH + HClO_4 \leftrightarrow HCOOH_2^+ + ClO_4^- \tag{5.15}$$

The chloride ions are then titrated according to the reaction

$$Cl^- + HCOOH_2^+ \rightarrow HCl + HCOOH \tag{5.16}$$

Unlike aqueous titration, titration in non-aqueous solvents is very dependent on temperature. This is because organic solvents have a substantially higher volumetric thermal expansion coefficient than water. For practical work, this means that it is important to measure the temperature when the titrant is standardized (t_1) as well as when the titration is done (t_2). If the temperatures are different, correction for the volume of titrant (V) is performed, and the following equation applies to the calculation of the corrected titration volume (V_c) when the titrant is 0.1 M perchloric acid dissolved in anhydrous acetic acid:

$$V_c = V[1 + (t_1 - t_2) \times 0.0011] \tag{5.17}$$

Box 5.6 demonstrates an example of volume correction for titration with 0.1 M perchloric acid in anhydrous acetic acid.

Box 5.6   Volume correction for titration with 0.1 M perchloric acid in anhydrous acetic acid

Standardization of the titrant was at 20 °C. Subsequent titration of API was at 25 °C and one of the titrations consumed 15.45 mL of titrant. Since the temperatures differ, the volume of titrant must be corrected:

$$V_c = 15.45 \text{ mL} \times [1 + (20 - 25) \times 0.0011] = 15.37 \text{ mL}$$

This is a significant difference in an assay of an API and highlights the importance of performing volume correction.

The titrant of perchloric acid in anhydrous acetic acid is standardized prior to use. According to the Ph. Eur., this is done by titration with potassium hydrogen phthalate as the primary standard. Preparation and standardization of 0.1 M perchloric acid is described in Box 5.7.

Box 5.7   Preparation and standardization of 0.1 M perchloric acid

PREPARATION

8.5 mL of perchloric acid is pipetted into a 1000 mL volumetric flask and 900 mL of glacial acetic acid is added and mixed. Thirty millilitres of acetic anhydride is added and the volumetric flask is filled to the calibration mark with glacial acetic acid. The titrant is allowed to stand for 24 hours.

STANDARDIZATION

0.3521 g of potassium hydrogen phthalate is weighed on an analytical balance and dissolved in 50 mL of anhydrous acetic acid. This solution is titrated with 17.23 mL of the perchloric acid solution to the endpoint using crystal violet as the indicator. The exact molarity of the perchloric acid solution is calculated as follows: taking 204.2 g/mol as the molar mass for potassium hydrogen phthalate, and based on 1 : 1 stoichiometry in the titration reaction:

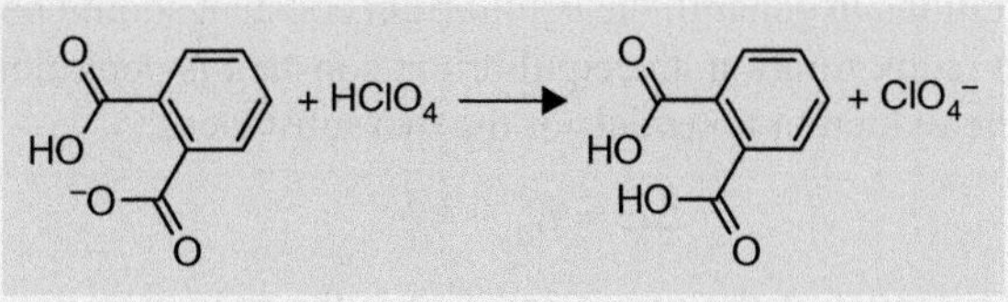

Mol potassium hydrogen phthalate ($=$ mol perchloric acid)

$$= \frac{0.3521\,\text{g}}{204.2\,\text{g/mol}} = 1.724 \times 10^{-3}\,\text{mol}$$

This number of moles corresponded to 17.23 mL ($= 0.017\,23$ L) of perchloric acid titrant, and the molarity of the titrant is calculated as follows:

$$\frac{1.724 \times 10^{-3}\,\text{mol}}{0.01723\,\text{L}} = 0.1001\,\text{M}$$

5.5   Redox Titrations

Acid–base titrations are based on the transfer of protons. Titrations can also be based on chemical reactions that transfer electrons. Such titrations are termed *redox titrations*. A redox titration can be described by the following equation:

$$\text{Ox}_1 + \text{Red}_2 \rightarrow \text{Red}_1 + \text{Ox}_2 \tag{5.18}$$

Table 5.3 *Standard reduction potentials (E_0)*

Oxidized form	Reduced form	E_0 (V)
$Ce^{4+} + e^-$	Ce^{3+}	1.61
$MnO_4^- + 5e^- + 8H^+$	$Mn^{2+} + 4H_2O$	1.51
$Fe^{3+} + e^-$	Fe^{2+}	0.55
$Br_2 + 2e^-$	$2Br^-$	1.05
$I_2 + 2e^-$	$2I^-$	0.54
$2H^+ + 2e^-$	H_2	0.00
$Fe^{2+} + 2e^-$	Fe	−0.44
$Ca^{2+} + 2e^-$	Ca	−2.89

The subscript 1 refers to the titrate and 2 refers to the titrant, respectively. The *reduction potential* of a given substance is an expression of the extent to which the substance may take up electrons. A high positive value for the reduction potential indicates:

- The substance is easily reduced.
- The substance is a powerful oxidizing agent.
- The substance easily removes electrons from other substances with a lower reduction potential.

Table 5.3 list values for standard reduction potentials (E_0) for typical redox pairs.

A substance with a higher reduction potential oxidizes a compound with a lower reduction potential. For titrations in general, the equilibrium constant should be high to ensure a complete reaction. In redox titration, the equilibrium constant is determined by the difference (ΔE) between the reduction potentials of the two substances:

$$\Delta E = E_0^T - E_0^A \tag{5.19}$$

E_0^T is the reduction potential of the titrant and E_0^A is the reduction potential of the analyte (titrate). ΔE is termed the *reaction potential*. In practice, ΔE should not be less than 0.1–0.2 V. Throughout titration, the potential changes gradually with addition of the titrant, and at the equivalence point the potential increases or decreases very sharply as a function of the added titrant. An example of this is shown in Figure 5.11 for the titration of iron (II) (Fe^{2+}) with cerium (IV) (Ce^{4+}), which reacts according to the following equation:

$$Fe^{2+} + Ce^{4+} \rightarrow Fe^{3+} + Ce^{3+} \tag{5.20}$$

For the titration of iron (II) with cerium (IV), the cell potential changes sharply at the equivalence point because the difference in reduction potential for Fe^{2+} and Ce^{4+} is large. For compounds with less difference in the reduction potential, the change in potential is less pronounced, as shown in Figure 5.12. The sigmoidal shape of the titration curves in Figures 5.11 and 5.12 follows from the *Nernst equation* (5.21) known from electrochemistry, which expresses the *half-cell potential* (E) for a half-reaction (5.22):

$$E = E^0 - \frac{RT}{nF} \ln \frac{a_B^b}{a_A^a} \tag{5.21}$$

$$aA + ne^- \rightarrow bB \tag{5.22}$$

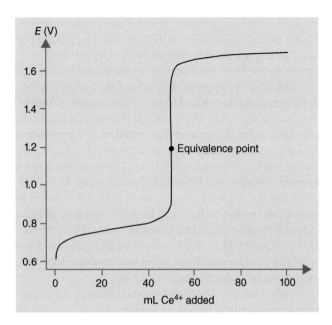

Figure 5.11   *Redox titration of 50 mL 0.1 M Fe²⁺ with 0.1 M Ce⁴⁺*

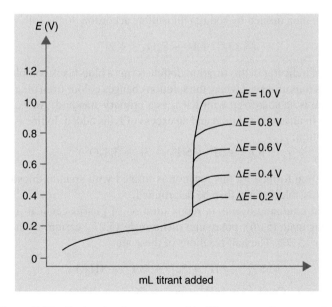

Figure 5.12   *Redox titration curves with different reaction potentials*

where E^0 is the *standard reduction potential*, R is the *gas constant* (8.314 J/K mol), T is temperature (K), n is the number of electrons in the half-reaction, F is the *Faraday constant* (96.49 kJ/mol), and a_B and a_A are the activities of the species involved.

Indicator and potentiometric endpoint detection are both used in redox titration. Colour indicators are compounds that are reduced or oxidized depending on the potential, and change colour from their oxidized to reduced form. Colour indicators should be selected so that the colour shift is within the range where the cell potential changes when the titration has reached the equivalence point. As for acid–base titration, it is important that the indicator is added in such small quantities that it does not consume significant amounts of titrate. One example of a colour indicator for redox titrations is *ferroin*, which is a mixture of ferrous sulfate and phenanthroline hydrochloride dissolved in water. Using colour indicators, it is normally required that the potential changes by at least 0.3–0.4 V in the equivalence point in order to make a clear endpoint detection. In the bottom case in Figure 5.12, where ΔE is only 0.2 V, indicator endpoint detection is inappropriate. In such cases, potentiometric endpoint detection is required. Most often, potentiometric endpoint detection is used in redox titrations, applying a platinum electrode as the indicator electrode and a silver–silver chloride or a calomel electrode as the reference electrode.

Sodium thiosulfate is often used as the titrant for redox titration. An excess of KI is added to the solution and the pharmaceutical ingredient oxidizes iodide to iodine according to the following reaction:

$$\text{Analyte}_{oxidized} + 2I^- \rightarrow \text{Analyte}_{reduced} + I_2 \tag{5.23}$$

Excess of iodine is then titrated by sodium thiosulfate according to the following reaction:

$$2S_2O_3^{2-} + I_2 \rightarrow S_4O_6^{2-} + 2I^- \tag{5.24}$$

Starch is used as an indicator in this titration. Iodine forms a blue-black complex with starch and the endpoint is thus observed where the solution changes colour from blue to colourless. Sodium thiosulfate is standardized with KIO_3 as a primary standard. An exact amount of KIO_3 is dissolved in an acidic solution and an excess of KI is added. Iodine is generated by the following reaction:

$$IO_3^- + 5I^- + 6H^+ \rightarrow 3I_2 + 3H_2O \tag{5.25}$$

The amount of iodine formed by this reaction is titrated with sodium thiosulfate and the concentration of this solution can then be determined.

Commonly used oxidizing agents in redox titration of pharmaceutical ingredients are potassium permanganate (5.26), potassium dichromate (5.27), cerium sulfate (5.28), and potassium bromate (5.29). The half reactions of these are:

$$MnO_4^- + 8H^+ + 5e^- \rightarrow Mn^{2+} + 4H_2O \tag{5.26}$$

$$Cr_2O_7^{2-} + 14H^+ + 6e^- \rightarrow 2Cr^{3+} + 7H_2O \tag{5.27}$$

$$Ce^{4+} + e^- \rightarrow Ce^{3+} \tag{5.28}$$

$$BrO_3^- + 6H^+ + 5e^- \rightarrow \tfrac{1}{2}Br_2 + 3H_2O \tag{5.29}$$

Potassium bromate is a primary standard. Potassium dichromate and cerium sulfate can be primary standards, but often they are standardized against sodium thiosulfate. This also applies to potassium permanganate, which is not a primary standard. Potassium permanganate is standardized according to the following reactions:

$$2MnO_4^- + 10I^- + 16H^+ \rightarrow 2Mn^{2+} + 5I_2 + 8H_2O \qquad (5.30)$$

$$2S_2O_3^{2-} + I_2 \rightarrow S_4O_6^{2-} + 2I^- \qquad (5.31)$$

A form of redox titration that is widely used in pharmaceutical analysis is *Karl–Fischer titration* for accurate determination of the water. In Ph. Eur. this assay is termed *Water: semi-micro determination* and in USP it is termed *Water determination*. A small amount of the API is weighed accurately and dissolved in anhydrous methanol. This titrate solution is then titrated with *Karl–Fischer reagent*. This titrant consists of iodine and sulfur dioxide dissolved in an anhydrous solvent such as methanol, and in the presence of pyridine. The following reaction takes place between water and Karl–Fischer reagent:

$$H_2O + C_5H_5N \cdot I_2 + C_5H_5N \cdot SO_2 + C_5H_5N \rightarrow 2C_5H_5N \cdot HI + C_5H_5N \cdot SO_3 \quad (5.32)$$

$$C_5H_5N \cdot SO_3 + CH_3OH \rightarrow C_5H_5N(H)SO_4CH_3 \qquad (5.33)$$

In titration reaction (5.32) oxidation of SO_2 by I_2 to SO_3 consumes the water originating from the API, and I_2 is reduced to HI. There is a large excess of pyridine, and SO_2, I_2, SO_3, and HI will be present as complexes with pyridine. These are marked by · in the reaction schemes. Reaction (5.33), which occurs due to the excess of methanol, is important because the sulfur trioxide/pyridine complex also react with water according to the following reaction:

$$C_5H_5N \cdot SO_3 + H_2O \rightarrow C_5H_5NHSO_4H \qquad (5.34)$$

Reaction (5.34) is not specific for water and is therefore undesirable. It is prevented by adding a large excess of methanol in the solution to be titrated. During the titration a colour change of the solution is observed. When all the water originating from the API have reacted with Karl–Fischer reagent, a further addition of the reagent results in the presence of unreacted reagent (brown colour) in the sample solution. This gives a colour change from yellow to brown. In most cases, however, Karl–Fischer titrations are performed with potentiometric endpoint detection. Applications of Karl–Fischer titration are discussed in more detail in Chapter 18.

5.6   Alternative Principles of Titration

In addition to acid–base and redox titrations, complexation reactions and precipitation reactions can also be used for titration purposes. Titrations based on complexation between titrate and titrant are termed *complexometric titrations*. Complexometric titrations are often used for assays of metal ions, and ethylenediaminetetraacetic acid (EDTA) or sodium edetate is used as the titrant. EDTA forms complexes with metal ions in the stoichiometric ratio 1 : 1. One example of complexometric titration is the assay of calcium carbonate in Ph. Eur., where calcium is titrated with 0.1 M sodium edetate under acidic conditions (Figure 5.13).

Precipitation titrations are often used for assay of inorganic ions such as Cl^-, Br^-, I^-, and Zn^{2+}. Endpoint detection can be either by colour indicator or by potentiometric endpoint

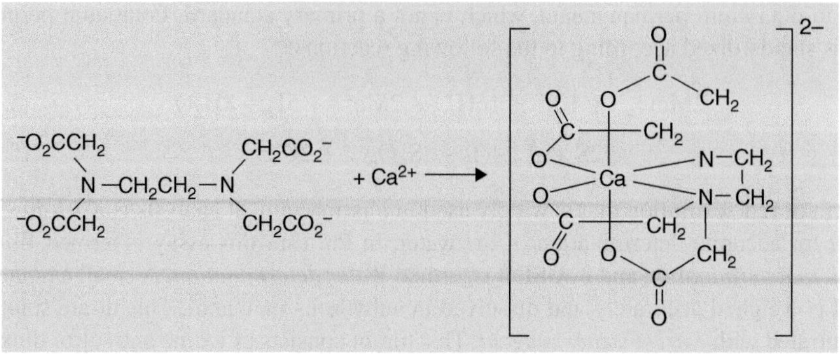

Figure 5.13 Complexation reaction of calcium with EDTA

detection. One example of precipitation titration is the assay of sodium chloride in Ph. Eur., where chloride ions are titrated with 0.1 M silver nitrate.

Another titration method occasionally encountered in pharmaceutical analysis is the *Kjeldahl analysis* for the determination of organically bound nitrogen. This method is based on acid–base titration and is used extensively in connection with the determination of total amounts of proteins. In Kjeldahl analysis, the sample is added to concentrated sulfuric acid and a catalyst (a Cu and Se salt), and the mixture is combusted by boiling at 340 °C (boiling point of H_2SO_4). Carbon in the sample is converted to CO_2, while organically bound nitrogen is transformed to ammonium (NH_4^+). Then NaOH is added to convert ammonium into ammonia (NH_3). Ammonia is distilled into a solution containing a surplus of HCl. The HCl reacts with ammonia and the remaining surplus of HCl is titrated with NaOH. This is an example of a *back-titration*. Based on this titration, the amount of ammonia can be determined and the amount of organically bound nitrogen is calculated.

6

Introduction to Spectroscopic Methods

Chapters 7, 8, and 9 will discuss ultraviolet-visible (UV-Vis) spectrophotometry, infrared (IR) spectrophotometry, and atomic spectrometry, which are highly important techniques for quantitation and identification in pharmaceutical analysis. These methods are all based on electromagnetic radiation, and therefore this chapter focuses on the fundamentals of electromagnetic radiation.

6.1   Electromagnetic Radiation

Electromagnetic radiation can be described as both *waves* and *particles* termed *photons*. The wave description characterizes the electromagnetic radiation in terms of *wavelength* and *frequency*, while the particle description is used as a model for light absorption. Electromagnetic radiation consists of regular electric and magnetic fluctuations. For many purposes it can be pictured as an electric field that undergoes sinusoidal oscillations as it moves through space. Figure 6.1 is a representation of a beam of *monochromatic* (single wavelength) *radiation*. The *wavelength* of the sinusoidal wave can be measured as the distance between any two points with the same phase, such as between maxima as shown in Figure 6.1.

The wavelength is commonly represented by λ. The SI unit of λ is the metre and the wavelengths in the UV-Vis range is most often given in nm (10^{-9} m). Electromagnetic

Introduction to Pharmaceutical Analytical Chemistry, Second Edition.
Stig Pedersen-Bjergaard, Bente Gammelgaard and Trine Grønhaug Halvorsen.
© 2019 John Wiley & Sons Ltd. Published 2019 by John Wiley & Sons Ltd.

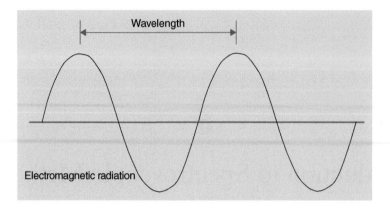

Figure 6.1   Representation of a beam of monochromatic radiation

radiation is also characterized by its *frequency*. Frequency is the number of oscillations per second and is commonly represented by v. The frequency is measured in *Hertz* (Hz), and a wave with a frequency of 1 Hz passes by at 1 cycle/s. The product of wavelength and frequency is the speed of light (c), and corresponds to 2.998×10^8 m/s in vacuum:

$$\lambda v = c \tag{6.1}$$

In other media than vacuum, the speed of light is c/n, where n is the *refractive index* of the medium. For visible wavelengths, n is larger than 1 for most materials, and light therefore travels more slowly through materials like water and glass. When light moves between different media like air, glass, and water, the frequency remains constant, while the wavelength changes.

The energy (E) of the radiation is proportional to the frequency according the following equation:

$$E = hv = h\frac{c}{\lambda} \tag{6.2}$$

where h is *Planck's constant* (6.626×10^{-34} J s). This means that the energy of electromagnetic radiation increases with increasing frequency. Since frequency and wavelength are inversely proportional, the energy of electromagnetic radiation decreases with increasing wavelength.

The *electromagnetic spectrum* comprises various types of electromagnetic radiation and their wavelength ranges are shown in Figure 6.2.

Short wavelengths result in high energy and X-rays belong to this high energy region of the spectrum. At the opposite end of the electromagnetic spectrum, the wavelengths are longer, resulting in low energy; radio waves belong to this region.

In the context of pharmaceutical analysis, only three small ranges in the electromagnetic spectrum are discussed in this book:

Ultraviolet (UV) radiation	200–380 nm
Visible (Vis) radiation	380–780 nm
Infrared (IR) radiation and near-infrared (NIR) radiation	780–15 000 nm

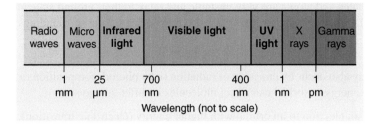

Figure 6.2   *The electromagnetic spectrum*

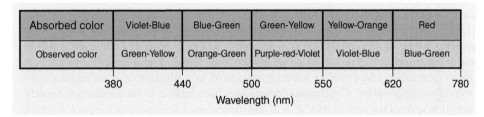

Figure 6.3   *Colours of visible light. When the compound absorbs light of a wavelength region, the colour of the compound is perceived as the complementary colour to the eye*

Visible light is electromagnetic radiation that can be perceived by the human eye. The colours of visible light are shown in Figure 6.3. The UV-Vis spectral range is used for UV-Vis spectrophotometry, which is discussed in Chapter 7. The infrared range is used in IR spectrophotometry (Chapter 8).

6.2   Molecules and Absorption of Electromagnetic Radiation

When molecules are exposed to electromagnetic radiation (light), either in the solution or gas phase, they may absorb part of the radiation. The absorption is dependent on the structure of the absorbing molecule and the wavelengths of the radiation. To describe the phenomenon of absorption of electromagnetic radiation by molecules, it is more convenient to describe light as particles or photons ('energy packages'). The energy of photons is described by Eq. (6.2).

Photons of short wavelength are rich in energy, while photons of higher wavelength are less energy rich. Electrons in molecules are normally present in the lowest energy state, which is called the *ground state*. When a photon passes near an electron, absorption becomes possible if the energy of the photon matches exactly one of the higher energy states of the electron. The energy of the photon is then transferred to the electron (present in an atom, ion, or molecule), converting it to a higher energy state, termed the *excited state*. Excitation of species M from the ground state to its excited state M* can be described by the following equation:

$$M + h\nu \rightarrow M^*$$
(6.3)

Excited molecules and atoms are very unstable and relax to their ground state after a very short time (10^{-6}–10^{-9} s). This process may also release heat to the surroundings:

$$M^* \rightarrow M + \text{heat} \tag{6.4}$$

Relaxation may also occur by emission of radiation or by photodecomposition to form new products. The energy absorbed can affect molecules in different ways:

- Transfer of an electron to an orbital with higher energy (electronic transition).
- Increase the vibrations of the molecule (vibrational transition).
- Increase the bonding rotations of the molecule (rotational transition).

The energy of rotational transition is smaller than the energy of vibrational transition and both of these energies are much lower than the electronic transition energy. For a particular molecule, a large number of combinations of transitions are possible, and each combination represents a specific amount of energy. This means that each possible combination will be equal to the absorption of radiation at one specific wavelength. These wavelengths are very close to each other, and therefore plotting absorption as a function of wavelength results in a *continuous spectrum*. Continuous spectra are characteristic of molecular compounds. Absorption spectra for acetylsalicylic acid and paracetamol are shown in Figure 6.4 for the UV region.

The wavelength of maximum absorbance is termed the *absorption maximum*. Absorption spectra are different for different chemical compounds and the absorption maximum is unaffected by the concentration. Absorption spectra may therefore be used for identification of a compound by the location of the absorption maximum. This is illustrated by the different absorption maxima of acetylsalicylic acid (275 nm) and paracetamol (245 nm) in Figure 6.4. In quantitative determinations the amount of absorbed radiation is measured

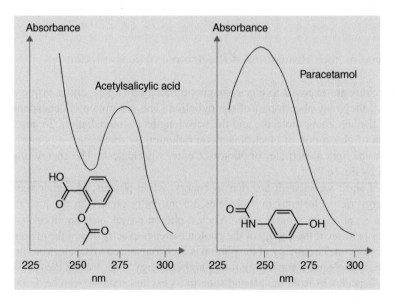

Figure 6.4   UV spectra of acetylsalicylic acid and paracetamol

and related to concentration. The more concentrated the solution, the more radiation is absorbed.

6.3   Absorbing Structures – Chromophores

The parts of molecules taking part in the absorption of light are termed *chromophores*. The absorption of electromagnetic radiation involves transfer of electrons to orbitals of higher energy. Two types of electrons can participate in this excitation process:

- electrons shared by several atoms participating in chemical bonding and
- unshared electrons localized to one atom such as in O, N, S, or Cl (halogens).

Electrons involved in single bonds, like C—C and C—H bonds, require high energy to be excited (radiation with wavelength < 200 nm). Electrons in single bonds therefore do not contribute to UV-Vis absorbance. Electrons in double and triple bonds demand less energy to be excited. Thus, compounds containing C=C, C=O, and C=N are chromophores, and conjugation of these bonds increases the absorption. In addition to double and triple bonds, atoms with unshared electrons contribute to UV absorption, as these electrons are also easily excited. The UV-Vis absorption of a molecule therefore results from the sum of the excitation of various electrons that are easily excited. Active pharmaceutical ingredients (APIs) often contain aromatic rings, ketones, aldehydes, carboxylic acids, amines, and nitro groups, and therefore the majority can be measured by UV spectrophotometry.

Compounds that absorb visible light are coloured. These compounds are often characterized by having a large number of conjugated double bonds and the electrons are easily excited by visible light. White light (sunlight) contains all the colours of the visible spectrum. When a compound absorbs certain wavelength regions of the white light, the eye detects the wavelengths that are not absorbed. An example of a compound absorbing visible light is riboflavin, shown in Figure 6.5. The compound absorbs light of 300–500 nm, which is the violet-green part of the spectrum, and the eye observes the complementary colours in the yellow-orange region.

6.4   Fluorescence

In some cases, the energy from excited molecules is partly emitted as electromagnetic radiation instead of heat. This is termed *fluorescence*. Figure 6.6 shows an absorption spectrum and a fluorescence spectrum of the same compound. Fluorescence is observed at slightly higher wavelengths than the radiation used to excite the molecule. This is because some of the energy used for excitation will be emitted as heat before the emission of radiation. Fluorescence emission will thus be of lower energy and appear at higher wavelengths. Fluorescence can be used for quantitative determinations because the fluorescence intensity is dependent on the concentration of the fluorescent substance. Fluorescence can also be used as a detection method in liquid chromatography.

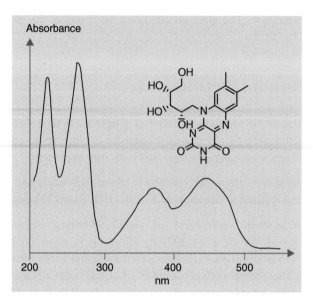

Figure 6.5 UV-Vis spectrum of the orange coloured substance riboflavin

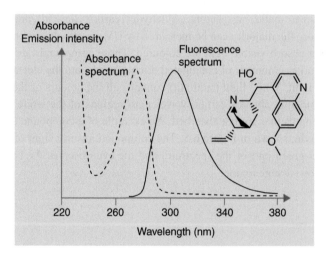

Figure 6.6 Absorption and fluorescence spectra of quinine

6.5 Atoms and Electromagnetic Radiation

Atomic absorption spectrometry (AAS) and atomic emission spectrometry (AES) are presented in Chapter 9. Both AAS and AES are primarily used for quantitative determinations. Both of these techniques are based on samples being decomposed into free atoms at high temperature (*atomization*). In AAS, free atoms absorb electromagnetic radiation and in AES

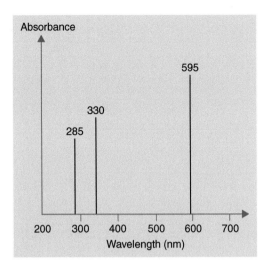

Figure 6.7   Absorption spectrum of sodium in the gas phase

free atoms emit electromagnetic radiation when they are exposed to high temperatures. This process is called *emission*.

While absorption of radiation by molecules results in continuous absorption spectra, this is not the case for free atoms. The continuous spectra of molecules are due to increased vibrations and rotations between the atoms in the molecules. This is not happening in free atoms as free atoms only absorb electromagnetic radiation by electrons being transferred to orbitals with higher energy (electronic transition). Consequently, free atoms only have a few possibilities of absorbing electromagnetic radiation. Each of these requires a specific energy, which results in absorption at a few very specific wavelengths. Figure 6.7 shows the absorption lines of sodium in the gas phase. The three major absorption lines at wavelengths 285, 330, and 590 nm are due to an electron in the 3s orbital being excited to the 5p, 4p, and 3p orbitals, respectively. Atomic spectra consisting of narrow absorption lines are termed *line spectra*. Emission of radiation from excited free atoms appears at the same wavelengths.

7

UV-Vis Spectrophotometry

7.1 Areas of Use

Ultraviolet–visible (UV-Vis) spectrophotometry is particularly suitable for identification and quantitative analysis of active pharmaceutical ingredients (APIs) and excipients and for quantitative analysis of APIs in pharmaceutical preparations of low complexity. For more complex samples, the method is not used directly, but is performed as UV-detection combined with liquid chromatography (high performance liquid chromatography – HPLC) separation. UV detection is also used in capillary electrophoresis and atomic absorption spectrometry. Only compounds showing UV-Vis absorbance are measurable; otherwise the

Introduction to Pharmaceutical Analytical Chemistry, Second Edition.
Stig Pedersen-Bjergaard, Bente Gammelgaard and Trine Grønhaug Halvorsen.
© 2019 John Wiley & Sons Ltd. Published 2019 by John Wiley & Sons Ltd.

compound must be converted to a substance that absorbs UV radiation. Compared with many other analytical techniques, UV-Vis spectrophotometry is rapid and easy to perform and spectrophotometers are relatively cheap in comparison with other instrumentation used in pharmaceutical analysis.

7.2   Quantitation

When *monochromatic* electromagnetic radiation (radiation of one wavelength only) with the intensity of I_0 passes through a solution of an analyte, some of the radiation is absorbed by the analyte molecules, while the rest passes through (Figure 7.1). When the intensity of the transmitted monochromatic radiation is I, the *absorbance* of radiation (A) is defined as

$$A = \log_{10}\left(\frac{I_0}{I}\right) = \log_{10}\left(\frac{1}{T}\right) \tag{7.1}$$

Here T is the *transmittance* (I/I_0). The transmittance thus describes the fraction of light that was not absorbed. *Beer's law* describes the relation between the absorbance (A) and the concentration (c_M) of the absorbing molecules:

$$A = \varepsilon b c_M \tag{7.2}$$

where A is the absorbance, ε is the *molar absorption coefficient* (or *molar absorptivity*), b is the *path length* of the radiation, and c_M is the *concentration* of the analyte in the solution (mol l^{-1}) (Figure 7.1). The molar absorption coefficient, ε, describes how well the compound absorbs light and is a constant for that particular substance at a defined wavelength. When b is expressed in centimetres and c in moles per litre (M), the unit of ε is cm^{-1} M^{-1}.

The specific absorbance ($A_{1\,cm}^{1\%}$) is defined as the absorbance of a solution containing 1 g of the substance dissolved in 100 ml of solvent (10 mg/mL = 1% (w/v)) measured at a defined wavelength at a path length of 1 cm and can be calculated as:

$$A_{1\,cm}^{1\%} = \frac{10\varepsilon}{M_r} \tag{7.3}$$

where M_r is the relative molar mass. The specific absorbance is also written as A (1%, 1 cm). Using the specific absorbance, Beer's law can be written as:

$$A = A_{1\,cm}^{1\%}\, bc \tag{7.4}$$

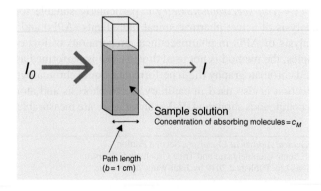

Figure 7.1   *Beer's law – absorption of radiation*

Box 7.1   Assay of phenindione in phenindione tablets by UV spectrophotometry according to BP

Content of phenindione, $C_{15}H_{10}O_2$
92.5–107.5% of the stated amount

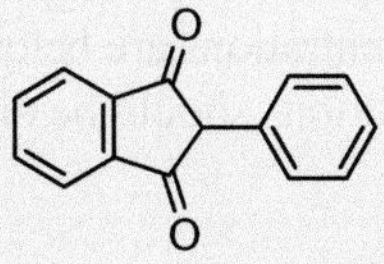

ASSAY

Weigh and powder 20 tablets. Shake a quantity of the powder containing 50 mg of phenindione with 150 mL of *0.1 M sodium hydroxide* for one hour, add sufficient *0.1 M sodium hydroxide* to produce 250 mL, filter, and dilute 5 mL of the filtrate to 250 mL with *0.1 M sodium hydroxide*. Measure the absorbance of the resulting solution at the maximum at 278 nm. Calculate the content of $C_{15}H_{10}O_2$ taking 1310 as the value of A (1%, 1 cm) at the maximum at 278 nm.

The stated amount of phenindione is 10 mg per tablet. As an example, the total mass of 20 tablets was 1.712 g (1712 mg). Thus, each tablet was on average 1.712 g/20 tablets = 0.0856 g = 85.6 mg (most of this was excipients). A 0.4184 g (418.4 mg) tablet powder was treated as described above. The absorbance of the resulting solution is 0.528.

The concentration of phenindione in the resulting solution is calculated using Beer's law as follows:

A (1%, 1 cm) is the absorbance of a solution containing 1% phenindione, where 1% = 1 g/100 mL = 10 mg/mL

$A_{1\ cm}^{1\ percent} = b \times 10$ mg/mL and $A = bc$, and the concentration is isolated by the relative calculation:

$$\frac{A}{A_{1\ cm}^{1\%}} = \frac{c}{10\ \text{mg/mL}} \Rightarrow c = \frac{A \times 10\ \text{mg/mL}}{A_{1\ cm}^{1\%}}$$

$$c = \frac{0.528 \times 10\ \text{mg/ml}}{1310} = 0.00403\ \text{mg/mL}$$

The tablet powder was originally dissolved in 250 mL (0.1 M sodium hydroxide) from which 5 mL was diluted to 250 mL:

$$\frac{0.00403\ \text{mg/mL} \times 250\ \text{mL} \times 250\ \text{mL}}{5\ \text{mL}} = 50.4\ \text{mg}$$

Thus, the 0.4184 g of tablet powder contained 50.4 mg of phenindione. The phenindione content in a tablet of average mass was

$$\frac{50.4\ \text{mg}}{418.4\ \text{mg}} \times \frac{1712\ \text{mg}}{20\ \text{tablets}} = 10.3\ \text{mg/tablet}$$

The content should be within 92.5–107.5% of the stated amount corresponding to 9.25–10.75 mg. This requirement is fulfilled.

where c is the concentration of the analyte in % (w/v). An example of quantitation by use of A (1%, 1 cm) without further calibration is given in Box 7.1. In pharmacopeia tests, the measurement conditions are well defined and calculations of concentrations are based on the given specific absorbance without further calibration. In all other cases, calibration with one or more standards is necessary to establish the exact value of A (1%, 1 cm) or ε.

Beer's law only applies to dilute solutions. In concentrated solutions, the absorbing molecules influence each other, as they are closer together. Typically, the deviation from Beer's law is small when analyte concentrations are below 0.01 M. For practical work, this means that relatively concentrated solutions should be diluted before the measurements.

7.3 Absorbance Dependence on Measurement Conditions

The absorbance of a given compound is dependent on wavelength, solvent, pH, and temperature. The dependence of temperature is relatively small and spectrophotometric assays are rarely temperature controlled. The absorbance is highly dependent on the wavelength, which appears from the absorption spectrum shown in Figure 7.2. The spectrum is recorded by irradiation of the sample and recording the absorbance, while scanning the wavelength. As the concentration of the sample and the light path are constant, the spectrum shows how ε or A (1%, 1 cm) changes with the wavelength. Theoretically, if the concentration of the analyte was 1 M, the curve would show ε as a function of λ and if the concentration of analyte was 1%, the curve would show A (1%, 1 cm) as a function of λ. This is only theoretical as such concentrated solutions should be diluted before measurement.

To achieve maximum sensitivity, quantitative measurements are performed at the *absorption maximum*, where ε (and $A_{1\ cm}^{1\ percent}$) have the highest values. For the substance displayed in Figure 7.2, the wavelength should be set to 290 nm. Beer's law applies to monochromatic radiation, that is, radiation of one wavelength. This is an ideal situation, which in practice cannot be achieved with the instruments used. In practice, the radiation transmitted to the

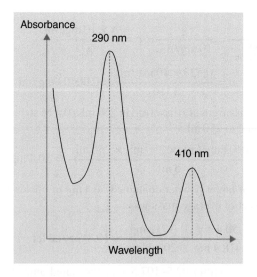

Figure 7.2 Absorption spectrum

species form a narrow wavelength range. The width of this range is termed *bandwidth*. *Polychromatic radiation* leads to deviation from Beer's law. The more the value of ε varies within the wavelength range that hits the sample, the larger is the deviation. As the variation of ε is smallest at the absorption maximum, this is another reason for performing quantitative spectrophotometric measurements at the absorption maximum. However, as long as the bandwidth of the incident light is substantially narrower than the absorption band of the absorbing structure, this will not cause significant deviations from Beer's law.

The absorption spectrum of a given compound in solution is very dependent on the solvent as it appears from Figure 7.3, showing that in polar solvents, like water, the spectrum is smoothed by the presence of the surrounding molecules, whereas in non-polar solutions, the vibrational structures are observed and even more structures are observed in the gas phase, where there is no interaction with surrounding molecules. Thus, the shape of the spectrum may change as well as the absorption maxima, depending on the solvent used to dissolve the analyte.

As many drug substances are acid or bases, pH of the test solution is important. Figure 7.4 shows the absorption maxima in acidic and alkaline solution of a drug substance (phenylephrine). As seen from this figure, both the wavelength and the specific absorption for phenylephrine are different in acidic and alkaline solutions.

If the absorbing analyte, e.g. a week acid, participates in concentration-dependent equilibria, the absorption can change by dilution. Participation in chemical reactions can also change absorption. Therefore, the analyte should be stable in the sample solution, and this cannot contain substances that react with the analyte.

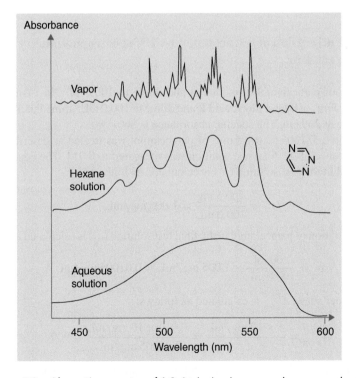

Figure 7.3   *Absorption spectra of 1,2,4-triazine in vapour, hexane, and water*

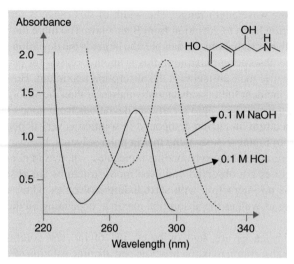

Figure 7.4 *Absorption of radiation is dependent on pH. The UV spectra of phenylephrine in 0.1 M NaOH and in 0.1 M HCl*

7.4   Identification

In the pharmacopoeias, UV spectrophotometry is used for identification of pharmaceutical ingredients, and most often as a supplement to identification by infrared (IR)

Box 7.2   Identification of paracetamol by UV spectrophotometry according to Ph. Eur.

Dissolve 0.1 g of paracetamol in methanol and dilute to 100.0 mL. To 1.0 mL of this solution add 0.5 mL of hydrochloric acid R and dilute to 100.0 mL; immediately measure the absorbance at 249 nm. The specific absorbance is 860–980.

As an example, 0.1005 g (100.5 mg) of paracetamol was treated as described above and the absorbance of the final test solution was measured to 0.914. The concentration of paracetamol in the first solution (c_1) is calculated as follows:

$$c_1 = \frac{100.5 \text{ mg}}{100.0 \text{ mL}} = 1.005 \text{ mg/mL}$$

The concentration of paracetamol in the final test solution (c_2) is calculated as follows:

$$c_2 = \frac{1.00 \text{ mL}}{100.0 \text{ mL}} \times 1.005 \text{ mg/mL} = 0.01005 \text{ mg/mL}$$

Based on Beer's law, $A_{1 \text{ cm}}^{1\%}$ is calculated as follows:

$$A_{1 \text{ cm}}^{1\%} = \frac{A \times 10 \text{ mg/mL}}{c} = \frac{0.914 \times 10 \text{ mg/mL}}{0.01005 \text{ mg/mL}} = 909$$

The calculated specific absorbance is within the required range of 860–920.

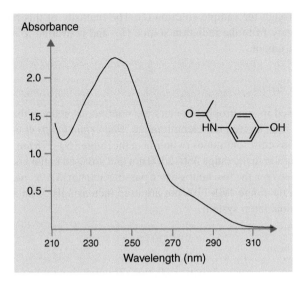

Figure 7.5 *UV spectrum for paracetamol*

spectrophotometry. The identification is based on the absorption maximum of the UV spectrum of the analyte, and is often combined with a calculation of the specific absorbance at the absorption maximum. An example of identification of paracetamol is given in Box 7.2 and the corresponding UV-spectrum is shown in Figure 7.5.

7.5 Instrumentation

Instruments for measuring absorbance in the UV-Vis range of the spectrum, termed *spectrophotometers,* consist of a *radiation source*, a *wavelength selector*, a *sample container* (called a *cell* or *cuvette*), and a *detector*. A schematic view of a UV spectrophotometer is shown in Figure 7.6.

The radiation source emits UV radiation, the wavelength selector (monochromator) isolates the wavelength that irradiates the sample, and the detector measures the intensity of

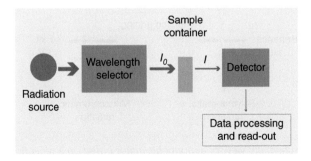

Figure 7.6 *Schematic view of a UV spectrophotometer*

radiation passing through the sample solution (I). The intensity of this radiation is compared with the intensity from the radiation source (I_0) and converted to absorbance units displayed by the instrument.

7.5.1   Radiation Sources

Radiation sources used in spectrophotometers are continuous sources that emit radiation of all wavelengths within a given wavelength range. *Deuterium lamps* emit radiation in the UV range. These lamps emit continuous radiation in the range 160–375 nm. *Tungsten lamps* emit continuous radiation in the range 350–2500 nm and are used in the visible range. Most instruments switch between the two lamps when passing 360 nm. *Xenon pulse lamp* sources emitting radiation in the range 190–1100 nm are used increasingly as an alternative to the deuterium/tungsten dual lamp system.

7.5.2   Monochromator

Beer's law only applies to monochromatic radiation. The function of the monochromator is to disperse the polychromatic radiation from the radiation source into its wavelength components and select a narrow band of wavelengths passing on to the sample. Monochromators used in UV–Vis spectrophotometers are *reflection gratings*. An outline of a *monochromator* based on a reflection grating is shown in Figure 7.7. Continuous radiation enters through a small slit (entrance slit) and reaches a concave mirror, where it is transformed into a beam of parallel rays (collimated). This beam reaches the *reflection grating*, which is a hard, thin plate with a large number of narrow parallel grooves in a reflective coating. For the UV–Vis region, there are typically between 300 and 2000 grooves/mm, with 1200–1400 being the most common. Continuous radiation that hits the grating is reflected and transmitted at various angles, depending on the wavelength. The radiation from the grating hits a new concave

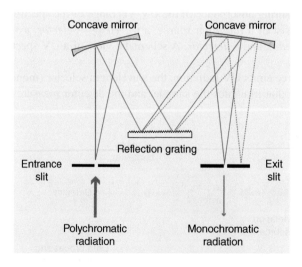

Figure 7.7   Schematic view of a monochromator based on a reflection grating

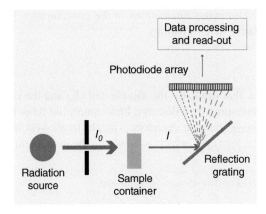

Figure 7.8   *A simplified diode array configuration*

mirror and travels on to the exit slit. A given small wavelength range passes through the slit and further on to the sample, while all other wavelengths (one such is visualized with a dotted line in Figure 7.7) are reflected at different angles from the grating and do not pass through the exit slit. The wavelength of the beam that irradiates the sample can be varied continuously by varying the angle of the reflection grating. When the grating is turned slightly, another small wavelength range will pass through the exit slit and another wavelength is thereby selected. Absorption spectra are recorded by scanning the relevant wavelength range while measuring the absorbance. During this recording, the angle of the reflection grating is turned continuously.

As an alternative to scanning instruments, a *photo diode array* instrument can be used (Figure 7.8). Continuous radiation is transmitted through the sample and hits the reflection grating. This reflects radiation at different angles according to the wavelength, as previously described. The reflected radiation strikes the photo diode array. A photo diode array is an array of a large number of small photo diodes that detect the intensity of the radiation that hits them. Each diode measures the radiation in a given wavelength interval. This means that the absorption of light can be detected and amplified to give an absorption spectrum in less than a second, in contrast to common spectrophotometers where recording of an absorption spectrum takes a longer time. Photo diode array instruments are often used as detectors in HPLC and are then named *diode array detectors* (DADs).

7.5.3   Sample Containers

The samples to be analysed are placed in a *sample container* (*cell* or *cuvette*) that is irradiated in the spectrophotometer. The sample cell windows must be transparent to the radiation used. For measurements in the UV region, sample containers are made of quartz, which is expensive, but the only material that allows transmittance of UV, as glass and plastic absorb UV radiation. In the visible region, glass or plastic containers can be used.

Cuvettes are often made with a square cross-section with an inner width of exactly 1.0 cm. Thus, the path length of radiation (b in Beer's law) is 1 cm. They can be supplied with a cap for analysis of samples in volatile solvents. The cap prevents evaporation during measurement and hence eliminates fluctuations in the concentration that may cause significant errors in quantitative determinations.

7.5.4 Detectors

The intensity of radiation that irradiates the sample cell (I_0) and the radiation transmitted through the cell (I) are measured by a detector. Photomultiplier tubes that convert light to electric current are the most common detectors. The electronic system in the instruments calculates the absorbance, $\log (I_0/I)$, which is shown on the instrument display.

7.5.5 Single-Beam and Double-Beam Instruments

Interaction between the radiation and the walls of a sample container in the form of reflection or absorption is inevitable and may result in a loss in intensity. In order to compensate for these effects, the intensity of the radiation transmitted through the sample cell is compared with the intensity of radiation that passes through an identical cell containing only solvent (*reference solution*). Single-beam instruments only contain one sample cell (Figure 7.6). This implies that first the instrument is set to zero absorbance with the reference solution. Then the reference solution is removed and the sample is loaded into the cell and measured. Each time the wavelength is changed, the absorbance has to be set to zero with the reference solution.

In a double-beam instrument (Figure 7.9), the electromagnetic radiation from the radiation source is split into two parallel beams. One beam passes through a cell for the reference solution, while the other beam passes through a cell with the sample solution. With this configuration, the absorbances in the reference cell and the sample cell are measured simultaneously, and the difference is the net absorbance of the sample solution. The instrument is adjusted to zero absorbance with two identical cuvettes containing the reference solution. Double-beam instruments thus compensate for variations in the characteristics of the radiation source, detector, and the electronic system. Measurements can thus be done faster and with higher precision and accuracy.

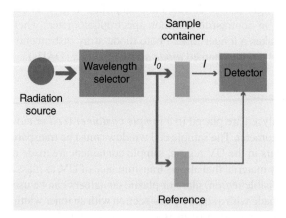

Figure 7.9   Schematic view of a double-beam spectrophotometer

7.6   Practical Work and Method Development

In the development and use of methods based on UV spectrophotometry, the following points are essential:

- Selection of solvent used to dissolve the sample
- Selection of wavelength used to read the absorbance
- Calibration and preparation of standards
- Handling of samples in the UV instrument

When the analysis is performed according to monographs in Ph. Eur., sample preparation and instrument setup are given in the monographs. When developing a new method based on UV spectrophotometry, however, all points must be considered.

The first step in method development is to select a suitable solvent. It is essential that the solvent dissolves the analyte and that particles from other substances in the sample are removed prior to absorbance measurement. Particles in the solution may cause *light scattering* and *reflection* of the radiation and inaccuracies in the absorbance reading. APIs are often organic compounds with basic functional groups and these will usually be very soluble in either acidic aqueous solutions (e.g. dilute hydrochloric acid) or in organic solvents such as ethanol or methanol. The solvent used should be transparent to the radiation in the wavelength range used (Table 7.1).

All solvents absorb radiation at low wavelengths in the UV region. At higher wavelengths, most solvents are transparent (do not absorb radiation). Above this wavelength, which is termed the *cut-off* value, the solvents can be used for UV spectrophotometry. Water and acetonitrile have low UV cut-off values and are used as solvents above 190 nm, while acetone has a high UV cut-off value and can only be used above 330 nm. The final decision should also include a verification of the stability of the analyte in the solvent. The solvent should not react with the analyte or catalyse decomposition reactions. Usually this is not a problem, but analyte stability should be verified experimentally prior to the final decision.

The absorbance readings should preferably be performed at the wavelength where the analytes have their *absorption maximum*. Quantitative measurements should be made at this wavelength for maximum sensitivity (Figure 7.10). At wavelengths where the analyte has strong absorbance the calibration curve is steep, and two samples with small differences in concentration result in significantly different absorbance readings. In this case, the *sensitivity*, which is defined as the slope of the calibration curve, is high, as shown in Figure 7.10. At wavelengths where the analyte has low absorbance, two samples with

Table 7.1   Solvents for UV spectrophotometry (list is not complete)

Solvent	Cut-off (nm)	Solvent	Cut-off (nm)
Water	190	Diethyl ether	210
Acetonitrile	190	1,4-Dioxane	220
n-Hexane	195	Dichloromethane	220
Methanol	205	Chloroform	240
Ethanol	210	Benzene	280
Cyclohexane	210	Acetone	330

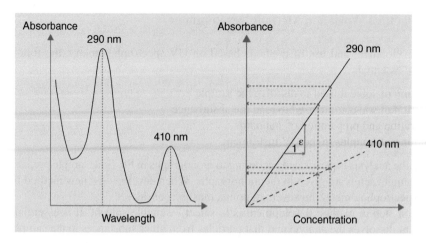

Figure 7.10   *Calibration curves at different wavelengths*

significantly different concentrations may give similar absorbance readings. In this case, the slope of the calibration curve is less and the sensitivity is low.

Published values for molar absorption coefficients, ε, or specific absorbance, A (1%, 1 cm), should be used with great caution because these depend on the experimental conditions. When developing new UV methods, a calibration curve based on standard solutions with exact concentrations of a *chemical reference substance* should be recorded. A chemical reference substance is the analyte substance of very high purity. Standard solutions must be prepared in the same way as the sample, so that the solvent, pH, and possible presence of other substances are identical for the sample and standards. Standard solutions for the calibration curve should be measured at the same wavelength as the samples. The calibration curve should be set up in the absorbance range 0.2–0.8 because this range offers the best precision. The samples must be diluted accordingly if the absorbance readings are above 0.8 absorbance units.

When samples and standards are dissolved in organic solvents, it is important that all sample containers and cuvettes are capped to prevent evaporation, as evaporation can cause significant changes of concentrations. Even with solvents like ethanol, evaporation is significant and can cause problems in quantitative determinations.

It is of particular importance that quartz cuvettes are clean to achieve the best possible optical quality. The walls should not be touched because impurities on the walls, such as grease stains, may result in significant measurement errors. Cuvettes should be wiped with lens paper soaked in methanol or ethanol. Drying of the cuvettes by heat may change their optical properties and should be avoided.

7.7   Test of Spectrophotometers

It is essential that the spectrophotometer is calibrated to ensure correct wavelength and absorbance reading. This is especially important when performing pharmacopeia analyses where calibration is not used.

7.7.1 Control of Wavelengths

According to Ph. Eur., the *wavelength scale* is verified using the absorption maxima of a holmium perchlorate solution. The test is performed by recording an absorption spectrum of holmium perchlorate solution (40 g/L of holmium oxide in perchloric acid), as shown in Figure 7.11, and comparing the absorption maxima read on the instrument with the maxima given in Table 7.2. The permitted tolerance for the wavelength is ± 1 nm in the ultraviolet range and ± 3 nm in the visible range. The reason for the lower tolerance in the UV range is that the absorption bands are narrower in this region.

The absorption wavelengths in Table 7.2 marked with (Ho) are maxima for holmium perchlorate, while the maxima marked with (Hg), (Hβ), and (Dβ) are emission lines from mercury, hydrogen, and deuterium lamps, respectively, used alternatively as radiation sources.

7.7.2 Control of Absorbance

For control of absorbance according to Ph. Eur., a solution of dried potassium dichromate is prepared with an exactly known concentration. The absorbance is read at the wavelengths shown in Table 7.3, and the specific absorbance calculated as shown in the example in Box 7.3. The calculated value of the specific absorbance should be within the maximum tolerance.

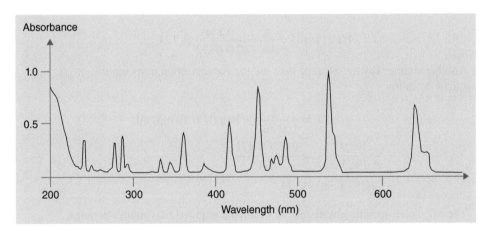

Figure 7.11   Holmium perchlorate absorption spectrum

Table 7.2   Absorption maxima for control of wavelength scale according to Ph. Eur

241.15 nm (Ho)	253.7 nm (Hg)	287.15 nm (Ho)	302.25 nm (Hg)
313.16 nm (Hg)	334.15 nm (Hg)	361.5 nm (Ho)	365.48 nm (Hg)
404.66 nm (Hg)	435.83 nm (Hg)	486.0 nm (Dβ)	486.1 nm (Hβ)
536.3 nm (Ho)	546.07 nm (Hg)	576.96 nm (Hg)	579.07 nm (Hg)

Table 7.3 *Specific absorbance of potassium dichromate solution for control of absorbance according to Ph. Eur*

Wavelength (nm)	Specific absorbance	Maximum tolerance
235	124.5	122.9–126.2
257	144.5	142.8–146.2
313	48.6	47.0–50.3
350	107.3	105.6–109.0
430	15.9	15.7–16.1

Box 7.3   Calculation of specific absorbance for control of absorbance scale according to Ph. Eur

A 59.92 mg/L (equivalent to 0.059 92 mg/mL or 0.005992% (*w/v*)) of potassium dichromate solution prepared as described in Ph. Eur. was analysed at 235 nm in a quartz cuvette with a light path of 1.00 cm. Prior to analysis, the spectrophotometer was set to zero absorbance when the quartz cuvette contained purified water.

The absorbance reading was 0.744. The specific absorbance is the absorption of a solution containing 1% (1 g/100 mL) of the substance:

$$A = A(1\%, 1\,\text{cm})bc$$

$$A(1\%, 1\,\text{cm}) = \frac{A}{bc} = \frac{0.744}{0.005992} = 124$$

Another simple way to calculate is to use the concentration units mg/mL or g/L and calculate by ratio:

$$1\% = \frac{1\,\text{g}}{100\,\text{mL}} = 10\,\text{g/L} = 10\,\text{mg/mL}$$

$$\frac{10\,\text{g/L}}{A(1\%, 1\,\text{cm})} = \frac{0.05992\,\text{g/L}}{0.744}$$

$$A(1\%, 1\,\text{cm}) = 124$$

The calculated specific absorbance is within the required maximum tolerance.

7.7.3   Limit of Stray Light

Stray light – or false light – refers to light that unintentionally reaches the detector. It can originate from the radiation source but follow other paths than that through the monochromator, or it can be light from the instrument surroundings. Errors from stray light are most serious at high absorbance values. In this case, only a small fraction of the incident light reaches the detector, and stray light will therefore constitute a large fraction of the light detected. According to Ph. Eur., stray light is controlled by recording the absorption spectrum of a 12 g/L solution of potassium chloride. The absorbance must increase steeply in the range 220–200 nm and must exceed 2.0 at 198 nm. An absorbance of 2.0 corresponds to a transmittance of 0.01 or 1%.

7.7.4   Resolution (for Qualitative Analysis)

Ph. Eur. also has a test for *resolution* that may be prescribed in monographs, where UV-absorption data are part of the identification of the compound. The test is based on recording an absorption spectrum of a solution of toluene in hexane 0.02% (*v/v*) and measuring the ratio of the absorbance at the maximum and minimum at 269 nm and 266 nm, respectively. The minimum ratio is stated in the monograph.

7.7.5   Spectral Slit-Width (for Quantitative Analysis)

The exit slit-width of the monochromator is given as the bandwidth of the radiation that exits the slit. Thus, the width of the exit slit is normally given as the *monochromator bandwidth*. For quantitative analysis this is typically 1.0 nm (although the physical width of the slit may be, for example, 0.3 mm). The energy reaching the detector increases with the exit slit-width, but the bandwidth must be small compared to the width of the absorption band of the analyte. On the other hand, decreasing the slit-width increases the resolution, which means that more details can be observed in an absorption spectrum. Therefore, a slit-width is chosen such that further reduction does not result in a change in the absorbance reading.

7.8   Fluorimetry

As described in Chapter 6, fluorescence is a process in which excited molecules emit the absorbed energy as radiation. Fluorescence is observed at slightly higher wavelengths than the radiation used to excite the molecules, as some of the absorbed energy is released before the radiation emission. Ph. Eur. describes fluorimetry as a relative procedure which uses measurement of the fluorescence intensity emitted by the analyte in relation to the fluorescence emitted by a standard solution of the same compound:

$$c_{\mathrm{x}} = \frac{I_{\mathrm{x}} c_{\mathrm{s}}}{I_{\mathrm{s}}} \tag{7.5}$$

where c_{x} is the concentration of the analyte, c_{s} is the concentration of the standard solution, I_{x} is the intensity of the light (fluorescence) emitted by the analyte, and I_{s} is the intensity of the light (fluorescence) emitted by the standard. It can be shown that the emission intensity at low concentrations can be described as

$$I = kI_0 c \tag{7.6}$$

where I_0 is the intensity of the excitation irradiation. This implies that at low concentrations the emission intensity is proportional to the analyte concentration. Blank samples must

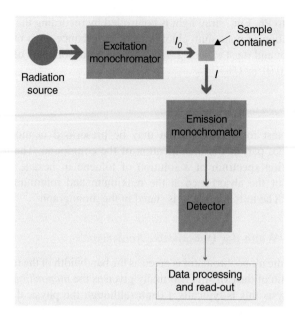

Figure 7.12   Schematic view of a fluorimeter

be run to correct for light scattering. In the context of pharmaceutical chemical analysis, fluorimetry is mainly used for detection in liquid chromatography (HPLC). A schematic overview of an instrument for fluorescence measurements, a *fluorescence spectrophotometer* or a *fluorimeter*, is shown in Figure 7.12. The excitation wavelength is selected in the excitation monochromator and the emitted fluorescence is isolated in the emission monochromator. The latter is positioned at an angle of 90° to the path of the excitation wavelength in order to avoid or minimize scattered light reaching the detector.

8

IR Spectrophotometry

8.1 IR Spectrophotometry

Infrared (IR) spectrophotometry is based on the fact that all atoms in organic molecules are in continuous vibration and rotation with respect to each other. When the frequency of a specific vibration is equal to the frequency of the IR radiation directed to the molecule, the molecule absorbs the radiation and the radiation is converted into energy of molecular vibration.

IR spectrophotometry follows the same basic principles for absorption of electromagnetic radiation as described for UV spectrophotometry and the principle is shown in Figure 8.1. Normally the wavelength region between 2500 and 15 000 nm is used for IR spectrophotometry. This region is termed the *middle infrared region*. Alternatively, the range from 800 to 2500 nm can be used. This region is termed the *near infrared* (NIR) *region* and is utilized in NIR spectrophotometry.

IR spectra show more detailed absorption bands than UV spectra (see Figure 8.2). IR spectra are like fingerprints of organic molecules and identification based on such spectra are considered highly reliable. The differences in complexity between UV and IR spectra are visualized in Figure 8.2 with salicylic acid as an example. In UV spectra, absorbance is plotted as a function of wavelength. However, by convention IR spectra plot *transmittance* as a function of *wavenumber* (see Figure 8.2). Transmittance is defined as follows:

$$T = \frac{I}{I_0} \tag{8.1}$$

Introduction to Pharmaceutical Analytical Chemistry, Second Edition.
Stig Pedersen-Bjergaard, Bente Gammelgaard and Trine Grønhaug Halvorsen.
© 2019 John Wiley & Sons Ltd. Published 2019 by John Wiley & Sons Ltd.

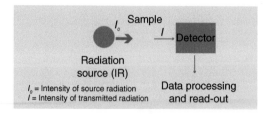

Figure 8.1   *Basic principle of IR spectrophotometry*

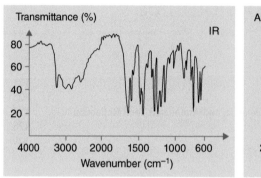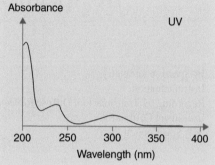

Figure 8.2   *IR and UV spectra of salicylic acid*

I_0 is the intensity of radiation directed to the sample and I is the intensity of radiation transmitted through the sample. Transmittance is often expressed as a percentage:

$$\%T = \frac{I}{I_0} \times 100\% \tag{8.2}$$

At a wavelength where the sample does not absorb any radiation, $I = I_0$ and the transmittance is 100%. In contrast, at a wavelength where all radiation is absorbed, the transmittance is 0%. The vertical transmittance axis in IR spectra is therefore scaled from 0% to 100%. Wavenumber is used on the horizontal axis. Wavenumber is defined as follows:

$$\text{Wavenumber} = \frac{1}{\text{Wavelength (cm)}} \tag{8.3}$$

A high value for the wavenumber corresponds to a low wavelength and vice versa. Wavenumbers are expressed in cm^{-1} and IR spectra are typically recorded in the spectral range from 4000 to 650 cm^{-1}.

Table 8.1   Characteristic infrared absorption bands

Binding	Functional group	Wavenumber (cm^{-1})	Intensity
C—H	Alkanes	2850–2970	Strong
		1340–1470	Strong
	Alkenes	3010–3095	Medium
		675–995	Strong
	Alkynes	3300	Strong
	Aromatic rings	3010–3100	Medium
		690–900	Strong
O—H	Alcohols, phenols	3650–3950	Variable
	H-bound to alcohols, phenols	3200–3600	Variable (broad)
	Carboxylic acids	3500–3650	Medium
	H-bound to carboxylic acids	2500–2700	Broad
N—H	Amines, amides	3300–3500	Medium
C=C	Alkenes	1610–1680	Variable
	Aromatic rings	1500–1600	Variable
C—N	Amines, amides	1180–1360	Strong
C—O	Alcohols, ethers, esters, carboxylic acids	1050–1300	Strong
C=O	Alcohols, ketones, esters, carboxylic acids	1690–1760	Strong
NO$_2$	Nitro substituents	1500–1570	Strong

Each absorption band in an IR spectrum is correlated to the molecular structure of the substance by the wavenumber that is characteristic of the absorbing bond. At wavenumbers above 1500 cm^{-1}, the most readily assigned bands are present. Table 8.1 shows some characteristic absorption bands. From IR spectra of unknown compounds, structural information can be derived using tables like Table 8.1. Below 1500 cm^{-1}, which is termed the *fingerprint region*, absorption is very complex, and reliable assignment of bands to particular functional groups can be challenging. Box 8.1 illustrates some of the absorption bands in the spectrum for salicylic acid in relation to Table 8.1.

In the context of pharmaceutical chemical analysis, IR spectrophotometry is primarily used for identification of pharmaceutical ingredients and active pharmaceutical ingredients (APIs) in pharmaceutical preparations. In such cases, the spectrum of the substance to be examined is compared with a spectrum of a chemical reference substance (CRS). Both spectra are recorded in the wavenumber range 4000–650 cm^{-1}. The transmission minima (absorption maxima) in the spectrum of the substance to be examined should correspond in position and relative size to those in the CRS spectrum. One example of such identification is illustrated in Box 8.2 for ibuprofen.

Box 8.1   Assignment of absorption bands to functional groups in the IR spectrum of salicylic acid

The IR spectrum of salicylic acid is shown below with the absorption band assignment.

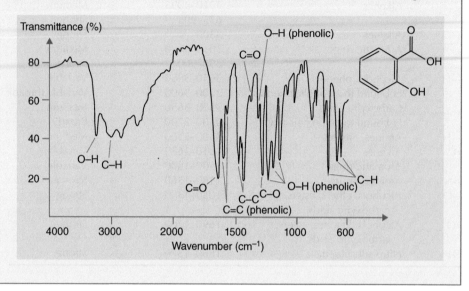

Box 8.2   Identification of ibuprofen API based on IR spectrophotometry

IR spectra are recorded for the ibuprofen API (sample) and for the ibuprofen CRS (reference) as shown below.

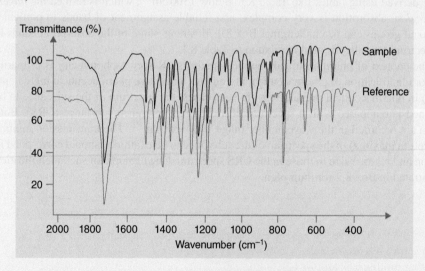

All transmission minima (absorption maxima) in the sample spectrum correspond in position and relative size to those in the reference spectrum. Therefore, the sample is positively identified as ibuprofen.

8.2 Instrumentation

The IR spectrophotometer comprises a *radiation source* (light source), a *monochromator* or *interferometer*, and a *detector*. Instruments equipped with a monochromator are termed *dispersive* instruments, which are very similar to UV spectrophotometers. The radiation source emits continuous IR radiation that is split into two equivalent beams (*double-beam* configuration) that pass through the sample holder (*sample channel*) and the reference holder (*reference channel*), respectively. The reference holder is usually empty and the sample is placed in the sample holder. Using an optical chopper (such as a rotating mirror) the IR radiation through the sample and the reference channels are alternately directed through the monochromator. The monochromator scans the wavelength and the radiation signals through the sample and reference channels are amplified and then compared electronically for recording of transmittance. The double-beam configuration is used to correct for variations in the intensity of the radiation source and to correct for fluctuations in the detector and the electronic systems. This is particularly important in the IR region because the intensity of radiation sources is relatively low. In addition, the double-beam configuration enables continuous subtraction (*background correction*) of absorbance from water vapour and carbon dioxide present in the instrument, which both have strong absorption bands in the IR spectrum between 1400 and 4000 cm^{-1} (Figure 8.3). Thus, the signal in the reference channel is subtracted from the signal in the sample channel at each wavelength, and this net absorbance (or transmittance) is plotted as a function of wavelength (or wavenumber). Therefore, no absorption bands from water and carbon dioxide are present in the net spectrum of the pharmaceutical substance to be examined.

The monochromator used in IR spectrophotometers is constructed in the same way as monochromators used in UV spectrophotometers. The major difference is the number of grooves per millimetre used in the optical grating. The grating used in UV spectrophotometers has 300–2000 grooves/mm, while the grating used in IR spectrophotometers has about 100 grooves/mm due to the higher wavelengths used in IR spectrophotometry.

The instruments described above are dispersive IR spectrophotometers where the beam from the radiation source is dispersed by the monochromator before the beam enters the detector. However, *Fourier transform IR (FTIR)* spectrophotometers have become very popular due to their superior speed and sensitivity. In the FTIR instruments, the monochromator is replaced by an *interferometer*. The interferometer produces interference signals (interferogram) that contain infrared spectral information generated after passage through the sample. A mathematical operation known as Fourier transformation converts the interferogram to the final IR spectrum. The operational principle of interferometers is very complex and is outside the scope of this textbook.

8.3 Recording by Transmission, Diffuse Reflectance, and Attenuated Total Reflection

UV spectra are recorded on diluted solutions of the analyte. In IR spectrophotometry, recording on dilute solutions is less convenient because most solvents (including water) have strong IR absorption bands. IR spectra recorded in solution will therefore contain absorption bands both from the pharmaceutical substance and the solvent. This complicates the interpretation of spectra. Therefore, IR spectrophotometry is preferably performed directly on the solid (or liquid) substance to be examined and the spectrum is recorded

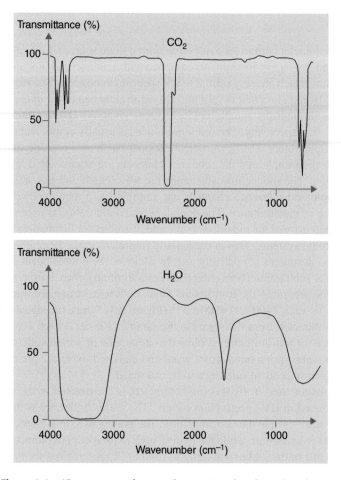

Figure 8.3   IR spectrum of water (bottom) and carbon dioxide (top)

either by transmission (or absorption), diffuse reflectance, or attenuated total reflection, as described in the following.

The principle of recording by *transmission* is illustrated in Figure 8.4. The IR radiation passes through the sample and the transmitted radiation (I) is measured by the detector. If the substance to be examined is a liquid, a thin film of the liquid is sandwiched between two plates of sodium chloride (NaCl) or potassium chloride (KCl). This sandwich is then inserted in the radiation path (sample channel) and the transmission measurement is performed. Sodium chloride and potassium chloride are both transparent to IR radiation above $400 \, cm^{-1}$, and does therefore not contribute to interference from absorption bands in the spectrum of the substance to be examined.

If the substance to be examined is solid, transmission is normally recorded on KBr discs. A small amount of the substance to be examined (1–2 mg) is finely ground and mixed with 300–400 mg of dry potassium bromide (KBr) in a mortar. The mixture is then compressed into a transparent disc at sufficiently high mechanical pressure. The disc is then placed in the sample holder which in turn is placed between the radiation

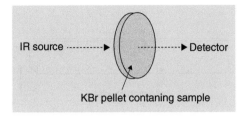

Figure 8.4 *Recording by transmission*

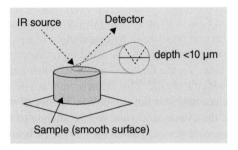

Figure 8.5 *Recording by diffuse reflectance*

source and the monochromator. Potassium bromide is transparent to IR radiation in the range above $400 \, cm^{-1}$ and has no absorption bands in the region traditionally used for IR spectrophotometry. The absorption bands recorded on the KBr disc thus originate from the analyte or impurities present in the KBr mixture. The best results are obtained when the KBr mixture is compressed in vacuo to exclude air and water. Residual water in the disc has absorption bands at 3450 and $1640 \, cm^{-1}$. Large particles in the disc contribute to band distortion due to scattering of the radiation. In order to minimize the effect of scattering, solid samples are ground to small particles of $2 \, \mu m$ or less in size.

A more recent approach is recording by *diffuse reflectance*, as illustrated in Figure 8.5. A small portion of the substance to be examined is ground to a fine powder and dispersed in KBr. Typically, this powder mixture contains 5% (*w/w*) of the substance. This powder is then loaded into a small metal cup. The powder is irradiated and the incident radiation is reflected from the base of the metal cup. IR radiation is absorbed during passage through the sample and back again, and this absorbance is measured by photocells. The resulting absorption spectrum is then converted by the instrument to a transmission spectrum. The resulting spectrum is thus very similar to the spectrum recorded by transmission through KBr discs. Diffuse reflectance is widely used in NIR spectrophotometry and can also be used to study different crystal forms of pharmaceutical ingredients (polymorphs).

Another recent and very popular approach is recording by *attenuated total reflectance* (*ATR*). The principle is illustrated in Figure 8.6.

The substance to be examined is placed directly in close contact with a *reflection crystal* such as diamond, germanium zinc selenide, or another suitable material having a high refractive index and that does not absorb IR radiation. In this case, the substance to be

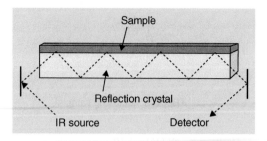

Figure 8.6   Recording by attenuated total reflectance (ATR)

examined is not dispersed in KBr. A close contact between the substance and the reflection crystal is obtained by mechanical pressure or by evaporation of a solution of the substance on the reflection crystal. An infrared beam is directed on to the crystal at a certain angle. The internal reflectance creates a wave that extends beyond the surface of the crystal into the sample held in contact with the crystal. This evanescent wave protrudes a few micrometres beyond the crystal surface and into the sample. In regions of the IR spectrum where the sample absorbs energy, the evanescent wave will be attenuated or altered. The attenuated energy from each evanescent wave is passed back to the IR beam, which then exits the opposite end of the crystal and is passed to the detector in the IR spectrophotometer. Using an FTIR spectrophotometer, the IR spectrum is collected in seconds.

8.4   Instrument Calibration

IR spectrometers should be calibrated regularly to verify that absorption bands are recorded at the correct wavenumber and to control the spectral resolution. The *wavenumber scale* is verified using *a polystyrene film*, which has transmission minima (absorption maxima) at the wavenumbers (in cm^{-1}) shown in Table 8.2. A number of absorption maxima for polystyrene are used to control the wavenumber scale and limits for satisfactory accuracy are given.

Control of *resolution performance* is based on a recorded spectrum of a polystyrene film. If resolution is too low, details in the spectra are lost, and consequently the value of the spectral information is reduced. The resolution performance test shown in Box 8.3 is according to the European Pharmacopoeia (Ph. Eur.).

Table 8.2   Transmission minima (cm^{-1}) and acceptance tolerances of a polystyrene film

Transmission minimum (and acceptance tolerance)	Transmission minimum (and acceptance tolerance)
3060.0 (±1.5)	1583.0 (±1.0)
2849.5 (±2.0)	1154.5 (±1.0)
1942.9 (±1.5)	1028.3 (±1.0)
1601.2 (±1.0)	

Box 8.3   Control of resolution performance

Record the spectrum of a polystyrene film approximately 35 μm in thickness. The figure below shows the difference x between the percentage transmittances at the transmission maximum A at 2870 cm^{-1} and at the first transmission minimum B at 2849.5 cm^{-1} must be greater than 18. The difference y between the percentage transmittances at transmission maximum C at 1589 cm^{-1} and at the transmission minimum D at 1583 cm^{-1} must be greater than 10.

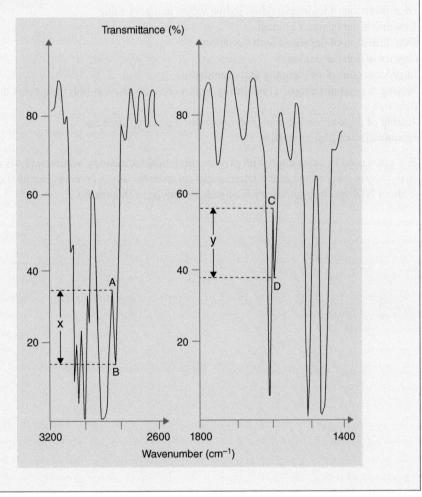

8.5   NIR Spectrophotometry

NIR spectrophotometry is a technique with wide and varied applications in pharmaceutical analysis. The NIR spectral range extends from about 780 nm to about 2500 nm (from about $12\,800\,\text{cm}^{-1}$ to about $4000\,\text{cm}^{-1}$). NIR spectrophotometry records spectra either in transmitted, scattered, or reflected mode. Measurements can be made directly on samples without any pretreatment. Physical as well as chemical information, both qualitative and quantitative, is available from NIR spectra. NIR spectrophotometry has a wide variety of applications for both chemical and physical testing, including:

- Identification of active substances, excipients, dosage forms, manufacturing intermediates, chemical raw materials, and packaging materials
- Quantitation of active substances and excipients
- Determination of hydroxyl value, iodine value, and acid value
- Determination of water content
- Determination of degree of hydroxylation
- Control of solvent content
- In-process control of blending and granulation
- Testing of crystalline form, crystallinity, polymorphism, pseudo-polymorphism, and particle size
- Testing of dissolution behaviour, disintegration pattern, hardness
- Examination of film properties

NIR is often used in connection with process analytical technology, where analysis is conducted directly and on-line in the pharmaceutical manufacturing process. Detailed discussion about NIR spectrophotometry is outside the scope of this textbook.

9

Atomic Spectrometry

9.1 Applications of Atomic Spectrometry

The main pharmaceutical use of atomic spectrometry is for control of elemental impurities in active pharmaceutical ingredients, excipients and pharmaceutical preparations. New harmonized monographs in Ph. Eur. and USP became official in 2018. In these regulations, instrumental methods replace the previous tests for heavy metals based on sulfide precipitation. Analysis of elements in dietary supplements and quantitation of zinc in insulin formulations are other examples of pharmaceutical use of atomic spectroscopy. The methods are characterized by high sensitivity, low detection limits, and high specificity.

Introduction to Pharmaceutical Analytical Chemistry, Second Edition.
Stig Pedersen-Bjergaard, Bente Gammelgaard and Trine Grønhaug Halvorsen.
© 2019 John Wiley & Sons Ltd. Published 2019 by John Wiley & Sons Ltd.

9.2   Atomic Absorption Spectrometry (AAS)

Atomic absorption is a process that occurs when an atom in the ground state absorbs electromagnetic radiation of a specific wavelength and is converted to an excited state. Atomic absorption spectrometry (AAS) is a technique for quantitative determination of elements based on absorption of electromagnetic radiation by an atomic vapour of the elements generated from the sample. The absorption is dependent on the number of atoms present in the atomic vapour. When the sample is heated to a high temperature, molecules are broken into free atoms (*atomization*). The volatilized atoms absorb electromagnetic radiation of a specific wavelength with an energy corresponding to the difference between the ground state and the excited state. According to *Beer's law*, the amount of radiation absorbed is proportional to the concentration:

$$A = \log \left(\frac{I_0}{I} \right) = abc \tag{9.1}$$

where I_0 is the intensity of the incident radiation and I is the intensity of the transmitted radiation, a is the *absorption coefficient* of the element at the specific wavelength, b is the path length, and c is the total concentration of the element in the test solution. Thus, the principle in AAS is similar to the principle of ultraviolet–visible (UV-Vis) spectrophotometry. A schematic view of an AAS instrument, where the sample is atomized in a flame is shown in Figure 9.1.

9.3   AAS Instrumentation

As illustrated in Figure 9.1, an AAS instrument consists of a radiation source (hollow cathode lamp), a sample introduction system, a sample atomizer (typically a flame), a monochromator, a detector, and a data-acquisition unit. The samples, which are usually acidic aqueous solutions, are aspirated into the flame where the sample evaporates and is volatilized and *atomized*. Radiation from the lamp is passed through the flame and the

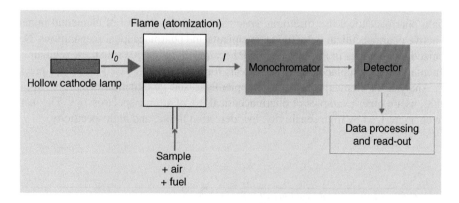

Figure 9.1   Principle of flame atomic absorption spectrometry

volatilized atoms absorb radiation of a specific wavelength. The monochromator selects the specific wavelength at which the atoms absorb radiation, and the detector records the intensity of the radiation line when a sample is present in the flame (I) and the intensity of the radiation line when the sample is not present (I_0). The signals are converted to absorbance by a data acquisition unit and shown at the display.

9.3.1   Radiation Sources

In contrast to the continuous absorption spectra of molecules, the absorption spectra of atoms in the gas phase are line spectra consisting of very narrow lines (Chapter 6). These lines are much narrower than the bandwidth of the exit radiation from the monochromator in a spectrophotometer, which is typically 1 nm. Thus, the energy absorbed by the atoms is very small, leading to a very small difference between I_0 and I and thereby resulting in absorbance close to zero. The emission lines from the lamp must therefore be comparable to the width of the lines in the absorption spectrum. This is achieved by use of *a hollow cathode lamp*, shown in Figure 9.2. The hollow cathode lamp is filled with an inert gas such as argon and contains an anode and a hollow cathode made from the analyte element. When a high voltage is applied between the anode and the cathode, the inert gas is ionized and positive ions are accelerated towards the negatively charged cathode. When positive ions strike the cathode, the cathode releases a vapour of atoms into the gas phase and these are excited by collision with high-energy electrons. The excited atoms immediately return to the ground state, while they release the energy by emitting a radiation spectrum of wavelengths that exactly matches the wavelengths that are able to excite the free atoms of that element in the flame (Figure 6.7). Consequently, a specific lamp is required for each element. For example, for the analysis of copper (Cu), a Cu-coated cathode is used. Some lamps, however, contain more than one element in the cathode.

Electrode-less discharge lamps (EDLs) are alternative elemental line radiation sources for AAS. Due to a different construction principle, they provide higher radiation intensity and are often used to replace the hollow cathode lamps, especially for volatile elements. The emission of these lamps consists of a spectrum of very narrow lines with half-widths of about 0.002 nm of the element.

9.3.2   Sample Introduction System

The most important step in atomic spectroscopic procedures is *atomization*, the process in which the sample is volatilized to produce the atomic vapour. Introduction of the sample solution and subsequent atomization in the flame is illustrated in Figure 9.3. In the first step,

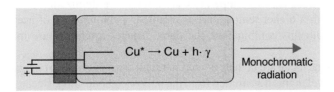

Figure 9.2   *Schematic view of the hollow cathode lamp*

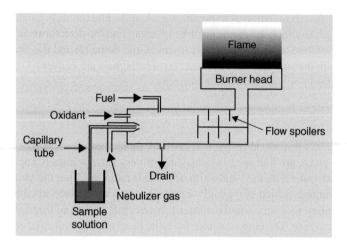

Figure 9.3 *Schematic view of the atomization device*

the sample solution is dispersed as a fine *aerosol* of small droplets. The formation of small droplets is termed *nebulization*. The aerosol is mixed with gaseous *fuel* and *oxidant* that carry it into the flame. The sample solution is drawn into the capillary tube of the *nebulizer* by a rapid flow of nebulizer gas (most often air) passing the tip of the capillary. The liquid sample is broken into fine droplets as it leaves the tip of the nebulizer, where after it is mixed with fuel (acetylene) and directed into the flame. The aerosol consists of droplets of different sizes and *flow spoilers* are used to block large droplets of liquid from entering the flame. The liquid from large droplets is drained at the bottom of the spray chamber and is led to a waste container. Only the very fine droplets of the aerosol, oxidant, and fuel are introduced to the atomizer.

9.3.3 Sample Atomizer

The atomizer in AAS is often a flame. The solvent of the aspirated sample evaporates in the base region of the flame and the resulting finely divided solid particles are carried to the centre of the flame, which is the hottest part. Here, the solid particles evaporate and the atomic vapour is formed. Finally, the atoms are carried to the outer edge of the flame where oxidation may occur before the atomization products disperse into the atmosphere. The atomization process is a consequence of the high temperature in the flame. The most common fuel/oxidizer combination is acetylene/air, which produces a flame with a temperature in the range 2100–2400 K. This temperature is sufficient for atomization of most elements. When a flame with a higher temperature is required, a combination of acetylene/nitrous oxide is used. With this combination, the flame reaches a temperature in the range of 2600–2800 K.

Most atoms in the flame are in their ground state and can absorb electromagnetic radiation of the specific wavelengths transmitted from the hollow cathode lamp. Because the velocity of the fuel/oxidant mixture through the flame is high, only a small fraction of the sample undergoes all these processes.

The flame replaces the cuvette in UV-Vis spectrometry and the path length of the flame is typically 10 cm. According to Beer's law, absorbance is proportional to the path length and long path lengths provide increased sensitivity.

9.3.4 Monochromator

As the hollow cathode lamp emits radiation at several wavelength lines and the flame emits continuous radiation, a *monochromator* is placed between the flame and the detector to select the wavelength of radiation directed to the detector. The spectral bandwidth of a monochromator in AAS is in the range 0.2–2 nm. To distinguish radiation from the hollow cathode lamp and the flame, the radiation from the lamp is pulsed or modulated by *beam chopping*, where a rotating chopper periodically blocks the light from the lamp. This makes the detector able to distinguish between the signal from the lamp and the background from the flame.

Instruments for AAS are *single-beam* or *double-beam instruments*, as described for UV spectrophotometers. The advantage of single-beam instruments is their simplicity and since all of the radiation is directed to the flame, single-beam instruments provide good sensitivity. A disadvantage of single-beam instruments is that these instruments cannot correct for variations in lamp intensity and variations due to the detector and the electronic system. These variations may affect the absorbance readings.

9.3.5 Electrothermal Atomizer

As an alternative to the flame atomizer, the sample can be atomized in an electrically heated *graphite furnace* (*electrothermal atomization*). In this system, the sample is introduced to a small graphite tube (about 3 cm in length) placed in a surrounding graphite furnace. A graphite furnace is shown in Figure 9.4. The sample, typically 5–50 µL, is introduced via an autosampler through a hole in the small graphite tube. The oven is temperature programmed in steps for *drying*, *ashing*, *atomization*, and *cleaning*. In the drying step, the temperature should be adequate to evaporate liquids and in the ashing step, the temperature is raised to remove matrix components by pyrolysis. A flow of inert gas removes the combustion products during these steps. The sample is then atomized by a fast rise in temperature. This atomizes the entire sample and retains the atomic vapour in the light path for an extended

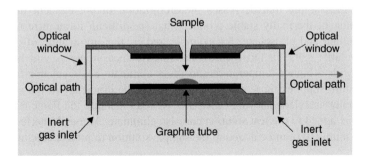

Figure 9.4 *Schematic view of the electrothermal atomizer*

Table 9.1 Comparison of methods based on atomic spectrometry

	Flame AAS	Furnace AAS	ICP	ICP-MS
Detection limits (μg/L)	10–1000	0.01–1	0.1–10	<ng/L
Linear range	10^2	10^2	10^5	10^8
Precision	0.1–1%	0.5–5%	0.1–2%	0.5–2%
Sample throughput	10–15 s/ element	3–4 min/ element	>50 elements/ min	>50 elements/ min
Sample volume	10 mL	Few μL	Few mL	Few mL

period, typically five seconds. This improves the sensitivity and therefore electrothermal atomization is mainly used to measure very low levels of element.

The advantage of electrothermal atomization is higher sensitivity as compared to flame atomization; detection limits are compared in Table 9.1. The low sample amount needed and the possibility of analysing samples with complicated matrixes, such as biofluids, are other advantages. A disadvantage is the longer analysis time, as a temperature program cycle often has a duration of 5–7 minutes, while analysis on a flame system takes only seconds. The graphite furnace technique also demands higher operator skills for developing temperature programs. Other types of atomization device include *cold vapour* and *hydride techniques*, which can be used for analysis of mercury and hydride forming elements (As, Sb, Bi, Se, and Sn), respectively. These techniques are beyond the scope of this book.

9.3.6 Interferences

Interference refers to any effect that changes the analyte signal while the analyte concentration remains the same. Interferences should either be removed or corrected. *Spectral interferences* are interferences owing to overlap of an analyte signal with signals from other elements or molecules or overlap with signals from the flame. The latter can be corrected by a deuterium lamp *background correction*. Overlap from other elements can be avoided by careful choice of the analyte wavelength and the use of a high resolution monochromator.

Errors may also occur if matrix components affect the atomization of elements. This is often referred to as *chemical interferences*. Chemical interference is caused by any component of the sample that decreases the atomization of the element to be determined. Formation of thermally stable products that are difficult to atomize reduces the atomization; sulfate and phosphate, for example, reduce the atomization of calcium by forming non-volatile salts. In those cases, quantitative measurements will be too low. Using flames with a higher temperature can often eliminate chemical interferences. A higher temperature supplies more energy and increases the efficiency of the atomization process. Replacing acetylene/air with acetylene/N_2O can increase the flame temperature. *Releasing agents* added to the test solution can also eliminate chemical interferences. For example, lanthanum (III) can be added to the sample solution to protect calcium from the interfering effects of sulfate as La (III) forms more stable compounds with sulfate than calcium. *Ionization interference* can be a problem if the ionization potential of the analyte element is low and the sample is therefore easily ionized, leaving fewer neutral atoms for absorption of radiation. This will result in decreased sensitivity. Adding an ionization suppressor, which is a more easily ionized compound than the analyte, circumvents the problem. This is desirable in low temperature flames like that in flame-photometers. In other cases, the absorption of the ion signal could be used instead of the atom signal.

9.3.7   Background Correction

Scatter and background from flame or graphite furnace atomization can result in errors in the absorbance readings. The background is measured in the wavelength range defined by the bandwidth selected by the monochromator (0.2–2 nm), whereas the atomic absorption takes place in a very narrow wavelength range (0.005–0.02 nm). Background correction can be performed by measuring the absorption when switching between the hollow cathode lamp and a deuterium lamp. The absorbance resulting from the hollow cathode lamp is the total absorption (element + background) while the absorbance from the deuterium lamp is due to the background. By subtraction of the background absorption from the total absorption, a correction is made for background absorption.

9.4   AAS Practical Work and Method Development

As the atomic spectrometric methods most often are used for determination of low concentrations of elements, errors due to contamination may occur. Use of plastic lab ware is recommended wherever possible and cleaning of lab ware by soaking in nitric acid may be necessary for trace analysis (expected concentrations in the low ng/mL level). High purity chemicals and purified water must be used. Sample preparation may require boiling with strong acids or treatment in a microwave oven. Determination of the *blank* value of the total analysis including the sample pretreatment is important. This is determined by performing the whole analytical procedure on a blank sample (e.g. water) not containing the analyte element. For elements with ubiquitous occurrence, like iron, zinc, nickel, and chromium, the blank value can exceed the concentration to be analysed in the sample, leading to serious errors. Thorough rinsing between sample analyses is recommended to avoid *carryover* from previous samples.

Before analysis, the instrument is rinsed by a dilute acid solution followed by setting the instrument to zero absorption. The instrument sensitivity is controlled by measuring a standard solution and calculating the *characteristic concentration*, which is defined as the concentration resulting in 1% absorption corresponding to an absorbance of 0.0044. Box 9.1 exemplifies a control of characteristic concentration. The specifications of the instrument given by the manufacturer list the values with which the instruments must comply. If the instrument does not comply, sample aspiration efficiency, lamp intensity, and the position of lamps and burner heads should be controlled and optimized.

Box 9.1   Calculation of characteristic concentration in AAS

According to the instrument manual, the characteristic concentration of Ca at 422.7 nm is 0.092 mg/mL.

Measuring a standard containing 5.0 mg/mL resulted in an absorbance of 0.345.

As the characteristic concentration is defined as the concentration resulting in an absorbance of 0.0044, it can be calculated by the ratio

$$\frac{X}{0.0044} = \frac{5.0\,\text{mg/mL}}{0.345}$$

$$X = \frac{5.0\,\text{mg/mL} \times 0.0044}{0.345} = 0.064\,\text{mg/mL}$$

As the characteristic concentration of the instrument was lower than the specification (0.092 mg/mL), an absorbance of 0.0044 was obtained with a lower concentration. Thus, the instrument was more sensitive than required and no further optimization was needed.

According to Ph. Eur., quantitative determination can be based on *direct calibration* or by the method of *standard addition*. In *direct calibration,* the *absorbances* (A) of not fewer than three reference solutions of known concentrations are recorded. Their concentrations should span the expected value of the test solution. For assay purposes, the optimal calibration levels are between 0.7 and 1.3 times the expected content of the element to be determined or the limit prescribed in the monograph. Each solution is introduced into the instrument using the same number of replicates for each of the solutions to obtain a steady reading. A calibration curve is prepared from the means of the readings obtained with the reference solutions by plotting the mean absorbance as a function of the element concentration. The concentration of the element in the test solution is determined from the curve.

Standard addition is an alternative to direct calibration. By standard addition the calibration is done directly in the test solution to avoid any experimental variations between samples and standards. In direct calibration, it is assumed that the atomization processes are the same for test solutions and reference solutions. This may not be the case if the viscosity of test solutions and reference solutions vary. Solutions of high viscosity are nebulized more slowly than solutions of lower viscosity. Furthermore, drop sizes of the aerosol may also vary; this is often the case, when samples contain organic solvents. In those cases, quantification by standard addition is recommended. Calibration by standard addition is described in Chapter 17.

9.5   Atomic Emission Spectrometry (AES)

Atomic emission is a process that occurs when electromagnetic radiation is emitted by excited atoms or ions. *Atomic emission spectrometry* (AES) is a technique for determination of the concentration of an element in a sample by measuring the intensity of one of the

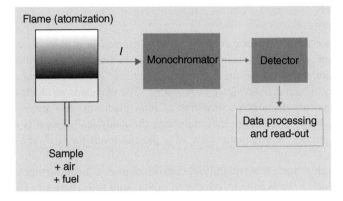

Figure 9.5   Principle of atomic emission spectrometry

emission lines of an atomic vapour of the element. The intensity of the emitted radiation (I) is proportional to the amount of element in the sample (c):

$$I = kc \tag{9.2}$$

The principle of AES is illustrated in Figure 9.5. The sample is brought into a *flame* or *plasma* as a gas or as an aerosol. The heat evaporates the solvent and breaks chemical bonds to create an atomic vapour. By heating the sample to high temperatures, the sample atomizes and a substantial part of the atoms is excited and ionized by collisional energy. The atoms and ions in the excited state are unstable and decay to lower states, resulting in emission of electromagnetic radiation. Emission lines resulting from the return of excited atoms to the ground state are separated in the monochromator and the intensity of the selected emission line is measured in the detector. Emission spectra contain several more lines than the corresponding absorption spectra. Quantitative determination of an element is performed by measuring the intensity of one of the emission lines characteristic for the element. A calibration curve is necessary to establish the relationship between the intensity of the signal and the concentration of the element.

9.6   Flame Photometry

AES based on flame atomization is termed *flame photometry* and is performed with flame photometers, which are relatively simple instruments. The flame is similar to the flame in AAS. Air is used as an oxidizer and common fuel gases are propane or butane. Interferences from other elements are less probable and wavelength selection can be made with simple *filters*. The advantage is that the instruments are relatively cheap and are easy to use. The number of free atoms excited increases sharply with increasing temperature. Flame photometry is only used for determination of elements that are easy to excite, such as *lithium*, *sodium*, and *potassium*. Because of its convenience, speed, and relative freedom from interferences, flame photometry has become the method of choice for these elements.

9.7 Inductively Coupled Plasma Emission Spectrometry

Most elements need a higher temperature than that created in a flame to be transferred to the excited state. This can be achieved by using inductively coupled plasma (ICP) as the excitation source. The technique is termed *inductively coupled plasma atomic emission spectroscopy*, which is abbreviated to *ICP-AES*. The plasma is an electrically neutral, highly ionized gas (usually argon) consisting of ions, electrons, and atoms sustained by a radiofrequency field. The energy that maintains the analytical plasma is derived from electromagnetic energy. A schematic view of a plasma torch for ICP-AES is shown in Figure 9.6.

The plasma torch is made from quartz and consists of three concentric tubes. A high flow of argon for producing the plasma is introduced in the outer tube (typically 15 L/min). The sample aerosol from the nebulizer is carried through the central tube, while an auxiliary cooling argon flow flows in the middle tube to prevent the plasma from overheating the inner tube. An *induction coil* surrounds the end of the torch. A radiofrequency generator (typically 1000–1500 W, 27 or 41 MHz) produces an oscillating current in the induction coil. This results in an oscillating magnetic field at the top of the torch. When the argon plasma is ignited by a spark, electrons are stripped from the argon atoms and accelerated in the circular paths of the magnetic field. Electrons collide with atoms and transfer energy to the entire gas, resulting in temperatures of 6000–10 000 K. The ICP appears as an intense, bright, tulip-shaped plasma. At the base, the plasma has a toroidal (donut shape); this region is called the induction region, in which the inductive energy transfer from the load coil to the plasma takes place.

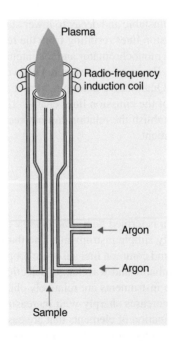

Figure 9.6 *Schematic view of the inductively coupled plasma torch*

Solvent from the nebulized sample evaporates forming anhydrous particles, broken down to individual molecules, followed by a further dissociation into atoms that are ionized in the plasma. The ionization energy of the plasma is adequate to ionize most elements in the periodic table, making ICP an ideal ionization source for multi-elemental determinations.

The detection limits for most elements are in the pg/mL to ng/mL range, which are significantly lower than those obtained with flame photometry. The emission is either observed across the plasma (radial view) or along the plasma (axial view). The latter increases the sensitivity with a factor of about 10.

By changing the wavelength settings of the monochromator, the emission from several elements can be measured. This is an obvious advantage as the content of several elements can be determined simultaneously. The range of the linear relationship between the intensity of emitted radiation and concentration is several orders of magnitude larger compared to the methods based on AAS.

9.8 Inductively Coupled Plasma Mass Spectrometry

Inductively coupled plasma mass spectrometry (ICP-MS) is a mass spectrometry method that uses ICP as the ionization source. The plasma is directed into a mass spectrometer (Chapter 15), which separates and measures ions according to their mass-to-charge ratio (m/z). The mass spectrometer can be a quadrupole or a triple quadrupole, or a time-of-flight instrument for high resolution applications.

The challenge in measuring with a mass spectrometer is that the instrument requires vacuum to avoid collisions between the introduced ions and air molecules. An important part of the ICP-MS instrument is therefore the interface that couples the ICP at atmospheric pressure and the mass spectrometer, which operates at vacuum conditions. A schematic view of an ICP-MS instrument is shown in Figure 9.7.

Ions from the ICP are transported from the plasma through the *sampling interface* consisting of two cones, a sampler, and a skimmer with small orifices (~ 1.0 mm). The purpose of the interface is to transfer ions from atmospheric pressure to the vacuum of the mass

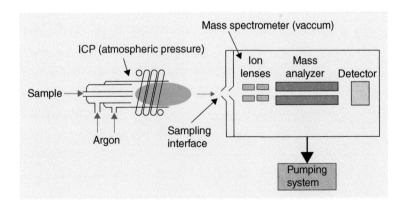

Figure 9.7   Schematic view of the ICP-MS instrument

spectrometer of 10^{-5} torr or less. A lens behind the skimmer cone with a high negative potential attracts the positive ions and separates these from electrons and molecular species. The ions are accelerated and focused by a magnetic ion lens into the mass analyser. The quadrupole mass analyser serves as a mass filter in the same way as discussed later in Chapter 15, and the different atomic ions with different m/z values are detected by a detector located after the quadrupole mass analyser.

ICP-MS is used for the determination of (atomic) elements in the same way as AAS and ICP-AES. ICP-MS is typically operated in the mass range of m/z 3–300. More than 90% of the elements in the periodic table have been determined by ICP-MS. The spectra produced by ICP-MS are remarkably simple compared to conventional ICP optical spectra and consist of a simple series of isotope peaks for each element present in the sample. These spectra are used to identify the elements present in the sample and for their quantitative determination. Usually, quantitative analyses are based on calibration curves in which the ratio of the ion count for the analyte to the count for an internal standard is plotted as a function of concentration. ICP-MS is a highly sensitive instrument and elements can be detected down to the sub-ppb level (Table 9.1). ICP-MS is the method of choice in industrial pharmaceutical quality control laboratories for determination of elemental impurities. The instruments are often placed in clean room facilities to avoid sample contamination from the surroundings.

10

Introduction to Chromatography

10.1   Introduction

Chromatography is a separation principle, which is indispensable in separation and detection of drug substances in complex mixtures. Separation takes place in a *column*, where the sample constituents interact with a *stationary phase* while they are transported down the column with a *mobile phase*. Due to different levels of interaction with the stationary phase, the different sample constituents (compounds) are transported with different velocities through the column and are thereby separated. A detection system (*detector*) measuring the individual compounds as a function of time is located at the outlet of the column. The detector signal plotted as a function of time results in a *chromatogram*, which appears as a number of peaks and is used to identify and quantify the individual compounds of the sample. If the mobile phase is a gas, the technique is termed *gas chromatography* (*GC*); when the mobile phase is a liquid, the technique is termed *liquid chromatography* (*LC*). This chapter discusses the fundamentals of chromatographic separation, which are valid for gas and liquid chromatography.

Introduction to Pharmaceutical Analytical Chemistry, Second Edition.
Stig Pedersen-Bjergaard, Bente Gammelgaard and Trine Grønhaug Halvorsen.
© 2019 John Wiley & Sons Ltd. Published 2019 by John Wiley & Sons Ltd.

10.2   General Principles

In *chromatography*, constituents of a complex sample are separated and measured individually by a detection system. Separation takes place in a tube containing the stationary phase, termed the *column*. In LC the stationary phase consists of packed solid particles, while in GC the stationary phase is a film on the column inner wall. The sample is introduced at the inlet of the column, and a liquid or a gas continuously flowing through the column brings the sample constituents through the column to a *detector*, which measures the separated sample constituents. The liquid or gas is termed the *mobile phase*. When moving downstream the column with the mobile phase, the different sample constituents *interact* with the stationary phase. Some sample constituents have little affinity for the stationary phase, and these will pass the column with a velocity close to the velocity of the mobile phase. Other sample constituents may have a higher affinity for the stationary phase, and these will be more delayed relative to the mobile phase. Thus, different chemical substances in a complex sample will interact to different degrees with the stationary phase and pass through the column with different velocities. The sample compounds are thereby separated in time and detected individually by the detector.

Figure 10.1 illustrates a chromatographic separation. In (a) the mobile phase is pumped through the column before introduction of the sample (the mobile phase is also pumped in (b) and (c)). In (b) a small volume of sample is introduced into the column inlet, and this is termed *injection*. In this case, the sample comprises three compounds, A, B, and C. In (c) A, B, and C are moving with the mobile phase through the column. Compound A has no interaction with the stationary phase; it moves with the same velocity as the mobile phase and is detected as the first component by the detector. Compound B has some interaction with the stationary phase and moves more slowly, while compound C is strongly interacting with the stationary phase and is therefore strongly *retained* in the system. The result is that compound C is detected later than compound B. The delay of compounds relative to the velocity of the mobile phase is termed *retention* and the process of moving the compounds

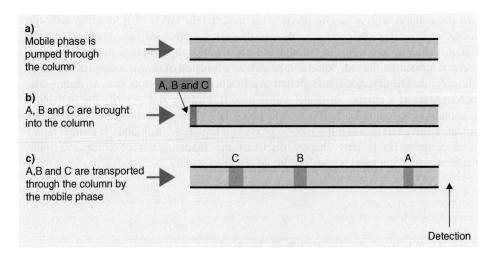

Figure 10.1   Chromatographic separation of components A, B, and C

through the column by the mobile phase is termed *elution*. When the differences in retention are sufficiently large, the different components elute from the column at different times and are separated and detected individually.

To separate different compounds by chromatography, their retention must be different. This is accomplished if they partition differently between the stationary phase and the mobile phase. The partition of a compound is described by the *distribution constant* (note that different textbooks use different terms for K_D, and in the current textbook the term distribution constant is used according to the European Pharmacopoeia (Ph. Eur.)):

$$K_D = \frac{C_S}{C_M} \tag{10.1}$$

C_S is the concentration of the compound in the stationary phase and C_M is the concentration in the mobile phase. The velocity by which each compound elutes through the column is determined by the portion of molecules present in the mobile phase, since molecules do not move through the column while they are in the stationary phase. Retention of a compound is therefore determined by its distribution constant. Compounds with large distribution constants have a large portion of molecules in the stationary phase and these compounds are strongly retained in the column. Compounds with a small distribution constant have a small portion of molecules in the stationary phase and these are less retained. If a compound is not retained by the stationary phase, the K_D value is zero.

At the molecular level, the elution of a compound is highly discontinuous. Each time a molecule is sorbed to the stationary phase, its elution is temporarily stopped while other molecules of the same kind pass with the mobile phase. Some of these are sorbed a moment later and are overtaken by the first molecule and so on. Each molecule thus follows a rapid, random alternation between the stationary and the mobile phase where the concentration of a compound in the stationary phase and in the mobile phase is determined by the distribution constant.

A schematic view of a chromatographic system is shown in Figure 10.2. The sample solution is injected into the flow of the mobile phase via the *injector* and at the column outlet the *detector* measures the individual compounds when eluted. This is plotted in a *chromatogram* showing the detector signal as a function of time, as illustrated in Figure 10.3.

The chromatogram illustrates the separation of compounds A, B, and C (Figure 10.2). The non-retained compound A has no affinity for the stationary phase and is eluted as the first peak in the chromatogram. Compounds B and C interact with the stationary phase as their distribution constants are larger. Thus, the compounds are eluting according to their distribution constants; the least retained is eluted first and the most retained is eluted last. The time from injection of the sample to the maximum detector response of a given

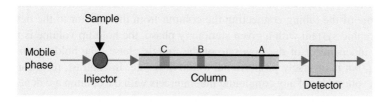

Figure 10.2   *Schematic view of the chromatographic system*

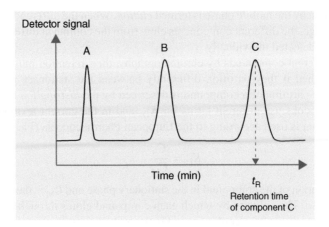

Figure 10.3   *Chromatogram of a sample containing compounds A, B, and C*

substance is termed *retention time* (t_R). Ideally the substances appear in the chromatogram as symmetric *Gaussian peaks*. In a given chromatographic system the retention time of a substance is characteristic for its physicochemical properties and is used for identification.

10.3   Retention

A simple chromatogram from the separation of two compounds is shown in Figure 10.4. The first compound is not retained by the stationary phase and passes the column with the same speed as the mobile phase. This compound is eluted at time t_M, which is termed the *hold-up time*. The second component is retained by the stationary phase and is eluted at retention time t_R. Retention is normally expressed as the retention time, but can also be expressed as the *retention volume* (V_R):

$$V_R = t_R F \tag{10.2}$$

F is the *volumetric flow rate of the mobile phase* measured in mL/min. In a similar way, the *hold-up volume* (V_M) is defined as

$$V_M = t_M F \tag{10.3}$$

The hold-up time is the retention time of a compound that is not retained by the column and the hold-up volume is thus the volume needed to elute a compound, which is not retained by the column. Thus, V_M consists of the volume of liquid between the particles in the column and the volume of the tubing connecting the column from the injector to the detector. In a chromatographic system with a given stationary phase, the hold-up volume is therefore constant and independent of the flow rate of the mobile phase. The hold-up time, t_M, is related to V_M, but is inversely proportional to the flow rate. In contrast, retention time t_R and retention volume V_R of any compound that interacts with the column are dependent on the choice of stationary and mobile phases, but are constant in the chosen system. Increasing the flow rate from 1.0 to 2.0 mL/min results in a reduction of retention times and hold-up time by 50%.

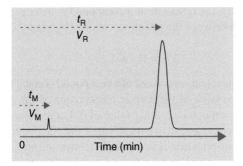

Figure 10.4   *Retention time, retention volume, hold-up time, and hold-up volume*

Box 10.1 exemplifies calculations related to retention volume, hold-up volume, and retention time.

Box 10.1   Calculation of retention volume and hold-up volume

In an LC system, the volumetric flow rate of the mobile phase is 0.40 mL/min. The active pharmaceutical ingredient atorvastatin has a retention time of 8.43 min in this system.

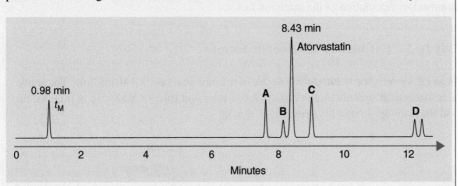

The retention volume is calculated as follows:

$$V_R = 8.43 \text{ min} \times 0.40 \text{ mL/min} = 3.37 \text{ mL}$$

A non-retained compound in the same system elutes at 0.98 min and the hold-up volume is calculated as follows:

$$V_M = 0.98 \text{ min} \times 0.40 \text{ mL/min} = 0.39 \text{ mL}$$

If the volumetric flow rate of the mobile phase is increased to 0.50 mL/min, the retention volume of atorvastatin is unaffected, but the retention time is decreased from 8.43 min to

$$t_R = \frac{3.37 \text{ mL}}{0.50 \text{ mL/min}} = 6.74 \text{ min}$$

As mentioned previously, the retention of a compound is determined by its distribution constant. The relationship between the retention volume (V_R) and the distribution constant (K_D) is given by

$$V_R = V_M + K_D V_S \tag{10.4}$$

The equation shows that the retention volume of a compound is equal to the hold-up volume (V_M) plus the additional volume of the mobile phase required for eluting the compound ($K_D V_S$). A non-retained compound has a K_D value of 0. Large values of K_D result in high affinity for the stationary phase and strong retention of the compound.

To provide a measure of retention it is common to use retention times. The *adjusted retention time* t'_R is the time a retained compound requires passing the column corrected for t_M:

$$t'_R = t_R - t_M \tag{10.5}$$

Retention times are affected by several factors such as the flow rate of the mobile phase and the column length. To avoid this dependency, retention can alternatively be expressed by the *retention factor* (k). The retention factor is defined as

$$k = \frac{t_R - t_M}{t_M} \tag{10.6}$$

The expression $t_R - t_M$ is the total time the compound stays in the stationary phase during elution, while t_M is the total time the compound stays in the mobile phase. Box 10.2 demonstrates calculation of the retention factor.

Box 10.2   Calculation of retention factor

In an LC system, the volumetric flow rate of the mobile phase is 0.40 mL/min. The active pharmaceutical ingredient atorvastatin has a retention time of 8.43 min in this system and the hold-up time in the system is 0.98 min.

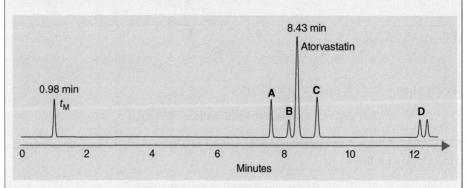

The retention factor for atorvastatin is calculated as follows:

$$k = \frac{8.43\ \text{min} - 0.98\ \text{min}}{0.98\ \text{min}} = \frac{7.45\ \text{min}}{0.98\ \text{min}} = 7.60$$

The molecules of atorvastatin continuously move in and out of the stationary phase and are present in the stationary phase for 7.45 min in total and in the mobile phase for 0.98 min in total.

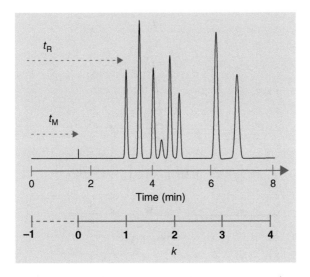

Figure 10.5   *Chromatogram illustrating retention time and retention factor*

Figure 10.5 shows a chromatogram from separation of eight compounds in which both retention times and retention factors are shown on the *x* axis. Because all eight compounds have different distribution constants in the chromatographic system, they are separated as eight individual *chromatographic peaks* in the chromatogram. The substances have retention times between 3.2 and 7.0 minutes and retention factors between 1.0 and 3.4.

10.4   Efficiency

The width of a chromatographic peak expresses how efficiently molecules are exchanged between the stationary and mobile phase during elution through the column. If the molecules of a certain compound are spread to only a small extent on their way downstream of the column, peak broadening is small and the final chromatographic peak is narrow. The narrower the chromatographic peaks are in relation to their retention times, the more efficient is the chromatographic column. Chromatographic peak broadening throughout the column is expressed by the parameter *N*, termed the *plate number* or the *number of theoretical plates*. The plate number is defined as

$$N = \left(\frac{t_R}{\sigma} \right)^2 \qquad (10.7)$$

where σ is the standard deviation of the peak. The ideal chromatographic peak is Gaussian, as shown in Figure 10.6, and for such peaks the width is 2σ at the points of inflection, which are at 0.607 times the height (*h*) of the peak.

In practice it is inconvenient to measure peak widths accurately at 0.607 times of the height, and in general the peak *width at half-height* (W_h) is measured instead (in most chromatographic software W_h is named $w_{\frac{1}{2}}$). The relationship between W_h and σ is given by

$$W_h = 2.354 \times \sigma \qquad (10.8)$$

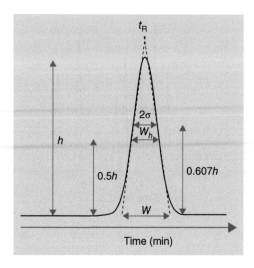

Figure 10.6   Parameters used to measure peak width

Box 10.3   Calculation of plate number

Atorvastatin is separated from other compounds (A, B, C, and D) in a chromatographic system. The retention time for atorvastatin is 8.43 min and the peak width at half-height (W_h) is 0.16 min.

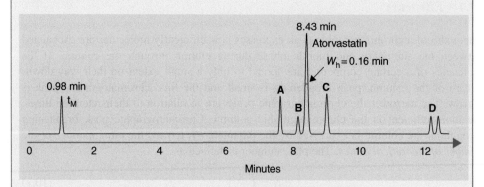

The plate number for atorvastatin is calculated as follows:

$$N = 5.54\left(\frac{8.43 \text{ min}}{0.16 \text{ min}}\right)^{2} = 15\,379 \approx 15\,000$$

The plate number can be measured on a regular basis to verify the chromatographic column performance.

Therefore, N is in practice estimated by the following equation:

$$N = 5.54 \times \left(\frac{t_R}{W_h} \right)^2 \qquad (10.9)$$

The narrower the peak is relative to the retention time, the greater is N. Therefore, the plate number is often used to express the performance of a chromatographic column. This is illustrated in Box 10.3.

The plate number (N) is proportional to the column length (L), and when other factors are equal, an increase in L results in an increase in N. A doubling of the column length results in doubling of the plate number. The relation between N and L is expressed as

$$H = \frac{L}{N} \qquad (10.10)$$

H is termed the *height equivalent to a theoretical plate (HETP)* or in short *plate height (H)*. Efficient chromatographic columns have small values of H and large N values. By using H instead of N, the efficiency of columns with different lengths can be compared.

In a chromatogram with a large number of peaks, the plate number is normally of similar value for all the compounds. The plate number calculated for one peak is therefore a general measure of the column efficiency. The plate number can be measured on a regular basis to check the efficiency of a column over time.

10.5   Selectivity

The selectivity between two peaks in a chromatogram is described by the *relative retention* (α), also called the *separation factor* (see Figure 10.7):

$$\alpha = \frac{k_2}{k_1} \qquad (10.11)$$

Here k_2 is the retention factor of the latter of the two eluting peaks and k_1 is the retention factor of the first eluting. Selectivity describes the ability of the chromatographic system to

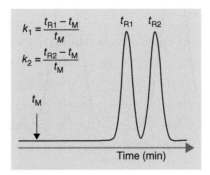

Figure 10.7   *Retention factors of two closely eluting peaks*

separate two specific compounds. The larger the value of α, the more efficient the system is in separating the two compounds. The separation factor is independent of column efficiency and column length but is affected by the chemical composition of the stationary and mobile phases. Thus, the separation factor can only be changed by changing the stationary or mobile phase of the system.

Separation factors have values larger or equal to 1. When $\alpha = 1$, the two compounds have identical retention times; they are *co-eluting* and are not separated. The larger the value of α, the larger the difference in retention times and the easier the separation becomes. As the retention factors, k_1 and k_2, describe the physicochemical properties of the compounds, the separation factor only describes the differences in these properties of the two compounds and the possibility of separating the compounds is based on α. However, α does not give information on the peak widths of the peaks.

10.6   Resolution

Although the compounds contained in a solution are eluted from the column as symmetrical Gaussian peaks with different retention times, it is not certain that they are separated completely. For the compounds to be separated completely, the detector response must reach a baseline between the two chromatographic peaks. The two peaks shown in Figure 10.8 are completely separated, and this is termed *base-line resolution* or *base-line separation*. Base-line separation of two different compounds requires that they have different retention times and that the two peaks are sufficiently narrow. The degree of separation between two chromatographic peaks is defined as the *resolution* (R_s):

$$R_S = 1.18 \times \frac{t_{R2} - t_{R1}}{W_{h1} + W_{h2}} \tag{10.12}$$

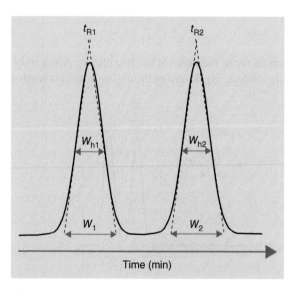

Figure 10.8   *Parameters needed for calculation of R_s*

Figure 10.8 shows the parameters needed for calculating R_s, where t_{R2} and t_{R1} are the retention times of peaks 2 and 1, respectively, and W_{h2} and W_{h1} are the corresponding peak widths at half-peak height. Thus, the resolution is highly dependent on the peak widths, as narrow peaks improve resolution. A value of $R_s = 1.0$ corresponds to a peak separation of 94% and base-line separation corresponds to an R_s value of 1.5. Box 10.4 demonstrates calculation of resolution.

Box 10.4   Calculation of resolution

Atorvastatin and a related impurity (related compound B) are separated with the following retention times and peak widths at half height:

Atorvastatin	$t_{R2} = 8.43$ min	$W_{h2} = 0.16$ min
Related compound B	$t_{R1} = 8.04$ min	$W_{h1} = 0.12$ min

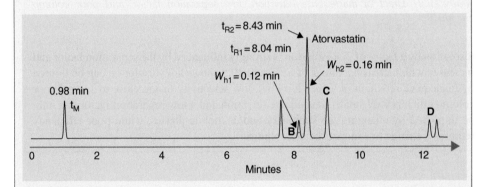

Resolution is calculated as follows:

$$R_S = 1.18 \times \left(\frac{8.43\,\text{min} - 8.04\,\text{min}}{0.12\,\text{min} + 0.16\,\text{min}} \right) \approx 1.4$$

The resolution is very close to 1.5, which corresponds to base-line separation.

To control the separation of closely eluting peaks, the following equation shows how the resolution between two peaks is related to the adjustable chromatographic variables retention factor (k), plate number (N), and separation factor (α):

$$R_S = \frac{1}{4} \times \sqrt{N} \, \frac{\alpha - 1}{\alpha} \, \frac{k}{k + 1} \tag{10.13}$$

In this equation k and N are the average values for the two peaks. The retention affects the resolution by $k/(k + 1)$, which approximates 1 as k increases. The optimum resolution for most separations falls in the range $k = 3$–10. The resolution increases with the square root of the plate number; thus the plate number must increase fourfold to increase

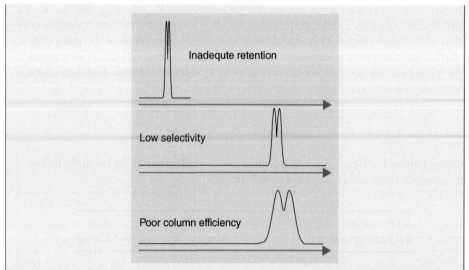

Figure 10.9   *Effect of inadequate retention, low separation factor, and poor column efficiency*

resolution by a factor of 2. Resolution is strongly influenced by the separation factor and increases with increasing values of α. Figure 10.9 shows how resolution can be limited by inadequate retention (k values too low), low selectivity (α too close to 1), and poor column efficiency (N small). Inadequate retention and a low separation factor can only be improved by changing the stationary and/or mobile phases, while poor efficiency requires a change to a more efficient column.

10.7   Peak Symmetry

Ideally Gaussian peaks are preferred in chromatography, but often chromatographic peaks show *peak tailing* or less often *peak fronting*. The result of these less ideal peak shapes is poorer separation. The symmetry factor (A_s) is used to describe symmetry of chromatographic peaks. The symmetry factor is calculated using the following equation:

$$A_s = \frac{w_{0.05}}{2 \times d} \tag{10.14}$$

Here $w_{0.05}$ is the width of the peak at 5% of the peak height and d is the distance between the perpendicular dropping from the peak maximum and the leading edge of the peak at 5% of the peak height, as illustrated in Figure 10.10. An A_s value of 1.0 signifies *symmetry* and when $A_s > 1.0$ the peak is *tailing*. When $A_s < 1.0$, the peak is *fronting*. Peak tailing and fronting are illustrated in Figure 10.10.

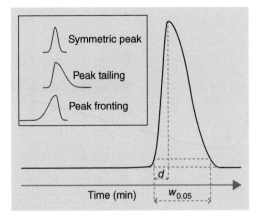

Figure 10.10   Parameters used to calculate peak symmetry and illustration of peak tailing and fronting

10.8   The Dynamics of Chromatography

When the sample molecules are present in the mobile phase, they interact with the stationary phase during the transport through the chromatographic system. During this process, chromatographic bands are inevitably prone to broadening, as illustrated in Figure 10.11.

During the chromatographic analysis, the sample molecules partition between the surface of the stationary phase and the mobile phase in a series of fast equilibria, while the mobile phase is pumped through the column. The flow rate of the mobile phase is an important parameter for the efficiency of the separation. This is given from the *van Deemter equation*:

$$H = A + \frac{b}{u} + C \cdot u \qquad (10.15)$$

The equation describes how the linear flow rate (u) of the mobile phase influences the plate height, H. The lower the value of H, the narrower are the peaks and the more efficient is the chromatographic system. A, B, and C are constants for a given column. Changing the column also changes the constants A, B, and C.

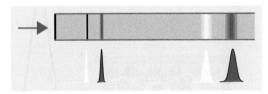

Figure 10.11   Chromatographic bands (peaks) are broadened during the transport through the column

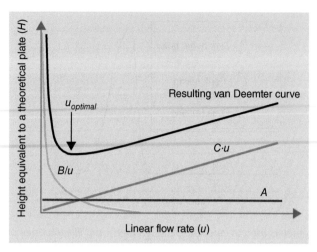

Figure 10.12 *van Deemter curve illustating separation efficiency (H) as a function of the linear flow rate of the mobile phase (u)*

The van Deemter equation is illustrated in Figure 10.12, in which the resulting van Deemter curve is plotted together with the three individual terms of the equation. The resulting van Deemter curve shows that the plate height is dependent on the linear flow rate and a minimum exists where separation efficiency is at an optimum. Thus, neither increasing nor decreasing the flow rate from this point will result in improved separation efficiency. The three individual terms in the van Deemter equation represent different contributions to band broadening of the chromatographic peaks.

The term A in the van Deemter equation is due to irregularity of stationary phase particles, which result in multiple paths for the sample molecules through the column. This type of band broadening (termed *Eddy diffusion*) is illustrated in Figure 10.13 and is independent of the mobile phase flow rate.

The second term in the equation (B/u) arises from *longitudinal diffusion* of the bands of sample molecules in the column. When a narrow band of a concentrated sample is injected

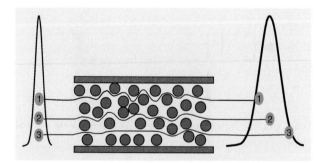

Figure 10.13 *Irregular particles of the column material result in different path lengths through the column and thereby band broadening*

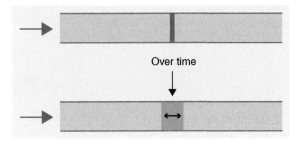

Figure 10.14   *Longitudinal diffusion – sample molecules diffuse from high concentration to low concentration, resulting in band broadening*

in the column, this is surrounded by a mobile phase not containing the sample. This will result in diffusion along the concentration gradient from the concentrated sample molecule bands to the mobile phase, as illustrated in Figure 10.14.

The sample molecules diffuse in all directions, but only diffusion in the direction of the length of the column (longitudinal diffusion) results in band broadening. The contribution from longitudinal diffusion is relatively high when the mobile phase flow rate is low, while the term B/u becomes less significant at high flow rates. The longitudinal diffusion is much higher in gases than in liquids, and therefore the B/u term is more dominant in GC than in LC.

The last term $(C \cdot u)$ arises from the equilibration time of the *mass transfer* between mobile and stationary phases and is proportional to the flow rate. The time required for the sample molecules to equilibrate between the stationary and mobile phases depends on the distance the molecules have to diffuse and the diffusion rate. If the distance is short, the equilibration is faster and therefore small and uniform stationary phase particles result in small distances and less band broadening. The faster the flow rate of the mobile phase, the less time for mass transfer. This is why the last term is proportional to the linear velocity of the mobile phase. If the equilibrium of the analyte molecule is fast compared to the flow rate, less band broadening takes place. The mass transfer term also decreases at elevated temperatures.

11

Separation Principles in Liquid Chromatography

11.1 Introduction

Different sample constituents (compounds) are separated in chromatography based on differences in their distribution between the stationary phase and the mobile phase. In gas chromatography (GC), the mobile phase is an inert gas, where little interaction takes place and retention and separation is mainly controlled by the choice of stationary phase. GC is discussed in detail in Chapter 13. In liquid chromatography (LC) (Chapter 12), compounds to be separated interact both with the stationary phase and the mobile phase. Thus, the

Introduction to Pharmaceutical Analytical Chemistry, Second Edition.
Stig Pedersen-Bjergaard, Bente Gammelgaard and Trine Grønhaug Halvorsen.
© 2019 John Wiley & Sons Ltd. Published 2019 by John Wiley & Sons Ltd.

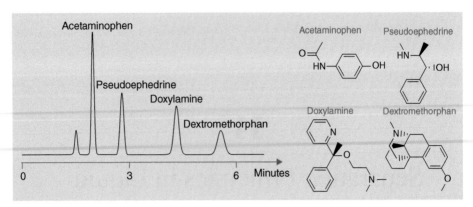

Figure 11.1 *Separation of acetaminophen, pseudoephedrine, doxylamine, and dextromethorphan by LC*

composition of the mobile phase plays a major role in LC, leaving more possibilities for optimization of the separation by variation of the mobile phase.

LC is used for *semi-volatile* and *non-volatile compounds*. Figure 11.1 shows an example of a chromatogram obtained by LC.

LC is used extensively for separation of drug substances in quality control of pharmaceutical ingredients and preparations, and for bioanalysis of pharmaceuticals. In LC, the stationary phase consists of particles with a large surface area packed into a column. Samples are injected into the mobile phase prior to the column inlet and are then transferred by the mobile phase to the column. Retention is caused by interactions of the compounds with the stationary phase and the compounds are transported with the mobile phase through the column. The individual compounds of the sample are separated in the column and detected at the outlet of the column. The detector is most often a UV detector or a mass spectrometer or a combination of these.

11.2 Reversed-Phase Chromatography

Reversed-phase (RP) chromatography is the most abundant separation principle in LC. In RP chromatography, the stationary phase is hydrophobic (non-polar) and the mobile phase is a mixture of water (aqueous buffer) and a polar organic solvent such as methanol or acetonitrile. The mechanism of RP chromatography is termed *partition chromatography* as the analyte partition between the stationary phase and the moving mobile phase.

11.2.1 Stationary Phases

Stationary phases for RP chromatography are typically based on silica particles where the surface has been chemically modified to form a hydrophobic surface. *Silica* (or silica gel, Figure 11.2) is a partially dehydrated form of colloidal polymerized silicic acid.

Silicic acid, H_2SiO_4, does not exist as the free monomer but is available in the form of a sodium silicate solution. When the sodium silicate solution is acidified the polymeric silica

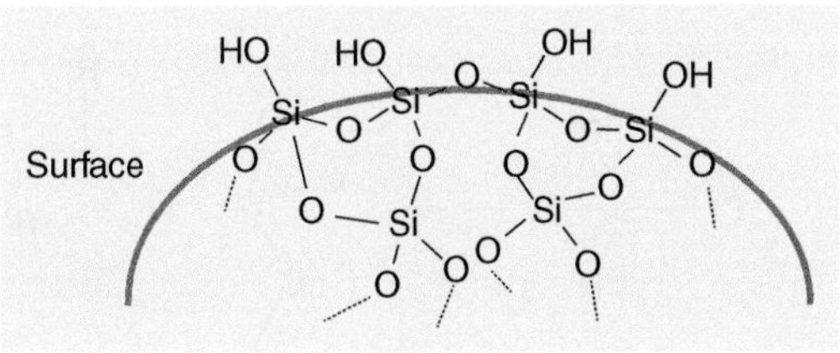

Figure 11.2 *Silica gel – the basis of stationary phase particles*

is formed. Silica used for chromatography undergoes an extensive purification process to remove metal impurities and is then pulverized, dried, and fractionated into particles of appropriate size. Silica has a large surface area in the range 200–800 m^2/g. The large surface area is due to the structure of the silica being a porous material similar to a sponge. The majority of the surface area is inside the pores. In practice, this means that silica is a porous skeleton that is manufactured with a well-defined pore diameter typically in the range 60–150 Å (1 Å = 10^{-10} m) when used for chromatography of small molecules. The pores must be wide enough to allow solvent and analyte molecules to enter freely. When the mobile phase flows through a column packed with silica particles, it enters the entire volume between the particles and the whole pore volume inside the particles. This provides a tremendous network of contacts between the stationary and the mobile phases and sample constituents are exposed to this surface area of the stationary phase. Column packing materials for RP chromatography are typically made of silica derivatized with reagents to form a hydrophobic (non-polar) surface. This is obtained by chemical reaction between the silanols on the surface of the silica particles and an organic silane reagent such as *chloro(dimethyl)octadecylsilane*. Common stationary phases for RP chromatography are shown in Figure 11.3 and are ranked after decreasing hydrophobicity. Octadecyl (—$C_{18}H_{37}$) silane is the most hydrophobic stationary phase and is often termed C_{18} or *ODS*. At the opposite end of the polarity scale, cyanopropyl (—C_3H_6—CN) is the least hydrophobic of these phases. RP chromatography is mainly performed using the C_{18} stationary phase.

The silane reagent has three alkyl groups connected to Si and is therefore bulky. Due to steric hindrance it is impossible to cover all silanols with the reagent. A significant percentage of the silanols can therefore still be present after chemical reaction as *free* or *residual silanols*. In order to minimize the number of free silanols the material can be treated with *trimethylchlorosilane*. This process is termed *end-capping*. However, even after end-capping some free silanols will still be present and accessible for polar interactions, for example polar interactions with amines.

Conventional silica-based stationary phases can be used with mobile phases in the pH range 2–8. With strongly acidic or alkaline mobile phases, silica-based stationary phases are prone to degradation. However, new specialized silica-based stationary materials that are stable in an extended pH range keep emerging. With acidic or alkaline mobile phases, stationary phases based on particles of pH stable organic polymers like *polystyrene–divinylbenzene* (PS-DVB) can be used as an alternative. These can be used with mobile phases in the pH range 1–13 and often provide a stronger retention of analytes compared to silica-based C_{18} materials. However, the column efficiency of the organic polymeric stationary phases is less than for silica-based materials.

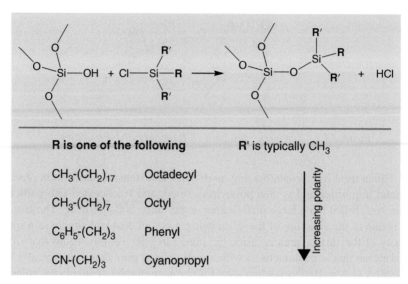

Figure 11.3 *Chemical reaction of silica with different chlorosilane reagents. Varying the substituents R' and R produces stationary phases with different properties*

11.2.2 Retention Mechanisms

Retention and separation in RP chromatography are mainly based on *hydrophobic interactions*. Non-polar sample constituents are therefore retained strongly on the stationary phase, while more polar constituents have less retention and elute earlier in the chromatogram. Octadecylsilane (C_{18}) stationary phases are the most hydrophobic phases and show the strongest retention for hydrophobic analytes. The main forces of interaction are *hydrophobic interactions* based on *van der Waal's forces*, which are relatively weak forces. The interactions principally take place between the hydrophobic hydrocarbon chains of the stationary phase and hydrophobic parts of the sample constituent molecules, as illustrated for naproxen in Figure 11.4. Interactions increase with molecular size.

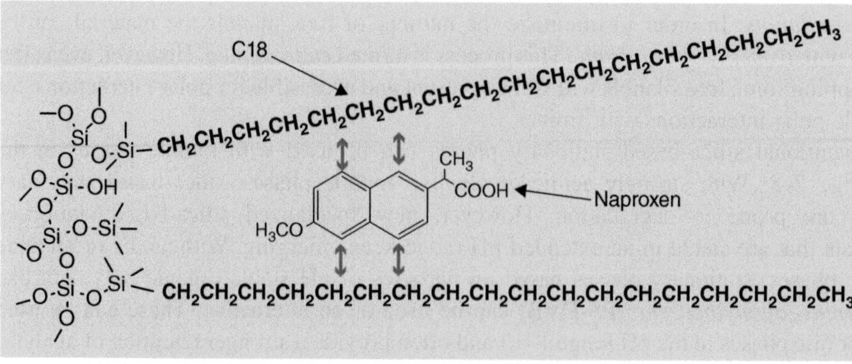

Figure 11.4 *Hydrophobic interactions between C_{18} (stationary phase) and naproxen (sample constituent)*

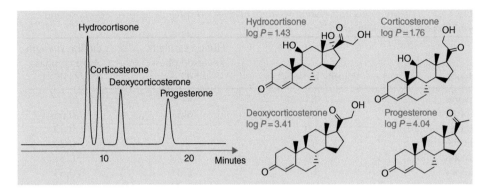

Figure 11.5 *Separation in order of increasing hydrophobicity by RP chromatography*

Separation is based on *partition chromatography*. Compounds partition between the non-polar stationary phase and the polar mobile phase. Hydrophilic (polar) compounds have low partition into the non-polar stationary phase, and elute from the column with short retention times. In contrast, hydrophobic (non-polar) compounds have more partition into the non-polar stationary phase and are stronger retained on the column. This is illustrated in Figure 11.5, where different steroids are separated in order of increasing hydrophobicity (increasing log *P* value).

Retention is also affected by the hydrophobicity and molecular size of the stationary phase, as illustrated in Figure 11.6, where retention times are longer with the C_{18} than with the C_8 stationary phase.

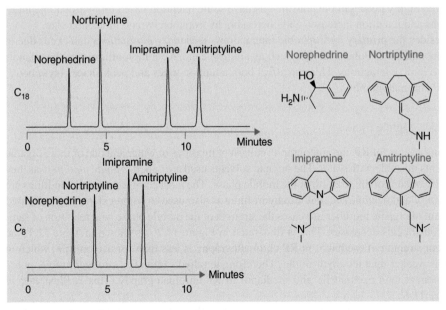

Figure 11.6 *Reversed-phase chromatography with the C_{18} and the C_8 stationary phase. The mobile phase is 80% methanol/20% 25 mM KH_2PO_4 (pH 6.0)*

Table 11.1 Polarity index and elution strength of selected solvents in reversed-phase and normal-phase chromatography

Solvent	Polarity index (P')	Elution strength Reversed phase (ε° (C_{18}))	Elution strength Normal phase (ε° (SiO_2))
Hexane	0.1	—	0
Isopropanol	3.9	8.3	0.6
Tetrahydrofuran	4.0	3.7	0.53
Chloroform	4.1	—	0.26
Dioxane	4.8	11.7	0.51
Acetone	5.1	8.8	0.53
Methanol	5.1	1.0	0.73
Acetonitrile	5.8	3.1	0.52
Water	10.2	—	High

In RP chromatography elution of the compounds from the column is promoted when the mobile phase becomes less polar, resulting in a higher *elution strength* of the mobile phase. The polarity of solvents used as the mobile phase can be described by the *polarity index*, P', which is a ranking of solvents according to their ability to dissolve polar compounds. A polar solvent like water has a high polarity index, while hexane (non-polar) has a low polarity index. The polarity index and elution strength of selected solvents are given in Table 11.1. The term ε° (C_{18}) is the strength in relation to elution from C_{18} stationary phases and applies to RP chromatography. The term ε° (SiO_2) is the elution strength from pure silica as a stationary phase and is discussed in Section 11.4.

Even highly polar sample constituents have some retention based on hydrophobic interactions and retention increases with increasing hydrophobicity and molecular size.

Besides the primary hydrophobic interactions, *secondary interactions* can occur due to residual silanols. Polar analytes, such as amines, can have a high affinity for the silanols due to ionic interactions. This can affect both retention times and peak shapes (symmetry) in RP chromatography.

11.2.3 Mobile Phases

Mobile phases for RP chromatography consist of mixtures of water and one or more organic solvents miscible with water. The organic solvents used are called *organic modifiers* as they modify or control the strength of the mobile phase. The most common organic modifiers are *methanol* and *acetonitrile*, but tetrahydrofuran is also used to some extent. The increased content of organic modifier increases the strength of the mobile phase, and retention of sample constituents decreases. This is illustrated in Figure 11.7. As shown in Table 11.1, the solvent strength of methanol in RP chromatography is less than for acetonitrile, which in turn is weaker than tetrahydrofuran. Therefore, tetrahydrofuran is a stronger mobile phase component than acetonitrile and methanol in RP chromatography. This is illustrated in Figure 11.8.

A mobile phase of 64% methanol in water has about the same eluting strength as 53% acetonitrile in water, or 40% tetrahydrofuran in water. These mobile phases are said to

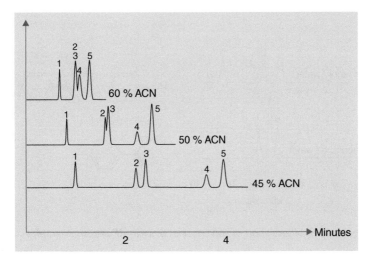

Figure 11.7 *Effect of percentage of acetonitrile in the mobile phase*

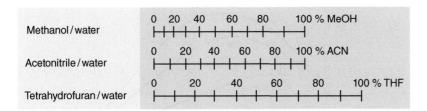

Figure 11.8 *Mobile phases with similar eluting strengths*

be *isoeluotropic* (they have similar elution strength). When performing a separation of a mixture of compounds using the isoeluotropic phases A, B, and C, the average retention of the compounds is about the same for the three phases. However, the order of elution between the substances can be somewhat different because the *selectivity* in the three systems can be different. Changing the organic modifier can thus be used to change the separation selectivity. This is exemplified in Figure 11.9 where the order of elution is changed when replacing methanol with acetonitrile in the mobile phase.

Methanol is cheaper and less toxic than acetonitrile. The main drawback when using methanol, however, is that methanol–water mixtures are more viscous than acetonitrile–water mixtures, as shown in Figure 11.10. As the mobile phase is pumped through the column containing the stationary phase particles, the back pressure is high. The pressure increases with the viscosity of the mobile phase, and therefore the pressure decreases when methanol is replaced with acetonitrile.

Retention of neutral sample constituents is only controlled by the content of organic modifier in the mobile phase and is not affected by pH. For basic and acidic compounds, on the other hand, retention is dependent on both the content of the organic modifier and pH in the mobile phase. Therefore, for basic and acidic compounds, a buffer system is used to

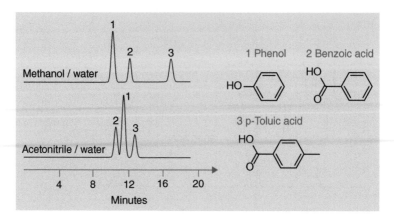

Figure 11.9   *Illustration of solvent selectivity in RP chromatography*

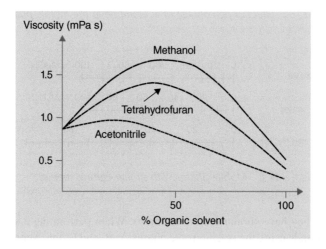

Figure 11.10   *Viscosity of mixtures of water and organic solvents*

control pH in the mobile phase. The buffer concentration in the mobile phase is typically in the range from 10 to 50 mM. For LC with UV detection, phosphate buffers are often preferred because phosphate buffers have good buffering properties and low UV absorbance. The latter is important in order to avoid background absorption from the mobile phase. A disadvantage of phosphate buffers is that they are poorly soluble in organic solvents and thus tend to precipitate at high concentrations of the organic modifier in the mobile phase. For LC coupled with mass spectrometry, volatile buffers of organic acids such as acetic acid or formic acid and their ammonium salts are used. These buffer components also have a better solubility in organic solvents.

For RP chromatography of basic and acidic drug substances, changing pH in the mobile phase can cause large changes in retention and separation. This is exemplified in Figure 11.11 for imipramine, amitriptyline, fenoprofen, and diclofenac. Imipramine and

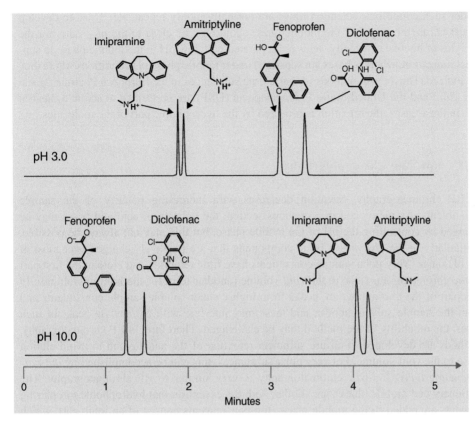

Figure 11.11 *Separation of acidic drugs (fenoprofen and ibuprofen) and basic drugs (imipramine and amitriptyline) at different pH values in the mobile phase*

amitriptyline are basic drug substances and with pH 3 in the mobile phase both substances are protonated and retention times are around two minutes. When pH in the mobile phase is increased to 10, imipramine and amitriptyline are deprotonated and prone to much stronger retention. Thus, retention times increase to about four minutes. Generally, basic substances experience less retention in their protonated form and retention decreases with decreasing pH. Fenoprofen and diclofenac are both acidic drug substances and with pH 3 in the mobile phase the molecules are protonated (uncharged). Under such conditions, the acidic drug substances experience strong retention, with retention times of about 3–3.5 minutes. When pH is increased to 10, the acidic drug substances deprotonate; as anions their retention times decrease to about 1.5 minutes. Thus, retention of acidic substances decreases with increasing pH in the mobile phase.

Retention varies greatly in the range around the pK_a value, where small changes in pH provide major changes in ionization and thereby retention. For weak bases, ionization is suppressed at high pH and therefore retention increases with increasing pH in the mobile phase. In the pH range where drugs are either fully ionized or where the ionization is completely suppressed, the retention is not affected by minor changes in mobile phase pH.

Under such conditions, retention times are normally highly repeatable. Thus, to develop robust LC methods, pH of the mobile phase should not be close to the pK_a value for the analytes of interest. Typically, acids are separated with low pH in the mobile phase to suppress ionization, whereas bases are separated under mobile phase pH conditions where they are ionized. Thus, most methods for basic drug substances in the European Pharmacopoeia (Ph. Eur.) and the United States Pharmacopoeia (USP) are performed at acidic to neutral pH. In these cases, the retention is provided by the hydrophobic part of the molecules.

11.3 Ion-Pair Chromatography

In RP chromatography, retention decreases with increasing polarity of the sample constituents. As mentioned in a previous section, the polarity of acid and bases may be changed by controlling the pH of the mobile phase, but this may not always be possible, particularly in the analysis of basic compounds that are positively charged over most of the pH range. Very polar sample constituents have little retention and elute in the first part of the chromatogram close to the *void* volume (another term for the hold-up volume). In this part of the chromatogram, peaks from highly water-soluble sample constituents and from the sample solvent appear, and these may interfere with the analyte peak. In such cases, the reliability of the method may be challenged. Therefore, in RP chromatography, methods are developed to ensure sufficient retention of the analyte and to avoid elution close to the void volume. For very polar substances, this can be accomplished by *ion-pair chromatography*. Ion-pair chromatography is very similar to RP chromatography. The stationary and mobile phases are similar, with the exception that hydrophobic ion-pairing reagents are added to the mobile phase. Ion-pair reagents consist of an ionic part, which interacts with the ionic analyte, and a hydrophobic part, which interacts with the hydrophobic stationary phase. The analytes and the ion-pair reagent are of opposite charge, and due to ionic interactions the formed hydrophobic ion pairs appear neutral to the surroundings and retention increase. The analytes have to be ionized in the mobile phase and therefore pH is adjusted accordingly, typically close to neutral pH where both carboxylic acids and amines are ionized. The ion-pairing process is a dynamic process where molecules are exchanging all the time, but when adding the ion-pairing reagent in excess, there is a high probability of ion-pair formation with the analyte.

Sulfonic acids and perfluorocarboxylic acids are common counter ions for ion-pair chromatography of basic drugs, and quaternary alkyl ammonium compounds are commonly used as counter ions for acidic drugs. Figure 11.12 shows the structure of three typical ion-pair reagents, *sodium octanesulfonate, heptafluorobutyric acid,* and *tetrabutylammonium hydrogen sulfate*. Octanesulfonate has a pK_a below 1, is negatively charged throughout the entire pH range, and forms ion pairs with positively charged basic drugs. Heptafluorobutyric acid behaves similarly and can also be considered negatively charged throughout the entire pH working range. Tetrabutylammonium hydrogen sulfate is a quaternary ammonium compound, is positively charged in the entire pH range, and forms ion pairs with negatively charged acidic drugs.

Figure 11.13 shows an example of the effect of ion pairing. Without an ion-pair reagent in the mobile phase (upper chromatogram), a polar basic drug substance and three related

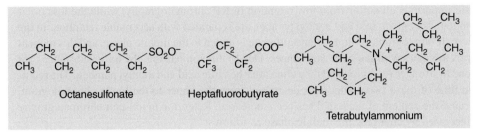

Figure 11.12 *Structures of (sodium) octanesulfonate, heptafluorobutyrate (heptafluorobutyric acid), and tetrabutylammonium (hydrogen sulfate)*

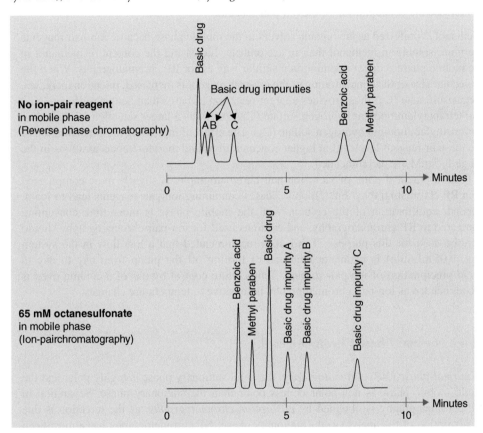

Figure 11.13 *Comparison of RP chromatography and ion-pair chromatography. The stationary phase is C_{18} in both experiments. The mobile phase in the upper chromatogram is 30% methanol mixed with 70% acetate buffer pH 3.5. The mobile phase in the lower chromatogram is 45% methanol mixed with 55% acetate buffer pH 3.5 containing 65 mM octanesulfonate*

basic impurities all suffer from low retention in RP chromatography, while the less polar compounds benzoic acid and methyl paraben are separated with acceptable retention. In the lower chromatogram, sodium octanesulfonate is added to the mobile phase, and this reagent forms ion pairs with the basic substances. Due to this ion pairing, the basic substances are prone to stronger retention, and they elute after benzoic acid and methyl paraben. The retention time of these two compounds decreased from the upper to the lower chromatogram, because the content of methanol has been increased. Retention in ion-pair chromatography is increased by changing the mobile phase as follows:

- Reduce the concentration of organic modifier.
- Increase the molecular size of the ion-pair reagent.
- Increase the concentration of the ion-pair reagent.

Methanol is preferred as the organic solvent in the mobile phase because ion-pair reagents are more soluble in methanol than in acetonitrile. Increasing the content of methanol in the mobile phase reduces retention in a similar way as for RP chromatography. When the molecular size or the concentration of the ion-pair reagent is increased, retention increases. Octanesulfonate (C_8) thus provides a larger retention of bases than pentanesulfonate (C_5) and tetrabutylammonium hydrogen sulfate (C_{16}) provides a larger retention of acids than tetramethylammonium hydrogen sulfate (C_4). The effect of increasing the concentration of the ion-pair reagent levels off at higher concentrations and therefore concentrations in the range 1–5 mM are recommended.

From a practical point of view, ion-pair chromatography is slightly more complicated than RP chromatography. First, mobile phases containing ion-pair reagents tend to foam. Second, equilibration of the column with the mobile phase is more time consuming compared to RP chromatography, and columns used for ion-pair chromatography should only be used for this purpose. Third, it is recommended that a low flow in the system (e.g. 0.05 mL/min) is maintained instead of turning off the pump from day to day to avoid precipitation of ion-pair reagent. Temperature control by use of a column oven is recommended as ion-pair chromatography is sensitive to temperature changes.

11.4   Normal-Phase Chromatography

In *normal-phase (NP) chromatography*, the solid stationary phase is highly polar and the liquid mobile phase is non-polar or less polar than the stationary phase. Separation in NP chromatography is obtained by *adsorption chromatography*, as the retention is due to adsorption of the analyte to the stationary phase. NP chromatography has almost been replaced by RP methods as the consumption of hazardous organic solvents in the latter methods are much lower. NP chromatography is mainly used for analytes, which cannot be analysed by RP chromatography, either because they are very hydrophobic compounds that are not soluble in the mobile phases used for RP chromatography or because they are very hydrophilic compounds that are not retained on the stationary phases used for RP chromatography.

11.4.1 Silica and Related Stationary Phases

A typical NP chromatographic system utilizes particles of pure silica as the stationary phase, while the mobile phase is a mixture of *n*-heptane and ethyl acetate or heptane–propanol. Sample constituents with no affinity to (no interaction with) the polar surface of silica are not retained, and pass the system with the speed of the mobile phase. Sample constituents with polar functional groups have a higher affinity to the polar silica particles and are retained in the system due to partial *adsorption*. Adsorption on to the stationary silica is a reversible interaction, and an increase in the polar component (ethyl acetate or propanol) of the mobile phase increases the competition for the adsorption sites on the surface and reduces the interaction of the sample constituents with silica. Thus, retention in NP chromatography can be decreased by increasing the polarity of the mobile phase.

Silica, or silica gel, as described in Figure 11.2, is the most abundant stationary phase used in NP chromatography. Silica has strong adsorption characteristics and many substances can be adsorbed on to silica. As illustrated in Figure 11.14, the surface of silica is covered with *silanol groups* (—Si—OH) and these are responsible for the adsorptive properties. Some silanols provide *hydrogen bonding interactions* with nearby silanols (*bridged*) while other silanols are isolated (*lone acidic*). In some cases, two silanols are attached to the same Si atom (*geminal silanols*). These structural differences between silanols result in different activities and different adsorptive properties. Performing NP chromatography where the silica is totally dry and the organic mobile phase is without water can be difficult to control because the most active silanols readily adsorb water and polar sample constituents. Therefore, small amounts of water or other polar solvents are added to the mobile phase to deactivate the most active silanols. The addition of minor amounts of an amine or an acid to the mobile phase has the same effect.

Silanols are weakly acidic and differences in their structure also result in differences in acidity. The pK_a value of pure silica is therefore in the range from 6 to 7. Furthermore, silica has an isoelectric point at about pH 2 due to increasing surface protonation of the polymer at low pH.

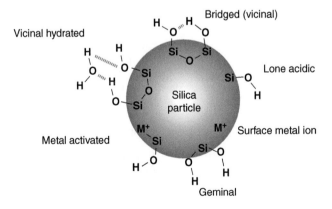

Figure 11.14   Silanol groups on the surface of silica

The silanols on silica can be derivatized with ligands containing other polar groups such as diol, —CN, and —NH$_2$. These polar materials can provide changes in selectivity and thus changes in the order of elution compared to that obtained on silica. *Alumina* and *magnesium silicates* are examples of other polar materials used as stationary phases in NP chromatography.

11.4.2 Molecular Interactions and Retention

The interactions between silanols on the silica surface and functional groups of the sample constituents are termed *polar interactions* and include dipole-dipole, hydrogen bonding, and ionic interactions. Sample constituents having functional groups with permanent dipole moment are retained by *dipole interactions*. Examples of functional groups that provide permanent dipoles are: $CN > NO_2 > C=O$, $CHO > COOR > halogen > OH > COOH >$ —O— $> NH_2$. Analytes having nitrile or nitro groups participate in the strongest dipole interactions. Hydrogen bond donors and acceptors interact by *hydrogen bonding interactions*. The more acidic the hydrogen bond donor and the more basic the hydrogen bond acceptor, the stronger is the interaction. Carboxylic acids and phenols are examples of strong hydrogen bond donors and amines are examples of strong hydrogen bond acceptors. Silanols have both hydrogen bond acceptor and donor properties. *Ionic interactions* can occur between basic sample constituents and the acidic silanol groups. Ionic interactions are undesirable in NP chromatography and give rise to very strong retention. These can be avoided by masking the most acidic silanol groups.

Sample constituents with a high number of polar functional groups will be prone to strong retention, while less polar substances will show less retention. This is visualized in Figure 11.15 showing decreased retention with decreasing polarity for different vitamin E variants.

Saturated hydrocarbons do not interact with silica and thus show no retention in NP chromatography. Aromatic hydrocarbons are slightly retained by silica due to weak

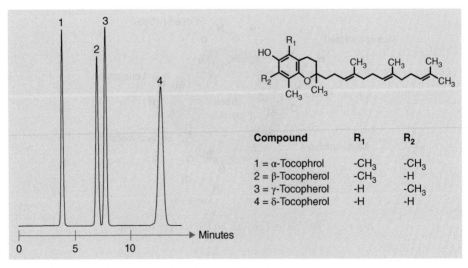

Compound	R$_1$	R$_2$
1 = α-Tocophrol	-CH$_3$	-CH$_3$
2 = β-Tocopherol	-CH$_3$	-H
3 = γ-Tocopherol	-H	-CH$_3$
4 = δ-Tocopherol	-H	-H

Figure 11.15 NP chromatography of vitamin E variants. The stationary phase is silica. The mobile phase is n-hexane/isopropanol 99.5 : 0.5 (v/v)

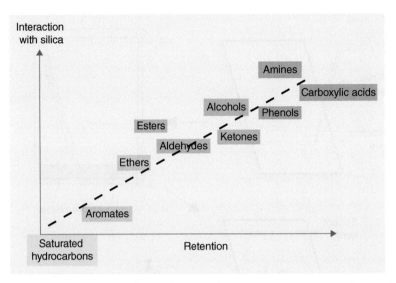

Figure 11.16 *Contribution of different functional groups to interaction and retention with silica*

dispersion interactions. Interaction and retention increases with the polarity of the functional groups. This is illustrated in Figure 11.16. Alcohols and phenols are of higher polarity than ketones and experience stronger retention because of hydrogen bonding interactions with silica. Primary amines and carboxylic acids provide strong hydrogen bonding interactions due to high hydrogen bond donor (acidity) and acceptor (basicity) properties, respectively, and are retained strongly by silica. Retention increases with the number of functional groups. A sample constituent with two hydroxyl groups is retained more strongly than a constituent with only one hydroxyl group.

11.4.3 Mobile Phases

In NP chromatography the mobile phase usually consists of a mixture of two or three solvents. The strength of the mobile phase is given by the polarity of the solvents, which can be ranked according to their solvent strength, $\varepsilon°$ (Table 11.1, $\varepsilon°$ (SiO_2)). Solvent strength increases with increasing polarity. Increased solvent strength of the mobile phase decreases retention of sample constituents in NP chromatography. As shown in Table 11.1, hydrocarbons have $\varepsilon°$ values close to 0, as hydrocarbons cannot suppress polar interactions between sample constituents and silica. The $\varepsilon°$ value of methanol is 0.73 and, together with water, methanol is the strongest mobile phase solvent in NP chromatography.

11.5 Thin-Layer Chromatography

Thin-layer chromatography (TLC) is an NP separation technique where the stationary phase particles are dispersed as a thin layer on the surface of an alumina foil or glass plate. Typical applications of TLC are identification of pharmaceutical ingredients and related substances (impurities), identification of active pharmaceutical ingredients in pharmaceutical

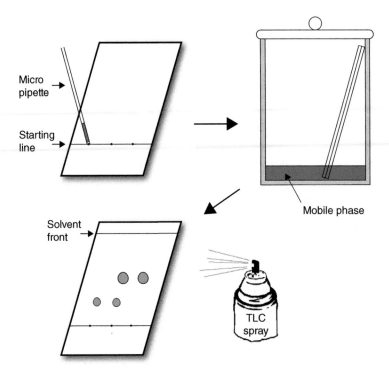

Figure 11.17 *TLC procedure: the sample is applied to the TLC plate, separation is performed in the development tank and sample constituents visualized*

preparations, and analysis of herbals used as drugs or food supplements. Separation in TLC is basically *adsorption chromatography* and is similar to NP chromatography.

An overview of the TLC procedure is given in Figure 11.17. The samples are prepared in a volatile solvent and microlitre volumes of those are placed as spots or bands on a starting line in the bottom region of the TLC plate. The stationary phase consists of a thin layer of silica particles. The sample solvent is evaporated and the TLC plate is inserted in a tank with the mobile phase in the bottom. The mobile phase is a mixture of organic solvents. As an example, USP recommends a mixture of chloroform, methanol, and water (180 : 15 : 1 (*v/v*)) as a general mobile phase for TLC when no other mobile phase is specified. Chloroform is a relative weak eluent with pure silica ($\varepsilon^\circ = 0.26$), whereas the presence of methanol ($\varepsilon^\circ = 0.73$) and water (ε° high) increases the strength of the mobile phase (Table 11.1).

When the tank is closed and the mobile phase moves up the TLC plate by capillary forces, the sample constituents move with the mobile phase, but are retained depending on their degree of adsorption to the stationary phase. When the mobile phase has moved close to the top of the plate, the TLC plate is removed from the tank and residual mobile phase on the plate is evaporated. The different sample constituents can then be visualized as individual spots on the TLC plate. Coloured compounds are immediately visible, while UV-absorbing compounds can be visualized by means of plates with fluorescent coating or spraying with colour-forming derivatizing agents.

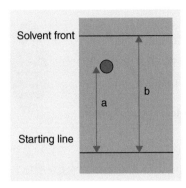

Figure 11.18 *Parameters for determination of R$_f$ values*

Retention of a given substance on the TLC plate is described by the *retention factor, R$_f$*. The R_f value is defined as

$$R_f = \frac{a}{b} \tag{11.1}$$

Here a is the distance from the point of application (starting line) to the centre of the spot and b is the total distance travelled by the mobile phase from the point of application, as shown in Figure 11.18. R_f values range between 0 and 1, and compounds with a strong interaction with the stationary phase are retained most and have low R_f values. Often R_f values are multiplied by 100 and thus range between 0 and 100.

The limitation of TLC is primarily related to the low sensitivity of the technique. Detection limits are significantly higher than for LC. Another limitation is that volatile substances cannot be analysed by TLC because they may evaporate during sample application and separation. While TLC was used extensively in previous editions of Ph. Eur. and USP, these methods are now being replaced by LC.

11.6 Hydrophilic Interaction Chromatography

Hydrophilic interaction chromatography (HILIC) is a chromatographic principle used for the separation of relatively polar analytes, for which RP chromatography does not provide sufficient retention. HILIC is therefore an alternative separation method for polar drug substances, peptides and sugars. The retention mechanism of HILIC is *mixed mode chromatography* as it can be described as *partition* chromatography of analytes between the water layer on the surface of the stationary phase particles and the mobile phase, combined with hydrogen bonding and electrostatic interactions, which are more related to *adsorption* chromatography.

The mobile phases used in HILIC contain a large amount (>80%) of organic solvent such as acetonitrile. The stationary phases are strongly polar and are either bare silica or silica where the surface has been functionalized with different groups, as illustrated in Figure 11.19.

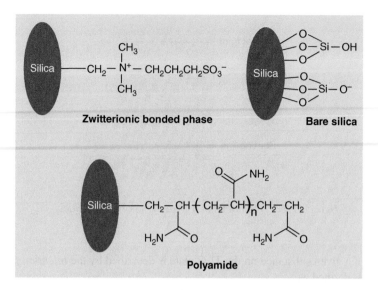

Figure 11.19 *Examples of HILIC stationary phases*

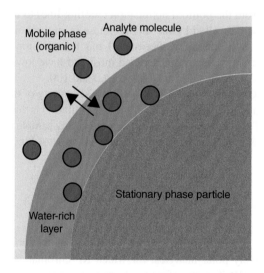

Figure 11.20 *Principle of hydrophilic interaction liquid chromatography (HILIC)*

Water in the mobile phase forms a water layer on the surface of the polar stationary phase particles, as illustrated in Figure 11.20. Sample constituents are retained (i) by partition into this water layer and (ii) by interactions with silica or functionalized silica. The less water present in the mobile phase, the stronger is the retention of the analytes. Thus, water is the strongest solvent in the HILIC mobile phase.

HILIC has become very popular for analysis of polar substances where RP chromatography is less successful. The high content of organic solvent improves compatibility with

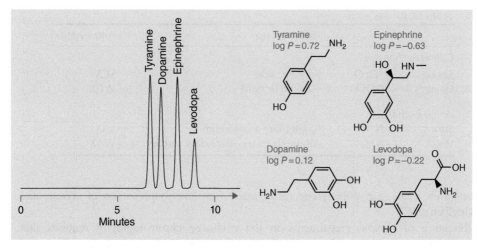

Figure 11.21 *Separation of amino acids by hydrophilic interaction liquid chromatography. The stationary phase is bare silica. The mobile phase is 0.1% ammonium acetate in acetonitrile/water 85 : 15 (v/v)*

electrospray ionization mass spectrometry and thus improves sensitivity in LC-MS methods (discussed later). Although HILIC is close to NP chromatography, the content of water in the mobile phase still makes it possible to use this mode for bioanalysis, which is more problematic when using NP chromatography. Figure 11.21 illustrates a separation of amino acids by HILIC.

11.7 Ion Exchange Chromatography

Ion exchange chromatography is a technique that allows the separation of ions and charged sample constituents based on their charge. Ion exchange chromatography can be used for almost any kind of charged molecules, including separation of large proteins, nucleotides, and amino acids. The ionic analytes are retained by ionic interactions with ion exchanger groups of the opposite charge bound to the stationary phase. The ionic groups $-SO_3^-$, $-COO^-$, $-NH_3^+$, and $-NR_3^+$ are common active groups for ion exchangers. Cations have affinity to negatively charged ion exchangers ($-SO_3^-$, $-COO^-$) and these ion exchangers are therefore termed *cation exchangers*. Positively charged ion exchangers ($-NH_3^+$, $-NR_3^+$) are likewise termed *anion exchangers*.

Ion exchangers can be classified into strong and weak ion exchangers, as illustrated in Table 11.2. Strong ion exchangers are charged throughout the entire pH range. *Strong cation exchangers* contain sulfonic acid groups ($-SO_3^-$), while *strong anion exchanges* contain quaternary ammonium groups ($-NR_3^+$) attached to the surface of the stationary phase. Weak ion exchangers are only charged within a certain pH range depending on the pK_a value of the functional group. *Weak cation exchangers* are based on carboxylic acid groups ($-COO^-$), while *weak anion exchangers* are aliphatic amino groups ($-NH_3^+$). The aliphatic amino group can be primary, secondary, or tertiary. Stationary phases are

Table 11.2 Ion exchangers

		Active group	Abbreviation
Cation exchangers			
Strong	$—SO_2O^-$	Sulfonic acid	SCX
Weak	$—COO^-$	Carboxylic acid	WCX
Anion exchangers			
Strong	$—N^+R_3$	Quaternary ammonium	SAX
Weak	$—NH_3^+$	Amine (primary, secondary, tertiary)	WAX

most often *resins*, e.g. polystyrene resins made from copolymerization of styrene and divinylbenzene.

Retention of sample constituents in ion exchange chromatography requires that the substances as well as the ion exchanger are charged. In general, the affinity of sample constituents depends on their charge and molecular size. Small and multiple charged substances have high retention, and retention decreases with decreasing charge and increasing molecular size. The retention of a given substance can be changed by changing the concentration or nature of the buffer in the mobile phase. The buffer ions in the mobile phase compete with the sample constituents for the ionic sites on the stationary phase. When the buffer concentration is increased or buffer ions with a stronger affinity to the ionic sites are used, the retention of sample constituents decreases. With cation exchangers, the eluting strength of the buffer cation increases in the order $Li^+ < H^+ < Na^+ < NH_4^+ < K^+ < Ag^+ < Mg^{2+} < Zn^{2+}$. With anion exchangers, the eluting strength of buffer anions increases in the order $OH^- < CH_3COO^- < HCOO^- < Cl^- < Br^- < H_2PO_4^- < $ oxalate $<$ citrate. However, this may change with the type of ion exchanger. Gradient elution in ion exchange chromatography can be performed by increasing the ionic strength of the mobile phase. An example of separation by ion exchange chromatography is illustrated in Figure 11.22 for the analysis of small inorganic anions.

11.8   Size Exclusion Chromatography

In *size exclusion chromatography* (SEC) the sample constituents are separated according to their molecular size. SEC is usually performed with a column tightly packed with extremely small porous polymer beads designed to have pores of different sizes. For separation of organic substances of low polarity, organic solvents are used in the mobile phase to keep the substances dissolved. With organic solvent in the mobile phase, the technique is termed *gel permeation chromatography* (GPC). For separation of water-soluble substances, aqueous mobile phases are used, and the technique is termed *gel filtration chromatography*.

Separation in SEC occurs when different sample constituents (molecules) penetrate differently into the pores of the matrix. The large molecules pass through the column outside the particles, at the same speed as the mobile phase. The volume of the mobile phase

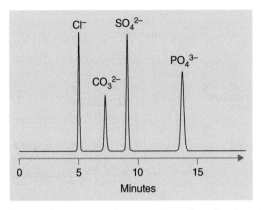

Figure 11.22 *Ion exchange chromatography separation of small anions. The stationary phase is an anion exchanger. The mobile phase is an aqueous solution of KOH, where the concentration of KOH is increasing from 22 to 40 mM during the separation*

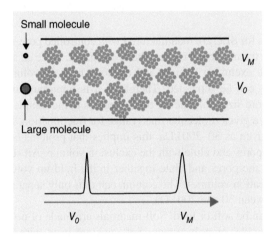

Figure 11.23 *Principle of size exclusion chromatography*

that carries the large molecules through the column is termed the *exclusion volume*, V_0 (Figure 11.23). The exclusion volume is equal to the volume of the mobile phase between the particles in the column. The smallest molecules are transported by the mobile phase into the smallest pores of the particles. Since the path length through all the pores of the packing material is much longer, the small molecules get a much longer retention time. Thus, sample constituents are separated in order of decreasing molecular size. The volume of the mobile phase that is used to elute the smallest substances is called the *total permeation volume*, V_M. Figure 11.24 shows an example of separation based on SEC. The separation is independent on the choice of mobile phase; the only demand is that the analytes must be soluble in the eluent.

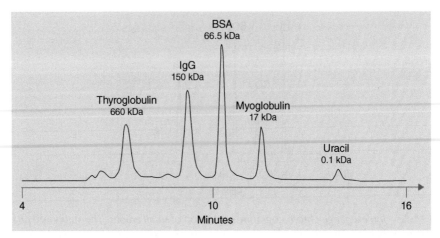

Figure 11.24 *SEC separation of thyroglobulin (660 kDa), IgG (150 kDa), BSA (66.5 kDa), myoglobin (17 kDa), and uracil (0.1 kDa). The stationary phase is termed Sepax Unix SEC-300. The mobile phase is 0.2 M dibasic sodium phosphate, pH 7 with phosphoric acid*

The column particles for SEC are manufactured with controlled pore sizes. The pore size is chosen dependent on the molecular range to be investigated. Since all sample constituents are eluted between the exclusion volume and the total hold-up volume, only a limited number of compounds can be separated. Therefore, it is important to choose a column with particles having pore sizes suitable for the molecules in the sample to be separated. A given SEC column has a given molecular mass range for fractionation. If the *fractionation range* for proteins is given as 30–200 kDa, this implies that proteins larger than 200 kDa are excluded from the pores and elute with the exclusion volume. All substances less than 30 kDa fully penetrate the pores and elute together in the hold-up volume independently of size (the total permeation volume). The column can thus only separate substances with molecular weights between 30 and 200 kDa.

Packing materials can be soft or hard. Soft materials are made of polysaccharides such as dextran, polyacrylamide, or polystyrene. Soft gels are used with gravity flow (flow driven only by gravity), where the mobile phase flows through the column due to gravity. When the mobile phase is pumped through the SEC system, rigid packing materials are used. Rigid packaging materials resists high pressure and are made of silica or a highly cross-linked organic polymer such as polystyrene–divinylbenzene. Silica is used with aqueous mobile phases, while polystyrene–divinylbenzene is often used with organic solvents as mobile phases.

11.9 Chiral Separations

Pairs of enantiomers are difficult to separate by chromatography because they have identical physicochemical characteristics. Chromatographic separation of two enantiomers is

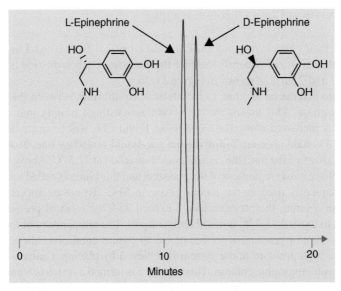

Figure 11.25   *Separation of L- and D-epinephrine using a chiral stationary phase. The station-ary phase is termed Shodex ORpak CDBS-453. The mobile phase is acetonitrile with 0.05% (w/v) acetic acid/0.2 M sodium chloride in water 5 : 95 (v/v)*

therefore only possible if a difference in their physicochemical characteristics is introduced. This can either be accomplished by an indirect approach based on derivatization or a direct approach using a chiral stationary or mobile phase.

In the indirect approach, the enantiomers are derivatized in the sample with an optically active reagent to form diastereomers. This is done prior to the chromatographic separation. Diastereomers have more than one chiral centre and have different physicochemical char-acteristics. Therefore, they have different distribution constants and can be separated in a traditional chromatographic system. When performing derivatization with chiral reagents, it is important to be aware of the purity of the derivatization reagent, which ideally should be 100% pure with respect to enantiomeric composition.

Direct chiral separation (without derivatization) of enantiomers is only possible if chiral-ity is introduced into the chromatographic system. This is achieved using chiral mobile or stationary phases. If chirality is in the mobile phase, diastereomeric complexes are formed in the mobile phase between sample constituents and a chiral mobile phase additive; the resulting diastereomeric complexes are separated on a common LC column. However, for most practical applications, a chiral stationary phase is used. This consists of polymer particles (silica, cellulose, or methacrylate) where the surface has been derivatized with chiral functional groups. The chiral functional groups in use are very different in nature and include, among others, proteins, polysaccharides, cyclodextrin, antibiotics, and helical methacrylate. An example of separation of two enantiomers using a chiral stationary phase in LC is illustrated in Figure 11.25.

11.10   Supercritical Fluid Chromatography

In *supercritical fluid chromatography (SFC)*, a *supercritical fluid* is used as the mobile phase. Supercritical CO_2 is frequently used for this purpose. Carbon dioxide is in the solid state at −78.7 °C and 2 bars, as shown in Figure 11.26.

As the pressure is reduced to 1 bar, CO_2 will be in equilibrium between the solid phase and the gaseous phase. This means that CO_2 sublimes without turning into a liquid. As the temperature is increased above the *triple point*, liquid CO_2 will be in equilibrium with gaseous CO_2 at a certain pressure following the gas–liquid boundary line. Moving up this line, two phases always exist until the *critical point* is reached at 31.3 °C. Above this temperature, only one phase exists regardless of the pressure, and this phase is called a *supercritical fluid*. CO_2 is frequently used as the mobile phase in SFC. To ensure supercritical fluid conditions in the column, the pressure has to exceed 73.9 bar (critical pressure) and the temperature has to exceed 31.3 °C (critical temperature). The temperature of the column is controlled in a column oven similar to that of GC to ensure conditions above the critical temperature. The high pressure in the system is achieved by placing a narrow capillary at the end of the chromatographic column. This capillary is termed a *restrictor*, and the strong restriction creates a high pressure in the column. When the supercritical fluid is leaving the restrictor, it turns into a gas, leaving the solutes in the gas phase for easy detection.

SFC provides increased speed and chromatographic resolution as compared to LC. The reason for this is that diffusion coefficients are higher in supercritical fluids than in liquids. On the other hand, SFC is less efficient than GC because diffusion coefficients are even higher in gasses than in supercritical fluids. Unlike gases, supercritical fluids can dissolve

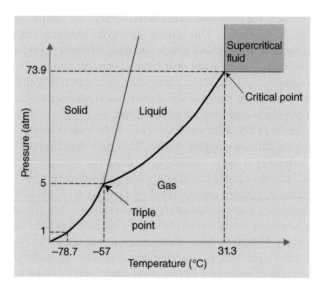

Figure 11.26   Phase diagram for carbon dioxide

non-volatile analytes. This means that SFC can be used to separate non-volatile analytes, as in LC. In GC elution strength is increased by increasing the temperature and in LC the elution strength is increased by gradient elution. In SFC, on the other hand, elution strength is increased during the chromatographic separation by increasing the density of the supercritical fluid. This can be achieved by gradually increasing the pressure of the supercritical fluid. In addition, organic modifiers added to the mobile phase, like methanol, can also increase the elution strength of the mobile phase.

The SFC instrument consists of a pumping system that pumps CO_2 into the column at a pressure above the critical pressure. Injection is typically performed by a loop-injector similar to in LC. GC-like columns are used for capillary SFC, whereas LC-like columns are used for packed-column SFC. Detection is usually performed with a flame ionization detector (FID), UV detector, or by mass spectrometry. SFC is an official method in Ph. Eur., but it is not in use very much.

12

High Performance Liquid Chromatography

12.1 Introduction

High performance liquid chromatography (*HPLC*) is a separation technique using high pressure to force the liquid mobile phase through the separation column. The separation column contains small, solid, stationary phase particles. When HPLC was introduced in the late 1960s the name high *pressure* liquid chromatography was used. Over time, the name was changed to high *performance* liquid chromatography (HPLC), focusing on the increasingly efficient separations obtained, and common practice is now to use the abbreviation *liquid chromatography* (*LC*). Systems capable of running at very high pressures have been introduced in recent years under the name *ultra high performance liquid chromatography* or *ultra high pressure liquid chromatography* (*UHPLC* or *UPLC*). A schematic overview of an LC system is shown in Figure 12.1.

The four main parts of the LC system, the solvent delivery pump (*LC pump*), the *injector*, the *column*, and the *detector* are all vital and indispensable units. The individual parts of the system are carefully connected by small volume *tubing* and tight *fittings* to avoid any dead volumes in the system.

Introduction to Pharmaceutical Analytical Chemistry, Second Edition.
Stig Pedersen-Bjergaard, Bente Gammelgaard and Trine Grønhaug Halvorsen.
© 2019 John Wiley & Sons Ltd. Published 2019 by John Wiley & Sons Ltd.

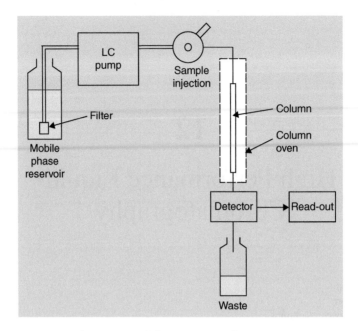

Figure 12.1 Schematic view of LC system

The mobile phase(s) are delivered by the pump and forced with a constant flow rate through the column, which is packed with the column material. The column is often placed in a thermostatic oven (*column oven*) to control the temperature. Samples are injected to the column by a manual injector or most often by an *autosampler*. The column may be protected by a short pre-column. After separation in the column, the individual sample constituents are detected in one or more detectors, which provide an electronic response of the compounds.

The broad applicability and the high degree of automation of LC are among the reasons why this technique has gained such a dominant position in pharmaceutical analysis. LC is used extensively both for chemical analysis of pharmaceutical ingredients and preparations, and for bioanalysis.

12.2 The Column

The column is the heart of the separation process and therefore it is important to keep the column in good condition. The separation columns are typically made of steel tubing and are filled with *packing material* containing the *stationary phase*. The packing material is held in place in the column by a *filter* at each end. The filter is either a porous frit or a net that allows the mobile phase to pass through. The pore size of the filter needs to be smaller than the diameter of the packing material in order to prevent the latter from leaking from the column.

The dynamics of the chromatographic separation process is described by the van Deemter equation. Increasing separation efficiency, expressed by the number of theoretical plates, N,

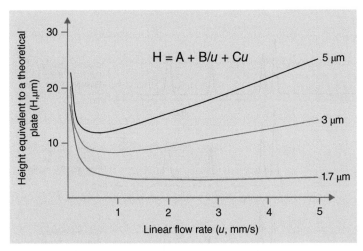

Figure 12.2 *Height equivalent to a theoretical plate as a function of the flow rate of the mobile phase for packing materials of different sizes*

corresponds to the decreasing plate height, and the van Deemter equation describes the theoretical plate height, H, as function of the linear velocity (u). According to this equation, the separation efficiency is highly dependent on the size and uniformity of the particles of the packing material and the separation efficiency is increased when the particle size is reduced. Thus, decreasing the particle size provides a larger number of theoretical plates. This is illustrated in Figure 12.2, showing van Deemter curves plotted for different particle sizes.

The van Deemter equation gives several reasons why smaller particles result in narrow peaks. Small particles reduce the different analyte pathways and thereby term A in the equation. At the same time the distance between analyte and stationary phase is reduced for small particles, leading to faster mass transfer of analyte between the phases; this reduces the term C in the equation. The minima of the van Deemter curves (and thereby optimum efficiency) shift towards higher linear flow rates when the particle size is reduced. Thus, higher flow rates can be used without losing efficiency. The practical implication of this is that efficient separations can be achieved on short columns with particles of about 2 µm in a very short time, as shown in Figure 12.3. However, the small particle size also provides a very high back pressure and it may be necessary to use UHPLC equipment.

The typical LC columns used in the European Pharmacopoeia (Ph. Eur.) and the United States Pharmacopeia (USP) are 15–25 cm long, with inner diameters of 4.6 mm, and packed with 5 µm particles. With the introduction of mass spectrometry as a routine LC detection technique and the improvement of column packings, column dimensions of 10–15 cm with a 2 mm internal diameter have become the common standard. For fast UHPLC analysis, smaller particles between 1.5 and 2.0 µm are used in columns of 3–5 cm in length. Using short columns with a small internal diameter result in major savings in consumption of the mobile phase. In Table 12.1, the reduction in mobile phase consumption is calculated for columns of equal length with different internal diameters and operated with the same linear flow of mobile phase. A further reduction of mobile phase consumption is obtained by reducing the column length. A reduction in column length results in a similar reduction in analysis time.

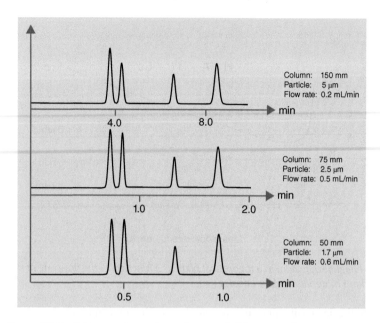

Figure 12.3   *LC separation using columns of different packing material size*

Table 12.1   *Saving of mobile phase as a result of the reduction in the internal column diameter keeping the linear flow rate (1.33 mm/s) in the column constant*

Internal diameter (mm)	Flow rate (mL/min)	Reduction in mobile phase (%)
4.6	1.0	—
4.0	0.69	31
3.0	0.43	57
2.0	0.19	81
1.0	0.05	95

Apart from a narrow particle size distribution of the column packing material, low dead volumes of the tubing and connections in the system is important. Tight end fittings of the column and short tubing with a low internal diameter from the injection device to the column and from the column to the detector maintain high quality separations. If non-optimal connections of tubing and fittings are made during column installation, the separation can be less efficient due to extra band broadening, even if the column itself is in perfect condition. Hence, measuring the efficiency of the 'column' is a test of the total system.

12.3   Scaling Between Columns

When operating a LC system, the flow rate of mobile phase is set as volumetric flow as measured in mL/min. In contrast, the van Deemter equation is based on the linear velocity of the mobile phase (u) measured in mm/s. This needs to be considered when changing between columns of different size, and the flow rate and injection volume has to be changed accordingly, as exemplified in Box 12.1. A column with an internal diameter of 4.6 mm is commonly operated with a flow rate of 1 mL/min and an injection volume of 20 μL, while a column with an internal diameter of 2.1 mm is commonly operated with a flow rate of 0.2 mL/min and an injection volume of 5 μL.

Box 12.1   Calculation of flow rate and injection volume when changing the column diameter

A column with an inner diameter of 4.6 mm used in a system with a flow rate of 1 mL/min and an injection volume of 20 μL is changed to a column with an inner diameter of 2.1 mm. The column volume is proportional to the column diameter. A column with an inner diameter of 4.6 mm contains a volume of

$$3.14 \times (0.23\ \text{cm})^2 \times 1\ \text{cm} = 0.17\ \text{cm}^3 = 0.17\ \text{mL/cm length}$$

Calculation of linear velocity: for general purposes, it can be estimated that the column particles occupy about 40% of the column volume. Thus, the volume of liquid in a 1 cm segment of the column is estimated to

$$0.17\ \text{mL} \times 0.6 = 0.10\ \text{mL} \approx 0.1\ \text{mL/cm}$$

If the flow rate is 1 mL/min, the linear velocity is

$$\frac{1\ \text{mL/min}}{0.1\ \text{mL/cm}} = 10\ \text{cm/min} = 1.7\ \text{mm/s}$$

To obtain the same linear velocity on the narrow column, the flow rate must be reduced by the factor

$$\left(\frac{2.1}{4.6}\right)^2 = 0.21$$

corresponding to a flow rate of

$$0.21 \times 1\ \text{mL/min} = 0.21\ \text{mL/min}$$

The sample volume (injection volume) of 20 μL on the 4.6 i.d. column is changed accordingly:

$$0.21 \times 20\ \text{μL} = 4.2\ \text{μL}$$

12.4   Pumps

Separation and detection occur when the mobile phase is pumped at a constant flow rate through the column. In a standard analytical LC system, the typical flow rate of the mobile phase is 0.2–2.0 mL/min. The small particle size of the column packing materials results in

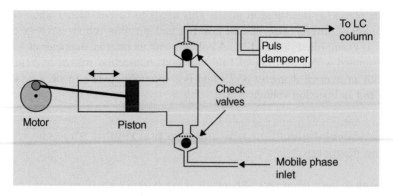

Figure 12.4 Schematic view of a piston pump

a back-pressure of 30–300 bar (3–30 MPa) in an LC system and up to 1500 bar (150 MPa) in an UHPLC system, depending on the flow rate. The pumps must therefore be capable of delivering the mobile phase at a constant flow rate against a high back-pressure. Any particle in the injected samples is trapped on the column inlet and results in gradually blocking the column, which leads to an increased back-pressure. LC pumps are equipped with a regulation mechanism that keeps the flow rate constant and a gradual blockage thus results in an increase in back pressure while the flow is kept constant.

The pumps deliver the mobile phase at a constant flow rate. The pumps can be constructed in different ways, but a *piston pump* is the most common and is sketched in Figure 12.4. The piston pump consists of a small steel *cylinder* with a volume of approximately 100 μL. A *piston* is moved back and forth in the cylinder by means of a motor. There are two ball valves (*check valves*) attached to the cylinder so that the mobile phase can only flow in one direction, into the cylinder from the mobile phase reservoir (mobile phase inlet) and out of the cylinder to the LC column. When the piston is moved back the lower ball valve opens while the upper ball valve closes, dragging the mobile phase into the cylinder. When the piston again is moved forward into the cylinder, the bottom ball valve closes while the top valve opens and the mobile phase is forced out of the cylinder and into the LC column. Since the mobile phase is forced into the column only when the plunger is pushed into the cylinder, the flow pulsates. This pulsation introduces extra noise in the detector signal and should be eliminated. A *pulse dampener* is therefore included in the system to ensure a smooth flow of the mobile phase. Other pump systems ensure a smooth flow by linking together two piston heads into a *double piston pump*, where one piston head delivers the mobile phase to the column while the second is filled up with the mobile phase.

When the pumping system delivers a mobile phase with a constant composition to the column, it is called *isocratic elution*. In a *gradient elution* system, the composition of the mobile phase is changed during the analytical run. In this case, two mobile phases are pumped from two separate reservoirs and mixed during the analysis. This can be done by *low pressure mixing* using a single pump equipped with a valve connected to up to four different mobile phase reservoirs. The mixing valve opens for only one pipeline at a time and in this way the solvent mixture can be controlled. Alternatively, *high pressure mixing* is performed by two pumps, each delivering a controlled amount of mobile phase. The

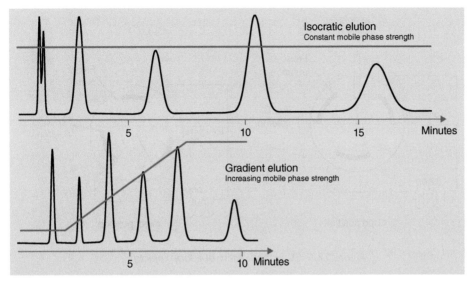

Figure 12.5 *Isocratic and gradient elution of a sample containing analytes with large differences in retention*

mixing is then performed at the high pressure side. Gradient elution can be compared to temperature programming in gas chromatography (GC).

Gradient elution is used to separate sample constituents with large differences in retention to avoid an unnecessary long analysis time. This is illustrated in Figure 12.5, where the early eluting substances show partially overlapping peaks in the upper chromatogram for isocratic elution, while the most retained substances elute as broad peaks with a long retention time. Using gradient elution (lower chromatogram), the composition of the mobile phase is continuously changed during the analysis, starting with a weak eluting composition of the mobile phase. In this way, the first eluting compounds are adequately retained, and because the eluting strength of the mobile phase is increased during chromatography, the late elution peaks are now eluted faster.

12.5 Injectors

The purpose of the *injector* is to bring a definite volume of sample solution into the mobile phase and to the column. The substances to be separated must be dissolved in a liquid that is miscible with the mobile phase. In addition, the sample liquid should not have a stronger elution strength than the mobile phase. Typical injection volumes are 5–100 μL. The injection systems are optimized to inject the solution under high pressure and must be leak-proof in the whole pressure range of the system. For manual injections a simple *loop injector* as shown in Figure 12.6 can be used.

The loop injector is a *six-port valve*. In the *load position* the mobile phase from the pump passes through the valve directly to the column. In this position it is possible to inject the sample into the *loop* using a syringe. When the valve is switched to the *inject position*, the

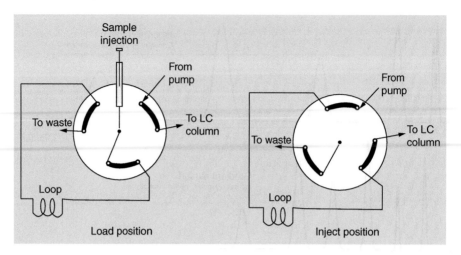

Figure 12.6 Schematic view of a loop injector

mobile phase from the pump passes through the loop and in this way brings the sample to the column. The loop is a piece of tubing with a total volume that should not exceed the capacity of the column used. In standard LC systems it is common to use a loop of 20 µL together with columns of 4.6 mm internal diameter. In the load position, the loop is filled with excess sample in order to remove the previous solvent (mobile phase) completely from the loop. After an injection, the injector is switched back to the load position. The channels in the injector are flushed with a cleaning solution between every sample to remove any residual sample solution.

The injection process is most often automated by use of an *autosampler*. Sample vials are placed in the autosampler and samples are injected automatically into the LC according to pre-programmed volumes and time intervals. Thus, large series of samples can be analysed automatically. The autosampler is temperature controlled, and sensitive samples can be stored there at a low temperature (typically 5 °C) prior to analysis.

12.6 Detectors

The purpose of the detector is to generate an electrical signal (response) for each compound eluting from the column. The detector response should be proportional to either the concentration or the mass of the compound in such a manner that quantitative analysis can be carried out based on the measurement of peak areas or peak heights. The main LC detectors used in pharmaceutical analysis are given in Table 12.2.

For chemical analysis of pharmaceutical ingredients and preparations by LC, the UV detector is standard. Fluorescence detectors and electrochemical detectors (ECDs) show much lower detection limits than the UV detector, but only for selected analytes. The mass spectrometer provides important information on the molecular structure and is a common detector with LC in bioanalysis. LC combined with mass spectrometry is abbreviated LC-MS and is described in Chapter 15. A refractive index (RI) detector may be used in case

Table 12.2   LC detectors and their typical performance

Detector	Lower limit of detection (ng)	Gradient elution
Ultraviolet (UV)	0.1–1.0	Yes
Fluorescence	0.001–0.01	Yes
Electrochemical (ECD)	0.01–1.0	No
Mass spectrometry (MS)	0.001–0.01	Yes
Refractive index (RI)	100–1000	No
Evaporative light scattering (ELSD)	0.1–1.0	Yes
Charged aerosol (CAD)	0.1–1.0	Yes

analytes do not absorb UV radiation. It is considerably less sensitive than UV detection and it is not applicable to methods using gradient elution.

12.6.1   UV Detectors

UV detectors are preferred for LC of pharmaceutical ingredients and preparations, mainly due to high operational stability and ease of use. The lower limit of detection is sufficient with pharmaceutical ingredients and preparations, but is insufficient in bioanalysis where drug substances are measured at low concentrations in biological fluids. UV detection can be used for any analyte absorbing UV or visible light. This requires the analyte to contain a *chromophore*.

According to Beer's law, the absorbance is proportional to the concentration of the substance in the mobile phase, the path length of radiation through the flow cell, and the molar absorption coefficient of the substance. Figure 12.7 shows a sketch of a traditional UV detector for LC.

The *radiation source* is most often a *deuterium lamp* that emits a continuum of light in the UV region (190–400 nm) and any wavelength can be selected by the *monochromator*. For optimal detection sensitivity, substances should be measured at a wavelength corresponding to their maximum UV absorbance. The mobile phase from the LC column (eluent) passes through a *flow cell* inside the UV detector and UV light with the selected wavelength is directed through the flow cell where the intensity is measured by a detector (*light sensor*).

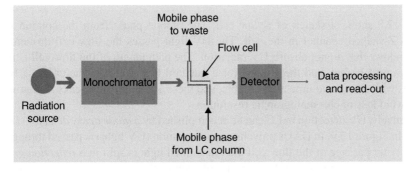

Figure 12.7   Schematic view of a UV detector

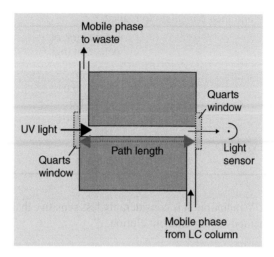

Figure 12.8   *Schematic view of the flow cell in a UV detector*

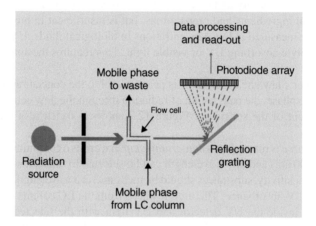

Figure 12.9   *Schematic view of a diode array detector (DAD)*

Figure 12.8 shows a sketch of a *flow cell*. The mobile phase from the column flows through a Z-shaped channel in the cell. The UV light passes the flow cell through two quartz windows that do not absorb UV radiation. The path length of the flow cell is in the range from 6 to 10 mm, and the volume for a standard flow cell is in the order of 10 μL. Flow cells with a lower volume are available, and these are used when peaks are narrow in order to avoid loss of chromatographic resolution.

Alternatively, UV detection in LC can be accomplished by a *diode array detector (DAD)*, as shown in Figure 12.9. In DADs polychromatic radiation (UV light) is passed through the flow cell. After passage of the flow cell the transmitted light is split in a *reflection grating* into individual wavelengths and the intensity of each of these is measured by an array of photodiodes (*photodiode array*). There may be up to several hundred diodes in the array to measure the entire UV region. DADs offer several possibilities. A full UV spectrum of a peak can be recorded during the elution of a given substance and this can be used to identify the substance. An important use of the DAD is for control of *peak purity* in chromatograms. If the absorption spectrum of a compound changes during the elution of a peak, it is an indication that more than one compound is present in the peak (co-elution) and separation is not complete.

12.6.2   Fluorescence Detectors

Some substances emit the temporary energy as *fluorescence* when they are irradiated. However, while most drug substances absorb radiation in the UV region, only a few substances emit fluorescence. The *fluorescence detector* is therefore more selective than the UV detector and has lower limits of detection. The detector is therefore useful for detection of low concentrations of fluorescent analytes. In some cases, reaction of analytes with fluorescent reagents and subsequent fluorescence detection is used to improve the sensitivity. Figure 12.10 shows a schematic diagram of a fluorescence detector. The lamp (radiation source) is usually a *xenon lamp* emitting an intense continuum of light in the whole UV region and partly in the visible region.

The wavelengths of *excitation* and *emission* light (fluorescence) are controlled by two monochromators. The molecules are excited using intense excitation light of a selected wavelength (*excitation wavelength*). Some of the absorbed energy is released as heat, and therefore the emitted light (*emission wavelength*) is shifted towards longer wavelengths. This is shown in Figure 12.11 for quinine as an example. The excitation wavelength is chosen from the UV absorption spectrum of the compound, preferably at the wavelength of the UV absorption maximum, but for selectivity reasons another wavelength can be chosen. The emission wavelength chosen can be varied as the emission spectrum also covers a range of wavelengths. Fluorescence is emitted in all directions but is normally measured perpendicular (90°) to the excitation radiation to reduce background noise.

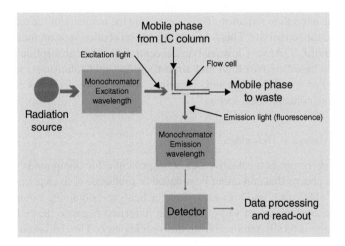

Figure 12.10   *Schematic view of a fluorescence detector*

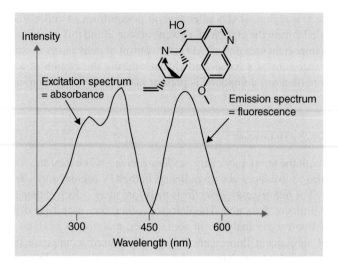

Figure 12.11 *Excitation and emission spectra of quinine*

For dilute solutions, the following equation expresses how the fluorescence (F) is dependent on a number of parameters:

$$F = \Phi I_0 \varepsilon bc \tag{12.1}$$

Fluorescence intensity F is proportional to the quantum yield Φ, the intensity of the excitation radiation I_0, the molar absorption coefficient of the compound ε, the path length of the detector b, and the concentration of the compound in the mobile phase c. The quantum yield is a compound-dependent constant in a given environment and describes the fraction of absorbed light that is re-emitted as fluorescence. As seen from Eq. (12.1), the measured intensity of fluorescence is proportional to the concentration of analyte in the mobile phase. As fluorescence is an emission method, the fluorescence intensity is directly proportional to the intensity of the excitation radiation. Thus, increasing the intensity of the excitation radiation will increase the sensitivity. This is in contrast to absorbance measurements, which are based on the ratio of I_0/I. At very low analyte concentrations (low absorption), absorbance measurements are based on two high and almost similar light intensities, while fluorescence is measured on a dark background. Therefore, fluorescence detectors are favourable for detection at low concentration levels.

12.6.3 Electrochemical Detectors

Electrochemical detection is a selective detection principle for compounds with electrochemically active groups that can either be reduced or oxidized, such as phenols, aromatic amines, ketones, aldehydes, and aromatic nitro- or halogen-containing compounds. Oxidation is in practice much easier to perform than reduction because the reduction mode requires that oxygen is totally removed from the mobile phase. The oxidation is performed at a given voltage, typically between +0.3 and +1.0 V. The higher the voltage, the more the substances are oxidized. The detector measures the current as a function of oxidation.

The detector is especially well suited for the detection of phenols, amines, and thiols. The ECD is not as robust as UV and fluorescence detectors, and contamination of the electrodes by impurities in the mobile phase and from samples may occur. Pulsed amperometric detection with Au or Pt electrodes expands the applicability of electrochemical detection to cover alcohols and carbohydrates. In this mode, carbohydrates can be detected with high sensitivity, but a high pH value is demanded (pH 12–13), which may not be compatible with the LC system.

12.6.4 Refractive Index, Evaporative Light Scattering, and Charged Aerosol Detectors

The *refractive index (RI) detector* measures the change in the refractive index of the mobile phase at the column outlet. An analyte with a refractive index different from the mobile phase is detected. However, the detector has low sensitivity and is easily influenced by changes in temperature. In the *evaporative light scattering detector* (*ELSD*) the column effluent is nebulized using nitrogen at an elevated temperature. Non-volatile compounds are detected in the aerosol as they scatter the light from a diode laser. Volatile substances cannot be measured by this technique. The *charged aerosol detector* (*CAD*) uses a similar principle for detection. The column effluent is evaporated and an aerosol of non-volatile compounds is formed. Nitrogen molecules passing a corona receive positive charges that are transferred to the aerosol particles. The positively charged aerosol particles are then measured by an electrometer. Both ELSD and CAD are universal detectors with the exception of volatile substances.

12.7 Mobile Phases

The mobile phases used in LC are contained in bottles, and from there they are withdrawn by the pump through polyethylene tubing, each equipped with a *filter* that prevents any particles from entering the pump. The following requirements for mobile phase solvents have to be considered:

- The solvents should not give any response in the detector used.
- The solvents must have a satisfactory degree of purity. A number of solvents are available in LC quality and sometimes specified for selected purposes (e.g. for gradient elution or for MS detection).
- The solvents should have low viscosity to provide as low a back pressure as possible in the LC system.
- The solvents should have low toxicity, preferable be inflammable and non-reactive, and they must be suitable for disposal after use.

Mobile phases for reversed phase LC often contain aqueous buffer solutions. The buffer salts are dissolved in water of *LC quality*, which is either produced in the laboratory using a water treatment system that removes common contaminants from tap water or obtained commercially. The aqueous solution is mixed with an organic modifier (typically acetonitrile or methanol) to the prescribed solvent strength. If the buffer salts give rise to particles

Table 12.3 UV cut-off for common solvents (1 cm path length)

Solvent	UV cut-off (nm)
Acetone	330
Acetonitrile	190
Dichloromethane	220
Ethanol	210
Heptane	195
Methanol	205
Tetrahydrofuran	210
Water	190

in the solution, it is filtered through a filter with a pore size of 0.45 μm. Before use, the mobile phase has to be degassed to remove dissolved air. *Degassing* can be performed by ultrasound treatment, by purging with helium, or by vacuum treatment. Vacuum treatment is most common and is now included in most instruments as a continuous online degassing system. Degassing is necessary in order to prevent bubble formation in the detector cell causing noise in the detector signal.

The UV detector is by far the most widely used LC detector. Analytes can be detected when they have a UV response above 195 nm. This requires the mobile phase to be transparent at the given wavelength. Table 12.3 gives an overview of the UV cut-off of common solvents (organic modifiers). The UV response increases as the wavelength decreases, and the UV cut-off value is the shortest wavelength that can be used with the solvent.

12.8 Solvents for Sample Preparation

A prerequisite for the samples to be analysed by LC is that they are soluble in the mobile phase and the mobile phase is therefore the first choice as a solvent for the sample. In general, the substances should not be dissolved in a solvent with a higher solvent strength than the mobile phase. The use of neat methanol (or acetonitrile) should be avoided in reversed phase LC, as the solvent strength of the sample solution will become too high. If substances dissolved in neat methanol are injected, methanol prevents the substances from being retained on the stationary phase in the column. The substances remain in the methanol plug until it has been diluted with the mobile phase and equilibrium has been re-established in the system. The result is that the substances are eluted from the column earlier and as broader peaks. If methanol must be used as a solvent, it must be injected only in small volumes in order to prevent peak broadening. If larger sample volumes are to be injected, it is preferable to have a lower solvent strength in the sample compared to the mobile phase. The substances can then under the correct conditions be concentrated at the inlet of the column during the injection, and spreading of extra bands because of dilution in the mobile phase can be avoided.

13

Gas Chromatography

13.1   Introduction

Gas chromatography (*GC*) is a separation technique used for separation and detection of volatile and semi-volatile substances. GC is an official method of analysis in the European Pharmacopoeia (Ph. Eur.) and the United States Pharmacopoeia (USP). In the area of pharmaceutical analysis, a common application is determination of volatile impurities in pharmaceutical ingredients and preparations. These are primarily residual solvents from production, but GC is also used for determination of the composition of essential oils and of fatty acids in vegetable fats and oils. Essential oils are added as excipients to pharmaceutical preparations because of their odour and taste characteristics. Essential oils contain a complex mixture of readily volatile components and capillary GC is therefore the method of choice. Vegetable fats and oils are used as excipients in ointments and creams. GC is used for the analysis of the composition of fatty acids in oils and to determine the content of sterols.

The ability of GC analysis to separate substances with large differences in volatility makes this method suitable for screening biological materials, such as blood or urine, in

Introduction to Pharmaceutical Analytical Chemistry, Second Edition.
Stig Pedersen-Bjergaard, Bente Gammelgaard and Trine Grønhaug Halvorsen.
© 2019 John Wiley & Sons Ltd. Published 2019 by John Wiley & Sons Ltd.

toxicological and doping analysis, where the purpose is to identify known or unknown substances that have been ingested. For such analyses, mass spectrometry is needed as the detection method. GC-MS is discussed in Chapter 15. Furthermore, GC is an important technique for analysis of pesticides and pollutants and is used in the food, oil, and perfume industries.

13.2   Basic Principle

Only *volatile* or *semi-volatile* substances can be separated and detected by GC. The analytes are separated as a function of different distribution ratios between the mobile and the stationary phases. The mobile phase is a gas and the stationary phase is a polymer film coated on the inner surface of a capillary column. The carrier gas transfers the volatile sample constituents along the column while they are distributed between the stationary phase and the carrier gas. Sample constituents with different distribution ratios are transferred through the column with different velocities and are thereby separated prior to detection.

A schematic overview of a *gas chromatograph* is presented in Figure 13.1. A sample volume of 0.1–1 μL is injected by a *microsyringe* into a heated *injection port*. The sample constituents immediately evaporate inside the injection port and are then transported along the *capillary column* with the mobile phase. The mobile phase is a clean gas delivered from a pressurized cylinder and is also named the *carrier gas*.

The capillary column (*GC column*) is a 10–50 m long, open tube with a typical internal diameter of 0.25 mm. The inside wall of the GC column is coated with a thin layer of polymer *stationary phase*. The sample constituents interact with the stationary phase as they pass through the GC column, and different compounds are separated based on differences in their affinity towards the stationary phase. At the end of the GC column, the separated compounds are measured by a *detector*, where the signal is recorded continuously during the separation and plotted as a chromatogram. A typical GC chromatogram from analysis of peppermint oil is presented in Figure 13.2. The individual peaks are narrow,

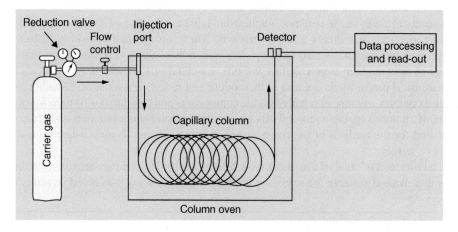

Figure 13.1   *Schematic illustration of a gas chromatograph*

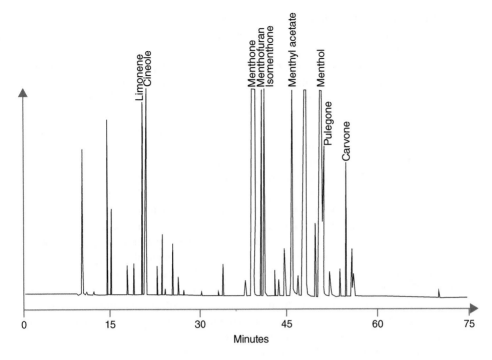

Figure 13.2   GC analysis of peppermint essential oil

which illustrates the high separation power and efficiency of GC. The entire GC column is located inside a temperature-controlled oven (column oven), as the retention in GC is highly dependent on temperature.

Only *volatile* or *semi-volatile* substances can be separated and detected by GC. The volatility of a substance is temperature dependent, and the vapour pressure of the substance increases with temperature. Therefore, retention is reduced in GC by increasing the column temperature. Highly volatile substances are separated at low temperatures, whereas less volatile compounds are separated at higher temperatures. In addition, the polarity and the thermal stability of the analytes are important for the applicability of GC. Polar compounds are difficult to separate by GC as they are prone to serious peak tailing. The thermal stability of the analyte also plays a role. Because GC is accomplished at elevated temperatures, the analytes need to be thermally stable up to 250–300 °C. Consequently, non-polar small molecule drug substances are ideal for GC, whereas more polar substances and large molecule substances are separated by liquid chromatography (LC). Most pharmaceuticals and biopharmaceuticals are indeed either relatively polar substances or large molecules, and therefore LC is used more than GC in pharmaceutical analysis.

13.3   Instrumentation

As shown in Figure 13.1, the carrier gas used for GC is contained in a high pressure steel cylinder, and is delivered through a reduction valve and into the GC instrument. The carrier gas is heated and passes through flow controllers into the injection port and flows through the column to the detector. All instrument settings, data acquisition, and data handling are performed by computer software.

The *reduction valve* reduces the carrier gas pressure from high pressure (up to 200 bar in the cylinder) to about 5 bar at the column inlet. The *flow controllers* are used to set the flow rate of carrier gas and to make sure that this flow remains constant over time. Operation with a constant flow of carrier gas is mandatory in order to obtain reproducible GC separations, retention times, and detector signals. The samples are dissolved in a volatile organic solvent and injected with a microsyringe into the heated injector. Common injection volumes are 0.1–1 µL. Injection port temperatures are typically in the range 200–300 °C. The temperature must be sufficiently high to evaporate the sample constituents immediately, for efficient transfer into the GC column. The injection port temperature is normally higher than the initial GC column temperature to obtain a rapid evaporation and transfer of sample constituents into the column. If the injection port temperature is too low, slow and incomplete transfer of sample constituents may occur. On the other hand, if the injection port temperature is too high, this may result in thermal degradation of sample constituents during injection.

A detector is placed at the column outlet and is typically heated to 250–350 °C. This is generally 25–50 °C higher than the final (and highest) column temperature used for separation. The detector is heated in order to avoid condensation of sample constituents at the column end. Heating the detector also prevents sample constituents of low volatility from condensing and contaminating the inside of the detector.

The column is placed in the GC oven with a fan that ensures efficient circulation and temperature control during operation. The column temperature is the main parameter that controls the retention in GC. During GC separation, the sample constituents are distributed between the carrier gas and the stationary phase, but they only move forward through the GC column when they are in the carrier gas phase. The higher the affinity of the sample constituents to the stationary phase, the stronger they are retained in the column. An increase in the column temperature will increase the volatility of the substances and in this way change the distribution in favour of the gas phase. Therefore, reducing retention in GC is obtained by increasing column temperature. Column temperatures above 300 °C increase the risk of analyte decomposition and are rarely used.

The oven can be operated in two different ways during separation, namely in the *isothermal mode* (*isothermal analysis*) and in the *temperature programmed mode* (*temperature programmed analysis*). In the isothermal mode, the temperature in the oven (and the column) is kept constant during the entire separation. In the temperature programmed mode, the temperature is gradually increased during the separation. This is illustrated in Figure 13.3.

Samples of low complexity and with constituents of similar volatility can be separated using the isothermal mode. However, samples are often complex and contain a broad variety of compounds of different volatility. In those cases, temperature programmed analysis is preferred. In temperature programmed analysis, the initial column temperature is low and typically in the range 40–80 °C during sample injection. At this stage only the most volatile sample constituents are carried through the GC column. Less volatile sample constituents are trapped at the inlet of the GC column. The oven temperature is then gradually increased, typically with a rate of 5–20 °C/min, and gradually less volatile sample constituents are released from the GC column inlet and carried to the detector by the carrier gas. Thus, even for samples containing compounds with major differences in terms

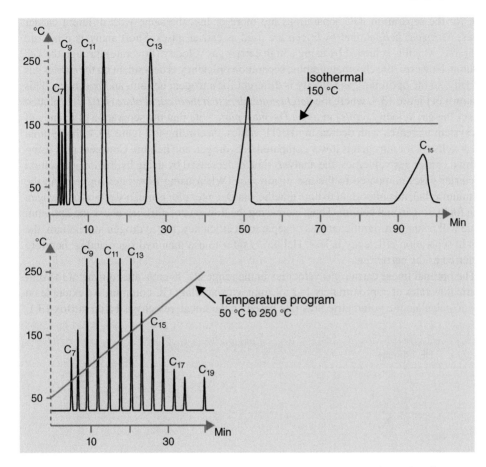

Figure 13.3   Isothermal versus temperature programming in GC analysis of n-alkanes

of volatility, very efficient GC separations can be obtained, as illustrated in Figure 13.3. In the chromatogram from the isothermal analysis, the first peaks are narrow, but not very well separated, while the less volatile compounds elute as broad peaks at extensively long retention times. In contrast, with a temperature programmed analysis, both the volatile and less volatile compounds were separated with high efficiency. Temperature programs always end at a high temperature (250–325 °C) to make sure that sample constituents of low volatility are removed from the column. Finally, the column temperature is reduced to the starting temperature again and a new sample can be injected. The principle of isothermal GC is comparable to isocratic elution in LC, whereas temperature programmed GC is comparable to gradient elution in LC.

13.4   Carrier Gas

The *carrier gas* is an inert transport medium for the sample constituents in GC. Thus, in contrast to LC, where the chemical composition of the mobile phase plays an important role for the separation, GC separations are more or less the same with different carrier gases. *Nitrogen, helium,* and *hydrogen* are used as carrier gases. Short analysis times are desirable, which is achieved by using a high carrier gas velocity (flow rate) through the GC column. However, the chromatographic separation efficiency is dependent on the carrier gas velocity, so the optimum gas velocity is different for nitrogen, helium, and hydrogen. This is shown in Figure 13.4, where the *height equivalent to a theoretical plate (HETP)* is plotted versus the gas velocity (cm/s) in a *van Deemter plot*. Note that the separation efficiency of the system increases with decreasing HETP values. According to Figure 13.4, the optimal linear velocity for nitrogen is lower compared to hydrogen and helium. Consequently, using optimal carrier gas velocities, the analysis time is decreased by using hydrogen and helium as carrier gases compared to the use of nitrogen. When using flow rates higher than the optimum value, it is important to note that the van Deemter plot is much steeper for nitrogen than for hydrogen and helium. Thus, increasing the flow rate of nitrogen above the optimum value will result in a significant loss of separation efficiency. For hydrogen and helium, the loss in separation efficiency is less. Helium is safer to use than hydrogen and is therefore preferred as the carrier gas.

The optimal linear carrier gas velocities in the range 20–40 cm/s are equivalent to volumetric flow rates of approximately 1–2 mL/min with capillary GC columns. An example on how to calculate the volumetric flow rate based on the linear velocity is given in Box 13.1.

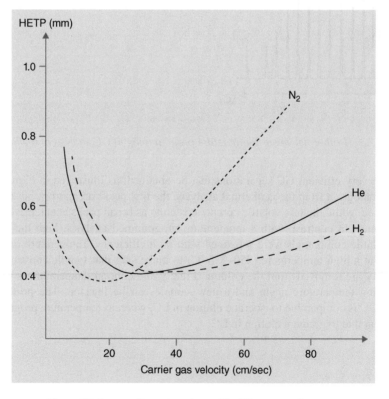

Figure 13.4 *van Deemter plots with different carrier gases*

The carrier gases used are of high purity, and in the GC system, the gas is further purified by moisture, hydrocarbon and oxygen traps before entrance to the column. This is extremely important in order to avoid deterioration of the stationary phase (caused by oxygen) and to avoid a high background in the detector response caused by organic impurities. The purities of GC carrier gases normally exceed 99.99%.

Box 13.1   Calculation of volumetric flow rate from linear gas velocity

A linear carrier gas velocity of 30 cm/s through a GC column with a 0.32 mm (0.032 cm) internal diameter corresponds to the following volumetric flow rate (mL/min).
First the cross-sectional area of the GC column is calculated:

$$\text{Cross-sectional area} = \pi \times (\text{radius})^2 = 3.14 \times \left(\frac{0.032 \text{ cm}}{2} \right)^2 = 8.0 \times 10^{-4} \text{ cm}^2$$

Then the volumetric flow rate is calculated:

Volumetric flow = Linear gas velocity × Cross-sectional area

$$= 30 \text{ cm/s} \times 8.0 \times 10^{-4} \text{ cm}^2 = 0.024 \text{ cm}^3 \times 60 \text{ s/min} = 1.4 \text{ mL/min}$$

13.5   Stationary Phases

The stationary phase in capillary GC is a non-volatile liquid coated as a thin film on the inside wall of the capillary column. The separation is based on the distribution of sample constituents between the carrier gas and the liquid stationary phase. Liquid stationary phases in GC are temperature stable liquids having very low vapour pressure. Many different stationary phases have been developed, but the preferred ones are either *polysiloxanes* or *polyethylene glycols*. These are liquids at room temperature, chemically stable up to 225–325 °C and have a low vapour pressure even at high temperatures. The stationary phase should be a liquid throughout the entire operational temperature range. Chemical stability at high temperatures is important because most GC procedures are accomplished in temperature programmed mode, where the final temperature often is 250–300 °C. Low vapour pressure is important in order to avoid loss of the stationary phase from the column during high temperature operation, also termed *column bleeding*.

The *operational temperature range* of a column is decided by the upper temperature limit, where thermal degradation and column bleeding of the stationary phase is a risk, and the lower temperature limit, where the stationary phase becomes too viscous and diffusion of sample constituents will be reduced. This will lead to peak broadening and poor chromatography.

Examples of polysiloxane stationary phases are shown in Figure 13.5. *Polydimethylsiloxane* is the most hydrophobic of the polysiloxanes, and this stationary phase is mainly used for separation of non-polar sample constituents.

The polarity of the stationary phase can be varied by substitution of some of the methyl functionalities with either phenyl or cyanopropyl functionalities. This provides

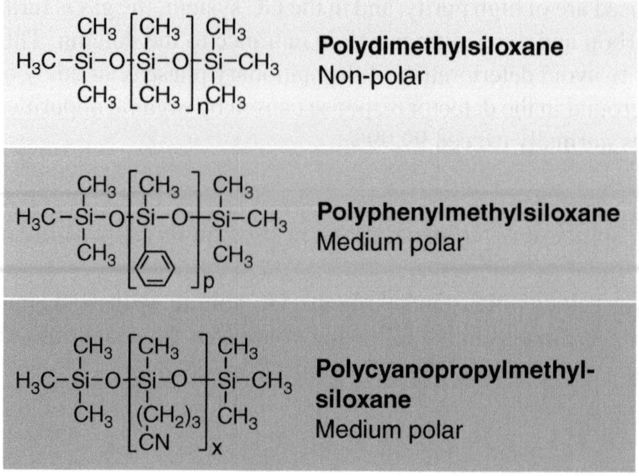

Figure 13.5 Basic skeletons of polydimethylsiloxane, polyphenylmethylsiloxane, and poly-cyanopropylmethylsiloxane. The numbers of repeating units (n, p, and x) can vary significantly from one stationary phase to another

medium-polar stationary phases as *polyphenylmethylsiloxane* or *polycyanopropylmethylsiloxane* (Figure 13.5). With an increasing degree of phenyl substitution, the polarity of the stationary phase will increase. When cyanopropyl functionalities are introduced, the polarity is increased even further. The level of substitution can vary, and a large number of different polysiloxane-based stationary phases are commercially available. Polysiloxanes are temperature stable and can be used over a wide temperature range. However, the introduction of polar functionalities reduces the temperature stability to some extent. Thus, while polydimethylsiloxane columns can be used up to 300–325 °C, polycyanopropylmethylsiloxane columns can only be used up to about 275 °C.

Polyethylene glycols represent another family of stationary phases. The general skeleton of polyethylene glycol polymers is shown in Figure 13.6. Polyethylene glycols are often named *Macrogol* or *Carbowax* followed by a number indicating the molecular weight. The temperature stability increases and the vapour pressure decreases with increasing molecular weight. These stationary phases are available with molecular masses in the range from 300 to 10^7 Da. The most popular is Carbowax 20M having a molecular mass of 20 000 Da. This can be used in the temperature range 60–250 °C. The hydroxyl groups terminating the polyethylene glycol chain are polar, and these are responsible for the intermediate polarity characteristics of these stationary phase. The polyethylene glycols provide different separation selectivity as compared to polysiloxane-based stationary phases.

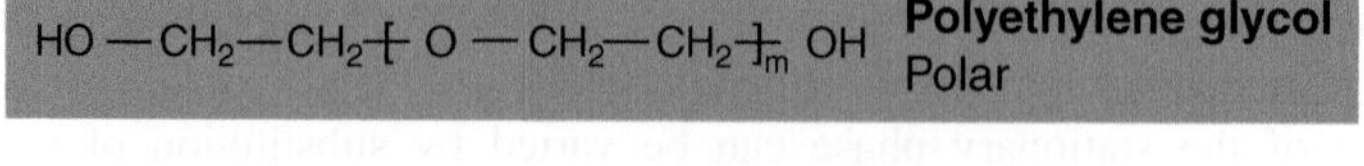

Figure 13.6 Basic skeleton of polyethylene glycol

13.6   Retention

As mentioned previously, the retention of substances in GC is controlled by temperature. In general, sample constituents are separated at substantially lower column temperatures than their respective boiling points. Increasing the column temperature increases the volatility of the substances to be separated, and the retention decreases. As a rule of thumb, the retention time is halved when the column temperature is increased by about 30 °C.

The solubility of the analyte into the stationary phase is another important parameter influencing the retention. The solubility depends on the interaction that takes place at the molecular level between the analyte and the liquid stationary phase. Strong intermolecular interactions result in increased retention. This can be expressed as the general rule 'like dissolves like', meaning that the stationary phase and the analytes should be of similar polarity for the analytes to be separated.

Box 13.2   Retention time (minutes) with different stationary phases

	Benzene 80°C	Cyclohexane 81°C	Ethanol 79°C
Polydimethylsiloxane (non-polar)	8.00	8.32	2.69
Polycyanopropyl-phenyl-dimethylsiloxane (medium polar)	8.69	8.10	3.47
Polyethylene glycol (polar)	6.46	2.27	6.46

Ethanol is very polar; benzene is less polar, while cyclohexane is non-polar.

On the non-polar polydimethylsiloxane, the retention of the very polar ethanol is low, as hydrophobic interactions are limited. The non-polar cyclohexane is strongly retained, while the more polar benzene is less retained.

On the medium polar dimethylpolysiloxane, the retention of ethanol increases due to dipole–dipole interactions between the —C≡N and —OH groups. This medium polar stationary phase contains 94% methyl functionalities, and therefore benzene and cyclohexane are still retained by hydrophobic interactions. The retention order of benzene and cyclohexane changes as compared to the non-polar stationary phase. Benzene is now more strongly retained due to π–π interactions between benzene itself and the benzene rings in the stationary phase.

On the highly polar polyethylene glycol, the retention of ethanol is strongest due to hydrogen bonding interactions between the —OH groups of ethanol and the —OH groups of polyethylene glycol. The retention of cyclohexane is now very low, whereas benzene with π-electrons is still retained strongly by the polar stationary phase.

A very non-polar stationary phase such as polydimethylsiloxane mainly interacts with sample constituents through *hydrophobic interactions*. Hydrophobic interactions are relatively weak and are based on van der Waals forces. Therefore, the non-polar stationary phases provide relatively weak retention. Thus, non-polar analytes like an homologous series of alkanes are eluted mainly in order of increasing boiling points.

The interactions become stronger as the polarity of the stationary phase is increased. Introduction of phenyl functionalities in the stationary phase give rise to *temporary dipole interactions*. Introduction of cyanopropyl functionalities increases polarity even more, and *dipole–dipole interactions* can now occur between the substances and the stationary phase. Thus, on medium-polar GC columns, the retention of a given substance is a complex function of its volatility and its ability to interact with the stationary phase based on temporary dipole and dipole–dipole interactions.

Stationary phases based on the strongly polar polyethylene glycols also facilitates *hydrogen bonding interactions*, especially for substances comprising alcohol functionalities, and thus will have stronger retention as compared to other substances. Thus, on polar GC columns, the retention of a given substance is a complex function of its volatility and its ability to interact with the stationary phase based on hydrogen bonding and dipole–dipole and hydrophobic interactions. An overview of these interactions is presented in Box 13.2, which illustrates the different selectivities of non-polar to highly polar stationary phases for three different substances with very similar boiling points (79–81 °C).

13.7   Columns

The columns used in GC are mainly *capillary columns*. For preparative chromatography or poorly retained gases *packed columns* may be used, but these will not be further discussed. For analytical purposes, capillary columns are used due to their high efficiency and resolution. Capillary columns are made of *fused silica*. As fused silica is very fragile, the outer surface of the capillary is coated with a layer of *polyimide* to improve the mechanical stability. This results in very flexible columns, which are easy to handle and install in the GC instrument. Narrow columns provide high resolution as a result of fast mass transfer. On the other hand, the sample capacity of narrow columns is limited so they are more easily overloaded. The internal diameters of the columns are in the range 0.1–0.53 mm, with 0.25 and 0.32 mm being the most popular ones.

The stationary phase is coated as a thin film on the inside wall of the column (*wall-coated column*, see Figure 13.7). The stationary phase can either be physically coated on the inner surface or be chemically bonded to the surface. The more stationary phase that is coated in the column, the greater is the *film thickness*. The film thickness is an important parameter affecting separations and increasing film thickness results in increased retention and sample capacity. Thick films reduce the risk of tailing as they shield the analytes from the silica surface, but at the same time, the risk of column bleeding increases. The film thickness is in the range from 0.05 to 10 μm, and 0.25 μm is often used. Columns with a thin film are often used for the separation of semi-volatile substances, whereas the thick-film columns are used for separation of volatile substances.

Capillary columns are open tubes, and the pressure drop along the columns is relatively small, which allows for use of fairly long columns. Typical column lengths are in the range

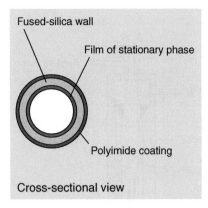

Figure 13.7   Capillary column (wall-coated) for GC

10–30 m (the columns are coiled). The number of theoretical plates, N, is proportional to the column length and the resolution is proportional to $\sqrt{N}$. This makes GC on capillary columns one of the most powerful separation techniques, and 50–100 peaks or even more can be detected in a single chromatogram. However, long columns also increase the analysis time. To fully specify a GC column, the internal diameter, the length, the film thickness, and the stationary phase and film thickness should be provided.

In recent years, *fast GC* has evolved using very narrow columns. With narrow columns, the van Deemter curve is flat and the GC system can be operated at very high linear gas velocities without reducing the chromatographic separation efficiency. Thus, complex separations can be performed in a couple of minutes using a rapid temperature program. Typically, the columns in fast GC have an internal diameter of 0.1 mm and the length is 10–15 m.

13.8   Injection

In GC analysis, samples are dissolved in a volatile solvent and volumes of 0.1–1 μL are injected with a microsyringe. The solvent and the sample constituents evaporate in the injector and the vapour is transferred to the column by the carrier gas. Introduction of a 1 μL liquid sample solution in the injector will after evaporation form about 0.5–1 mL of vapour. The volume of the injector must therefore be sufficiently large to allow the sample solution to expand to this gas volume, and the sample volume must not overload the column. Otherwise, peak broadening will occur, resulting in poor separation. Moreover, the injection system must prevent the formed gas to enter directly into the capillary column, as the internal volume of the capillary column is so small that 1 mL of gas fills a large part of the column and causes band broadening.

In order to prevent band broadening, the injector provides two different injection possibilities, namely *split injection* and *splitless injection*. Figure 13.8 shows a schematic overview of a *split/splitless injector*. The sample solution is injected with a microsyringe into the injector. The syringe needle penetrates a silicon membrane (*septum*) and the tip of the

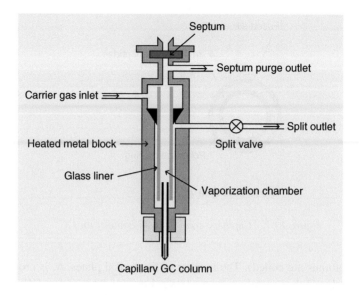

Figure 13.8   Split/splitless injector for capillary GC

syringe needle is inserted into a glass tube, the *glass liner*, inside the injector. The sample solution is injected and evaporates in the glass liner. The inner volume of the glass liner is large enough to allow expansion of the sample into vapour. The heated carrier gas enters the injector, from where there are three outlets. The capillary GC column is attached to one outlet. Another outlet termed the *septum purge outlet* is located just under the septum to prevent any impurity from the septum reaching the column. The third outlet is termed the *split outlet* and this outlet is used during split injection to divert part of the sample volume to waste. Gas flow rates through the three outlets are controlled by software to determine the amount of sample introduced to the column.

The principle of split injection is that only a small controlled proportion of the evaporated sample is transferred to the column, while the rest is diverted to waste through the split outlet. The flow rate ratio between gas to the split outlet and gas to the capillary GC column is termed the *split ratio*. With carrier gas flow of 1 mL/min through the column and 50 mL/min through the split outlet, the split ratio is 50 : 1, and only about 2% of the sample enters the column. In this case, 98% is vented through the split purge outlet to waste. Split ratios are normally in the range from 10 : 1 to 100 : 1. Since only a small proportion of the sample volume is used for separation and detection, split injection is best suited for sample solutions containing relatively high concentrations of analytes and for dirty samples. In the latter cases, most of the matrix is diverted to waste and potential column contamination is reduced.

In a splitless injection, the total sample volume is introduced into the column. The sample evaporates in the injector and the sample constituents are transferred as vapour with the carrier gas to the GC column inlet, where sample constituents are condensed and focused. In the splitless mode, the split outlet is therefore closed during injection. If 1 μL sample expands to 1 mL of gas and the carrier gas velocity through the column is 1 mL/min, it takes approximately 60 seconds to transfer the entire sample to the column. To prevent

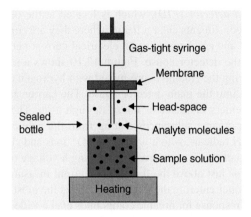

Figure 13.9   *Static headspace injection of residual solvent*

spreading of the analytes throughout the capillary column during 60 seconds of sample transfer, the analytes must be trapped at the column inlet. This is accomplished by *solvent trapping,* which is obtained when the solvent used for dissolution of the sample constituents condenses in the stationary phase in the first part of the column. The sample constituents are normally less volatile than the solvent and are trapped in a film of condensed solvent. For this solvent effect to occur, the temperature of the column has to be 20–50 °C below the boiling point of the solvent. After the sample constituents have been trapped in a narrow zone of solvent in the column inlet, the oven temperature is gradually increased by temperature programming, and the solvent as well as the sample constituents will gradually elute and be separated based on volatility and interactions with the stationary phase. With this technique low limits of detection can be obtained because the full injection volume is used for detection.

The determination of residual solvents in pharmaceutical ingredients and preparations is performed using *static headspace injection.* The sample is pulverized and suspended in water in a headspace vial (sealed bottle), as shown in Figure 13.9. The headspace vial is heated and residual solvent evaporates to the air volume above the sample solution, which is termed the *headspace.* The headspace vial is closed with a membrane that prevents residual solvent from leaking to the external atmosphere. A volume of 0.5–2.0 mL of the headspace is collected with a gas-tight syringe and the sample is injected into the GC.

13.9   Detectors

Several different detectors are used in GC. In this context, the principles of flame ionization detection, nitrogen–phosphorus detection, and electron capture detection are discussed, since these are important in pharmaceutical analysis. In addition, mass spectrometers (MS) are widely used as detectors in GC, which, when GC and MS are coupled, is termed *GC-MS.* GC-MS can be used for identification and quantitation of volatile and semi-volatile compounds. GC-MS is discussed in detail in Chapter 15.

In the *flame ionization detector* (*FID*), which is located at the outlet of the capillary GC column, the sample constituents enter a flame, where they are combusted. During this process, ions are formed and are detected as an electrical current between two electrodes (cathode and anode) in the detector house. Figure 13.10 shows a schematic view of an FID. The carrier gas leaving the column is pre-mixed with hydrogen at the entrance of the detector, while air is led into the main detector house. The carrier gas/hydrogen mixture containing the analytes enters the detector house through a jet. The flame burns on the top of the jet, and when organic substances eluting from the column burn in the flame, carbon atoms produce CH radicals, which produce CHO^+ ions and electrons in the flame. The amount of charged particles is measured by applying a voltage of 300 V between the flame tip and the *collector* just above the flame. The current measured is proportional to the amount of organic matter entering the flame. The FID is the most commonly used GC detector. It gives a linear response for organic compounds over a wide concentration range. The lower limit of detection is about 10^{-9} g. The FID is a *universal detector* and responds to all organic compounds.

The *nitrogen–phosphorus detector* (*NPD*) measures organic substances containing nitrogen or phosphorus. The detector is also named the *alkali flame ionization detector* (*AFID*) or the *thermionic detector* (*TID*). The detector is similar to the FID, but with a crystal of alkali metal salt placed just above the flame. Usually a *rubidium salt* is used. Organic compounds containing nitrogen or phosphorus react with alkali metals during combustion under the formation of anions such as cyanide and phosphorus-containing anions as well as electrons. The response of nitrogen-containing organic compounds can be up to 10^6 times higher than the response of organic substances without nitrogen. The

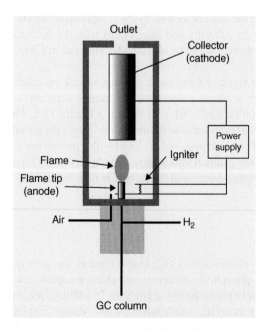

Figure 13.10   Schematic view of a flame ionization detector (FID)

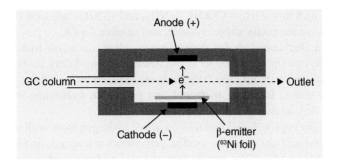

Figure 13.11 *Schematic view of an electron capture detector*

lower limit of detection for nitrogen and phosphorus-containing organic substances is about 10^{-10} g.

The *electron capture detector* (*ECD*) shown in Figure 13.11 is a selective detector suited for the determination of compounds with a high affinity for electrons. The detector contains a radioactive foil of ^{63}Ni, emitting β-electrons, which by collision with the carrier gas molecules generate a plasma of electrons. Thus, a constant current in the detector provides a background signal that is changed when electron capturing compounds enter the detector. Compounds containing halogens or nitro groups have a high affinity for electrons, and the lower limit of detection is at the fg level. The detector is among others used for substances derivatized with polyhalogenated reagents. A number of derivatization reagents containing fluorine are available and derivatized substances can be detected at very low concentrations with the ECD.

13.10 Derivatization

Many drug substances contain polar functional groups that lower the volatility of the substances due to hydrogen bonding or ionic intermolecular forces. Furthermore, the polar substances often interact with surfaces in the GC system by adsorption, resulting in serious peak tailing and poor reproducibility. This can be suppressed or eliminated by *derivatization*. The purpose of derivatization is to create more volatile and thermally stable derivatives of polar substances, thereby improving their GC behaviour and detectability. The derivatization blocks polar intra- and intermolecular forces, thus enhancing volatility. By *silylation* active H-atoms are replaced by *trimethylsilyl* ligands, as shown in Figure 13.12.

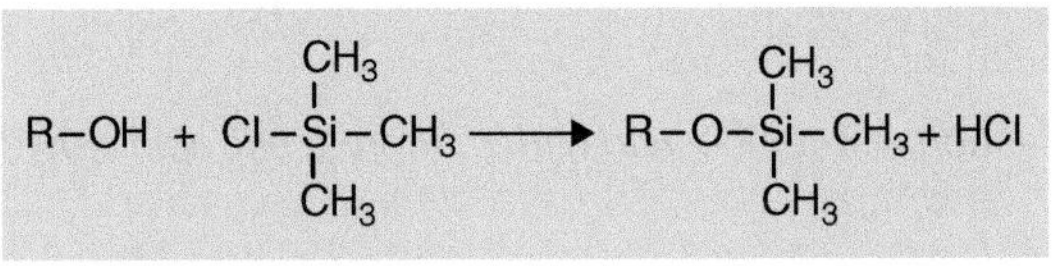

$$R-OH + Cl-\underset{\underset{CH_3}{|}}{\overset{\overset{CH_3}{|}}{Si}}-CH_3 \longrightarrow R-O-\underset{\underset{CH_3}{|}}{\overset{\overset{CH_3}{|}}{Si}}-CH_3 + HCl$$

Figure 13.12 *Silylation of a hydroxyl group*

Functionalities such as —OH, —COOH, —NH$_2$, and —NH— are well suited for sily-lation. Derivatives are thermally stable, volatile, and suitable for GC.

Alkylation is another approach for derivatization where an active hydrogen atom is replaced by an alkyl group. An example is the determination of fatty acids in oils. Fatty acids exist as triglycerides in oils. The fatty acids are released by hydrolysis and separated by GC as methyl esters. By using methanol and potassium hydroxide, the reaction is performed in one step, as shown in Figure 13.13.

Acylation is a third approach and replaces an active hydrogen atom with an acyl group. Both anhydrides and acid chlorides are used as derivatization reagents, and functionalities such as —OH, —NH$_2$, and —NH— can be derivatized. The derivatization reagents usually contain fluorine atoms because fluorine atoms provide stable and volatile derivatives suitable for GC. Commonly used reagents are *trifluoroacetic acid anhydride*, or anhydrides of *pentafluoropropionic acid* or *heptafluorobutyric acid*. Amino acids are non-volatile due to the strong intermolecular ionic interactions, but can be analysed by GC after acylation and methylation, as shown in Figure 13.14.

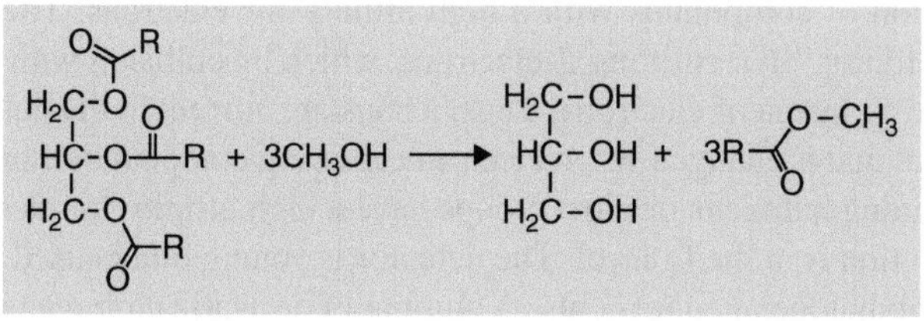

Figure 13.13   *Methylation of fatty acids in oils*

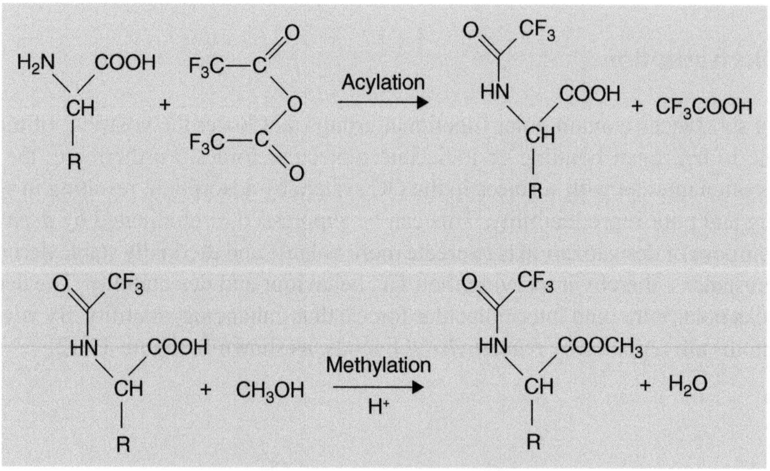

Figure 13.14   *Derivatization of α-amino acid with trifluoroacetic acid anhydride and methanol*

14

Electrophoretic Methods

14.1 Introduction

Electrophoresis is a liquid phase separation technique based on migration of charged species in an electrical field. The driving force is a voltage gradient applied between two electrodes. Charged analytes migrate towards the electrode with the opposite charge and sample constituents are separated based on their charge-to-size ratio. In the simplest case, charged sample constituents migrate in a buffer solution. This separation principle is termed *zone electrophoresis*. Electrophoresis is mainly used for testing of biopharmaceuticals (peptides and proteins) but can also be used as an alternative to liquid chromatography (LC) for small-molecule drug substances. Electrophoresis is an official method of analysis in the European Pharmacopoeia (Ph. Eur.) and the United States Pharmacopoeia (USP).

Introduction to Pharmaceutical Analytical Chemistry, Second Edition.
Stig Pedersen-Bjergaard, Bente Gammelgaard and Trine Grønhaug Halvorsen.
© 2019 John Wiley & Sons Ltd. Published 2019 by John Wiley & Sons Ltd.

When the electrophoretic separation is performed in a gel, the procedure is termed *gel electrophoresis* and this procedure has been further developed for protein separation into different techniques, namely *sodium dodecyl sulfate polyacrylamide gel electrophoresis (SDS-PAGE)*, *isoelectric focusing (IEF)*, and *Western blotting*. Alternatively, the electrophoretic separation is performed in a capillary; this procedure is termed *capillary electrophoresis (CE)*. CE is an effective tool for quantitation.

14.2 Principle and Theory

Charged compounds in solution migrate when an *electrical field* is imposed on the solution, as shown in Figure 14.1. The term *migration* is used for molecular movement in electrical fields. Positively charged compounds (*cations*) migrate towards the negative electrode (*cathode*) as a result of electrostatic attraction, while negatively charged compounds (*anions*) migrate towards the positive electrode (*anode*).

The velocity (*v*) of a charged compound is given by

$$v = \mu_e E \tag{14.1}$$

where μ_e is the *electrophoretic mobility* of the compound and E is the applied electrical field. The electrical field is expressed as volts per centimeter (V/cm). Thus, applying a potential difference of 10 000 V between two electrodes, placed at a distance of 50 cm from each other, will result in an electrical field of 200 V/cm. When charged molecules migrate in solution they are exposed to frictional forces. When the positive force of the electrical field is balanced by and is equal to the negative frictional force, the electrophoretic mobility μ_e can be expressed as

$$\mu_e = \frac{z}{6\pi r \eta} \tag{14.2}$$

where z is the *charge* of the molecules, r is the *molecular radius*, and η is the *viscosity* of the solution. From the two equations, it is given that the velocity of a given compound increases with increasing charge, decreasing molecular radius (size), decreasing solution viscosity, and increasing electrical field. Thus, small molecules travel faster than large molecules

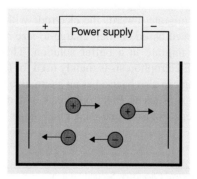

Figure 14.1 *Charged compounds in solution migrate under application of an electrical field*

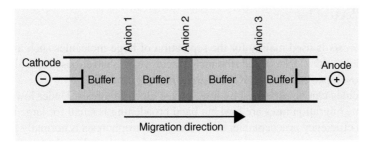

Figure 14.2   *Schematic view of zone electrophoresis in a capillary*

and multiple charged molecules travel faster than singly charged molecules. As different sample constituents may have different velocities based on their charge and size, they can be separated according to their *charge-to-molecular radius ratio*.

For small molecules with a molecular mass less than 10^3 Da, the molecular radius is considered proportional to the molecular mass, and separation is principally related to molecular mass and charge. However, for larger molecules like peptides and proteins, this is not the case because the three-dimensional structure plays a major role. In the following discussion, the term *size* is used for simplicity rather than molecular radius, and charged molecules therefore migrate with a velocity determined by their *charge-to-size ratio*.

A schematic overview of the electrophoresis process in solution is given in Figure 14.2. The solution where the electrophoretic separation takes place is usually a buffer and this is termed the *separation buffer* or *electrophoresis buffer*. Separation of charged sample constituents in a buffer solution is termed *zone electrophoresis*. Zone electrophoresis can be performed as *high voltage electrophoresis*, with an electrical field in the range of 50–200 V/cm, or as *low voltage electrophoresis*, with an electrical field of 2–10 V/cm. High voltage electrophoresis is used for small molecules. Band broadening due to diffusion is a challenge with small molecules in free solution, but diffusion is reduced when migration is fast. Therefore, high electrical fields are used for zone electrophoresis of small molecules and such separations are done with CE. This is termed *capillary zone electrophoresis* (CZE) and is shown in Figure 14.2.

A prerequisite for electrophoresis is that the analytes are charged. Thus, pH is important and therefore electrophoresis is performed in buffers.

During electrophoresis, current will pass the separation buffer between the anode and the cathode, and this inherently generates heat. The velocity of molecules in electrophoresis is dependent on temperature, and therefore the heat generated may affect the reproducibility of the separations. Excessive heating of the electrophoresis system should therefore be avoided, which is done by using separation buffers with low ionic strength and conductivity.

In *gel electrophoresis*, the separation takes place in a gel containing the separation buffer. After sample loading, the electrical field is applied across the gel. The sample constituents migrate along the gel and different sample constituents migrate with different velocity, resulting in individual bands across the gel. In *CE*, the sample is injected into the inlet of a narrow capillary filled with separation buffer. The electrical field is applied across the capillary and the sample constituents migrate with different velocities through the capillary, where they are separated and detected close to the outlet of the capillary.

14.3   Gel Electrophoresis

Gel electrophoresis is used mainly for the separation of large molecules such as proteins, deoxyribonucleic acids (DNAs), and ribonucleic acids (RNAs). Gel-based techniques have been further developed into *SDS-PAGE, Western blotting*, and *IEF*.

Large molecules can be separated by low voltage electrophoresis. Under low electrical field conditions, migration rates are low, but band broadening is small for large molecules and separation efficiency is acceptable. Low voltage electrophoresis is normally performed as *slab-gel electrophoresis*, where the separation is performed on thin plates coated with a gel. Although zone electrophoresis occurs in the gel, separation of molecules according to their size will take place due to the *sieving effect* of gels. This separation principle is termed *gel electrophoresis* and is illustrated in Figure 14.3.

In gel electrophoresis, small molecules can move almost freely through the pores of the gel and are mainly separated based on charge-to-size ratios. Larger molecules, however, will be more restricted and will mainly be separated in order of increasing size. The sieving effect is dependent on the pore size of the gel and can provide separation of molecules with similar charge-to-size ratios. This is highly advantageous for large biomolecules such as peptides and proteins, where the differences in charge-to-size ratios are small, but where

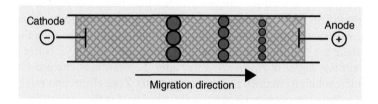

Figure 14.3   *Principle of gel electrophoresis*

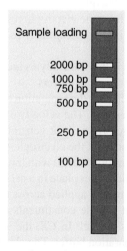

Figure 14.4   *Slab gel electrophoresis of DNA fragments containing 100, 250, 500, 750, 1000, and 2000 base pairs*

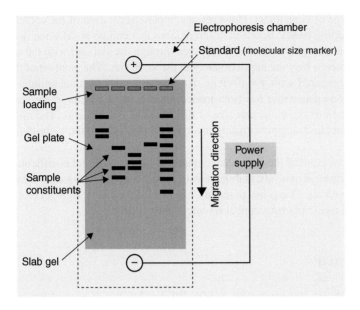

Figure 14.5   *Setup for slab-gel electrophoresis with samples and standard*

the differences in size can be large. An example of a slab-gel electrophoresis separation is shown in Figure 14.4.

The setup for slab-gel electrophoresis is schematically illustrated in Figure 14.5. The setup comprises an electrophoresis chamber (containing the slab-gel or gel plate), electrodes, and a power supply. The slab-gel used as the separation medium is normally a cross-linked polymer with a composition and a porosity that is scaled to the size and nature of the molecules to be separated. For the separation of proteins or small nucleic acids, the slab-gel is typically composed of acrylamide and a cross-linker, producing a network of *polyacrylamide*. For the separation of larger nucleic acids (larger than a few hundred bases), the preferred matrix is purified *agarose*. Agarose is composed of long unbranched chains of neutral carbohydrate without cross-links, providing a gel with large pores that allows for the separation of macromolecules. The gel is placed in the *electrophoresis chamber* and the *electrodes* are connected to the *power supply*.

Samples are loaded into *wells* in the slab-gel, and the electrical field is applied across the gel. Upon application of the electrical field, the sample constituents migrate in the slab-gel. The constituents migrate towards the anode if they are negatively charged or towards the cathode if they are positively charged. Different molecules migrate with different velocities through the slab-gel depending on size. The separated molecules in the slab-gel are stained after separation to make them visible (*visualization*). Silver staining by silver nitrate and Coomassie staining with the dye Coomassie blue are frequently used for staining of proteins. If the target molecules are fluorescent, a photograph can be taken of the slab-gel under a UV lamp.

Several samples can be loaded on the same slab-gel into adjacent *wells* and the different samples are developed in individual *lanes*, as illustrated in Figure 14.5. Each lane shows the separation of the constituents from the original sample as distinct bands, in principle one band per type of constituent. However, incomplete separation of the constituents may result in overlapping bands. Bands in different lanes that end up at the same distance from the starting point contain molecules that passed through the slab-gel with the same speed, which usually means they are approximately of the same size. The position of the different bands can be compared with a *molecular size marker* (standard) containing a mixture of molecules of known size that has been loaded into one of the wells in the slab-gel. This enables estimation of the molecular size of the separated components. The migration distance of a certain band is approximately inversely proportional to the logarithm of the size of the molecule.

Reliable quantitation of the individual compound bands in the gel is difficult, and in the area of pharmaceutical analysis, slab-gel electrophoresis is mainly used for quality control of biopharmaceuticals. This is discussed in Chapter 21. In addition, slab-gel electrophoresis is commonly in use in pharmaceutical research laboratories.

14.4 SDS-PAGE

A variant of slab-gel electrophoresis is *SDS-PAGE*. SDS-PAGE is used for separation of proteins. The principle is illustrated in Figure 14.6.

SDS-PAGE is performed in a slab gel after denaturation (unfolding) of the proteins in the sample by addition of 2-mercaptoethanol, which breaks the disulfide bonds. The negatively charged surfactant, sodium dodecyl sulfate (SDS), is then added to the protein sample. SDS molecules bind to the proteins by hydrophobic interactions, whereby the proteins are

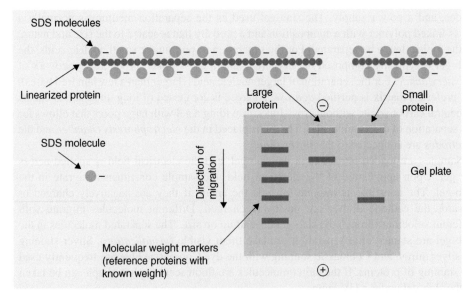

Figure 14.6 *Principle of sodium dodecyl sulfate polyacrylamide gel electrophoresis (SDS-PAGE)*

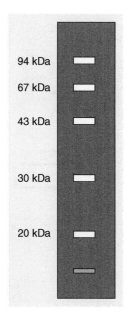

Figure 14.7   *SDS-PAGE separation of proteins of 20, 30, 43, 67, and 93 kDa*

linearized and covered with a negative charge. In most proteins, the binding of SDS to the polypeptide chain imparts an even distribution of charge per unit mass, corresponding to a ratio of SDS-to-protein of approximately 1.4 : 1 (*w/w*). Therefore, different proteins will have approximately the same charge-to-size ratio. When the protein–SDS complexes are loaded into the gel and the electrical field is applied, the proteins will migrate in the gel and will be separated based on size due to the sieving effect of the gel. Since the shape of denatured proteins is like long rods instead of a complex tertiary shape, the rate at which the resulting SDS-coated proteins migrate in the gel is dependent only on its size and not on its charge or shape. An example of an SDS-PAGE separation of proteins is shown in Figure 14.7.

14.5   Western Blotting

Another variant of protein separation using gel electrophoresis is *Western blotting*. The principle is illustrated in Figure 14.8. Western blotting is used for separation and detection of proteins (*antigens*) by interaction with *antibodies* for specific detection of the antigens. In the first step, the proteins are separated using gel electrophoresis, most often by SDS-PAGE. For very complex samples, it is also possible to use *two-dimensional gel electrophoresis* in which the proteins from a single sample are spread out in two dimensions according to the isoelectric point (pI) and size. In the first dimension, proteins are separated by IEF (discussed below), and in the second dimension, the individual proteins are separated by gel

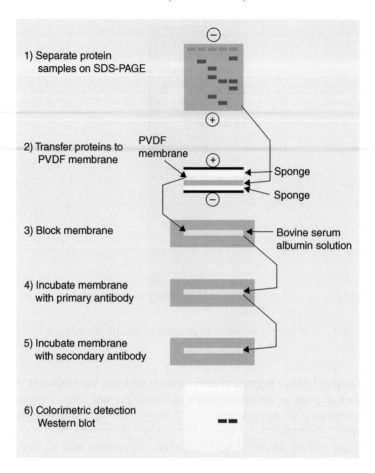

Figure 14.8   Principle of Western blotting

electrophoresis. The separated proteins are transferred or *blotted* on to a nitrocellulose or polyvinylidene difluoride (PVDF) membrane. During blotting, the proteins are transferred from the electrophoresis gel and on to the membrane (for detection), while maintaining their organisation from the gel. Blotting is facilitated in an electrical field. In the third step, the membrane is *blocked* to prevent any non-specific binding of the antibodies used for detection to the surface of the membrane. Blocking of non-specific binding is achieved by placing the membrane in a dilute solution of protein (such as bovine serum albumin) or a solution of surfactant. The protein in the dilute solution (or the surfactant) attaches to the membrane in all places where the target proteins have not attached. When the antibody is added for the final detection (fourth step), only the binding sites of the specific target protein are available. During detection, the membrane is 'probed' for the protein of interest with a modified antibody, which is linked to a reporter enzyme; when exposed to an appropriate substrate, this enzyme drives a colorimetric reaction and produces a colour for colorimetric detection.

14.6 Isoelectric Focusing

IEF is used for separation of proteins according to their *isoelectric point* (pI). The pI of a protein is the pH value at which the protein carries no net electrical charge (the sum of positive ($-NH_3^+$) and negative ($-COO^-$) charges is zero). IEF takes advantage of the fact that the net charge of a protein changes with the pH of its surroundings. A protein that is in a pH region below its pI is positively charged and migrates towards the cathode in an electrical field. As it migrates through a separation buffer of increasing pH (in a gel), the net charge of the protein decreases until the protein reaches a pH region that corresponds to the pI of the protein. At this point the protein has no net charge and the protein no longer migrates. As a result, the proteins become focused into sharp bands, with each protein positioned at a point in the pH gradient in the gel corresponding to its pI. This is illustrated in Figure 14.9.

The target molecules to be focused and separated are loaded into a gel with a pH gradient. The pH gradient in the gel is established prior to adding the sample, by subjecting a solution of small molecules such as polyampholytes with varying pI values to electrophoresis in the gel. Gels with large pores are used in IEF to eliminate any sieving effects, in order to avoid differences in migration rates for proteins due to different sizes. IEF can resolve proteins that differ in pI value by as little as 0.01. IEF is capable of high resolution and is the first step in two-dimensional gel electrophoresis, as discussed above.

14.7 Capillary Electrophoresis

14.7.1 Principle and Instrumentation

CE is used for separation of both small and large molecules. Typically, the separation principle is zone electrophoresis and separation is based on charge-to-size. A schematic

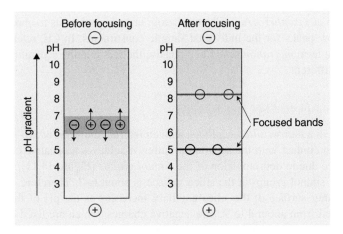

Figure 14.9 *Principle of isoelectric focusing*

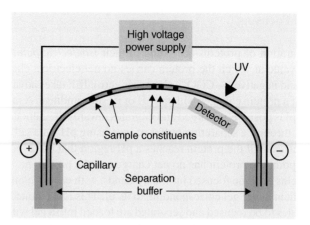

Figure 14.10   Schematic overview of capillary electrophoresis

overview of an apparatus for CE is shown in Figure 14.10. CE instruments are equipped with auto-samplers and are software controlled; large sequences of samples can be analysed automatically.

The separation takes place in a capillary, typically made of fused silica with an internal diameter of 50 or 75 μm. The capillary is filled with *separation buffer*, is typically 30–70 cm long, and is placed between two vials containing separation buffer. The *anode* (positive charge) is located in the source vial and the cathode (negative charge) is located in the destination vial. The anode and the cathode are connected to a *high voltage power supply* (10–30 kV).

In the basic setup for CE, samples are introduced at the anodic end of the capillary. When the voltage is applied across the electrodes, the positive analytes migrate towards the cathode electrode and are detected near the outlet of the capillary. UV detection is common. The UV absorption of the sample constituents is measured and plotted as a function of time. This plot is termed an *electropherogram*. Electropherograms are analogous to chromatograms, as they both show peaks for the individual sample constituents. In CE, retention time is replaced with the term *migration time* to emphasise the fact that the separation principles are completely different.

14.7.2   Electro-osmotic Flow and Mobility

CE is performed in a narrow *silica capillary* with an internal diameter of 50–75 μm. When the capillary is in contact with the separation buffer, it achieves a negative charge on the inner wall surface due to deprotonation of the *silanol groups* (Figure 14.11). The average pK_a value of the silanol groups at the silica surface is about 6–7. Therefore, the negative charge on the inner surface of the silica capillary increases as the pH of the separation buffer is increased from about 4 to 9. The negative charges, which are fixed on the capillary surface, are balanced by mobile cations (typically Na^+) from the separation buffer to maintain charge neutrality. In this way a *double layer* is built at the surface, creating a poten-

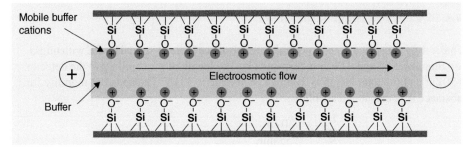

Figure 14.11   *Inner surface of capillary with electrical double layer and electro-osmotic flow (EOF)*

tial difference reaching from the capillary wall and into the buffer. The potential is known as the *zeta potential*. The buffer cations are hydrated, which means that they are bound to the surrounding water molecules, which in turn are bound to the other water molecules in the separation buffer via hydrogen bonds. When the electrical field is applied across the capillary, the hydrated cations migrate towards the cathode and in this way they drag the bulk separation buffer in the capillary in the same direction. This is termed *electro-osmosis*. The flow of separation buffer generated by electro-osmosis is termed *electro-osmotic flow* (*EOF*), and is strongly dependent on pH in the separation buffer. At pH > 7, a relatively high EOF is obtained (in the range of 10 cm/min), but even at low pH, such as in 10 mM phosphoric acid, an EOF is observed (0.5–1.0 cm/min). The EOF decreases with increasing buffer strength as the increase in buffer ions suppresses the zeta potential and thus the surplus of cations close to the surface. The reduced surplus of cations at the surface results in a reduced drag on the separation buffer. With a separation buffer at pH 7, the EOF is significant and even anions move towards the detector, although they have a certain electrophoretic migration in the opposite direction (towards the anode).

The voltage applied across the capillary initiates the *electrophoretic migration* of the charged sample constituents as well as the *EOF*. The observed mobility (*apparent mobility*, μ_{app}) for a given sample constituent is a function of the electrophoretic mobility μ_e and the *electro-osmotic mobility* μ_{eo}:

$$\mu_{app} = \mu_e + \mu_{eo} \tag{14.3}$$

Charged sample constituents are separated due to their different electrophoretic mobility. Neutral constituents migrate with EOF and are not separated. The apparent mobility can also be written as

$$\mu_{app} = \frac{\dfrac{L_d}{t_m}}{\dfrac{V}{L_t}} \tag{14.4}$$

where L_d is the length of the capillary from the injection to the detector, L_t is the total length of the capillary, V is the voltage, and t_m is the migration time of an ion for distance L_d. An example of calculation of the electrophoretic, electro-osmotic, and apparent mobility based on an electropherogram is given in Box 14.1.

Box 14.1 Calculation of mobility

A basic drug substance A and a neutral drug substance B are separated by CE with migration times of 2.55 and 4.06 min, respectively. The length of the capillary to the detector is 30 cm and the total length is 37 cm. The voltage is 30 kV. The apparent mobility for substance A is calculated as follows:

$$\mu_{app} = \frac{\dfrac{L_d}{t_m}}{\dfrac{V}{L_t}} = \frac{\dfrac{30\ cm}{2.55\ min}}{\dfrac{30\,000\ V}{37\ cm}} = 14.5 \times 10^{-3} cm^2/min\ V$$

Since substance B is neutral, this follows the velocity of the electro-osmotic flow through the capillary, and the electro-osmotic mobility is equal to the apparent mobility of substance B:

$$\mu_{eo} = \frac{\dfrac{L_d}{t_m}}{\dfrac{V}{L_t}} = \frac{\dfrac{30\ cm}{4.06\ min}}{\dfrac{30\,000\ V}{37\ cm}} = 9.1 \times 10^{-3}\ cm^2/min\ V$$

The electrophoretic mobility of substance A is therefore

$$\mu_e = \mu_{app} - \mu_{eo} = 14.5 \times 10^{-3}\ cm^2/min\ V - 9.1 \times 10^{-3}\ cm^2/min\ V = 5.4 \times 10^{-3}\ cm^2/min\ V$$

14.7.3 Dispersion

The bulk flow (EOF) in the capillary is generated at the same time at any part of the inner surface of the capillary and therefore results in a flat flow profile. Thus, the velocity of the separation buffer is the same at any point in the capillary. This should be seen in comparison to LC where a parabolic flow is generated due to the pumping of the mobile phase from one end of the system. Radial dispersion takes place across the *parabolic flow* profile and this type of dispersion is eliminated in CE (Figure 14.12). Stationary phase related peak broadening effects are absent in CE because separations are performed in open capillaries (no stationary phase). Therefore, a high number of theoretical plates can be generated in a fairly short capillary.

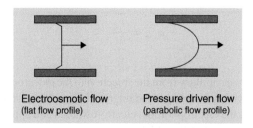

Electroosmotic flow
(flat flow profile)

Pressure driven flow
(parabolic flow profile)

Figure 14.12 Flow profiles in CE (electro-osmotic flow) and LC (pressure driven flow)

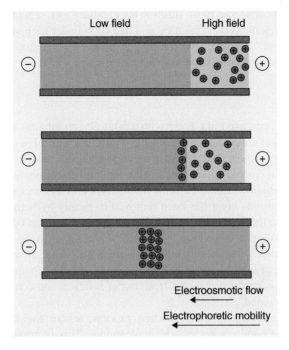

Figure 14.13   Stacking of cations at low pH

The two primary causes of dispersion in CE are *longitudinal diffusion* and *electrodispersion*. Diffusion of the analyte molecules is only of importance in the longitudinal direction of the capillary and the diffusion is low because narrow capillaries are used and because the separation medium is a liquid. Electrodispersion is caused by a difference in mobility between the analyte ions and the ions of the separation buffer. A lower mobility of analyte ions compared to buffer ions results in tailing peaks and a higher mobility of analyte ions results in fronting peaks.

Tuning of the sample conductivity relative to the conductivity of the separation buffer can lead to a local concentration of the introduced sample – a concentration step that can improve the signal-to-noise ratio by a factor between 10 and 100. This principle is termed *stacking* (Figure 14.13). Stacking can be obtained when the ionic strength in the sample is one-tenth or less than that of the separation buffer. The electric field is inversely proportional to the conductivity and a low concentration in the sample results in a high electrical field. The electrophoretic mobility of the sample ions is high in the sample zone. When the ions in the sample zone reach the boundary to the surrounding separation buffer, they experience the lower field in the separation buffer and the migration rate decreases. In this way the sample ions are concentrated at the boundary between the sample zone and the separation buffer. Subsequently the concentrated sample zone is separated in the separation buffer.

Considering the reverse situation where a high concentration of ions is present in the sample, the mobility of ions is low until they reach the boundary to the surrounding separation buffer where the electric field is higher. The ions then speed up, resulting in band broadening. Thus high salt concentrations in the sample ruin CE separations.

CE separation comprises three steps: flushing the capillary with separation buffer, injection of sample, and simultaneous separation and detection under high voltage conditions. Close to the destination vial end of the capillary, the detector continuously measures the passage of separated sample constituents during electrophoresis.

14.7.4 Capillaries

The fused silica capillaries used in CE are with a 50 or 75 μm internal diameter, are typically 30–70 cm long, and are coated on the outside with a thin layer of *polyimide*. This layer makes the capillaries extremely flexible and prevents breakage (similar to gas chromatography (GC) columns). To be able to perform on-capillary UV detection the polyimide layer, which absorbs UV radiation, has to be removed to form a *detection window* of 2–4 mm length. When the polyimide layer has been removed (typically by burning) the capillary becomes very fragile at the detection window. Before the capillary can be used for CE, it is flushed with a dilute solution of sodium hydroxide, water, and with the separation buffer. This procedure is repeated prior to every new sequence of samples. Between each CE injection within a sequence of samples, the capillary is flushed with the separation buffer. In this way, every separation is performed with a fresh buffer solution, and contamination due to previous samples is avoided.

Electrophoresis is a temperature-dependent process, where the migration rate of molecules through the capillary varies with temperature. This is important to consider, because the current generated during application of voltage generates heat. This is termed *Joule heating* and if the temperature becomes too high in the separation buffer inside the capillary, air bubbles may be formed. This results in loss of conductivity and the electrophoresis is interrupted. Temperature control around the capillary is therefore essential in order to obtain reliable and repeatable separations. For this reason, CE instruments are equipped with a liquid or air-based *cooling system* to carefully control the temperature of the capillary.

14.7.5 Sample Introduction

The sample is normally introduced into the anodic end of the capillary by *hydrodynamic injection*. This injection principle is illustrated in Figure 14.14. The capillary end is placed in the sample vial and pressure is applied to the vial for a short time (typically 1–10 seconds). This forces a small plug of sample solution into the capillary (injection). After injection the capillary end is moved from the sample vial and back to the source vial, and voltage is applied to facilitate the electrophoretic separation. The sample volume injected by hydrodynamic injection is controlled by the magnitude and duration of the pressure. Increasing either the injection time or the injection pressure increases the volume of sample injected. To maintain a high separation efficiency, the injected plug of sample should

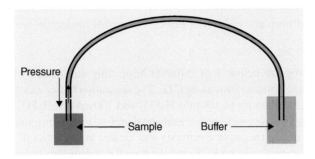

Figure 14.14 *Hydrodynamic injection*

not exceed 1% of the capillary volume. With a capillary of 50 μm inner diameter and 50 cm length, the total volume of the capillary is 1 μL. In this capillary, the volume of injected sample should be less than 10 nL. In other words, the sample volumes in CE are very small.

14.7.6 Detection

Detection is most often performed *on-capillary* across a detection window close to the cathodic end of the capillary. In the detection window, the capillary is transparent and radiation from an UV lamp is transmitted across the capillary. The UV detectors used for CE are often *diode array detectors* and full UV spectra can be recorded during electrophoresis. Using UV spectrophotometry for on-capillary detection provides less sensitivity as compared to LC, because the effective light path length is limited to the small internal diameter of the capillary (50 or 75 μm). CE can also be coupled to mass spectrometry (*CE-MS*). CE-MS enables detection at lower concentration levels and also provides important structural information. CE-MS is currently not a common technique, but the use for characterization of biopharmaceuticals is increasing.

14.7.7 Applications

In the area of pharmaceutical analysis, CE is mainly used for quality control of biopharmaceuticals. This is discussed in Chapter 21. However, CE can also be applied as an alternative to LC for separation of small molecule drug substances, and one such application is shown in Box 14.2, where small molecule substances were separated based on their charge-to-size ratio by CZE, which is zone electrophoresis performed as CE. In addition, CE is commonly in use in pharmaceutical research laboratories. CE can be used for very effective separations of chiral substances (not discussed in this textbook).

Box 14.2   Capillary zone electrophoresis of small molecule basic drug substances

In the electropherogram below, four different basic drug substances were separated based on their charge-to-size ratios using CZE. The separation buffer was 100 mM phosphate buffer pH 2.38 (mixture of 100 mM H_3PO_4 and 100 mM NaH_2PO_4). At this pH, all the basic drug substances were protonated and therefore they all migrated in the electrical field. The cathode (negative electrode) was located at the outlet of the capillary. At pH 2.38, there were almost no EOF, and the transfer of drug substances through the capillary was mainly due to electrophoretic migration. From a theoretical point of view, the drug substances are separated according to the charge-to-size ratio. From a practical point of view, three of the drug substances were singly charged at pH 2.38 and migrated in order of increasing molecular weight (increasing molecular weight normally results in increasing size). Pheniramine was protonated at two sites (exact calculation of net charge not feasible), and due to a larger charge-to-size ratio this compound migrated prior to the three drugs with a single charge. The capillary used was of 50 μm internal diameter and the length was 50 cm. Injection was by hydrodynamic injection, detection was by UV spectrophotometry, and the capillary electrophoresis was conducted for 15 minutes at 30 kV to complete the separation of the four different basic drug substances.

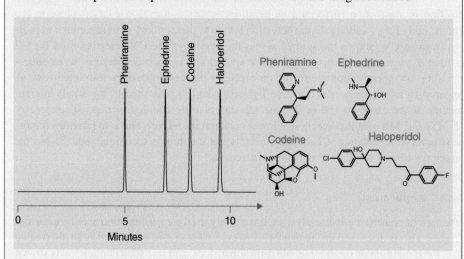

Drug substance	Molecular mass (g/mol)	Charge at pH 2
Pheniramine	240.3	$\approx +2$
Ephedrine	165.2	+1
Codeine	299.4	+1
Haloperidol	375.9	+1

15

Mass Spectrometry

15.1 Introduction

Mass spectrometry (MS) is an official method of analysis in the European Pharmacopoeia (Ph. Eur.) and the United States Pharmacopoeia (USP). It is the most powerful detector for chromatography as it provides qualitative as well as quantitative information. MS coupled with liquid chromatography (LC) is abbreviated to *LC-MS*, and MS coupled with gas

Introduction to Pharmaceutical Analytical Chemistry, Second Edition.
Stig Pedersen-Bjergaard, Bente Gammelgaard and Trine Grønhaug Halvorsen.
© 2019 John Wiley & Sons Ltd. Published 2019 by John Wiley & Sons Ltd.

chromatography (GC) is abbreviated to *GC-MS*. A major advantage is the ability to distinguish different substances with similar retention times.

LC-MS is a very important technique for therapeutic drug monitoring (TDM), illicit drug analysis, doping analysis, and for drug analysis in pharmaceutical research. GC-MS is used in the same application areas when the analytes are volatile or semi-volatile. MS is used in all phases of drug development. MS is indispensable for identification and quantitation of metabolites and degradation products in biological samples and analysis of peptides and proteins.

MS is a very sensitive analytical technique and can detect compounds down to the pico- or femtogram level. This is especially important for LC-MS and GC-MS analysis of blood and urine samples, where drug substances and metabolites can be present in very low concentrations and where the amount of sample can be limited.

MS is a technique for measuring masses of atoms, molecules, or fragments of molecules and is performed in a *mass spectrometer*. A mass spectrometer is built from three components, an *ion source*, a *mass analyser*, and a *detector*. The analytes are ionized in the ion source and are separated according to their *mass-to-charge ratio* (*m/z*) in the mass analyser. The abundance of each ion with a different *m/z* value is finally measured by the *detector*. The result is displayed as a *mass spectrum* showing the number of ions detected at each value of the *m/z* ratio. Schematic views of LC-MS and GC-MS systems are presented in Figure 15.1.

In LC-MS, the chromatographic system is linked to the mass spectrometer by an *interface*. In this interface, the sample constituents in the mobile phase are *ionized* and *vapourized* before they enter the mass spectrometer (see Figure 15.1). The interface

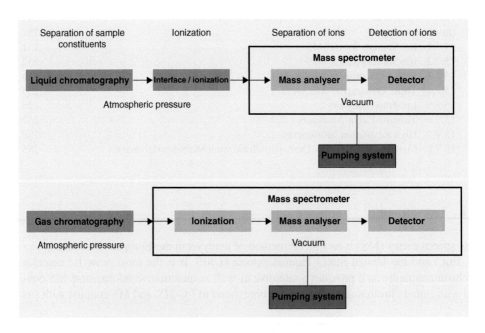

Figure 15.1   Schematic views of LC-MS and GC-MS

is at atmospheric pressure and also serves to remove most of the mobile phase. Thus, the bulk of mobile phase from the LC system is prevented from entering the mass spectrometer.

GC-MS is fundamentally different from LC-MS. In GC-MS, the sample constituents enter the mass spectrometer in the carrier gas directly inside the vacuum region of the mass spectrometer and an interface is not necessary.

15.2 Basic Theory of Mass Spectrometry

Mass spectra display all isotopes of the molecule. A basic understanding of the abundance of isotopes is therefore necessary to interpret a mass spectrum. Atomic masses are measured in *atomic mass units* (*u*), where 1 u is defined as one-twelfth of the mass of the ^{12}C isotope. The atomic mass unit, u, is the official SI unit, but the unit *Dalton*, which is the same as u (1 u = 1 Da), is also used.

The mass of the ^{12}C isotope is exactly 12.0000 Da. The mass of other atoms is measured relative to the mass of the ^{12}C isotope. The ^{1}H isotope, for example, is 11.9068 times lighter than the ^{12}C isotope, so the mass of ^{1}H is therefore

$$\frac{1}{11.9068} \times 12.0000 \, \text{Da} = 1.0078 \, \text{Da} \tag{15.1}$$

The ^{35}Cl isotope is 2.91407 times heavier than the ^{12}C isotope, so the mass of ^{35}Cl is therefore

$$2.91407 \times 12.0000 \, \text{Da} = 34.9688 \, \text{Da} \tag{15.2}$$

Chlorine has two isotopes, ^{35}Cl and ^{37}Cl, so the mass of ^{37}Cl is 36.965896. The natural abundances of the two isotopes are 75.77% and 24.23%, respectively. All elements frequently found in drug substances exist as more than one isotope, except for fluorine, iodine, and phosphorus, which are *monoisotopic*. An overview of isotopes of selected elements and their natural abundances are given in Table 15.1. The *monoisotopic mass* of a certain molecule is calculated by the sum of the masses of the most abundant isotopes. For simplicity, the word mass will be used as a synonym for monoisotopic mass throughout this chapter. Calculation of monoisotopic mass is exemplified in Box 15.1, where the difference from molecular weight is clarified.

Atomic and molecular masses as given above with three or four decimal places are *exact masses* or *accurate* masses. MS with *high resolution* (HR) measures exact masses, whereas in *low resolution* MS the masses are often rounded to the nearest integer, and these masses are termed *nominal masses*. Measured in nominal masses, the molecular mass of chlorambucil ($^{12}C_{14}{}^{1}H_{19}{}^{35}Cl_2{}^{14}N\,{}^{16}O_2$) is therefore

$$14 \times 12 + 19 \times 1 + 2 \times 35 + 14 + 2 \times 16 = 303 \, \text{Da} \tag{15.3}$$

For simplicity, mainly nominal masses will be used in this chapter.

Table 15.1 *Masses and abundance of stable isotopes*

Isotope	Exact mass (Da)	Nominal mass	Abundance (%)
^1H	1.007825	1	99.988
^2H	2.014102	2	0.012
^{12}C	12.00000	12	98.93
^{13}C	13.003354	13	1.07
^{14}N	14.003074	14	99.632
^{15}N	15.000108	15	0.368
^{16}O	15.994915	16	99.757
^{17}O	16.999133	17	0.038
^{18}O	17.999160	18	0.205
^{19}F	18.998405	19	100
^{31}P	30.973763	31	100
^{32}S	31.972074	32	94.93
^{33}S	32.971461	33	0.76
^{34}S	33.967865	34	4.29
^{36}S	35.967091	36	0.02
^{35}Cl	34.968855	35	75.77
^{37}Cl	36.965896	37	24.231
^{79}Br	78.918348	79	50.69
^{81}Br	80.916344	81	49.31
^{127}I	126.904352	127	100

Box 15.1   Monoisotopic mass and molecular weight of chlorambucil

The elemental composition of chlorambucil is $C_{14}H_{19}Cl_2NO_2$.

The monoisotopic mass is the exact mass of the molecule and is calculated from the sum of the most abundant isotopes ^{12}C, ^1H, ^{35}Cl, ^{14}N, and ^{16}O:

$$(14 \times 12.0000 + 19 \times 1.0078 + 2 \times 34.9689 + 14.0031 + 2 \times 15.9949)\ Da$$

$$= 303.0789\ Da$$

Mass spectrometers measure monoisotopic masses.

The molar mass is the average mass of the compound to be used in molar calculations. Molar weights take into consideration the average abundances of the different isotopes, which are the masses given in the periodic table:

$$(14 \times 12.011 + 19 \times 1.008 + 2 \times 35.453 + 14.007 + 2 \times 15.999)g/mol$$

$$= 304.217\ g/mol$$

15.3   Ionization

The first step in the MS analysis is ionization of the sample constituents. The ionization methods are different in GC-MS and LC-MS. In GC-MS, the sample constituents eluting from the GC column are bombarded with electrons inside the mass spectrometer under vacuum conditions. This is a *hard ionization* method termed *electron ionization* (EI) (or *electron impact ionization*), in which sample constituents are often degraded to *fragments*. On the other hand, samples in LC-MS are most often exposed to *electrospray ionization* (ESI). This is a *soft ionization* method, in which the original sample molecules are maintained. The difference in ionization methods thus results in different types of mass spectra.

EI is used in GC-MS. Electron bombardment leads to some of the molecules being ionized, in most cases by loss of a single electron per molecule. These ions are termed *molecular ions*. The mass of molecular ions is equivalent to the mass of the original molecules because the mass of an electron that is lost (or taken up) is insignificant. However, molecular ions are unstable and most ions momentarily split into smaller fragments when chemical bonds in the molecular ions are broken. This process is termed *fragmentation*. Some of the fragments remain ionized (*fragment ions*), while others lose their charge. In cases where virtually all molecular ions decompose to fragment ions, no molecular ions are found in the resulting mass spectrum. An example of a mass spectrum based on EI is shown in Figure 15.2.

In LC-MS, the chromatographic system is linked to the mass spectrometer by an *interface*. In this interface, sample constituents are *ionized* and *vapourized* before they enter the mass spectrometer. In the *ESI* interface, the mobile phase is converted to a fine spray and exposed to an electric potential of 2–5 kV, resulting in a fine aerosol of charged droplets.

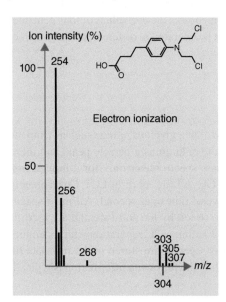
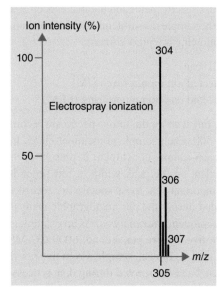

Figure 15.2   *Mass spectrum of chlorambucil in the m/z range 250–310 by electron ionization (GC-MS) and electrospray ionization (LC-MS)*

The analyte molecules are then extracted into the mass analyser in the vacuum region of the mass spectrometer. ESI forms positive or negative ions depending on applied voltage. Ions are produced based on acid–base chemistry, by proton uptake, or release from the analyte molecule, resulting in $[M + H]^+$ ions and $[M-H]^-$ ions, respectively. Strictly speaking, these ions are not 'molecular ions' due to the gain or loss of a proton. They are, however, often referred to as molecular ions and will also be named this throughout the textbook. ESI is a *soft ionization* technique, which leaves the ions stable without, or only limited, fragmentation. An example of a mass spectrum obtained by ESI is presented in Figure 15.2.

Figure 15.2 shows mass spectra of chlorambucil as an example. The *y*-axis in a mass spectrum shows the abundance or intensity of the different ions. The *x*-axis displays the *m/z* value. Mass spectra are plotted independently of concentration by scaling the spectra relative to the most intense ion. The signal for the most intense ion is termed the *base peak* and the intensity of the base peak is scaled to 100%. The other fragment ions are then scaled relative to the base peak. The base peak in the EI mass spectrum appears at *m/z* 254. The EI mass spectrum shows only a small signal for the molecular ion at *m/z* 303, as the molecular ion is extensively fragmented. In the ESI mass spectrum, the base peak is the protonated molecule, which appears as $[M + H]^+$ at *m/z* 304.

15.4 The Mass Spectrometer as a Chromatographic Detector – Data Acquisition

The mass spectrometer is in most cases used as a detector in combination with LC-MS or GC-MS. In both cases, sample constituents separated by chromatography elute directly into the mass spectrometer and the mass spectrometer measures the *m/z* values of ions generated from the sample constituents. LC-MS systems can be operated in different modes, of which the following are predominant:

- Full scan
- Selected ion monitoring (SIM)
- Selected reaction monitoring (SRM)

In the *full scan* mode, mass spectra are recorded within a given *m/z* range such as from *m/z* 50 to 500. Each sample constituent elutes from the column as a narrow peak, and therefore each sample constituent is present in the mass spectrometer only for a limited time dependent on the peak width (a few seconds in GC, a little more in LC). Consequently, it is important that mass spectra are recorded several times per second. All mass spectra recorded during the chromatographic analysis are stored by the software and are used for the subsequent data analysis. As an example, in a 20 minute GC separation with a sampling rate of five spectra per second, 6000 GC-MS mass spectra are stored in the raw data file from analysis of each sample. From this raw data file, the following information is typically extracted and interpreted during data processing:

- Total ion current chromatogram (TIC)
- Extracted ion chromatogram (EIC)
- Mass spectra

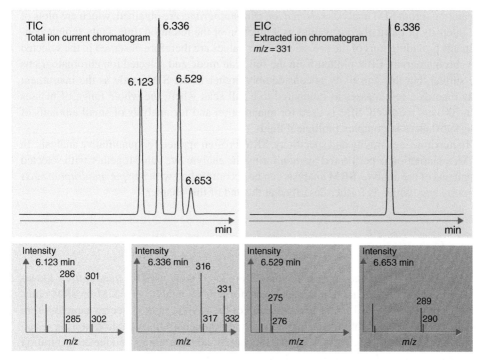

Figure 15.3   *GC-MS total ion current chromatogram and extracted ion chromatogram at* m/z *311. Below, mass spectra of the four separated sample constituents*

A simplified example of a TIC is displayed in Figure 15.3, where four different sample constituents have been separated by GC. In the TIC, the total ion current is plotted as a function of the retention time. The TIC is very low when pure carrier gas flows from the GC into the mass spectrometer. Each time a sample constituent elutes into the mass spectrometer, the total ion current increases for a short period of time, giving rise to a peak in the TIC. Thus, the TIC is a universal chromatogram and shows individual peaks for all the constituents of the sample that have been separated in the column. A major advantage of the full scan mode is that mass spectra are collected and stored for all the sample constituents separated in the chromatographic system. Thus, unknown peaks in the chromatogram can be identified based on their mass spectra.

For locating known compounds of particular interest, EICs are more suitable. In an EIC, the signal intensity of a selected m/z value is extracted from the TIC and plotted as a function of the retention time, as illustrated in Figure 15.3, where the ions of m/z 331 have been extracted, while signals for all other m/z values have been excluded.

By recording the full mass spectra, a lot of structural information is available for each of the separated sample constituents, but the drawback is that the sensitivity of the mass spectrometer is limited in the full scan mode, and it may therefore be difficult to detect compounds at very low concentration levels. To address the issue of sensitivity, the mass spectrometer can be operated in *SIM mode*. In SIM mode, the mass spectrometer is tuned to measure only a single m/z value or a few selected m/z values during the chromatographic separation. From SIM analysis *selected ion chromatograms* are obtained, which are plots of the intensity of particular *m/z* values as a function of the retention time. Only sample constituents providing ions of the pre-selected *m/z* values are therefore observed in the selected ion chromatogram. EICs plotted from the full scan mode and selected ion chromatograms are similar, but the sensitivity is considerably higher in the SIM mode as the instrument only records a few masses as compared to a full scan where the whole range of masses (50–500) are recorded. SIM is used for quantitation and for analysis of small amounts of drug substances in complex biological fluids.

To maximize sensitivity and specificity, SRM is often applied for quantitative analysis. In SRM, quantitation is performed by monitoring the analyte *m/z* value together with selected fragments of the analyte. SRM analysis can be accomplished with a *triple quadrupole mass spectrometer*, which is further described at the end of this chapter.

15.5 Quantitation by MS

LC-MS and GC-MS based analytical methods are intensively used for *quantitation*. Quantitation is based on the fact that the peak area for a given analyte in a SIM or SRM chromatogram is proportional to the concentration of the analyte in the injected sample. From an analysis of standard solutions with known analyte concentrations a calibration curve can be established by plotting peak areas as a function of concentrations and the concentration of samples calculated by comparison of a sample of the peak area with the calibration curve.

One great advantage of MS detection is that even though the resolution of the chromatographic separation does not provide baseline resolution, data acquisition in the SIM or SRM mode compensates for this, as only the selected *m/z* value is recorded. This is illustrated in Figure 15.4. Cannabidiol and tetrahydrocannabinol acid elute from the LC system with very similar retention times, but because they are detected at different masses by the MS, both compounds can be quantified independently based on the chromatograms in Figure 15.4.

15.6 Identification by MS

Mass spectrometric detection is highly suitable for *identification*. Due to the differences between LC-MS (soft ionization) and GC-MS (hard ionization), somewhat different identification principles are used.

15.6.1 GC-MS and LC-MS

Mass spectra of a given substance acquired in different laboratories and with different GC-MS instruments are normally very similar when EI at 70 eV is used. Using such hard ionization, the analytes are fragmented independently of instrument type and setup, and the mass spectra are like fingerprints and are ideal for comparison. Peaks in a GC chromatogram can often be identified by comparing its corresponding mass spectrum with a *reference spectrum*. Mass spectra at 70 eV have therefore been recorded and published for a very large number of different drug substances, drug metabolites, drugs of abuse, and

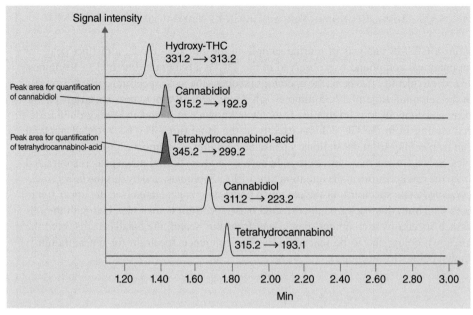

Figure 15.4 *Areas of co-eluting peaks can be separated as the analytes have different* m/z *values*

other chemical compounds. These spectra are stored in *computer-based libraries*. Mass spectrometers used for GC-MS are delivered with such libraries of reference spectra connected online to the instrument. Searching the library for suggestions of compound identity is therefore fast and simple. An example of the use of a reference spectrum for GC-MS identification of an unknown substance is shown in Box 15.2. The search for library reference spectra often gives a list of several suggestions that match the spectrum of the unknown compound to a certain level. The list is ranked according to the quality of the match. The match is normally based on the 5–10 most abundant ions in the mass spectrum. The suggestion of highest match quality should be controlled to verify that all major fragments in the reference spectrum are also present in the spectrum of the unknown compound. For reliable identification, an unknown compound should also be identified by retention time. With reference to the example in Box 15.2, fencamfamin should therefore be purchased as a chemical reference substance and analysed by GC-MS as standard solution under identical conditions to check the retention time. Provided that the main peak of the reference solution has a retention time of 17.35 minutes, which is identical to the retention time of the unknown compound, the latter has been positively identified as fencamfamin.

Data bases for LC-MS with ESI are more limited as the soft ionization mode only leads to production of the protonated molecular ion and fragmentation patterns are much more dependent on instruments and sample conditions. Identification in LC-MS based on nominal masses is therefore mainly based on the molecular ion and deduction from fragmentation patterns. However, using high resolution instruments capable of distinguishing *m/z* values with up to four decimal places, suggestions for identities of unknown compounds can be obtained from data base searches.

Box 15.2 Identification of fencamfamin in a urine sample by GC-MS

After a GC-MS analysis of a urine sample in the full scan mode, a distinct peak for an unknown compound was observed in the TIC at a retention time of 17.35 minutes. In order to identify this peak, the operator first opens the mass spectrum collected exactly at the retention time of 17.35 minutes. This represents the raw mass spectrum for the peak and contains ions originating from the unknown compound and background material eluting from the GC column exactly at the same time. The background material can be bleeding from the stationary phase and column contamination from earlier samples. Second, a raw mass spectrum right before (or after) the unknown peak is subtracted from the raw spectrum at the retention time of 17.35 minutes, and by this the *background corrected* mass spectrum for the unknown compound is generated (see the upper figure below). Third, this background corrected mass spectrum is then matched with the reference spectra by a computer search, and from this search the database suggested the unknown compound to be fencamfamin. The reference spectrum for fencamfamin is shown below (lower figure).

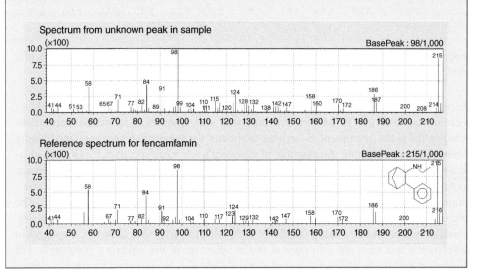

15.6.2 Structural Information from Isotopes

As already mentioned, isotopes are very important in mass spectral interpretation. Carbon exists naturally as two isotopes, namely the ^{12}C isotope with nominal mass 12 and the ^{13}C isotope with nominal mass 13. The ^{12}C isotope dominates and constitutes 98.9%, while the ^{13}C isotope represents only 1.1%. The more carbon atoms in a given substance, the larger is the fraction of molecules containing one ^{13}C. Chlorambucil (Figure 15.2) contains 14 carbon atoms, and $14 \times 1.1\% = 15.4\%$ of these molecular ions contains a ^{13}C atom. This is why the EI mass spectrum shows a peak at m/z 304 and the ESI mass spectrum shows a peak at m/z 305, respectively. This peak corresponds to 15.4% of the intensity of the 303 and 304 peaks, respectively.

In clean mass spectra without interference from other substances, the ^{13}C isotope can be used to estimate the number of carbon atoms in the molecular ion as well as in fragment ions. If the intensity of an ion with mass m is I_m (this contains only ^{12}C atoms) and the

---

Box 15.3 Calculation of the number of carbon atoms in chlorambucil ($C_{14}H_{19}Cl_2NO_2$) based on the ratio between ^{12}C and ^{13}C

From a magnification of the mass spectrum of chlorambucil (see below) we read the following intensities for the molecular ion $(M + H)^+$ with only ^{12}C isotopes (I_m) at m/z 304 and the molecular ion with one ^{13}C isotope (I_{m+1}) at m/z 305:

$$I_m = 100.0\%$$
$$I_{m+1} = 15.0\%$$

This intensity ratio corresponds to the following number of carbon atoms:

$$n = \frac{15.0\%}{0.011 \times 100.0\%} = 13.6 \approx 14$$

The calculation demonstrates that, within the uncertainty of the measurement, the molecular ion contains 14 carbon atoms.

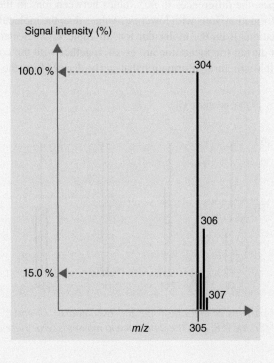

intensity of the corresponding ion with one ^{13}C atom at mass $(m + 1)$ is I_{m+1}, the following equation is valid:

$$I_{m+1} = n \times 0.011 \times I_m \tag{15.4}$$

Here n is the total number of carbon atoms in the ion. This number can then be determined by the following reorganization of Eq. (15.4):

$$n = \frac{I_{m+1}}{0.011 \times I_m} \tag{15.5}$$

An example of how to calculate the number of carbon atoms is shown in Box 15.3. Multiplication with 0.011 in the equation is linked to the 1.10% natural abundance of ^{13}C as given in Table 15.1.

In addition to the ^{13}C isotope, isotope patterns of chlorine and bromine are distinct. As shown in Table 15.1, the abundances of the chlorine isotopes ^{35}Cl and ^{37}Cl are in the ratio $76\% : 24\% \approx 3 : 1$ and the abundances of the bromine isotopes, ^{79}Br and ^{81}Br, are in the ratio $\approx 1 : 1$. The isotope patterns of these are therefore very characteristic, as illustrated in Figure 15.5, and all molecular ions and fragment ions containing Cl or Br are relatively easy to locate. The calculation for the isotope patterns of molecules containing one and two chlorine atoms are given in Box 15.4.

15.6.3   Structural Information from Fragmentation

In addition to isotopes, the differences in m/z values between ions in the mass spectrum can give valuable information. Ions with lower m/z values than the molecular ion have been formed by fragmentations from the molecular ion. In Table 15.2 selected possible losses from molecular ions during fragmentation are given, together with the corresponding m/z differences and possible structural information that can be derived from the actual m/z shifts.

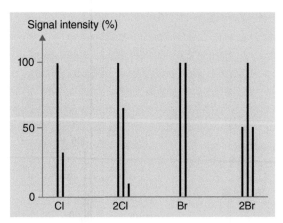

Figure 15.5   *Isotope patterns of chlorine (one and two atoms ^{35}Cl and ^{37}Cl) and bromine (one and two atoms ^{79}Br and ^{81}Br). As the difference in masses is two, there are two m/z units between each peak*

Box 15.4   Calculation of isotope patterns for molecules containing one and two chlorine atoms

Chlorine has two naturally abundant isotopes, ^{35}Cl and ^{37}Cl, with nominal masses of 35 and 37, respectively. The ^{35}Cl isotope is predominant and is at the 75.77% level, while the ^{37}Cl isotope is 24.23%.

Diazepam with the chemical formula $C_{16}H_{13}ClN_2O$ contains one chlorine atom and has the monoisotopic mass 284.07 corresponding to the nominal mass 284. The monoisotopic mass is calculated based on the most abundant isotopes and thus the monoisotopic mass is calculated for ^{35}Cl. According to the abundances, the probability that the isotope is ^{35}Cl is 76% and the probability that the isotope is ^{37}Cl is 24%, so a third of the molecules will contain ^{37}Cl. Thus, a signal at 2 *m/z* units higher than M will appear in the mass spectrum and the size will be about one-third of the signal of M. In ESI, the molecular ion is protonated $[M + H]^+$ and is observed at *m/z* 285, while the signal of ^{37}Cl is seen at *m/z* 287. (In EI the signals will appear at *m/z* 284 and 286, if the molecular ion is not totally fragmented.)

In a molecule containing two chlorine atoms, like chlorambucil ($C_{14}H_{19}Cl_2NO_2$), three peaks for the molecular ion will be observed in ESI: the signal at *m/z* 304 corresponds to molecular ions having two ^{35}Cl isotopes, *m/z* 306 corresponds to one ^{35}Cl and one ^{37}Cl isotope, while *m/z* 308 corresponds to two ^{37}Cl isotopes. Based on the isotope abundances, the probabilities of the different combinations of isotopes are:

M:	Two ^{35}Cl	$0.7577 \times 0.7577 = 0.5741$
M + 2	One ^{35}Cl and one ^{37}Cl	$0.7577 \times 0.2423 + 0.2423 \times 0.7577 = 0.3672$
M + 4:	Two ^{37}Cl	$0.2423 \times 0.2423 = 0.0587$

If M is the base peak set to 100%, the intensity of M + 2 is (0.3672/0.5741) × 100% = 64% and the intensity at M + 4 is (0.05871/0.5741) × 100% = 10%. A compound containing two chlorine atoms will therefore appear as an isotope pattern of M, M + 2, M + 4 with the intensities 100 : 64 : 10, as shown in Figure 15.5. For chlorambucil the signals would appear at *m/z* 303, 305, and 307 with EI and at *m/z* 304, 306, and 308 with ESI (Figure 15.2).

Thus, in a spectrum where fragment ions are found 18 m/z values lower than the molecular ion, loss of water has occurred. This may be indicative for a primary alcohol. Fragment ions with high m/z values close to the molecular ion can often be explained by losses included in Table 15.2, whereas fragment ions with low m/z values can be difficult to explain because they are formed due to multiple fragmentations. In the EI mass spectrum in Figure 15.2, the peak at m/z 254 corresponds to the loss of CH_2Cl from the molecular ion ($C_{14}H_{19}Cl_2NO_2 - CH_2Cl = C_{13}H_{17}ClNO_2$). This fragment ion contains 13 carbon atoms and an average of $13 \times 1.1\% = 14.3\%$ of the ions with the formula $C_{13}H_{17}ClNO_2$ contain one ^{13}C atom. Therefore, the mass spectrum shows a peak at m/z 255. This fragment is not

Table 15.2 *Typical fragmentations and losses from the molecular ion (M) with electron ionization*

m/z Difference	Loss	Interpretation
$M - 1$	H	
$M - 2$	H_2	
$M - 15$	CH_3	α-Cleavage
$M - 16$	O	Aromatic nitro compound, N-oxide
$M - 16$	NH_2	Primary amide
$M - 17$	OH	Carboxylic acid, tertiary alcohol
$M - 18$	H_2O	Primary alcohol
$M - 19$	F	Fluoride
$M - 20$	HF	Fluoride
$M - 26$	C_2H_2	Unsubstituted aromatic hydrocarbon
$M - 27$	C_2H_3	Ethyl ester
$M - 28$	C_2H_4	Aromatic ethyl ether, n-propylketone
$M - 28$	CO	
$M - 29$	C_2H_5	Ethyl ketone, α-cleavage
$M - 29$	HCO	Aliphatic aldehyde
$M - 30$	CH_2O	Aromatic methyl ether
$M - 30$	NO	Aromatic nitro compound
$M - 31$	OCH_3	Methyl ester, aromatic methyl ester
$M - 32$	CH_3OH	Methyl ether
$M - 33$	SH	Thiol
$M - 34$	H_2S	Thiol
$M - 35/37$	Cl	Chlorinated compound
$M - 36/38$	HCl	Chlorinated aliphatic compound
$M - 41$	C_3H_5	Propyl ester
$M - 42$	CH_2CO	Aromatic acetate, $ArNHCOCH_3$
$M - 42$	C_3H_6	Aromatic propyl ether
$M - 43$	C_3H_7	Propyl ketone, α-cleavage
$M - 43$	CH_3CO	Methyl ketone
$M - 45$	OC_2H_5	Ethyl ester
$M - 45$	COOH	Carboxylic acid
$M - 46$	NO_2	Aromatic nitro compound
$M - 46$	C_2H_5OH	Ethyl ether
$M - 60$	CH_3CO_2H	Acetate
$M - 127$	I	Iodated compound

seen in the ESI spectrum due to the softer ionization. Boxes 15.5 and 15.6 show examples of losses due to fragmentation.

Many drug substances contain nitrogen, and for these compounds the *nitrogen rule* applies. How to apply the nitrogen rule is dependent on the ionization method. For EI the rule is as follows:

Box 15.5   Loss of halogens during fragmentation

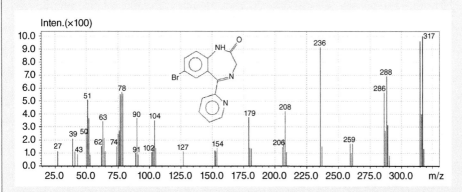

EI spectrum (GC-MS) of bromazepam

A number of drugs contain halogen atoms (F, Cl, Br, I). According to Table 15.2 such compounds may give a characteristic release of the halogens corresponding to $(M - 19)^+$ for F, $(M - 35)^+$ for Cl, $(M - 79)^+$ for Br, and $(M - 127)^+$ for I. An example of the release of bromine is shown in the EI spectrum for bromazepam. The molecular ion for bromazepam is found at m/z 315, which is an odd nominal mass. This is consistent with the nitrogen rule, since bromazepam contains three nitrogen atoms. The compound contains a bromine atom, and the typical isotope distribution of bromine can be seen clearly (compare with Figure 15.5); the heavier bromine isotope ^{81}Br gives a peak at mass 317 with about the same intensity as the peak at 315. Normally, bromine is easily released from the molecular ion as the C—Br bond in general is relatively weak compared to C—F and C—Cl bonds. For bromazepam a peak is observed at $(M - 79)^+$, corresponding to m/z 236. The fragment ion at m/z 236 no longer contains bromine, and no bromine isotope pattern is observed at this fragment. The signal at m/z 236 is large and shows that the release of bromine from the molecular ion in this case is a dominant process during fragmentation.

Box 15.6 Fragmentation of *p*-amino-benzoic acid

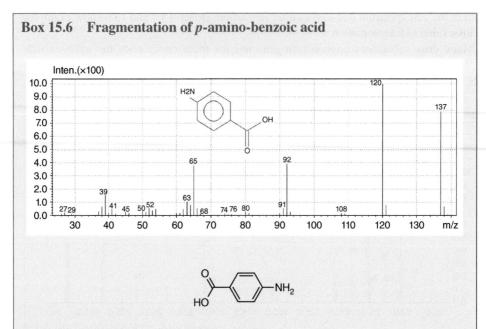

EI spectrum (GC-MS) of *p*-amino-benzoic acid

For *p*-amino-benzoic acid, a strong signal is observed for the molecular ion at *m/z* 137. Note that the molecular ion has an odd nominal mass. This follows the nitrogen rule, since the compound contains a single nitrogen atom. The next fragment of interest is found at *m/z* 120, 17 *m/z* units from the molecular ion. According to Table 15.2, this corresponds to the loss of an OH radical $(M - 17)^+$, which is frequently observed for carboxylic acids (and tertiary alcohols). The $(M - 45)^+$ fragment at *m/z* 92 corresponds to the loss of a COOH-radical.

- If a compound has an odd number of nitrogen atoms then the molecular ion $M^{•+}$ has an *odd* nominal *m/z* value.
- If a compound has an even number of nitrogen atoms, or no nitrogen atoms, the molecular ion $M^{•+}$ has an *even* nominal *m/z* value.

For ESI (and other ionization methods generating $(M + H)^+$ or $(M - H)^-$ ions) the rule should be applied as follows:

- If a compound has an odd number of nitrogen atoms then the molecular ion $(M + H)^+$ or $(M - H)^-$ has an *even* nominal *m/z* value.
- If a compound has an even number of nitrogen atoms, or no nitrogen atoms, the molecular ion $(M + H)^+$ or $(M - H)^-$ has an *odd* nominal *m/z* value.

If the elemental composition of the analyte is known, the number of rings (R) + double bonds (DB) can be calculated from the following equation:

$$R + DB = c - \frac{h}{2} + \frac{n}{2} + 1 \qquad (15.6)$$

Here c is the number of carbon atoms, h is the number of hydrogen atoms, and n is the number of nitrogen and phosphorous atoms. This equation may help in predicting unknown structures from mass spectra. One example is given in Box 15.7.

Box 15.7 Calculation of the number of rings and double bonds for methadone

The gross formula for methadone is $C_{21}H_{27}NO$. The structural formula is shown below. The number of rings and double bonds are calculated as follows:

$$R + DB = 21 - \frac{27}{2} + \frac{1}{2} + 1 = 9$$

This is in accordance with the molecular structure shown below, with seven double bonds and two rings

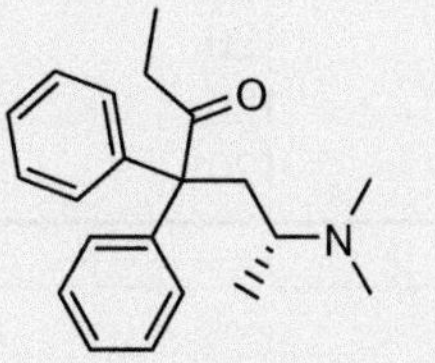

15.6.4   Structural Information from Accurate Masses (High Resolution MS)

Use of high resolution MS for identification provides accurate mass determinations with four or five decimals. By combining accurate mass determinations with search in different data bases, suggestions for elemental composition, structure formulas, and in some cases also fragmentation patterns may be available. The added information from measuring exact masses is illustrated in Box 15.8.

High resolution mass spectrometry can thus determine the elemental composition of both molecular ions and fragment ions.

Different mass spectrometers have different mass resolutions. The ability to separate two peaks of similar mass can be expressed by the *resolution* or by the *resolving power,* as defined by the following equations:

$$\text{Resolution} = \frac{\Delta m}{m} \tag{15.7}$$

$$\text{Resolving power} = \frac{m}{\Delta m} \tag{15.8}$$

Here m is the smaller value of m/z for two adjacent peaks with no more than 10% overlap in their m/z signals and Δm is their m/z difference. An example on how to calculate resolving power and resolution is shown in Box 15.9.

Box 15.8   Several ions with the same nominal mass but with significant differences in their exact mass

A molecular ion with a nominal m/z value of 120 can correspond to several different elemental compositions with very small differences in the accurate mass (see the table below) and knowledge about the correct elemental composition may be crucial for correct identification. Elemental compositions can be determined if the mass determinations are highly accurate. This is done by high resolution mass spectrometry, where the mass spectrometer is tuned for highly accurate mass determinations. If the accurate mass of the molecular ion is determined to 120.070 ($\pm$0.0005), it can be concluded that this corresponds to the elemental composition of the ion $C_7H_8N_2^+$.

Ion	Nominal mass	Exact mass
$C_5H_4N_4^+$	120	120.044
$C_7H_8N_2^+$	120	120.069
$C_9H_{12}^+$	120	120.096
$C_8H_8O^+$	120	120.058

Box 15.9   Calculation of resolving power and resolution

A mass spectrometer is capable of separating m/z 120.044 and m/z 120.058 with approximately 10% overlap, as illustrated in the figure below.

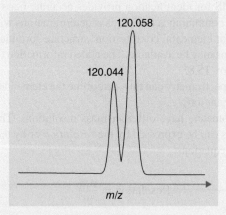

This corresponds to the following resolving power:

$$\text{Resolving power} = \frac{m}{\Delta m} = \frac{120.044}{120.058 - 120.044} = 8575 \approx 8600$$

The resolving power of the mass spectrometer is thus about 8600. This corresponds to the following number for resolution:

$$\text{Resolution} = \frac{\Delta m}{m} = \frac{120.058 - 120.044}{120.044} = 1.17 \times 10^{-4}$$

Resolving power is a big number and resolution is a small number. Resolution and resolving power are given for the m/z value where they have been calculated, as the resolving power varies with the mass. Simple instruments have *unit resolution*, meaning that they can only measure differences of 1 Da, corresponding to a resolving power of $300/1 = 300$ at m/z 300. High resolution instruments often show resolutions of 10^{-3}, e.g. an Orbitrap instrument (see the instrument section) with a resolving power of 70 000 at m/z 200 can distinguish two peaks with a mass difference of 0.0029. An instrument with a resolution 10^{-3} throughout the m/z scale has a resolving power of $100/0.001 = 100 000$ at m/z 100, while the resolution at m/z 1000 is 10 000.

15.6.5 Structural Information for Peptides and Proteins

Peptides and proteins are analysed by LC-MS with ESI. In large molecules containing many carbon atoms like peptide and proteins, the probability of the presence of a ^{13}C atom is much larger than for small molecules, and the molecular ion produced by ESI will not be the largest signal in the mass spectrum. Furthermore, several amino acids in the biomolecule can be protonated, resulting in multiple-charged ions. Thus, the monovalent molecular ion does not appear directly from the mass spectrum. This is illustrated in Figure 15.6 in the electrospray mass spectrum of the insulin monomer.

The peaks in the mass spectrum represent molecules with a different number of protons, n, and thereby different charges, e.g. $[M + 4H]^{4+}$ at m/z 1452.8334. To find the molecular mass, M, of the neutral protein, the charge of each of the species, m_n, must be calculated. A peak resulting from a molecule with the charge n will appear at m/z:

$$m_n = \frac{M + n(1.008)}{n} = \frac{M}{n} + 1.008 \rightarrow \frac{M}{n} = m_n - 1.008 \qquad (15.9)$$

1.008 is the mass of a proton. The peak with one more proton has $n + 1$ protons and a charge of $n + 1$. The mass-to-charge ratio of this peak is

$$m_{n+1} = \frac{M + (n + 1) \times 1.008}{n + 1} = \frac{M}{n + 1} + 1.008 \rightarrow \frac{M}{n + 1} = m_{n+1} - 1.008 \qquad (15.10)$$

Combining the equations and solving them for n gives the charge for n:

$$n = \frac{m_{n+1} - 1.008}{m_n - m_{n+1}} \qquad (15.11)$$

In Box 15.10 calculation using Eq. (15.11) reveals that the signal at m/z 1162 corresponds to the charge state +5 ($[M + 5H]^{5+}$), while the m/z values of 1453 and 1937 correspond

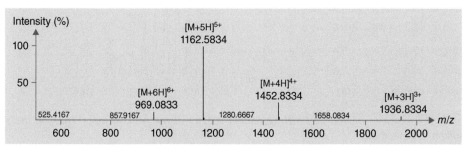

Figure 15.6   *Electrospray mass spectrum of the insulin monomer ($C_{257}H_{383}N_{65}O_{77}S_6$, molecular mass = 5807.57 Da)*

to $[M+4H]^{4+}$ and $[M+3H]^{3+}$, respectively, and the signal at m/z 969 corresponds to $[M+6H]^{6+}$.

Box 15.10   Calculation of charge and molecular mass of protein

Applying Eq. (15.11) for the two most distinct peaks in the mass spectrum in Figure 15.6 leads to

$$m_n = 1452.8334 \text{ and } m_{n+1} = 1162.5834$$

$$n = \frac{m_{n+1} - 1.008}{m_n - m_{n+1}} = \frac{1162.5834 - 1.008}{1452.8334 - 1162.5834} = 4.00$$

Thus, the peak at m/z 1453 has the charge +4, and the molecular mass of the neutral protein is

$$\frac{M}{n} = m_n - 1.008 \;\rightarrow\; M = (m_n - 1.008)n = (1452.8334 - 1.008) \times 4 = 5807.3$$

Figure 15.7 shows a zoom on the largest peak at m/z 1162 (in Figure 15.6) recorded with an ion trap instrument with unit resolution and an Orbitrap instrument with high resolution. The high resolution instrument can distinguish the small differences between the signals from molecules containing different numbers of ^{13}C atoms, although the differences are small as they are divided by the charge. For a molecule with a charge of 1, the increase in m/z for each ^{13}C atom would be 1. For a molecule in the charge state +2, the increment would be only 0.5 (1/2), while in a molecule in charge state +4, it would be 0.25 (1/4). The difference between the m/z values in the present peak is 0.2, corresponding to a charge state of +5. Thus, the peaks in the spectrum correspond to increasing numbers of ^{13}C isotopes in the molecule. In large molecules, the probability of only one ^{13}C isotope present in the molecule is much less than the probability of the presence of several ^{13}C isotopes.

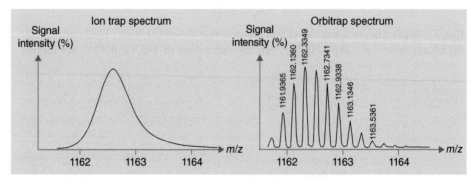

Figure 15.7   *Unit resolution and high resolution spectra of the peak at m/z 1162.6 in the insulin mass spectrum*

15.7   Instrumentation

The first step in mass spectrometry is the ionization. In LC-MS, ESI and atmospheric pressure chemical ionization (APCI) takes place outside the vacuum region of the instrument. In GC-MS, EI and chemical ionization (CI) occurs under vacuum conditions in an ion source inside the mass spectrometer. In both LC-MS and GC-MS, the ions are separated in the mass analyser according to their *m/z* values and detected by the detector. This is summarized in Figure 15.8.

MS is performed under vacuum conditions, with a very low pressure inside the instrument. The reason for this is to prevent the ions of interest from colliding with air molecules. This ensures that the ions reach the detector. To maintain a low pressure in the mass spectrometer, the instrument must be closed, and air penetrating the system must constantly be pumped out by means of a powerful *pumping system*. The pumping system is connected both to the ion source (in the case of EI and CI), to the mass analyser, and to the detector. This ensures that the pressure inside the instrument does not exceed 10^{-4}–10^{-8} torr (10^{-7}–10^{-11} bar).

15.7.1   Ion Sources for GC-MS

In GC-MS, separations are performed on capillary columns and the outlet of the column is placed directly inside the ion source of the mass spectrometer. Thus, EI (or CI) takes place when the compounds leave the column. The ion source is a small chamber inside the mass spectrometer in front of the mass analyser. Usually, the carrier gas (helium or hydrogen) has a flow rate of 1–3 mL/min, and this gas enters the mass spectrometer. However, the capacity of the pumping system is so powerful that this gas flow does not influence the pressure significantly.

In EI, the sample constituents are bombarded with electrons. EI is a *hard ionization* method. The EI ionization takes place in the *ion source* located inside the mass spectrometer, where the sample constituents cross a beam of electrons released from a small wire (*filament*) of rhenium or tungsten. The electrons released from the filament are accelerated

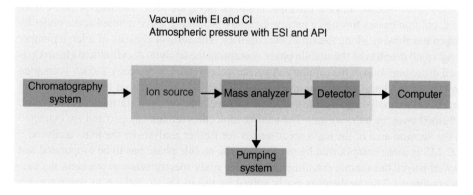

Figure 15.8   *Schematic view of mass spectrometer*

in an electric field with a potential of 70 eV. When sample molecules interact with the electrons, an electron will be released from some of the molecules as a consequence of electrical repulsion (minus charges on the electrons repel each other):

$$M + e^- \rightarrow M^{\bullet+} + 2e^- \tag{15.12}$$

The positive $M^{\bullet+}$ ion formed during the EI is termed the molecular ion. In addition to the charge, the symbol for the molecular ions is usually supplied with a small dot to indicate that they are radical ions, meaning that the molecular ion contains an unpaired electron. (Normally, all the electrons in a molecule are paired, but when one electron is removed, one of the remaining electrons is unpaired). From the ion source, molecular ions and fragment ions are accelerated into the mass analyser.

If EI does not provide sufficient molecular ions, the electrons that bombard the sample contain too much energy. The energy of the electrons can be reduced by adding a *reagent gas* like methane into the ion source. This is known as CI. CI takes place in the ion source in the vacuum region of the mass spectrometer and the reagent gas is supplied in relatively large amounts compared to the amount of sample constituents (1000–10 000 times more reagent gas than sample). Electrons released by the filament in the ion source therefore almost exclusively ionize the reagent gas. With methane as the reagent gas, CH_5^+ and $C_2H_5^+$ are formed. These ions are highly reactive and react with the analyte (XH) according to the following reactions:

$$CH_5^+ + XH \rightarrow XH_2^+ + CH_4 \tag{15.13}$$

$$C_2H_5^+ + XH \rightarrow XH_2^+ + C_2H_4 \tag{15.14}$$

$$C_2H_5^+ + XH \rightarrow X^+ + C_2H_6 \tag{15.15}$$

15.7.2   Ion Sources for LC-MS

In LC-MS, the chromatography is performed at high pressure, while the mass spectrometer is operated under vacuum conditions. Therefore, coupling the two systems is a technical challenge, which is solved by an *interface*. The most common one is the *ESI* interface. In this interface, the sample constituents are continuously delivered in a volatile mobile phase from the LC system. ESI takes place at atmospheric pressure outside the vacuum region of the mass spectrometer, and the principle is shown in Figure 15.9. The mobile phase from the LC column passes through a narrow capillary producing a fine *aerosol* at the outlet by nitrogen gas flowing along the tip of the capillary. The aerosol consists of a large number of very small droplets of the mobile phase containing the analyte. A cylindrical electrode is placed in continuation of the capillary. An electrical potential of typically 2–5 kV is coupled between the capillary tip and the cylindrical electrode. This potential transfers electrical potential to the small drops. As the mobile phase is volatile, the liquid droplets evaporate and are flushed away by a *drying gas*. The analyte molecules remain charged and are extracted into the vacuum area of the mass spectrometer for further analysis by the mass analyser.

LC-MS is more complicated by the fact that the mobile phase has to be evaporated and removed before the sample constituents enter the mass spectrometer, to maintain the vacuum conditions. This problem is partly solved by use of the drying gas to remove much

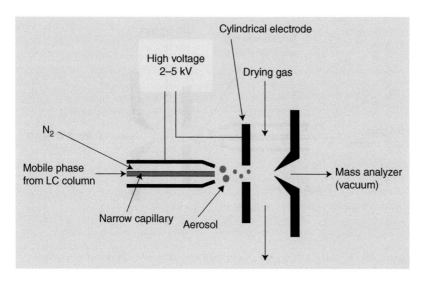

Figure 15.9 Schematic view of electrospray ionization

of the mobile phase and partly by using narrow-bore LC columns. In addition, powerful pumps are used in LC-MS to keep the vacuum inside the instrument.

ESI can form positive as well as negative ions. The positive ions $[M+H]^+$ are formed by protonation, which is favoured by acidic conditions, while negative ions $[M-H]^-$ are formed by release of a proton, which is favoured under alkaline conditions. The ions will appear in the mass spectrum at m/z $[M+1]$ and $[M-1]$, respectively. Thus, only molecules able to release a proton are observed in the negative ionization mode, which limits the use to acidic compounds like carboxylic acids and phenols. In the positive mode, amines are easily ionized but most polar molecules with functional groups such as alcohols, carbonyl, amides, and even carboxylic acids are ionized as well.

Some compounds that are not basic may be ionized by *adduct formation* with sodium or potassium, and will be observed at m/z values of 23 and 39 units above the values of the neutral molecule, respectively. Adduct formation can also happen unintentionally.

Due to the low energy ionization, fragmentation is limited. Dependent on the instrument settings, however, some substances undergo *in-source fragmentation* by collision with gas molecules on their way into the mass analyser. In this case, fragments of the molecular ion will also be present in the mass spectrum.

The pH of the mobile phase is adjusted to favour ionization of the analytes. For basic drugs, this is accomplished by acidic conditions in the LC mobile phase. Volatile buffer components, such as ammonium salts of formic acid and acetic acid should be used in the mobile phase to ease an efficient transport away with the drying gas. Non-volatile components like phosphate buffers should be avoided in the mobile phase.

As an alternative to ESI, *APCI* is used in LC-MS. The principle of APCI is shown in Figure 15.10. The mobile phase containing the analytes flows through a heated capillary and is evaporated at the end of the capillary and transported away by a drying gas. A needle

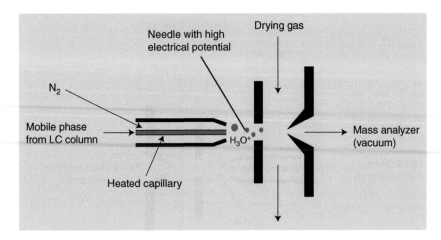

Figure 15.10 *Schematic view of atmospheric pressure chemical ionization*

connected to a high electric potential in the order of several kilovolts ensures that nitrogen and small amounts of water present at the end of the capillary are ionized according to the following reactions:

$$N_2 + e^- \rightarrow N_2^{\bullet+} + 2e^- \tag{15.16}$$

$$N_2^{\bullet+} + 2N_2 \rightarrow N_4^{\bullet+} + N_2 \tag{15.17}$$

$$N_4^{\bullet+} + H_2O \rightarrow H_2O^{\bullet+} + 2N_2 \tag{15.18}$$

$$H_2O^{\bullet+} + H_2O \rightarrow H_3O^+ + OH^\bullet \tag{15.19}$$

H_3O^+ and $OH^\bullet$ (OH radical) can then react with the analyte (M) according to the following reactions:

$$H_3O^+ + M \rightarrow (M+H)^+ + H_2O \tag{15.20}$$

$$OH^\bullet + M \rightarrow (M-H)^- + H_2O \tag{15.21}$$

Protonated molecular ions $(M+H)^+$ or deprotonated molecular ions $(M-H)^-$ are measured by the mass spectrometer. APCI is a relatively soft ionization technique and mainly molecular ions are formed. APCI can be used as an alternative to ESI and is particularly applicable if the analytes do not contain acidic or basic groups. For this type of compounds, it can be difficult to get sufficient signals with ESI.

15.7.3   Single Quadrupole Analysers

The produced ions are separated according to their m/z values in the mass analyser. The most simple and versatile mass analyser is termed *quadrupole*. The quadrupole mass analyser consists of four parallel rods to which are applied both a constant voltage and a radio frequency oscillating voltage (Figure 15.11). The electric field deflects ions in complex trajectories as they pass through the mass analyser, allowing only ions with one particular m/z ratio to reach the detector. Other ions collide with the rods and are lost before they reach

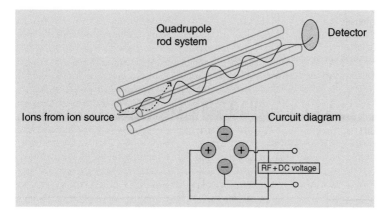

Figure 15.11 *Schematic view of quadrupole mass analyser*

the detector, as illustrated with a dashed line in Figure 15.11. Rapidly varying voltages select ions of different m/z values to reach the detector, and in this way the instrument can scan a wide m/z range in a very short time (full scan) or rapidly switch between different preset m/z values (SIM mode). Quadrupole MS instruments are very common detectors for chromatography. They have unit resolution and can be operated either in full scan or SIM mode.

15.7.4 Triple Quadrupole Analysers

Alternatively, mass spectrometers can be equipped with three quadrupoles in series, and such instruments are termed *triple-quadrupole mass spectrometers*. The principle of triple-quadrupole mass spectrometers is illustrated in Figure 15.12 and these instruments can be used for quantitative determinations by SRM or *multiple reaction monitoring* (MRM).

In a *triple-quadrupole mass spectrometer*, three mass analysers are coupled in series prior to the detector. Sample constituents enter the first mass analyser (Q1), which is locked to an m/z value that is characteristic for the substance to be determined. This m/z value is most often the molecular ion of the substance. The ions that are selected in the first

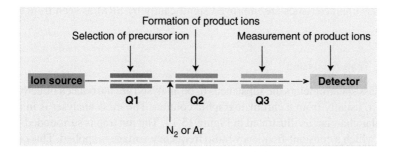

Figure 15.12 *Schematic view of triple quadrupole mass analyser*

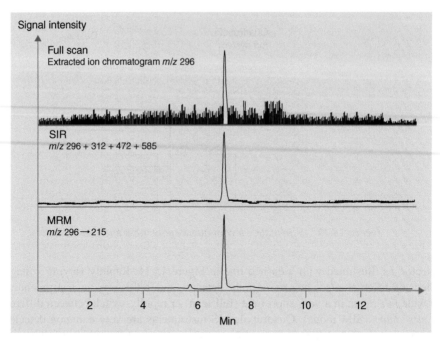

Figure 15.13 *Comparison of full scan (extracted ion chromatogram), SIM, and SRM from analysis of diclofenac in rat plasma*

mass analyser are termed *precursor ions*. All other *m/z* values cannot pass Q1 and this gives rise to the high selectivity of SRM. The precursor ions then pass to the next mass analyser (Q2), where they react with either N_2 or Ar. Here the ions fragment to *product ions* and the most favourable product ions are selected in the third mass analyser (Q3) and measured by the detector. Product ions are usually very specific to a given substance, and detection of product ions at given masses therefore provides a highly reliable identification. At the same time, the measurements are very sensitive because the ion background from other substances is effectively eliminated. Due to the high sensitivity and selectivity, triple quadrupole instruments are used extensively in analysis of drug substances and metabolites in biological samples like plasma. Figure 15.13 illustrates the sensitivity and selectivity of SRM compared to a full scan and SIM in analysis of diclofenac in rat plasma. Signal-to-noise characteristics are improved significantly by SRM as seen from the chromatograms.

15.7.5 Ion Trap Analysers

In *ion trap* mass spectrometers the substances to be analysed are introduced directly into the mass analyser, usually from a chromatographic column. The mass analyser is in principle a small circular chamber as illustrated in Figure 15.14. The ion trap is surrounded by a ring electrode, to which a constant-frequency radio frequency voltage is applied. This causes the ionized substances to move in stable trajectories inside the ion trap. Increasing the amplitude of the radio frequency voltage expels ions of a particular *m/z* value by sending them into

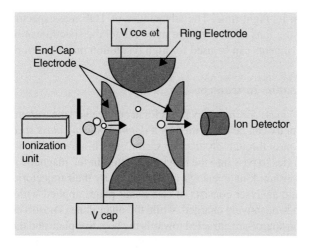

Figure 15.14 *Schematic view of ion trap mass analyser*

unstable trajectories that pass through the exit holes in the end caps. Ions expelled through the lower end cap are captured by the detector and registered with high sensitivity. Ion trap instruments can isolate ions of interest and reject all other ions. The isolated ions can then be fragmented by collision with a gas, resulting in a mass spectrum of the fragment ions (MS/MS). This procedure for further fragmentation of fragment ions can be repeated several (*n*) times and is termed MS^n. Using MS/MS increases sensitivity as much of the background is removed. This is often referred to as *MS/MS in time* in contrast to fragmentation in the triple quadrupole, which is referred to as *MS/MS in space*.

15.7.6 Time-of-Flight Analysers

In *time-of-flight* (TOF) mass spectrometry, ions formed in the ion source are accelerated in pulses by means of an electric potential imposed on a back plate right in the back of the ion source. Positive ions are accelerated by a positive potential applied to the back plate. Ions fly from the ion source and into a field-free flight tube without any magnetic or electrostatic field, as illustrated in Figure 15.15. All ions are accelerated to the same kinetic energy ($1/2 \times mv^2$), which means that the heavier ions will fly more slowly than light ions. At the other end of the flight tube is a detector. The light ions will hit the detector at the other end of the flight tube faster than heavier ions (lower flight time). The mass of each ion is thus

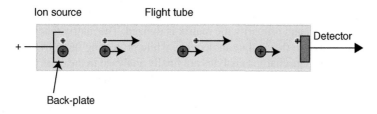

Figure 15.15 *Schematic view of time-of-flight mass spectrometry*

determined based on its flight time. The advantages of TOF mass spectrometry is that it can be used to measure ions in a very large mass range, the spectra can be taken up very quickly, and the instruments can be used for high resolution mass spectrometry.

15.7.7   High Resolution Instruments

High resolution (HR) mass spectrometry provides accurate mass measurements. High resolution mass spectrometry can be performed by a *double focusing mass spectrometer*. After ionization, the ions pass through an electric sector, which allows only ions of a narrow range of kinetic energies to pass into the magnetic sector. In the magnetic sector, ions with different mass but constant kinetic energy are separated by their trajectories in a magnetic field. The electrostatic analyser consists of two metal plates applied with an electrostatic field. The inner plate is negatively charged, while the outer plate is positively charged. Positive ions from the ion source are attracted towards the negative plate and the deflection will depend on the strength of the electrostatic field applied to the plates and on the m/z value of the ions. The electrostatic analyser focuses ions with the same mass, even if these have a slightly different speed and direction from the ion source. The electrostatic field can be varied in order to scan ions with different masses through the electrostatic analyser. Following this, the ions enter a magnetic field, which causes deflection of their trajectories depending on the magnetic field strength and the m/z values of the ions. By varying the magnetic field, different ions pass the magnetic analyser and are registered by the detector. The focusing in the electrostatic analyser makes the mass determination in the subsequent magnetic analyser much more accurate, and resolving powers of 10 000 corresponding to a resolution of 0.001 at m/z 100 can be obtained.

Another high resolution instrument is the *Orbitrap* MS. The Orbitrap mass analyser provides resolving powers of 140 000–480 000 at m/z 100 and m/z accuracy of 1–5 ppm. The principle of the m/z separation in the analyser is beyond the scope of this textbook. The soft ionization by ESI combined with high resolution make the Orbitrap instruments very suitable for identification and quantitation of large molecules like peptides and proteins, where high resolution is needed for reliable identification.

15.7.8   Matrix-Assisted Laser Desorption/Ionization Mass Spectrometry

Matrix-assisted laser desorption/ionization mass spectrometry (MALDI-MS) is a soft MS ionization technique that is primarily used for the analysis of large biomolecules, like proteins, peptides, and sugars. Ionization is accomplished with a laser beam. Typically, a small solution containing the analyte is mixed with a solution of a matrix component such as 2,5-dihydroxybenzoic acid. The liquid is evaporated, providing a dry spot of fine crystals of the matrix and the analyte. The laser beam is directed towards this spot. The matrix absorbs the laser energy, and primarily the matrix is ionized. The matrix then transfers part of its energy and charge to the analyte molecules, while protecting them from the disruptive energy of the laser. The charged analytes are finally analysed in the mass spectrometer; typically TOF analysers are used for this purpose. MALDI-MS is an official method of analysis in Ph. Eur., and the use of this technique is expected to increase in the future with the introduction of more drugs based on biological macromolecules. MALDI-MS is not combined with chromatography.

16

Sample Preparation

16.1 When is Sample Preparation Required?

Many samples are complex mixtures of different chemical and biochemical substances. In pharmaceutical analysis, the samples are typically pharmaceutical preparations like tablets or ointments, or biological fluids like blood and urine. In pharmaceutical preparations, the active pharmaceutical ingredient (API) is normally mixed with various excipients, and this may complicate the identification and quantitation of the API. Blood and urine are also

Introduction to Pharmaceutical Analytical Chemistry, Second Edition.
Stig Pedersen-Bjergaard, Bente Gammelgaard and Trine Grønhaug Halvorsen.
© 2019 John Wiley & Sons Ltd. Published 2019 by John Wiley & Sons Ltd.

very complex mixtures consisting of numerous components and the drug substances of interest (analytes) are often present in very low concentrations. Low concentration in this context is from a high pg/mL to a low μg/mL level. Analysis of drugs in complex mixtures is challenging for several reasons:

- Complex mixtures can contain one or more substances that provide a false response (interference) during the analytical measurement.
- The analyte concentration may be below the detection limit of the analytical instrument.
- Samples may be incompatible with the instrument or may contain substances that contaminate the analytical instrument.

Many pharmaceutical preparations are not readily compatible with analytical instruments as they are solid or semi-solid, and the API has to be extracted into a solvent before introduction to the analytical instrument. Analysis of biological samples is in general far more complicated than analysis of pharmaceutical preparations. Biological samples contain numerous components, and there is always a risk that some of these matrix components may cause interference. In some cases, the concentration of drugs is below the quantitation limit, and samples have to be pre-concentrated prior to detection. Finally, substances in biological samples can also damage the analytical instrument. Clogging of liquid chromatography (LC) columns by proteins from blood samples is an example of this.

The solution to all of these challenges is sample preparation. Very often, sample preparation involves *extraction,* which is defined as a process where the analyte is selectively transferred from the sample phase into another phase (extraction phase). Thus, extraction involves a phase transfer.

16.2 Main Strategies

The following four sample preparation approaches are frequently used in pharmaceutical analysis:

- Protein precipitation
- Liquid–liquid extraction (LLE)
- Solid–liquid extraction (SLE)
- Solid phase extraction (SPE)

Protein precipitation is widely used for plasma and serum samples to remove proteins prior to analysis by LC. Thus, protein precipitation is mainly used in *bioanalysis,* and the technique is therefore described in detail in Chapter 20.

LLE is used for liquid samples, including blood, urine, and liquid pharmaceutical preparations. The sample solution for LLE is aqueous. The aqueous sample is mixed with an organic solvent immiscible with water, resulting in a two-phase system. This two-phase system is shaken vigorously and, during this shaking, the analytes are transferred from the aqueous sample to the organic phase, which is termed *extract.* After shaking, the mixture is centrifuged, and the two phases separate again. The extract (organic phase) containing the analyte is then collected for the final analysis, typically performed by LC or gas chromatography (GC). LLE is an extraction technique as the analyte is transferred from one liquid phase (sample) to another liquid phase (extract).

For solid samples, sample preparation can be accomplished by SLE. Typical examples of solid samples are tablets or capsules. The solid sample to be analysed is powdered and an organic solvent or an aqueous solvent is added. The mixture is shaken and the analyte is extracted from the solid matrix and into the solvent. The solvent containing the analyte (extract) is then collected for the final analysis. Thus, in SLE, the analyte is transferred from the solid phase (sample) into a liquid phase (extract).

A very common sample preparation technique is SPE. In SPE, a small column packed with a sorbent (stationary phase) is used, which resembles the stationary phases used for LC. In SPE, the sample is normally a liquid in itself, often a biological fluid. The sample is applied to the column and drawn through the column, and in this process the analytes of interest are retained in the column by different types of interactions with the sorbent. After washing to remove matrix components, the analytes are eluted from the column with a solvent that breaks the analyte–sorbent interactions, and this eluate (extract) is collected for the final analysis. Thus, in SPE, the analyte is transferred from one liquid phase (sample) to another liquid phase (eluate).

16.3   Recovery and Enrichment

Before going more into the different extraction techniques, the terms recovery and enrichment have to be defined. These are essential parameters to evaluate the efficiency of the extraction. *Recovery* is the percentage of analyte collected during extraction and expresses the efficiency of the extraction process. Recovery (R) is given by the following equation:

$$R = \frac{n_{E,final}}{n_{S,initial}} \times 100\% \tag{16.1}$$

where $n_{E,final}$ is the amount of analyte collected in the final extract and $n_{S,initial}$ is the original amount of analyte present in the sample. If the recovery is 100%, all analyte molecules are transferred from the sample and into the extract. If recovery is 25%, only one-quarter of the analyte molecules are transferred into the extract. Extraction methods are preferably developed to provide close to 100% recovery. For practical applications, the following equation can be used to calculate recoveries:

$$R = \frac{C_{E,final} V_E}{C_{S,initial} V_S} \times 100\% \tag{16.2}$$

where $C_{E,final}$ is the concentration of analyte in the final extract, V_E is the volume of the extract, $C_{S,initial}$ is the concentration of analyte initially present in the sample, and V_S is the volume of the sample. Recoveries are used to characterize sample preparation methods based on extraction. For quantitative analytical methods where several analytes are measured simultaneously, the method should specify the recovery for each of the analytes because different analytes can be extracted with different efficiencies. Recoveries are determined during method development and validation. In addition, if the recovery for a certain analyte is less than 100%, the recovery has to be determined every time the method is calibrated with calibration samples for quantitation. This is important in order to correct the analytical results for the fact that the extraction is not complete. The precision on the recovery is equally important as the recovery percentage. An example of how to determine the recovery for an extraction method is given in Box 16.1.

Box 16.1   Determination of extraction recovery

A new drug substance under development was extracted from plasma by LLE and the extract was analysed by LC-MS. Because the method was new, the recovery for the LLE procedure had to be established. This was accomplished as follows:

1 mL of plasma was extracted with 4 mL of an organic solvent. After extraction and separation of the two phases, the organic phase was isolated and evaporated to dryness. The residual was dissolved in 0.25 mL of the mobile phase.

To measure the recovery the following solutions were prepared:

Solution 1: Drug substance spiked into drug-free plasma, concentration $= 1.0\,\mu g/mL$
Solution 2: Drug dissolved in the mobile phase, concentration $= 4.0\,\mu g/mL$

Solution 1 was extracted and analysed by LC-MS, measured peak area $= 11\,263$
Solution 2 was analysed directly by LC-MS, measured peak area $= 12\,467$
First the concentration of drug in solution 1 is calculated, which equals $C_{E,final}$ in Eq. (16.2):

$$C_{E,final} = \frac{11\,263}{12\,467} \times 4.0\,\mu g/mL = 3.6\,\mu g/mL$$

Then the recovery is calculated based on Eq. (16.2):

$$R = \frac{C_{E,final}V_E}{C_{S,initial}V_S} \times 100\% = \frac{3.6\,\mu g/mL \times 0.25\,mL}{1.0\,\mu g/mL \times 1.00\,mL} \times 100\% = 90\%$$

In this case, 90% of the analyte molecules were transferred into the extract and were available for the final analysis by LC-MS.

During extraction, the volume of the extract is often smaller than the volume of the sample ($V_E < V_S$) and in such cases the analytes may be *pre-concentrated* or *enriched* during the extraction process. The level of enrichment can be expressed by the *enrichment factor* (*E*), defined by the following equation:

$$E = \frac{C_{E,final}}{C_{S,initial}} = \frac{V_S R}{V_E \times 100\%} \tag{16.3}$$

Enrichment is important when analysing drugs at very low concentration levels (pg/mL or low ng/mL level), to ensure that the extract contains a concentration that is above the limit of quantitation for the analytical method and instrument. Enrichment may be an issue in bioanalysis where analyte concentrations can be low, and under certain conditions enrichment factors up to 10–20 times can be achieved. Enrichment is normally not an issue for the analysis of pharmaceutical preparations because API concentrations are relatively high. Box 16.2 illustrates an example on how to calculate enrichment.

Box 16.2   Determination of enrichment

In Box 16.1 the recovery was calculated for a new extraction method. Based on the data from this experiment, the enrichment can be calculated using Eq. (16.3):

$$E = \frac{C_{E,\text{final}}}{C_{S,\text{initial}}} = \frac{3.6\,\mu g/mL}{1.0\,\mu g/mL} = 3.6$$

This value means that the concentration of drug in the final extract was 3.6 times higher than in the original sample.

16.4   Liquid–Liquid Extraction

The principle of *LLE* is illustrated in Figure 16.1. In short, the aqueous sample is mixed with an organic solvent immiscible with water, resulting in a two-phase system. This two-phase system is shaken vigorously for a certain period of time and small droplets of the organic phase mix with the aqueous sample. During this shaking, the analytes are transferred from the aqueous sample to the organic phase, which is termed *extract*. After shaking, the two phases separate again. This is often accelerated by centrifugation. The extract (organic phase) containing the analyte is then collected for the final analysis, typically by LC or GC.

16.4.1   Procedure

LLE is frequently used for sample preparation of biological fluids (mainly blood and urine) and is also in use for liquid pharmaceutical preparations. For LLE of blood samples (whole blood, plasma, serum), sample volumes are typically limited to 100–1000 µL. For such samples the typical LLE procedure starts (step 1) with pipetting the sample into a *centrifuge tube*. For LLE of pharmaceutical preparations, larger sample volumes are available, typically in the range 10–100 mL, and in such cases LLE is up-scaled and performed in a *separation funnel*. Separation funnels are known from the organic chemistry laboratory. In step 2, pH in the sample is adjusted by adding a buffer solution or a strong acid or base to ensure an efficient extraction of the analyte. In step 3, the *extraction solvent* is added to the sample. It is important that the solvent is immiscible with water to form a two-phase system with the sample. Then, in step 4 the centrifuge tube or the separation funnel is shaken for a certain period of time to ensure efficient mixing of the two phases. During this, the analyte

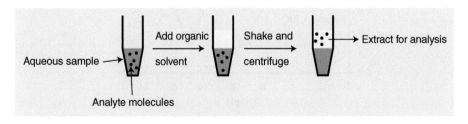

Figure 16.1   Principle of LLE

is transferred from the aqueous sample and into the small droplets of organic solvent formed by shaking. After this, the centrifuge tube or the separation funnel is allowed to stand for the two phases to separate again (step 5). Especially with biological fluids, this phase separation can be difficult, but centrifugation can accelerate the phase separation. After centrifugation, the organic phase (extract containing the analyte) can be collected (step 6), and basically it is ready for the final analytical measurement. However, often the organic phase has to be evaporated and the residue has to be reconstituted into a liquid that is compatible with the analytical instrument. As an example, extracts for LC are reconstituted in the mobile phase.

16.4.2   Theory

The fundamentals of liquid–liquid partition were discussed in Section 3.4. To summarize, Eqs. (16.4) to (16.6) describe the distribution ratio (D) under LLE conditions for neutral, acidic, and basic analytes, respectively:

$$\text{Neutral analytes}\quad D = K_D \tag{16.4}$$

$$\text{Acidic analytes}\quad D = K_D\frac{[\text{H}^+]}{[\text{H}^+] + K_a} \tag{16.5}$$

$$\text{Basic analytes}\quad D = K_D\frac{K_a}{[\text{H}^+] + K_a} \tag{16.6}$$

For a neutral analyte, with no acidic or basic functional groups, the extraction recovery is independent of pH in the aqueous sample and no pH adjustment is required. For neutral analytes, recovery is determined by the partition ratio (K_D) only. A high partition ratio gives a high recovery.

For acidic and basic analytes, pH in the sample plays a major role, and pH adjustment in the sample is normally required prior to extraction. For efficient extraction of acidic analytes, Eq. (16.5) shows that pH in the sample should be below the pK_a value of the analyte. This is exemplified in Figure 16.2, where the distribution ratio is plotted as a function of pH for the acidic drug substance ibuprofen. From this figure it is clear that the highest distribution into the organic phase is obtained at pH values at least two units below the pK_a

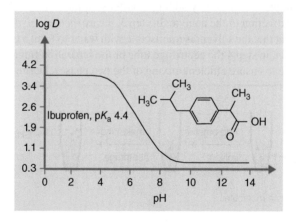

Figure 16.2   *Distribution ratio (as log D) as a function of pH for an acidic analyte (ibuprofen)*

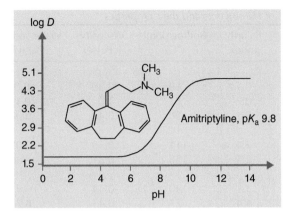

Figure 16.3 *Distribution ratio (as log D) as a function of pH for a basic analyte (amitriptyline)*

value for the acidic analyte, where the acidic analyte is protonated and thereby is neutral and prone to extraction into the organic phase.

Figure 16.3 shows a similar plot for a basic analyte (amitriptyline as an example). In this case, the highest distribution into the organic phase is obtained with a pH in the aqueous phase at least two units higher than the pK_a value. This is in agreement with Eq. (16.6). Thus, to effectively extract basic analytes into organic solvents, the sample should be alkaline. Under alkaline conditions in the sample, the protonated amine is neutralized and extraction into the organic solvent is more efficient.

16.4.3 Extraction Solvents

The organic solvent is very important in LLE. From a theoretical point of view, the organic solvent determines the value of K_D in Eqs. (16.4) to (16.6). From a practical point of view, the solvent should be selected with care and based on the 'like dissolves like' principle of organic chemistry. For very non-polar analytes, which are hydrocarbon-like substances with no polar functional groups, non-polar solvents are preferred. Typical non-polar extraction solvents are aliphatic hydrocarbons like *n*-heptane or cyclohexane, as mentioned in Table 16.1. For analytes of low polarity, but with an aromatic character, aromatic solvents like toluene or *p*-xylene can be more efficient. For the more polar analytes, a more polar solvent should be selected, but still not too polar because the solvent should be immiscible with water to form a two-phase system. For polar analytes, solvents with high *dipole moments* are very efficient. Examples of solvents with strong dipole properties are dichloromethane and ethyl acetate.

Basic analytes are amines, which are characterized by high *hydrogen bond basicity*. This implies that amines are prone to hydrogen bonding and that they serve as hydrogen bond acceptors based on the free electrons of nitrogen. Therefore, the ideal solvent for extraction of basic analytes is a solvent with high *hydrogen bond acidity*. Chloroform is one example of an LLE solvent with high hydrogen bond acidity. For acidic analytes, which are characterized by their hydrogen bond acidity, the ideal LLE solvent provides high hydrogen bond basicity. Typical examples of LLE solvents with high hydrogen bond basicity are diethyl ether and methyl-tertbutyl ether (see Table 16.1).

Table 16.1   Examples of LLE solvents and their properties

Solvent	Polarity index	Boiling point (°C)	Viscosity (cPoise)	Water solubility (%)(w/w)	Density (g/mL)
Non-polar					
n-Hexane	0.1	69	0.30	0.014	0.655
Cyclohexane	0.2	81	1.0	<0.1	0.779
n-Heptane	0.1	98	0.42	0.01	0.684
Aromatic					
Toluene	2.4	111	0.59	0.05	0.867
p-Xylene	2.5	138	0.81	Insoluble	0.861
Dipolar					
Dichloromethane	3.1	40	0.44	0.9	1.327
Ethyl acetate	4.4	77	0.43	8.7	0.894
Hydrogen bond acidity					
Chloroform	4.1	61	0.53	0.8	1.498
Hydrogen bond basicity					
Methyl-tertbutyl ether	2.5	55	0.27	5.1	0.741
Diethyl ether	2.9	35	0.24	7.5	0.713

As discussed above, polar analytes need to be extracted with relatively polar organic solvents and non-polar analytes need to be extracted with more non-polar solvents. The polarity of solvents can be described by the *polarity index*, as illustrated in Table 16.1. For the most non-polar aliphatic hydrocarbons, the polarity index is very close to 0.0. As the polarity increases, the polarity index also increases. Water has a polarity index of 10.2. Organic solvents with a polarity index below 4 or 5 are normally immiscible with water and form a two-phase system with aqueous samples. Organic solvents with a polarity index above 5 are miscible with water and can therefore not be used for LLE.

The choice of solvent for LLE does not solely focus on 'like dissolves like' principles and analyte solubility. Other properties of the solvents, such as density and viscosity, are also crucial. Solvents with low *viscosity* are preferred because they disintegrate more easily into small droplets and mix with the aqueous sample during extraction, which is beneficial for the transfer of analyte into the organic phase. Solvents with a lower *density* than water will float on top of the aqueous sample after extraction. This is an advantage if the extract is to be collected by a pipette after extraction from a small centrifuge tube. Examples of low density solvents are *n*-hexane and toluene. If a solvent with a higher density than water is used, it will be recovered below the aqueous sample after LLE. Examples of solvents with a higher density than water are chloroform and dichloromethane. The *volatility* of the solvent and the *solubility in water* are also crucial. Any evaporation of solvent after extraction is much faster with a solvent with a low boiling point, and volatile solvents are normally preferred if they have to be evaporated after extraction. Finally, as seen in Table 16.1, most organic solvents used in LLE are slightly soluble or less in water. The non-polar solvent *n*-hexane shows negligible water solubility, whereas ethyl acetate dissolves in water in substantial amounts. Thus, during LLE, small amounts of organic solvent leak into the aqueous sample and small amounts of water leak into the organic solvent. However, this leaking is at trace levels and has no practical consequences.

16.4.4 Parameters Affecting Extraction Recovery

Equations (16.5) and (16.6) demonstrate that the distribution ratio (D) is very close to the partition ratio (K_D) if the extraction is carried out at optimal pH (low pH for acids and high pH for bases). Under such conditions, the expected recovery (R) is linked to the partition ratio by the following equation:

$$R = \frac{K_D V_{\text{Organic}}}{K_D V_{\text{Organic}} + V_{\text{Aqueous}}} \times 100\% \qquad (16.7)$$

V_{Aqueous} is the sample volume and V_{Organic} is the volume of the organic phase. From Eq. (16.7), the basic parameters affecting extraction recovery can be identified as the partition ratio, sample volume, and volume of the organic phase. From the equation, recovery is expected to increase with an increasing partition ratio, a decreasing sample volume, and an increasing volume of the organic phase. This is exemplified in Box 16.3. Increasing the partition ratio is in practice accomplished by selection of a more suitable organic solvent. Because recoveries are dependent on the partition ratio, very polar analytes ($\log P < 1$) are less suited for LLE.

Box 16.3   Calculation of theoretical recoveries

A 1.0 mL plasma sample was extracted with a 1.0 mL organic solvent, and the partition ratio for the analyte in this particular system was 10. The recovery is calculated based on the following equation:

$$R = \frac{10 \times 1.0}{10 \times 1.0 + 1.0} \times 100\% = 91\%$$

A similar extraction is conducted where the volume of the organic solvent is increased to 5.0 mL:

$$R = \frac{10 \times 5.0}{10 \times 5.0 + 1.0} \times 100\% = 98\%$$

Clearly, the recovery in LLE can be increased by increasing the volume of organic solvent.

Alternatively, the organic solvent can be optimized to increase the partition ratio. Suppose that the extraction is carried out with 1 mL of a solvent providing a partition ratio of 30:

$$R = \frac{30 \times 1.0}{30 \times 1.0 + 1.0} \times 100\% = 97\%$$

Clearly, recovery in LLE can also be increased by increasing the partition ratio (selecting a more suitable solvent). If both approaches are combined, namely increasing the volume of organic solvent and the partition ratio, the recovery is calculated as

$$R = \frac{30 \times 5.0}{30 \times 5.0 + 1.0} \times 100\% = 99\%$$

In the latter case, the extraction is essentially quantitative or exhaustive, which implies that the total amount of analyte is transferred from the sample and into the extract.

During development of LLE methods, *exhaustive extraction* (100% recovery) is preferred. In some cases, this can be accomplished in a single extraction, as illustrated in Box 16.3. This requires a high partition ratio and implies that the analyte is relatively non-polar ($\log P > 2$). For more polar analytes ($\log P < 2$), the volume of the organic phase can be increased as illustrated in Box 16.3 to compensate for a poor partition ratio. However, another way of increasing the recovery, which is even more effective, is to extract the aqueous sample with several fresh portions of organic solvent, and combine the extracts after extraction. This is termed *multiple extractions*. Recovery (R) after n extractions can be calculated based on the following equation:

$$R = \left(1 - \left(\frac{V_{\text{Aqueous}}}{V_{\text{Aqueous}} + K_{\text{D}} V_{\text{Organic}}} \right)^n \right) \times 100\% \tag{16.8}$$

As illustrated by the equation, extraction recovery is increasing with an increasing number of extractions. The impact of multiple extractions is exemplified in Box 16.4.

Box 16.4   Calculation of theoretical recoveries for multiple extractions

A new drug substance is extracted from test oral solutions by LLE. The partition ratio is 5. First, a single extraction is performed with a 10.0 mL sample and a 10.0 mL organic solvent.

The expected extraction recovery after a single extraction is calculated according to Eq. (16.7):

$$R = \frac{K_{\text{D}} V_{\text{Organic}}}{K_{\text{D}} V_{\text{Organic}} + V_{\text{Aqueous}}} \times 100\% = \frac{5 \times 10.0 \text{ mL}}{5 \times 10.0 \text{ mL} + 10.0 \text{ mL}} \times 100\% = 83\%$$

Suppose that the same sample is extracted with two portions of 5.0 mL each as an alternative procedure. The expected recovery is calculated according to Eq. (16.8):

$$R = \left(1 - \left(\frac{V_{\text{Aqueous}}}{V_{\text{Aqueous}} + K_{\text{D}} V_{\text{Organic}}} \right)^n \right) \times 100\%$$

$$R = \left(1 - \left(\frac{10.0 \text{ mL}}{10.0 \text{ mL} + 5 \times 5.0 \text{ mL}} \right)^2 \right) \times 100 = 92\%$$

If a third extraction is performed with 5.0 ml organic solvent and all three portions of solvent are combined prior to analysis, the expected recovery is

$$R = \left(1 - \left(\frac{10.0 \text{ mL}}{10.0 \text{ mL} + 5 \times 5.0 \text{ mL}} \right)^3 \right) \times 100 = 98\%$$

These calculations clearly demonstrate the effectiveness of multiple extractions.

16.4.5   Liquid–Liquid Extraction with Back-Extraction

In LLE discussed so far, analytes have been transferred from aqueous sample and into an organic solvent (single extraction). In such cases, the organic extract is analysed directly, or the organic phase is evaporated and reconstituted in another liquid prior to analysis. With very complicated samples, a single extraction may not be sufficient because some of the matrix compounds are also extracted into the organic solvent. These matrix components can be endogenous substances from biological fluids or excipients of low water solubility from pharmaceutical preparations. In such cases, a second extraction step termed *back-extraction* may be performed, where the analytes are extracted into a new aqueous phase that is used for the final chemical analysis. Thus, the organic extract from the first extraction is subsequently extracted with a new aqueous phase in a second step. The advantage of back-extraction is improved sample clean-up, and the final extract is now aqueous. Aqueous extracts can be injected directly in LC and LC-MS. A typical scheme for back-extraction is as follows (for a basic drug substance): the sample is made alkaline and the basic drug substance is first extracted into an organic solvent. Subsequently, the organic extract is transferred into a new extraction tube and an acidic aqueous solution is added as a back-extraction solution. As the basic analyte is protonated (positively charged) in the acidic solution, it is effectively back-extracted from the organic solvent into the back-extraction solution.

16.5   Solid–Liquid Extraction

Solid–liquid extraction is abbreviated as *SLE* and is used to extract analytes from solid samples. In pharmaceutical analysis, SLE is mainly used for extraction of APIs from solid pharmaceutical preparations such as tablets and capsules. Figure 16.4 illustrates the principle for SLE. In short, the solid sample to be analysed is powdered and an organic solvent or an aqueous solvent is added. The mixture is shaken for a certain period of time and the analyte is extracted from the solid matrix and into the solvent. The solvent containing the analyte (extract) is then collected for the final analysis.

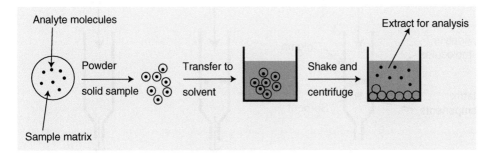

Figure 16.4   Principle of SLE

SLE is similar to LLE because the analytes are transferred from one phase to another; in this case from a solid sample into an extraction solvent. Thus, the theoretical considerations discussed for LLE also apply to SLE.

A typical SLE procedure starts with milling or powdering of the solid sample to homogenize and disintegrate the sample into small particles. In this way the analyte becomes more accessible to the solvent (step 1). This is important in order to achieve high recovery. After disintegration of the sample, a certain amount of sample is transferred to the extraction vessel (step 2). The extraction solvent is added (step 3), and the whole system is shaken, vibrated, or sonicated for a certain period of time (step 4). Shaking or sonication is important to speed up the extraction. During the extraction, the analyte molecules are released from the solid sample and dissolved into the extraction solvent. After extraction, the content of the vessel is filtrated in order to separate undissolved residuals of the solid sample from the extraction solvent (step 5). The filtrate is then used for the final chemical analysis.

SLE differs from LLE in two ways. Extraction of analytes from a solid sample may be difficult and often the same sample is extracted several times with new portions of solvent, and the individual portions are combined into the final extract (multiple extractions). Solvents miscible with water can be used in SLE as long as the major matrix constituents are not dissolved. Therefore, methanol, ethanol, and acetonitrile can be used as extraction solvents in SLE. Even water or aqueous solutions of base, acid, or buffer are used for SLE. Alkaline solutions are used to extract acidic analytes, while basic analytes are extracted using acidic solutions.

16.6 Solid Phase Extraction

16.6.1 Fundamentals

A very common sample preparation technique is *SPE*. The principle for SPE is illustrated in Figure 16.5. In SPE, a small column is packed with a *sorbent* (stationary phase), which

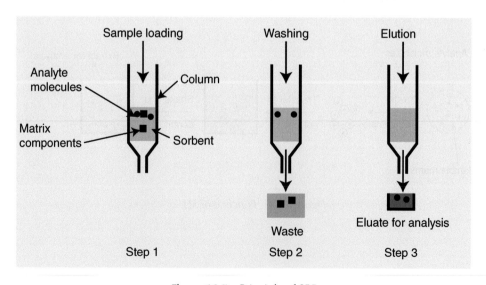

Figure 16.5 Principle of SPE

resembles the stationary phases used for LC. The mechanisms of interaction are comparable to the interactions in LC. The sample is in most cases a biological fluid, such as plasma or urine. The procedure involves three individual steps. In step 1, the sample is applied to the column and drawn through the column. This is termed *sample loading*, and the analytes of interest are retained in the column by different type of interactions with the sorbent. In most cases, the analytes are retained by hydrophobic interactions. In step 2, which is termed *washing*, the column is washed to remove matrix components, but without washing the analytes from the column. Finally, in step 3, which is termed *elution*, the analytes are eluted from the column with a solvent that disrupts the analyte–sorbent interactions, and the *eluate* (extract) is collected at the outlet of the column for the final analysis.

16.6.2   The Solid Phase Extraction Column

Figure 16.6 shows a schematic drawing of an *SPE column*. The column body is typically made of polypropylene. A certain mass of sorbent particles is packed in the column and immobilized by two polyethylene *filters* on each side of the bed. The size of the sorbent particles is typically in the range 40–50 µm. Thus, the SPE particles are much larger than the LC stationary phase particles, which are typically in the range 3–5 µm. The reason for using larger particles in SPE is that the liquid flow through the column is obtained by a vacuum system and smaller particles would result in a large back-pressure, which would make it impossible for liquid to flow through the column by means of a vacuum. Because of the large sorbent particles, biological fluids with relatively high viscosity are easily drawn through the column. In addition, the high separation efficiency of small LC stationary phase particles is not required in SPE, as the separation in SPE is very rough.

Columns for SPE are commercially available with a variety of physical dimensions and sorbent chemistries. The typical SPE column for bioanalysis contains 30–500 mg of sorbent. A broad range of different sorbents are available, based on *silica particles* or *organic polymeric particles*. Functional groups facilitate interactions with the analytes and are responsible for retention, and these are attached to the surface of the particles in large numbers. Silica-based particles have for many years been very popular. Those are highly porous particles with an average pore size of 60 Å. Small molecule drugs and related substances can diffuse into the pores, and the surface in the pore structures account for approximately 98–99% of the total surface of the particle. Thus, small molecule drugs can be retained very efficiently. On the other hand, macro-molecules, such as proteins present in plasma and serum, are prevented from entering the pore structures and are only exposed to the particle surface (1–2% of the total surface). Therefore, macro-molecules are poorly retained in the column and easily washed from the column. In this way, proteins are not present in the final eluate.

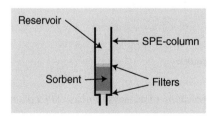

Figure 16.6   *SPE column*

16.6.3   Conditioning

Prior to the first step in the SPE procedure the column is *conditioned*. SPE columns are delivered from the manufacturer in a dry state. The functional groups are inactive and must be solvated to retain the analytes. *Solvation* is carried out by carefully drawing a polar organic solvent such as methanol through the column. Figure 16.7 illustrates the effect of solvation. After solvation, the functional groups have been 'raised' and retain the analytes much more efficiently. Following this, excess methanol located in the *bed volume* (= volume between the particles) needs to be removed to avoid loss of analyte, as methanol is a strong eluent in SPE, and if methanol is present in the bed volume during sample loading, the analytes can be poorly retained and partly lost from the column. Excess methanol is typically removed by drawing water or aqueous buffers through the column (reversed-phase SPE). Conditioning is very important with silica-based sorbents, whereas some organic polymer-based phases are claimed to be applicable without conditioning.

16.6.4   Equipment

The SPE columns are available from several manufacturers. They are intended for single use only and are discarded after use. SPE can be accomplished with a simple *vacuum manifold*, as shown in Figure 16.8. The purpose of the vacuum manifold is to generate a vacuum at the column outlet to draw the different liquids through the column. Individual valves for each column allow the vacuum to be activated and deactivated. The vacuum is activated when a liquid is to pass the column, whereas the vacuum is deactivated in between to prevent the column drying out. Small vials are placed under each column inside the vacuum manifold to collect the final eluate from each sample. In this way, several samples can be extracted simultaneously.

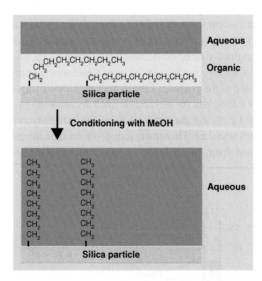

Figure 16.7   Conditioning and solvation of a reversed-phase SPE column

Figure 16.8   SPE vacuum manifold. Source: With permission from Agilent

16.6.5   Reversed-Phase Solid Phase Extraction

Reversed-phase SPE is typically used for extraction of relative non-polar analytes from biological fluids. Extraction is based on *hydrophobic interactions* between the analyte and the sorbent and is very similar to reversed-phase LC. Retention of the analyte increases with increasing hydrophobicity of the analyte. The interactions are based on *van der Waals forces*, mainly between hydrogen–carbon bonds in the sorbent and hydrogen–carbon bonds in the analyte molecules. Hydrophobic interactions are promoted in an aqueous solution, whereas they are suppressed in an organic solution.

Several different sorbents are available for reversed-phase SPE, as illustrated in Figure 16.9, where some of the most common silica-based materials are shown. C_{18} columns are the most common hydrophobic reversed-phase columns. They retain many compounds and are used for a variety of applications. However, as C_{18} columns can retain a broad range of compounds, the selectivity of this material is limited. In order to increase the selectivity, the use of less hydrophobic phases like C_8 columns or modified phases is an alternative.

An alternative to reversed-phase silica-based materials is polymer-based sorbents. An example of a polymer used as a sorbent in SPE is shown in Figure 16.10. The sorbent is a co-polymer built from two different monomers, namely the hydrophobic monomer *divinylbenzene* and the more polar monomer *N-vinylpyrrolidone*. The divinylbenzene moieties provide the sorbent hydrophobicity, which are responsible for hydrophobic interactions with the analytes. The *N*-vinylpyrrolidone moieties in addition provide polar functions, and these can retain more polar compounds. An important advantage of the polymer-based sorbents is that they do not change chemical properties even if they dry out during the procedure, and for some polymers initial conditioning is not required.

As mentioned above, hydrophobic interactions are promoted in aqueous solution, whereas they are suppressed in organic solution. This means that a relatively non-polar drug present in an aqueous biological fluid is an ideal case for reversed-phase SPE. The aqueous sample can be loaded directly on to the column. The non-polar nature

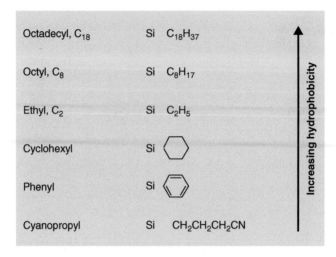

Figure 16.9   Different sorbents for reversed-phase SPE

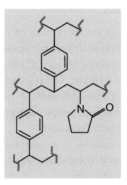

Figure 16.10   Example of polymeric sorbent for SPE

of the analyte results in strong retention by hydrophobic interactions with the sorbent. The aqueous nature of the biological fluid promotes hydrophobic interactions between the analyte and the sorbent. For acidic or basic analytes, retention is even better if the compounds are loaded in their uncharged state. For basic analytes, alkaline conditions in the sample can therefore improve the retention of the analyte on the column. This can be accomplished by diluting the sample with an alkaline buffer.

For *washing*, pure water or other 100% aqueous solutions are used to remove matrix components from the column. This will remove water-soluble compounds (polar compounds) with low affinity to the sorbent. By keeping the washing solution 100% aqueous, strong retention of the analytes is maintained. In some cases, if the analytes are very well retained, aqueous washing solutions containing 10% or 20% methanol or other water miscible organic solvents can be used, but care should be taken to avoid elution of the analytes from the column. Small amounts of methanol in the washing solution remove more interference from the column and improve clean-up.

In the final *elution* step, an eluent such as pure methanol or mixtures of methanol and water are used. Methanol provides organic conditions in the column and the hydrophobic interactions are suppressed. The key point during elution is that the hydrophobic interactions between the sorbent and analyte should be suppressed efficiently in order to elute the analytes from the column in the lowest volume of eluent possible. The reason for using small volumes of eluent is to avoid dilution of the eluate. If the analytes are basic or acidic compounds, pH in the elution solute can be adjusted to ionize the analytes and reduce hydrophobic interactions.

16.6.6   Secondary Interactions

With silica-based sorbents, the functional groups are attached to the surface of the silica particles. However, between the functional groups, *residual silanols* are present, as illustrated in Figure 16.11.

Residual silanols are acidic and in aqueous medium they can retain analytes due to ionic interactions. These interactions are termed *secondary interactions*. Secondary interactions can be advantageous, because analytes are retained by hydrophobic interactions as well as secondary interactions, and are therefore strongly retained in the column. In such cases, stronger washing solutions can be used without eluting the analyte, which provides more clean-up. For example, pure acetonitrile can be used during washing, and this solvent will remove many non-polar and more polar matrix components, while the analyte is still retained by secondary interactions. To elute analytes retained by secondary interactions, acidified methanol can be used. Methanol suppresses the hydrophobic interactions, while the acid reduces the ionization of the silanols and suppresses the secondary interaction. Because both types of interactions are suppressed simultaneously, the analyte is effectively eluted from the column.

16.6.7   Ion Exchange Solid Phase Extraction

Analytes containing ionic groups, like acids and bases, can be extracted from aqueous samples by *ion exchange SPE*. The principal type of interactions is *ionic interactions* between

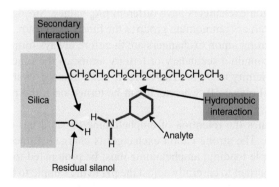

Figure 16.11   Secondary interactions in SPE

Sulfonylpropyl	$-\overset{\displaystyle |}{\underset{\displaystyle |}{Si}}- CH_2CH_2CH_2SO_3^-$	Strong cation exchanger
Carboxymethyl	$-\overset{\displaystyle |}{\underset{\displaystyle |}{Si}}- CH_2COO^-$	Weak cation exchanger ($pK_a = 4.8$)
Trimethylaminopropyl	$-\overset{\displaystyle |}{\underset{\displaystyle |}{Si}}- CH_2CH_2CH_2\overset{+}{N}-(CH_3)_3$	Strong anion exchanger
Diethylaminopropyl	$-\overset{\displaystyle |}{\underset{\displaystyle |}{Si}}- CH_2CH_2CH_2\underset{\displaystyle H}{\overset{+}{N}}(CH_2CH_3)_2$	Weak anion exchanger ($pK_a = 10.7$)

Figure 16.12 *Examples of sorbents for ion exchange SPE*

ionic groups on the sorbent and ionic groups on the analytes. Cationic analytes are positively charged and are retained by *cation exchangers*, which are negatively charged. Many APIs are amines and can be retained by cation exchangers with functional groups based on carboxylic or sulfonic acids. In a similar way, anionic analytes, like carboxylic acids, are negatively charged and can be retained by *anion exchangers*, which are based on functional groups of positively charged amines or quaternary ammonium ions.

Figure 16.12 gives an overview of some common ion exchangers used for SPE. The sorbents are characterized as *strong ion exchangers* or *weak ion exchangers*. A strong ion exchanger is ionized in the entire pH range and will act as an ion exchanger regardless of the pH in the SPE column. In contrast, a weak ion exchanger is only ionized in parts of the pH range, so weak ion exchangers can be turned on and off depending on the pH in the column. Strong cation exchangers typically use sulfonic acid functionalities as the ionic group. Sulfonic acids have very low pK_a values and are negatively charged in almost the entire pH range. Weak cation exchangers are often based on carboxylic acid functionalities as the ionic group. Carboxylic acids are weak acids, with pK_a values around 4.5–5.0. If the pH in the SPE column is above 3–5, the ion exchanger is negatively charged and turned on, while the ion exchanger is turned off when the pH is well below 3 and the carboxylic acid is uncharged.

In a similar way, anion exchangers have different pK_a values. Strong anion exchangers typically contain a quaternary ammonium group as the functional group, which is ionized in the entire pH range. Strong anion exchangers are therefore always turned on. Weak anion exchangers normally contain a secondary or tertiary amine as the functional group, with the pK_a value in the vicinity of 9–10. Therefore, a weak anion exchanger can be turned off by adjusting the pH above 10, whereas it can be turned on if the pH is lowered well below 10.

Figure 16.13 illustrates the retention of amphetamine (basic drug) on a strong cation exchange SPE column. The strong cation exchanger is charged (turned on) in the entire pH range. During sample loading, amphetamine must be protonated to be retained in the column. Thus, it is important to carefully adjust the pH in the sample to 8 (or lower), which is well below the pK_a value of amphetamine ($pK_a \approx 10.1$). Generally it is recommended to

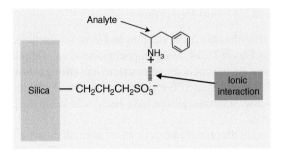

Figure 16.13 *Retention of amphetamine on a strong cation exchange SPE column*

adjust the pH at least two units below the pK_a value for basic analytes to be sure that the compounds are fully ionized. For pH adjustment of samples, buffers are recommended due to pH stability.

The same principle holds for retention of acidic analytes on cation exchangers. The acid must be ionized during loading, which requires a pH in the sample to exceed $pK_a + 2$.

To ensure strong retention of the analyte during sample loading, it is also important to have control over the content of other ions (matrix ions) in the sample. The reason for this is that matrix ions at a very high concentration, or with very high *affinity* for the ion exchanger, can compete with the analyte for the active sites on the sorbent. If this happens, the analyte may partly be flushed through the column during loading. This in turn will give low extraction recoveries. A high content of matrix ions in the sample is especially challenging when extracting urine samples. Ionic matrix components are abundant in urine and can vary substantially from sample to sample. In such cases, the sample can be diluted with water prior to extraction in order to reduce the impact of matrix ions.

After loading, the SPE column is *washed* in order to remove major matrix components. Buffer solutions are preferred for this purpose. Careful control of pH is mandatory in order not to turn off the ionization of the analyte during washing and thereby lose the analyte. This in turn results in low extraction recoveries. Methanol can be used for washing as it does not disturb the ionic interactions between the analyte and the sorbent. Washing with methanol can be efficient for removing neutral matrix compounds with low polarity, as they may be retained in the column by hydrophobic interactions with the hydrocarbon chain of the functional group (see Figure 16.12).

In the final step, the analyte is *eluted* from the SPE column. In the example from Figure 16.13 with amphetamine, the analyte can be eluted with an eluent at pH 12. Under alkaline conditions amphetamine is deprotonated, and the ionic interactions are immediately suppressed. Addition of methanol to the eluent is probably an advantage as amphetamine also can have some hydrophobic interaction with the hydrocarbon chains of the ion exchanger. Methanol will also increase the solubility of amphetamine in the eluent. In general, elution in the ion exchange SPE is based on turning off the ionic interactions. With strong ion exchangers, the ionic interactions are turned off by turning off the charge on the analyte. With weak ion exchangers, on the other hand, the ionic interactions are turned off by turning off the ionization of the ion exchanger. In all of these cases, turning off the ionic interactions involves selection of appropriate pH conditions in the eluent.

16.6.8 Mixed-Mode Solid Phase Extraction

In *mixed-mode SPE*, both hydrophobic interactions and ionic interactions are used for retention. This is illustrated in Figure 16.14, using amphetamine as an example. The silica-based sorbent contains C_8 groups for hydrophobic interactions and strong cation exchanger groups for ionic interactions. Polymer-based SPE sorbents are also available; an example is shown in Figure 16.15. In this case, sulfonate groups have been included in the polymer to provide cation exchange properties.

During sample loading, hydrophobic interactions are normally used as the primary source of retention. This is because hydrophobic interactions are less affected by variations in the sample matrix than ionic interactions. For washing, a buffer is used first to turn on the ionic interactions between the analyte and the sorbent. The pH in this solution should promote ionization of both the sorbent and the analyte. Washing with this buffer will remove polar matrix components. The column can subsequently be washed with pure methanol because the analyte is also retained by ionic interactions, which are not suppressed by methanol. By washing with methanol, neutral matrix components of low polarity can be removed. This is the reason why mixed-mode SPE results in very clean extracts from biological fluids.

For elution of the analytes in mixed-mode SPE, both the hydrophobic interactions and the ionic interactions should be turned off. The hydrophobic interactions are suppressed by using methanol in the eluent. The ionic interactions are suppressed by adding either a base or an acid to the eluent. For basic analytes, base is added to the eluent. High pH in

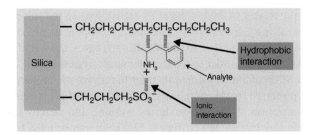

Figure 16.14   *Retention of amphetamine on a mixed-mode SPE column*

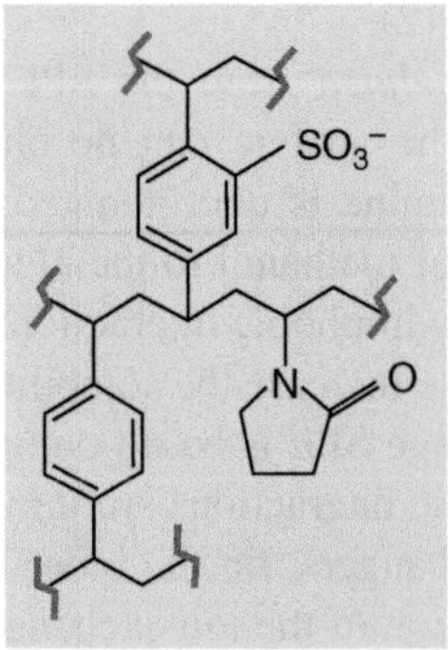

Figure 16.15   *Polymeric-based sorbent for mixed-mode SPE*

the eluent will suppress the ionization of the analyte, so typically methanol mixed with ammonia is used.

16.6.9   Normal-Phase Solid Phase Extraction

Normal-phase SPE can be used to isolate polar analytes present in non-aqueous samples. Normal-phase SPE is much less in use in pharmaceutical analysis than reversed-phase SPE. In normal-phase SPE, the analytes are retained in the column by *polar interactions* with the sorbent. The polar interactions are similar to the interactions in normal-phase LC, namely *hydrogen bond interactions, dipole/dipole interactions, induced dipole/dipole interactions*, and *π–π interactions*. An analyte should contain one or more polar functional group to take part in polar interactions. Thus, compounds containing hydroxyl groups, amines, carbonyl groups, aromatic rings, double bonds, or heteroatoms like oxygen, nitrogen, sulfur, or phosphorus can be extracted by normal-phase SPE.

Figure 16.16 presents some sorbents frequently used in normal-phase SPE. The functional groups are hydroxyl groups, amino groups, or cyanopropyl groups. Even pure silica particles, without any external groups attached, are used as sorbent for normal-phase SPE. In the latter case, the silanols on the surface of the silica particles serve as the functional groups.

Normal-phase SPE often relies on hydrogen bond interactions between the analyte and the sorbent. Hydrogen bond interactions occur when hydrogen interacts with electronegative atoms like oxygen or nitrogen. Strong hydrogen bond interactions usually occur between hydroxyl- and amino- groups. The polar interactions are turned on in a non-polar environment, whereas they are turned off in a polar environment. This means that the sample should be dissolved in a non-polar or low-polarity organic solvent during sample loading. An example of such a non-polar solvent is *n*-hexane. For elution of the analyte, change to a more polar solvent is required, which effectively competes with the analyte in terms of hydrogen bond interactions with the sorbent. Methanol is often used for this purpose.

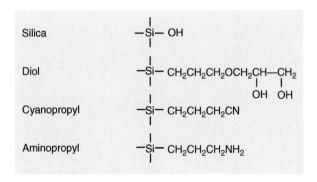

Figure 16.16   *Examples of sorbents for normal-phase SPE*

17

Quality of Analytical Data and Validation

17.1 Instrumental Signals

For quantitation of a given analyte in a certain sample, an instrumental signal is measured for the analyte. The instrumental signal normally increases linearly with concentration (amount) of the analyte in the sample, given by the following equation:

$$\text{Analyte concentration} = a \times [\text{instrumental signal}] + b \tag{17.1}$$

Introduction to Pharmaceutical Analytical Chemistry, Second Edition.
Stig Pedersen-Bjergaard, Bente Gammelgaard and Trine Grønhaug Halvorsen.
© 2019 John Wiley & Sons Ltd. Published 2019 by John Wiley & Sons Ltd.

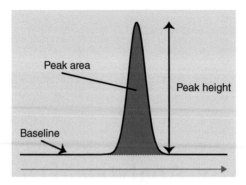

Figure 17.1   *Peak height and peak area of a chromatographic peak*

To quantify the analyte, *a* and *b* in Eq. (17.1) must be determined. The classical way to determine the relation is to measure the *instrumental signal* for a series of *standard solutions* containing exactly known concentrations of the analyte and to plot the signals of these versus concentration. The resulting linear relationship is termed a *standard curve* or a *calibration curve* and the entire procedure is termed *calibration*. When the data with standard solutions are treated mathematically by linear regression analysis, the values of *a* and *b* are calculated. The *slope* of the curve represents *a*, and *b* is the *interception* with the *x* axis. Simple standard solutions are prepared in pure water or another solvent such as the liquid chromatography (LC) mobile phase. Alternatively, the solutions used for calibration can be prepared in a sample solution similar to the real samples, and these are termed *calibration samples*. Calibration samples can be drug-free blood or urine samples spiked to an exactly known concentration with the analyte.

In high performance liquid chromatography (HPLC) and gas chromatography (GC), the instrumental signals are derived from chromatograms. A very simple chromatogram is shown in Figure 17.1, comprising a single peak for a single analyte. The signal, observed before and after the analyte elutes, is an almost straight line representing almost a zero detector response, which is termed the *baseline*. When the analyte enters the detector, the detector response increases until a maximum value is reached and then it returns to the baseline. Ideally the peak is a symmetrical Gaussian curve. The instrumental signal used for quantitation is the *peak area*, which is calculated by the instrument software. Previously, the *peak height* was also used for quantitation. If conditions are properly controlled both peak heights and peak areas vary linearly with the analyte concentration. The peak area is independent of peak broadening effects. Therefore, peak areas are preferred rather than peak heights as the measured instrumental signals in LC and GC as the peak area can be measured with a high degree of *accuracy* and *precision*.

17.2   Calibration Methods

Calibration is normally based on linear calibration equations (Eq. 17.1), and if more than one analyte is to be quantified, a calibration equation has to be established for each analyte. The most important calibration methods are:

- External standard method
- One point calibration
- Internal standard (IS) method
- Standard addition
- Normalization

17.2.1 External Standard

In the *external standard method*, a *chemical reference substance (CRS)* of the analyte is used to produce the *external standard solutions* for calibration. The CRS is a highly purified and well-defined quality of the analyte. The external standard solutions and the samples are analysed under identical conditions. The external calibration curve is normally and preferably linear. If analyte concentrations are expected to vary substantially from sample to sample (typical for bioanalysis), the external calibration curve is established within the expected concentration range. In such cases, the expected concentration range is typically covered by five or six external standard solutions. Quantitation based on the external standard method is exemplified in Box 17.1, where acetylsalicylic acid is determined in diluted samples of a new pharmaceutical product formulation.

Box 17.1   Quantitation of acetylsalicylic acid in a new pharmaceutical formulation based on calibration with the external standard method

The concentration of acetylsalicylic acid in diluted samples of a new pharmaceutical product formulation was determined. The analysis was by HPLC and the expected concentration range was from 0.1 to 1 µg/mL. External standard solutions containing 0.10, 0.25, 0.50, 0.75, and 1.0 µg/mL of acetylsalicylic acid were prepared in water and were analysed by the same procedure as the samples. The peak area (instrumental signal) for acetylsalicylic acid in the external standard solutions was read from the corresponding chromatograms:

Concentration of acetylsalicylic acid (µg/mL)	Peak area
0.10	11 952
0.25	29 906
0.50	60 335
0.75	87 988
1.0	117 520

The peak area (y) was plotted against concentration (x, µg/mL), yielding the following linear regression line as an external calibration curve calculated by the method of least squares:

$$y = 116\,971x + 715.15$$

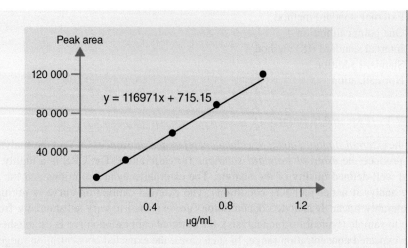

One of the new formulation samples was analysed under the same conditions and provided a peak area for acetylsalicylic acid of 105 432. The concentration x of acetylsalicylic acid in the sample was

$$105\,432 = 116\,971x + 715.15$$

$$x = \frac{105\,432 - 715.15}{116\,971} = 0.90 \, \mu g/mL$$

17.2.2 One Point Calibration

In the test of purity (assay) of pharmaceutical ingredients and assays for quantitation of the active pharmaceutical ingredient (API) in finished products, analyte concentrations are expected to be almost the same from sample to sample, because pharmaceutical ingredients are of high purity (very close to 100% w/w) and concentrations in pharmaceutical preparations are expected to be close to the nominal value. In such cases it is common practice to prepare a single external standard solution only, at the same concentration as expected for the sample. The approach is termed *one point calibration* and is the preferred pharmacopoeia method. The prerequisite of using one point calibration is that the analytical method is robust and validated, which is the case for pharmacopoeia methods.

The sample and the external standard solution are measured, and based on the instrumental signals, the concentration of analyte in the sample solution ($[A]_{sample}$) is determined according to the following equation:

$$\frac{S_{sample}}{S_{standard}} = \frac{[A]_{sample}}{[A]_{CRS}} \tag{17.2}$$

S_{sample} is the instrumental signal for the analyte in the sample, $S_{standard}$ is the instrumental signal for the analyte in the external standard solution, and $[A]_{CRS}$ is the concentration of CRS in the external standard solution. The concentration of the analyte in the sample solution is calculated by rearrangement of Eq. (17.2):

$$[A]_{sample} = [A]_{CRS} \frac{S_{sample}}{S_{standard}} \qquad (17.3)$$

Quantitation based on one point calibration is exemplified in Box 17.2.

Box 17.2   Quantitation (purity determination) of pharmaceutical ingredient of methotrexate based on one point calibration

The purity of the pharmaceutical ingredient methotrexate was determined by HPLC. The following procedure was applied based on one point calibration.

A *test solution* (sample) of methotrexate was prepared by dissolving 10.1 mg of API in 100.0 mL mobile phase. The *Test solution* was used to measure the purity of the API.

A *Standard preparation* (external standard solution) was prepared by dissolution of 10.2 mg of methotrexate CRS ($\sim$100% purity) in 100.0 mL mobile phase. The *Standard preparation* was used to calibrate the HPLC assay.

Both the *Test solution* and the *Standard preparation* were analysed by HPLC, providing the following peak areas (instrumental signal):

Peak area *Test solution* (S_{sample}) = 11 605
Peak area *Standard preparation* ($S_{standard}$) = 12 006

The concentration of methotrexate in the *Standard preparation* ($[A]_{CRS}$) was

$$\frac{10.2 \text{ mg}}{100.0 \text{ mL}} = 0.102 \text{ mg/mL}$$

The concentration of methotrexate in the *Test solution* ($[A]_{sample}$) was

$$[A]_{sample} = [A]_{CRS} \frac{S_{sample}}{S_{standard}} = 0.102 \text{ mg/mL} \times \frac{11\,605}{12\,006} = 0.0986 \text{ mg/mL}$$

The purity of the pharmaceutical substance was calculated as

$$Purity = \frac{0.0986 \text{ mg/mL} \times 100.0 \text{ mL}}{10.1 \text{ mg}} \times 100\% = 97.6\% (w/w)$$

17.2.3   Internal Standard

Use of *internal standard* (IS) compensates for analytical errors due to sample losses and variable injection volumes. To compensate for these errors, the same concentration of the *internal standard* is added to all standard solutions and to all sample solutions prior to any sample treatment. After sample processing and chromatography the peaks of the analyte and the internal standard are identified and integrated as shown in Figure 17.2. The ratio of standard peak areas to IS peak areas ($Area_{Standard}/Area_{IS}$) is then used for construction of the calibration curve, and the ratio of sample peak areas to IS peak areas ($Area_{Sample}/Area_{IS}$) is

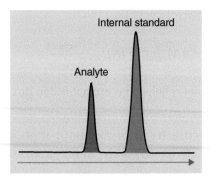

Figure 17.2 Chromatogram of analyte and internal standard

used for quantitation of the analyte. Thus, the calibration is similar to external calibration, except that peak area ratios are used instead of peak areas. The internal standard can improve precision when the dominant sources of error are related to sample preparation or instrumental variations. Such errors affect both the internal standard and the analyte peak to the same extent, and the peak area ratios are unaffected. For this method to work well, it is important to choose a suitable internal standard. The internal standard should match the analyte closely. Ideally an internal standard should:

- Be recovered to a similar extent as the analyte in the sample preparation procedure.
- Be separated from all other sample constituents during LC or GC.
- Have a detector response that is similar to the analyte.
- Have a concentration that gives a similar peak height or area as the analyte.
- Not be present in the sample itself.

Consequently, the physicochemical properties of the internal standard should be as similar as possible to those of the analyte. The ideal internal standard in LC-MS and GC-MS is an *isotopically labelled analogue* of the analyte. A labelled internal standard matches the analyte and is accurately determined at another mass. An example is shown in Figure 17.3, where atovaquone-D5 (five deuterium atoms) is used as the internal standard for atovaquone. Labelled internal standards are expensive and often compound analogues are used as internal standards. Internal standards are intensively used in bioanalysis.

For calibration with an internal standard, a series of standard solutions or calibration samples containing known concentrations of the CRS are prepared. The concentrations should approximate the concentrations of the real samples. This is similar to the external standard approach. In addition, a constant amount of internal standard is added to all

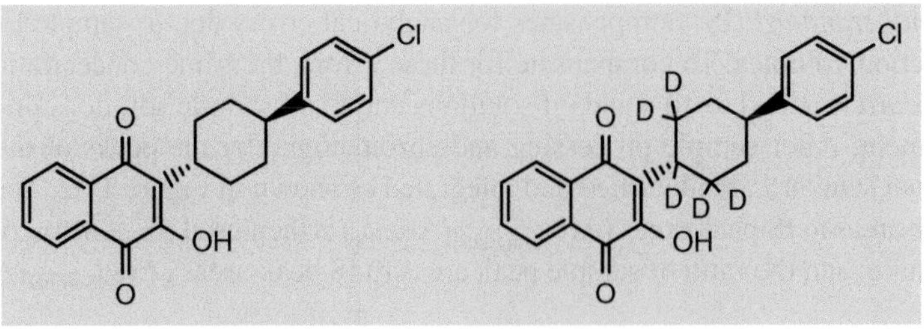

Figure 17.3 Atovaquone (analyte) and atovaquone-D5 (internal standard)

standard solutions and samples. All solutions are treated equally in the sample preparation procedure and are subsequently analysed. The peak areas of the analyte and the internal standard are read from the chromatograms and the ratios of analyte peak area to internal standard peak area in all solutions are calculated. The peak area ratios of standard solutions are plotted as a function of the concentration for the calibration curve. From the calibration curve, peak area ratios obtained for the samples can be converted to concentration. An example of this is shown in Box 17.3.

Box 17.3   Quantitation of hydrocortisone in a new ointment formulation based on calibration with internal standard

The concentration of hydrocortisone in extracts of a new ointment formulation is determined. The analysis is by HPLC, and the expected concentration range is from 100 to 250 µg/mL. Standard solutions containing 100, 150, 200, and 250 µg/mL of the chemical reference standard of hydrocortisone are prepared and an internal standard is added to a final concentration of 150 µg/mL. After sample preparation, 10 µL of standard solutions and sample solutions are injected into the chromatographic column. The peak areas of the reference standard and the internal standard are measured from the chromatograms and the ratio of analyte/IS peak areas are calculated:

Standard concentration (µg/mL)	Standard peak area	IS peak area	Ratio
100	111 068	164 302	0.676
150	159 143	163 997	0.975
200	218 063	164 204	1.328
250	271 241	164 289	1.651

The calibration curve is a plot of concentration versus peak area ratios:

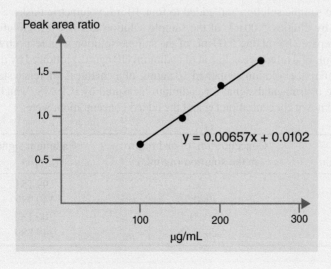

$y = 0.00657x + 0.0102$

The regression line $y = 0.00657x + 0.0102$, where x is the concentration in mg/mL and y is the peak area ratio used for calculation of analyte concentration in sample solutions.

In one of the sample solutions the peak area of the analyte is 175 432 and the peak area of the internal standard is 164 105. Concentration of hydrocortisone is calculated as follows:

$$\text{Peak area ratio} = \frac{175\,432}{164\,105} = 1.069$$

$$1.069 = 0.00657x + 0.0102$$

$$x = \frac{1.069 - 0.0102}{0.00657} = 161$$

The concentration is 161 µg/mL.

17.2.4 Standard Addition

Standard addition is a calibration method where calibration is performed directly in the sample by adding increasing amounts of CRS to aliquots of the sample. The sample is measured before and after addition of exact amounts of CRS. In this way, a calibration curve is constructed in the sample and any interference from the sample will influence standards and samples equally. Standard addition is often used in determination of elemental impurities and for trace amount elements in biological fluids. An example of a quantitative determination based on standard addition is shown in Box 17.4.

Box 17.4   Quantitation of trace levels of cadmium in zinc stearate (pharmaceutical ingredient) based on the standard addition method

Traces of cadmium were determined in a solution of the pharmaceutical ingredient zinc stearate. This was performed by ICP-MS using standard addition. The expected concentration level was about 1 ng/mL. To determine the unknown concentration, four standard addition solutions were prepared from the sample solution. Volumes of 5.00 mL were taken from the sample solution and added to four 10-mL volumetric flasks. Solution 1 was prepared by diluting 5.00 mL of the sample solution to 10.00 mL. Solutions 2, 3, and 4 were prepared by mixing 5.00 mL of the sample solution with respectively 0.25, 0.5, and 0.75 mL of a reference solution of cadmium (10 ng/mL) followed by dilution to 10 mL. The reference solution contained 10 ng/mL of a chemical reference standard of cadmium. The instrumental signal for cadmium measured by ICP-MS, which was the sum of the unknown concentration (x) and the added concentration, were:

Solution	Concentration of cadmium in the solution (ng/mL)	Cadmium signal (cps)
1	x	99 151
2	$x + 0.25$	149 987
3	$x + 0.5$	200 132
4	$x + 0.75$	249 889

The standard curve is a plot of concentration versus detector response:

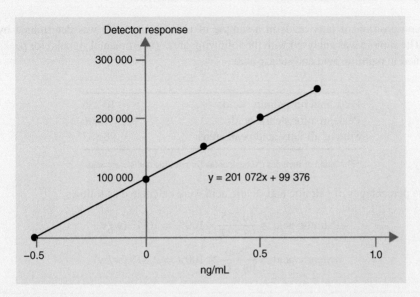

The calibration curve calculated according to the method of least squares was

$$y = 201\,072x + 99\,376$$

where y is the Cd signal and x is the concentration of added Cd (ng/mL). The unknown concentration in solution 1 is found at $y = 0$:

$$0 = 201\,072x + 99\,376$$

$$x = \frac{99\,376}{201\,072} = 0.494 \text{ ng/mL}$$

Since solution 1 was a $1:1$ (v/v) dilution of the original sample solution, the cadmium content in the sample was 0.988 ng/mL.

17.2.5 Normalization

Normalization is a technique where a certain analyte is quantified relative to the sum of components present in the sample. Normalization is used in chromatographic analysis. The quantitative results are obtained by expressing the area of a given peak (A) as a percentage of the sum of all peak areas of the components (ΣA_i) in the mixture:

$$A(\%) = \frac{A}{\sum A_i} \times 100\% \qquad (17.4)$$

An example of normalization is given in Box 17.5.

Box 17.5   Quantitation of palmitic and stearic acid in refined maize oil (pharmaceutical ingredient) based on normalization

The composition of fatty acids in a sample of refined maize oil was determined by GC. The sample was analysed with the following areas (instrumental signals) for peaks assigned to palmitic acid and stearic acid:

Peak area for palmitic acid:	10 235
Peak area for stearic acid:	2806
Sum of all fatty acid peak areas:	98 887[a]

[a]This number includes the peak areas for palmitic and stearic acid.

The percentages of palmitic and stearic acid were calculated as follows:

$$\%\text{Palmitic acid} = \frac{10\,235}{98\,887} \times 100\% = 10.4\%(\text{w/w})$$

$$\%\text{Stearic acid} = \frac{2806}{98\,887} \times 100\% = 2.84\%(\text{w/w})$$

17.3   Analytical Procedures

The *analytical procedure* refers to the way of performing the analytical measurement. It should describe in detail the steps necessary to perform each analytical test. The quality control system of any laboratory encompasses *standard operation procedures* (*SOP*) for all activities performed in the lab, including analytical procedures. Descriptions of analytical procedures should include detailed information as outlined in Table 17.1.

The section *Principle and scope* is a description of the target analyte and the sample type (e.g. assay of paracetamol in tablets or citalopram in serum). Also, this section gives a description of the principles of the analytical measurement (e.g. LC or IR

Table 17.1   General description of an analytical procedure

Principle and scope
Apparatus and equipment
Operating parameters
Chemicals
Reagents and standards
Sample preparation
Procedure
System suitability
Calculations
Data reporting

spectrophotometry). The section *Apparatus and equipment* gives information about all equipment and components required for the analytical measurement. The section *Operating parameters* lists optimal settings and ranges critical to the analytical measurement and to equipment (e.g. flow rate of the mobile phase in HPLC or wavelength in UV spectrophotometry). *Chemicals* is an overview of the quality and origin of the chemicals used. In the section *Reagents and standards* a description of reagents, standards, their source, purity, potencies, storage conditions, and directions for safe use is given. Samples have to be prepared prior to the analytical measurement, which is described step-by-step in the section *Sample preparation*, and finally the section *Procedure* gives a step-by-step description of the method and the analytical measurement. Prior to use, analytical instruments need to be checked for proper performance, which is described in section *System suitability*. Finally, *Calculations* and *Data reporting* describe the mathematical transformations, formulas, and format used to report the analytical results.

17.4   Validation

Validation is the process to demonstrate that an analytical procedure is suitable for its intended purpose. New analytical methods must be validated before they can be applied. The validation process should document that the method is fit for purpose. Furthermore, revalidation may be necessary if the synthesis of the drug substance is changed, if the composition of the finished product is changed, or if the analytical procedure is changed.

The *International Conference on Harmonization of Technical Requirements for Registration of Pharmaceuticals for Human Use (ICH)* is an international collaboration of expert working groups that publishes quality guidelines, adopted by the regulatory bodies of the European Union (European Medicines Agency (EMA)), USA (Food and Drug Administration (FDA)) and Japan. The ICH harmonized guideline, *Validation of Analytical Procedures: Text and Methodology Q2(R1)*, and ICH harmonized *Guideline for Elemental Impurities Q3D* are examples of guidelines relevant for pharmaceutical analysis.

According to the Q2 Validation guideline, 'the discussion on validation of analytical procedures is directed to the four most common types of analytical procedures:

- *Identification tests*
- *Quantitative tests for impurities*
- *Limit tests for impurities*
- Quantitative tests (assays) of the active moiety in samples of drug substance or drug product or other selected component(s) in the drug product'.

The guideline defines and describes the validation characteristics shown in Figure 17.4. Figure 17.5 lists the validation characteristics considered most important for validation of

Figure 17.4   Analytical characteristics used in validation of an analytical procedure

Validation characteristics	Identification tests	Impurities		Assays
		Quantitative tests	Limit tests	
Accuracy	–	+	–	+
Precision Repeatability Intermediate precision	– –	+ +	– –	+ +
Specificity	+	+	+	+
Detection Limit	–	+/–	+	–
Quantitation Limit	–	+/–	–	–
Linearity	–	+	–	+
Range	–	+	–	+

Figure 17.5   Characteristics applicable to analytical procedures

the different types of analytical procedures. Robustness is not included in the table, but should also be considered in the development of analytical procedures.

17.4.1   Specificity

The *specificity* of an analytical method is defined as 'the ability to assess unequivocally the analyte in the presence of compounds that may be expected to be present'. Typically these might include impurities, degradation products, matrix, etc. Specificity and *selectivity* are used in common for this characteristic (correctly speaking, however, selectivity can be graduated (more or less selectively), while a completely selective method is specific). Analytical methods with high selectivity provide measured instrumental signals solely from the target analyte. High (or sufficient) selectivity is vital for analytical methods. If an analytical method lacks specificity, one or several matrix components in the sample may contribute to the instrumental signal and the measurements may not provide an exact result.

Specificity is tested as part of the validation procedure for identification tests, determination of impurities, and assay. For identification methods (qualitative analysis), the ability to select between compounds of closely related structures that are likely to be present in real samples should be demonstrated. This should be confirmed by obtaining positive results (identification) from samples containing the analyte, coupled with negative results from samples that do not contain the analyte.

For quantitative methods, demonstration of specificity requires that the method is unaffected by the presence of matrix components. For quantitative methods of pharmaceutical ingredients and pharmaceutical preparations, this can be done by spiking the pharmaceutical ingredient or preparation with appropriate levels of impurities or excipients and demonstrate that the assay result is unaffected by the presence of these extraneous materials. For quantitative methods aimed at biological fluids, specificity is tested by analysing a certain number of drug-free samples from independent sources. For these samples, the instrumental signal for the analyte should be zero. Box 17.6 demonstrates an example of testing specificity.

Box 17.6   Test of Specificity

A new API has to be measured in oral solutions based on HPLC and prior to this the analytical method has to be validated. Specificity has to be tested as part of this validation and specificity is tested by analysing oral solutions without API (placebo preparation). The chromatogram from one of these samples is shown in the figure below (lower chromatogram). The upper chromatogram shows a similar analysis of oral solution spiked with the new API. The retention time of the latter is 2.553 minutes and at this retention time there is no signal in the drug-free sample. Thus, there are no peaks for excipients at the retention time of the API and interferences are avoided.

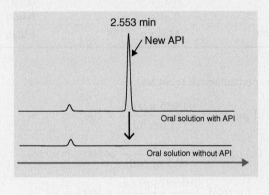

17.4.2   Accuracy

The *accuracy* of an analytical method is defined as 'the closeness of agreement between the value which is accepted either as a conventional true value or an accepted reference value, and the value found'. This is sometimes termed *trueness*. Accuracy should be established across the specified range of the analytical procedure. In the case of the assay of a pharmaceutical ingredient, accuracy may be determined by application of the analytical procedure to a substance of known purity, for example a certified reference substance or a substance with a similar high quality. In an assay of API in a pharmaceutical preparation, accuracy may be determined by application of the analytical procedure to synthetic mixtures of the preparation (drug-free) to which known amounts of the drug substance (e.g. a CRS) have been added. Another possibility of establishing accuracy is to compare the results with another independent well-characterized method for which accuracy has been documented.

Accuracy should be reported in either of two ways, either by the difference between the mean measured value and the accepted true value or the percentage of recovery by the assay of the known added amount of analyte in the sample. Box 17.7 demonstrates an example of accuracy testing.

Box 17.7   Test of Accuracy

The range of the method for acetylsalicylic acid in Box 17.1 is from 0.1 to 1.0 μg/mL. To test accuracy, three standard solutions are prepared, containing 0.1, 0.5, and 1.0 μg/mL of CRS. The purity of the chemical reference standard is 100% and the concentrations of the standard solutions are therefore the accepted true value. The concentrations of the standard solutions are determined three times. The results are:

0.1 μg/mL	0.5 μg/mL	1.0 μg/mL
0.0985	0.496	0.995
0.0974	0.499	0.997
0.0999	0.498	0.999
Mean = 0.0986	Mean = 0.498	Mean = 0.997

Accuracy reported as percentage of recovery:

$$0.1 \ \mu g/mL \ : \ \frac{0.0985 \ \mu g/mL}{0.1 \ \mu g/mL} \times 100\% = 98.5\%$$

$$0.5 \ \mu g/mL \ : \ \frac{0.498 \ \mu g/mL}{0.5 \ \mu g/mL} \times 100\% = 99.6\%$$

$$1 \ \mu g/mL \ : \ \frac{0.997 \ \mu g/mL}{1.0 \ \mu g/mL} \times 100\% = 99.7\%$$

17.4.3   Precision

The precision of an analytical procedure is defined as 'the closeness of agreement (degree of scatter) between a series of measurements obtained from multiple sampling of the same homogenous sample under the prescribed conditions'. Precision is usually expressed as the standard deviation (s) or the relative standard deviation (% RSD) of the mean (x) of a series of measurements:

$$\%RSD = \frac{s}{x} \times 100\% \qquad (17.5)$$

Precision is normally considered at three different levels:

- Repeatability
- Intermediate precision
- Reproducibility

The *repeatability* is the precision under the same operating conditions over a short interval of time. Repeatability is also termed *intra-assay precision*. Normally, the same operator with the same equipment carries out the measurement repeatedly within one day within the same laboratory. *Intermediate precision* expresses within-laboratories variations of measurements, as on different days, or with different analysts or equipment. The intermediate precision is also called *ruggedness*. *Reproducibility* expresses the precision of a procedure between different laboratories in a collaborative study.

The precision of the analytical procedure is determined by assaying a sufficient number of aliquots of a homogenous sample to be able to calculate statistically valid estimates of standard deviation or RSD. The assays are independent analysis of sample aliquots that have been carried through the complete analytical procedure from sample preparation to final measurement. Precision is assessed at different concentration levels covering the range of the analytical method. Recommended data for assessing repeatability are a minimum of nine determinations covering the specified range for the procedure or a minimum of six determinations at 100% test concentration. Box 17.8 shows an example of testing repeatability (within-run precision).

Box 17.8   Test of Repeatability

The concentration of hydrocortisone in a sample solution (500 μg/mL expected concentration) is determined seven times. The results of the assays are

496, 499, 498, 495, 499, 497, and 496 μg/mL
Standard deviation: 1.57 μg/mL
Mean concentration: 497 μg/mL

$$\text{Relative standard deviation} = \frac{1.57\ \mu g/mL}{497\ \mu g/mL} \times 100\% = 0.32\%$$

17.4.4   Detection Limit

The *detection limit* (*DL*) of an analytical method is defined as 'the lowest amount of analyte in a sample that can be detected, but not necessary quantitated as an exact value' under given experimental conditions. The detection limit is usually expressed as analyte concentration in the sample. The terms *limit of detection* (*LOD*) and *lower limit of detection* (*LLOD*) are used as alternatives to detection limit.

Several approaches for the determination of detection limits are used, depending on whether the procedure is non-instrumental or instrumental. For non-instrumental methods visual evaluation may be used. The detection limit is determined by establishing the minimum level of an analyte that can reliably be detected by visual evaluation.

For instrumental methods, calculation of the detection limit can be based on the *signal-to-noise ratio* (*S/N*). The measured detector signal is affected by small and rapid fluctuations defined as noise, as illustrated in Figure 17.6 (magnified chromatogram of a small analyte peak). On both sides of the analyte peak, baseline fluctuations are observed

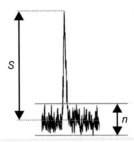

Figure 17.6   Definition of S and N. S/N is close to 3 is this illustration

as noise. The detection limit (*DL*) is defined as the concentration providing an instrumental signal two, three, or three-point-three times the noise level, thus providing a signal-to-noise ratio (*S/N* ratio) of 2, 3, or 3.3. The latter is the factor used in the ICH guideline:

$$DL = 3.3 \times \frac{\sigma}{S} \tag{17.6}$$

where σ is the standard deviation of the response (representing the noise level) and S is the slope of the calibration curve, which is used to convert the signal to a concentration. The noise level may be estimated in several ways. In the case of chromatography, the noise level can be estimated by visual inspection or measured by software, as in Figure 17.6, and an example is given in Box 17.9 on how to establish the detection limit based on the signal-to-noise ratio.

Box 17.9   Establishment of Detection Limit

A standard solution containing citalopram at a concentration of 0.16 ng/mL was injected into LC-MS system. The signal (peak height) (*S*) read from the chromatogram was 14 intensity units and the noise level (*N*) was 2 intensity units.

The detection limit is set to a signal-to-noise level of 3 (*S/N* = 3). Since *N* = 2 intensity units, the detection limit corresponds to a citalopram solution with a peak height *S* of $2 \times 3 = 6$ intensity units. The concentration of citalopram providing a peak height of 6 intensity units is

$$\text{Detection limit} = 0.16 \text{ ng/mL} \times \frac{6 \text{ intensity units}}{14 \text{ intensity units}} = 0.069 \text{ ng/mL} \approx 0.07 \text{ ng/mL}$$

The detection limit of the analyte is 0.07 ng/mL.

In other methods, the noise level can be estimated based on the standard deviation of the blank by analysing an appropriate number of blank samples and calculating the standard deviation of the responses. Alternatively, an appropriate number of analyses of a standard with a concentration close to the detection limit may be used. The residual standard deviation of a regression line or the standard deviation of Y-intercepts of regression lines from a calibration curve may also be used to estimate σ.

17.4.5   Quantitation Limit

The *quantitation limit (QL)*, also termed the *limit of quantitation (LOQ)* or the *lower limit of quantitation (LLOQ)* of an analytical method, is defined as 'the lowest amount of analyte in a sample, which can be quantitatively determined with suitable precision and accuracy'. This parameter should be estimated for assays for low levels of analytes and particularly for the determination of impurities and/or degradation products. Quantitation should not be performed below this limit. The quantitation limit is defined as the concentration of analyte providing a signal-to-noise ratio of 10:

$$QL = 10 \times \frac{\sigma}{S} \qquad (17.7)$$

and σ (the noise) can be estimated in the same way as described for the detection limit.

   The quantitation limit may also be derived from the precision of the analytical method. Normally, precision is concentration dependent, and values for RSD tend to increase at lower analyte concentrations, as illustrated in Figure 17.7. Based on this, the quantitation limit can be set to the concentration where the RSD is suitable for the purpose. This approach is mainly used in bioanalysis where the requirement for precision is lower than for assays for pharmaceuticals preparations, and often is at the level 15–20%. The limit is established by analysing a number of samples with decreasing amounts of analyte. The calculated RSD (%) is plotted against the analyte concentration, as shown in Figure 17.7, and the quantitation limit can be read from the graph. Box 17.10 demonstrates calculations of detection and quantitation limits based on precision.

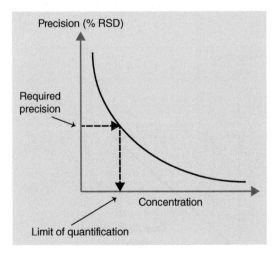

Figure 17.7   *Relationship between concentration and precision (relative standard deviation)*

Box 17.10   Establishment of Detection and Quantitation Limit

An HPLC method for testing impurities in the drug substance cisplatin is developed. A standard solution of 0.05 µg/mL is analysed six times, resulting in peak areas of

3349   3380   3378   3340   3317   3331

The mean is 3349.1 and the standard deviation is 25.43.

The detection limit is calculated as

$$DL = 3.3 \times \frac{\sigma}{S} = 3.3 \times \frac{25.43}{3349/0.05 \ \mu g/mL} = 1.25 \times 10^{-3} \mu g/mL = 1.25 \ ng/mL$$

(In the lack of a calibration curve, the mean divided by the standard concentration is used for the slope.)

The quantitation limit is calculated as

$$QL = 10 \times \frac{\sigma}{S} = 10 \times \frac{25.43}{3349/0.05 \ \mu g/mL} = 3.80 \times 10^{-3} \mu g/mL = 3.80 \ ng/mL$$

The advantage of this calculation is that variations in peak integration are included in σ.

17.4.6   Linearity and Range

The *linearity* of an analytical method is defined as 'its ability (within a given range) to obtain test results, which are directly proportional to the concentration (amount) of analyte in the sample', as shown in Figure 17.8. Linearity should be established across the *range* of the analytical procedure. The range is defined as 'the interval between the upper and lower

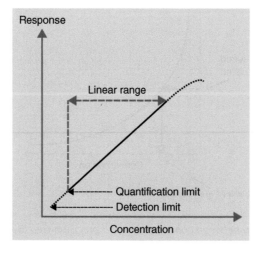

Figure 17.8   Illustration of linearity

levels of analyte concentration that have been demonstrated to be determined with a suitable level of precision, accuracy, and linearity'. Linearity should be evaluated by inspection of a plot of an instrumental signal as a function of analyte concentration or content. If there is a linear relationship, test results should be evaluated by appropriate statistical methods, for example by calculation of a regression line by the method of least squares. The square of the correlation coefficient and the regression line are typically reported.

For establishing linearity, standard solutions or calibration samples covering the entire range are analysed (see Boxes 17.1 and 17.3). For assay (quantitation) of pharmaceutical ingredients and preparations, linearity should be established in the range of 80–120% of expected concentration, while for quantitation of impurities the linearity should be established from 50% to 120% of acceptance criteria.

For bioanalytical methods, linearity should be established from the quantitation limit and to the highest recommended therapeutic concentration. The range of the procedure is validated by verifying that the analytical procedure provides precision, accuracy, and linearity in compliance with requirements within the entire range. Box 17.11 shows an example of testing linearity.

Box 17.11   Evaluation of linearity for assay of tetracaine hydrochloride in ophthalmic solution

The content of tetracaine hydrochloride is to be measured in an ophthalmic solution by LC. The ophthalmic solution contains 10 mg/mL of tetracaine hydrochloride, and is diluted by a factor of 100 before LC. Therefore the concentration of tetracaine hydrochloride in the sample is expected to be

$$\frac{10 \text{ mg/mL}}{100} = 0.10 \text{ mg/mL} = 100 \text{ μg/mL}$$

Standard solutions in the range 80–120 μg/mL (80–120% of expected concentration) are used to test linearity and five different concentrations are used to cover the range:

Tetracaine hydrochloride concentration (μg/mL)	Peak area
80	80 102
90	90 037
100	100 539
110	109 842
120	120 317

Linearity across the range of the analytical procedure is established by visual inspection of a plot of peak area as a function of analyte concentration, as shown below.

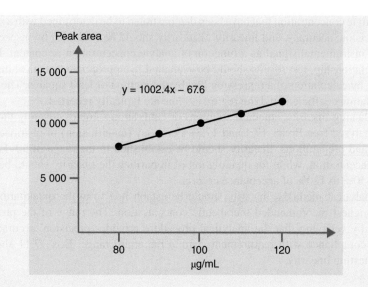

The regression line calculated by the method of least squares is

$$y = 1002.4x - 67.6$$

where y is the peak area and x is the analyte concentration (µg/mL). The square of the correlation coefficient $R^2 = 0.9997$ shows excellent linearity.

17.4.7 Robustness

The *robustness* of an analytical method is defined as 'a measure of its capacity to remain unaffected by small, but deliberate variations in method parameters and provides an indication of its reliability during normal usage'. The evaluation of robustness should be considered during the development of an analytical method and depends on the type of method. In the case of LC, examples of typical variations are:

- Influence of variations of pH in the mobile phase
- Influence of variations in the mobile phase composition
- Different columns (different lots/or suppliers)
- Temperature of the column
- Flow rate of the mobile phase

Thus, the performance of the method is tested with small changes in the parameters listed above, and the analytical results should be unaffected.

17.4.8 Test Methods in Ph. Eur. and USP

The test methods given in monographs and general chapters of the European Pharmacopoeia (Ph. Eur.) and the United States Pharmacopoeia (USP) have been validated in accordance with official procedures and requirements for analytical method validation. Unless otherwise stated in the monograph or the general chapter, validation of the test methods by the analyst is therefore not required.

17.5   System Suitability

If measurements are susceptible to variations in analytical conditions, these should be suitably controlled. This is done by *system suitability* testing. System suitability is an integrated part of many analytical procedures. The purpose of system suitability testing is to establish the performance of a particular instrument or procedure. In Ph. Eur. and USP, system suitability tests are listed in the monographs. For analytical methods based on LC or GC, system suitability testing is essential because performance of chromatography systems can vary from day to day. Typically, a standard solution of the analyte or a standard solution of the analyte and a related substance is injected into the system, and parameters such as RSD, resolution, symmetry factor, number of theoretical plates, or retention factor are checked. Typical criteria for system suitability are exemplified in Box 17.12.

Box 17.12   Typical system suitability criteria for the LC system

- The RSD of peak areas or peak heights is less than 1% (for drug substance assay).
- The resolution (R_s) is greater than 2 between the analyte peak and the related substance peak.
- The symmetry factor for the analyte peak is in the range 0.8–1.5.
- The number of theoretical plates for the analyte peak is larger than 5000 (for LC).
- The retention factor (k) for the analyte peak is larger than 2.

   System suitability is performed on daily basis to check that the analytical instruments are running properly and performance is acceptable.

17.5.1   Adjustment of Chromatographic Conditions

Chromatographic procedures in monographs of Ph. Eur. and USP contain necessary information that enables the procedure to be reproducible in a laboratory and typically includes system suitability requirements. In some instances, the chromatographic system fails to meet those requirements. Because this situation can arise, adjustments to the operating conditions might be necessary to meet the system suitability requirements. The maximum variations allowed for each chromatographic parameter can be found in general chapters in the pharmacopoeias. Users are allowed to implement these adjustments to bring the system into a suitable performance without a full validation of the analytical procedure. Typical criteria are summarized in Box 17.13 (the summary is not complete).

Box 17.13   Adjustment of chromatographic conditions in Ph. Eur

LIQUID CHROMATOGRAPHY: ISOCRATIC ELUTION

Composition of the mobile phase: the amount of the minor solvent component may be adjusted by ±30% relative or ±2% absolute; no other component is altered by more than 10% absolute.

pH of the aqueous component of the mobile phase: ±0.2 pH, unless otherwise prescribed, or ±1.0 pH when non-ionizable substances are to be examined.

Concentration of salts in the buffer component of a mobile phase: ±10%.

Flow rate: ±50%; a larger adjustment is acceptable when changing the column dimensions (see Ph. Eur.).

Column parameters

Stationary phase:

- No change of the identity of the substituent of the stationary phase permitted (e.g. no replacement of C_{18} by C_8).
- *Particle size*: maximum reduction of 50%; no increase permitted.

Column dimensions:

- *Length*: ±70%.
- *Internal diameter*: ±25%.

Temperature: ±10 °C, where the operating temperature is specified, unless otherwise prescribed.

Detector wavelength: no adjustment permitted.

Injection volume: may be decreased, provided detection and repeatability of the peak(s) to be determined are satisfactory; no increase permitted.

LIQUID CHROMATOGRAPHY: GRADIENT ELUTION

Adjustment of chromatographic conditions for gradient systems requires greater caution than for isocratic systems.

Composition of the mobile phase/gradient elution: minor adjustments of the composition of the mobile phase and the gradient are acceptable provided that:

- The system suitability requirements are fulfilled.
- The principal peak(s) elute(s) within ±15% of the indicated retention time(s).
- The final composition of the mobile phase is not weaker in elution power than the prescribed composition.

pH of the aqueous component of the mobile phase: no adjustment permitted.

Concentration of salts in the buffer component of a mobile phase: no adjustment permitted.

Flow rate: adjustment is acceptable when changing the column dimensions (see the formula below).
Column parameters
Stationary phase:

- No change of the identity of the substituent of the stationary phase permitted (e.g. no replacement of C_{18} by C_8).
- *Particle size*: no adjustment permitted.

Column dimensions:

- *Length*: $\pm 70\%$.
- *Internal diameter*: $\pm 25\%$.

Temperature: $\pm 5\,^{\circ}C$, where the operating temperature is specified, unless otherwise prescribed.
Detector wavelength: no adjustment permitted.
Injection volume: may be decreased, provided detection and repeatability of the peak(s) to be determined are satisfactory; no increase permitted.

GAS CHROMATOGRAPHY

Column parameters
Stationary phase:

- *Particle size*: maximum reduction of 50%; no increase permitted (packed columns).
- *Film thickness*: -50% to $+100\%$ (capillary columns).

Column dimensions:

- *Length*: $\pm 70\%$.
- *Internal diameter*: $\pm 50\%$.

Flow rate: $\pm 50\%$.
Temperature: $\pm 10\%$.
Injection volume and split volume: may be adjusted, provided detection and repeatability are satisfactory.

18

Chemical Analysis of Pharmaceutical Ingredients

Introduction to Pharmaceutical Analytical Chemistry, Second Edition.
Stig Pedersen-Bjergaard, Bente Gammelgaard and Trine Grønhaug Halvorsen.
© 2019 John Wiley & Sons Ltd. Published 2019 by John Wiley & Sons Ltd.

18.1   Pharmaceutical Ingredients, Production, and Control

All active pharmaceutical ingredients (APIs) and excipients to be used in the manufacture of pharmaceutical preparations are termed *pharmaceutical ingredients*. A large number of pharmaceutical ingredients exist, and examples with classification are given in Table 18.1. This textbook primarily focuses on pure chemical ingredients, multi-chemical ingredients, and biological macromolecules (biopharmaceuticals). These are the main ingredients in pharmaceutical preparations and chemical analysis plays a predominant role in their quality control. For the other ingredients given in Table 18.1, other techniques such as microscopy, physical testing, biological tests, and radioactivity measurements are used. These techniques are outside the scope of this textbook.

Pure chemical ingredients are pure chemical substances with a well-defined structure, which are typically organic compounds produced by industrial synthesis. In most cases, a number of synthetic steps are involved to produce the pharmaceutical ingredient, as exemplified in Figure 18.1 for ibuprofen. During synthesis, side reactions can occur, giving rise to undesirable by-products and impurities in the pharmaceutical ingredient. Such substances

Table 18.1   Classification of pharmaceutical ingredients

Raw material	Example	API/excipient
Pure chemical ingredients	Ibuprofen	API
	Benzoic acid	Excipient
Multi-chemical ingredients (organic)	Polyethylene glycol	Excipient
	Soya oil	Excipient
Multi-chemical ingredients (inorganic)	Bentonite	Excipient
	Talc	Excipient
Gasses and liquids	Oxygen	API
	Water	Excipient
Radioactive substances	Technetium	API
Biological macromolecules	Insulin	API
Products of human or animal origin	Coagulation factor	API
	Human plasma	Excipient
Plant materials	Herbal drugs	API

Synthesis scheme for ibuprofen

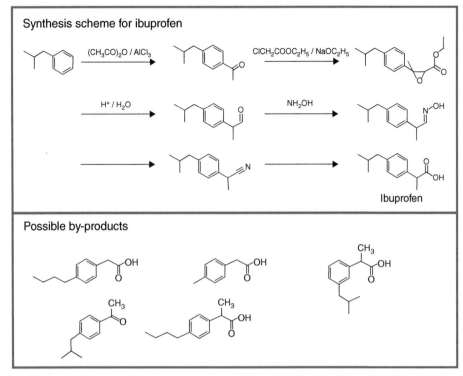

Ibuprofen

Possible by-products

Figure 18.1 *Synthesis scheme for ibuprofen (top frame) and possible by-products (bottom frame)*

are termed *related substances*. Ibuprofen examples of by-products are shown in Figure 18.1. The syntheses of pharmaceutical ingredients are normally performed in organic solvents, and the ingredients are re-crystallized from the organic solvent. The pharmaceutical ingredients can therefore contain traces of organic solvents. Such traces are termed *residual solvents*. To remove related substances and residual solvents, the products from synthesis are purified before use. The purification process removes related substances and residual solvents, but traces can remain. These traces may reduce the quality of the pharmaceutical ingredient and can represent a safety issue for the patient. Pharmaceutical ingredients can also be contaminated with small amounts of substances derived from:

- Production equipment
- Chemical degradation
- Packing

The purpose of chemical analysis of pharmaceutical ingredients is to measure such impurities and check that levels are below the limits set by the medicinal authorities. Pure chemical ingredients are not limited to organic compounds only, but also comprise inorganic compounds such as sodium chloride, calcium phosphate, and sodium fluoride (Table 18.2).

Pharmaceutical ingredients belonging to the group of *organic multi-chemical ingredients* have a more diverse composition. *Synthetic polymers* are manufactured from pure

Table 18.2 Overview of pure chemical ingredients and multi-chemical ingredients

Type of substance	Example	API/excipient
Pure chemical ingredients		
Organic compounds	Paracetamol	API
	Methyl parahydroxybenzoate	Excipient
Inorganic compounds	Sodium chloride	Excipient
	Sodium fluoride	API
Multi-chemical ingredients (organic)		
Polysaccharides	Cellulose	Excipient
Synthetic polymers	Polyethylene glycols/macrogols	Excipient
Fatty oils	Soya-bean oil	Excipient
Hydrocarbons	Vaseline	Excipient

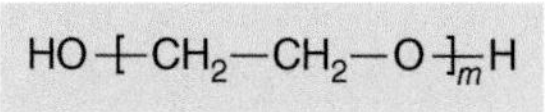

$$HO-\!\left[CH_2-CH_2-O\right]_m\!H$$

Figure 18.2 The general chemical structure of polyethylene glycol, with molecular mass between 5 and 1000

chemicals, an example of which is polyethylene glycol (PEG), as shown in Figure 18.2. PEGs are produced by reaction between water and ethylene oxide under pressure in the presence of a catalyst. *Fatty oils* and *polysaccharides* can be produced by isolation from plant materials. An example of this is soya-bean oil. Soya-bean oil is produced by extraction from the seeds of *Glycine max* or *G. soya*.

Pharmaceutical ingredients within the group of organic multi-chemical ingredients may vary in composition depending on production. Remnants of solvents from the production, impurities from packaging and production, and degradation products can be present, all of which may reduce the quality. For synthetic products, traces of starting materials, by-products from the synthesis, and degradation products are potential impurities. For pharmaceutical ingredients isolated from plant materials, various other undesired substances can be present, such as herbicides, insecticides, and mycotoxins. Chemical analysis of pharmaceutical ingredients is intended to measure such undesired substances and keep control, as well as characterize the overall composition of organic multi-chemical ingredients.

18.2 Pharmacopoeia Monographs

The Ph. Eur. and the United States Pharmacopeia (USP) contain requirements (*standards*) and test methods for identification and purity assessment of common pharmaceutical ingredients. Patent protected ingredients are tested in a similar manner, but are not included in official pharmacopoeias. USP and the British Pharmacopoeia (BP) also include standards and test procedures for pharmaceutical preparations.

The major part of Ph. Eur. comprises individual monographs for pharmaceutical ingredients. In addition to these monographs, a number of general chapters are published, as listed in Table 18.3. Chapter 2 in Ph. Eur. is a collection of all the methods of analysis used in the

Table 18.3   Main contents of the Ph.Eur

Volume I

1. General notices
2. Methods of analysis
3. Materials for containers and containers
4. Reagents
5. General texts
General monographs
Monographs
        Dosage forms
        Vaccines
        Radiopharmaceutical preparations
        Sutures
        Herbal drugs
        Homoeopathic preparations

Volume II

Monographs (individual monographs for pharmaceutical ingredients)

---

individual monographs. Each method of analysis is described, including principle, equipment, operational procedure, system suitability testing, and qualification/calibration. One example of such a description is shown in Figure 18.3 for the determination of the melting point. In all individual monographs where a certain method of analysis is prescribed, reference is given to the description in Chapter 2. Thus, when determination of a melting point is prescribed in an individual monograph, reference is given to Section 2.2.14.

Chapter 4 describe all reagents, standard solutions for limit tests, buffer solutions, and volumetric solutions used in the individual monographs, including their preparation, standardization, and purity requirements. Examples are given in Figure 18.4 for anhydrous acetic acid (reagent), lead standard solution (0.1% Pb), phosphate buffer solution pH 7.0 (buffer), and 0.1 M perchloric acid (volumetric solution). Volumetric solutions are those used for titration. Reagents included in Chapter 4 are followed by the letter *R* in the individual monographs. Thus, when anhydrous acetic acid is prescribed in an individual monograph, it is written *anhydrous acetic acid R*.

Volume II of Ph. Eur. contains the individual monographs (standards) for pharmaceutical ingredients with all the tests needed to ensure identity and high purity. The monographs for pharmaceutical ingredients are structured as follows:

- Definition
- Production (optional)
- Characters
- Identification
- Test
- Assay
- Storage
- Impurities

An example of a monograph is shown in Figure 18.5 for paracetamol. Paracetamol is an API and a well-defined chemical compound (pure chemical ingredient). At the top of the monograph the English *trade name* for the compound is given ('PARACETAMOL'). In

2.2.14. MELTING POINT - CAPILLARY METHOD

The melting point determined by the capillary method is the temperature at which the last solid particle of a compact column of a substance in a tube passes into the liquid phase (i.e. clear point). The melting point determined by this method is specific to the methodology (e.g. heating rate) described in this chapter. Similarly, whenever the use of certified reference materials is required, their certified values refer to the described analytical procedure.

When prescribed in the monograph, the same apparatus and method are used for the determination of other factors, such as meniscus formation or melting range, that characterize the melting behavior of a substance.

Equipment. The equipment consists of a metal heating block with 1 or more compartments for capillary tubes, or of a suitable glass vessel containing a liquid bath (e.g. water, liquid paraffin or silicone oil) and fitted with a suitable means of heating and stirring. The equipment is equipped with a temperature sensor or a suitable certified thermometer allowing readings at least to the nearest 0.1 °C.

Samples are introduced into the equipment in glass capillary tubes. The dimensions are chosen according to the manufacturer's requirements, typically with an external diameter of 1.3-1.5 mm and a wall thickness of 0.1-0.3 mm. In some equipment glass slides are used instead of capillary tubes.

The equipment is capable of heating samples at a rate of 1 °C/min or less. The accuracy of the equipment is at most ± 0.5 °C.

Detection can be performed either visually or instrumentally. In the case of instrumental detection, this is generally performed by image recording and subsequent analysis or by a photodetector that measures the transmitted or reflected light from the sample.

Method. The substance is previously treated as described in the monograph. Coarse crystals are to be avoided as they might lead to false results. If necessary, samples are crushed into a fine powder. Unless otherwise prescribed, dry the finely powdered substance *in vacuo* over anhydrous silica gel R for 24 h. Introduce a sufficient quantity into a capillary tube to give a compact column as described by the instrument manufacturer (e.g. 4-6 mm in height). Raise the temperature of the apparatus to about 5 °C below the presumed melting point. Allow the temperature to stabilize and then introduce the capillary tube into the instrument. Finally, adjust the rate of heating to about 1 °C/min unless otherwise prescribed.

In the case of instrumental detection, follow the instrument manufacturer's requirements for the determination of the melting point. For visual detection, record the temperature at which the last particle of the substance to be examined passes into the liquid phase.

Samples can be measured in parallel if the instrument allows multiple sample processing.

System suitability. Carry out a system suitability test before the measurements for example by choosing a suitable reference material with a melting point close to that expected for the substance to be examined.

Qualification / Calibration of the equipment. The qualification / calibration is carried out periodically according to the instrument manufacturer's requirements, using at least 2 certified reference materials. These are selected to cover the temperature range that is used on the equipment. Use capillary tubes with the same dimensions as those used for sample measurement.

Guidance on how to compare results obtained from certified reference materials with values from the certificates can be found on the European Reference Materials (ERM) website (Application note 1).

Figure 18.3   Description of method of analysis for the melting point (Ph. Eur.)

Acetic acid, anhydrous, $C_2H_4O_2$. (M_r 60.1). *1000300.* [64-19-7].
Content: minimum 99.6 per cent *m/m* of $C_2H_4O_2$.
Colourless liquid or white or almost white, shining, fern-like crystals, miscible with
  or very soluble in water, in ethanol (96 per cent), in glycerol (85 per cent), and in
  most fatty and essential oils.
d_{20}^{20}: 1.052 to 1.053.
bp: 117 °C to 119 °C.
A 100 g/L solution is strongly acid (2.2.4).
A 5 g/L solution neutralised with dilute ammonia R2 gives reaction (b) of acetates
  (2.3.1).
Freezing point (2.2.18): minimum 15.8 °C.
Water (2.5.12): maximum 0.4 per cent. If the water content is more than 0.4 per cent
  it may be adjusted by adding the calculated amount of acetic anhydride R.
Storage: protected from light.
Lead standard solution (0.1 per cent Pb). *5001700.*
Dissolve lead nitrate R equivalent to 0.400 g of $Pb(NO_3)_2$ in water R and dilute to
  250.0 mL with the same solvent.
Phosphate buffer solution pH 7.0. *4003700.*
Mix 82.4 mL of a 71.5 g/L solution of disodium hydrogen phosphate
  dodecahydrate R with 17.6 mL of a 21 g/L solution of citric acid monohydrate R.
0.1 M Perchloric acid. *3003900.*
Place 8.5 mL of perchloric acid R in a volumetric flask containing about 900 mL of
  glacial acetic acid R and mix. Add 30 mL of acetic anhydride R, dilute to
  1000.0 mL with glacial acetic acid R, mix and allow to stand for 24 h. Determine
  the water content (2.5.12) without addition of methanol and, if necessary, adjust
  the water content to 0.1-0.2 per cent by adding either acetic anhydride R or
  water R. Allow to stand for 24 h.
Standardisation. Dissolve 0.170 g of potassium hydrogen phthalate RV in 50 mL of
  anhydrous acetic acid R, warming gently if necessary. Allow to cool protected
  from air, and titrate with the perchloric acid solution, determining the end-point
  potentiometrically (2.2.20) or using 0.05 mL of crystal violet solution R as
  indicator. Note the temperature of the perchloric acid solution at the time of the
  titration. If the temperature at which an assay is carried out is different from that at
  which the 0.1 M perchloric acid has been standardised, the volume used in the
  assay becomes:
$V_c = V \cdot [1 + (t_1 - t_2) \cdot 0.0011]$
t_1 is temperature during standardisation, t_2 is temperature during the assay, V_c is
  corrected volume, and V is observed volume.
1 mL of 0.1 M perchloric acid is equivalent to 20.42 mg of $C_8H_5KO_4$.

Figure 18.4 *Descriptions of anhydrous acetic acid, lead standard solution, phosphate buffer
solution pH 7.0, and 0.1 M perchloric acid (Ph. Eur.)*

E. (1RS)-1-(3,4-dimethoxybenzyl)-6,7-dimethoxy-1,2,3,4-tetrahydroisoquinoline (tetrahydropapaverine),

F. 2-(3,4-dimethoxyphenyl)-N-[2-(3,4-dimethoxyphenyl)-ethyl]acetamide.

01/2017:0049

PARACETAMOL

Paracetamolum

$C_8H_9NO_2$ \hspace{2em} M_r 151.2
[103-90-2]

DEFINITION

N-(4-Hydroxyphenyl)acetamide.

Content: 99.0 per cent to 101.0 per cent (dried substance).

CHARACTERS

Appearance: white or almost white, crystalline powder.

Solubility: sparingly soluble in water, freely soluble in alcohol, very slightly soluble in methylene chloride.

IDENTIFICATION

First identification: A, C.

Second identification: A, B, D, E.

A. Melting point (*2.2.14*): 168 °C to 172 °C.

B. Dissolve 0.1 g in *methanol R* and dilute to 100.0 mL with the same solvent. To 1.0 mL of the solution add 0.5 mL of a 10.3 g/L solution of *hydrochloric acid R* and dilute to 100.0 mL with *methanol R*. Protect the solution from bright light and immediately measure the absorbance (*2.2.25*) at the absorption maximum at 249 nm. The specific absorbance at the maximum is 860 to 980.

C. Infrared absorption spectrophotometry (*2.2.24*).

  Preparation: discs.

  Comparison: paracetamol CRS.

D. To 0.1 g add 1 mL of *hydrochloric acid R*, heat to boiling for 3 min, add 1 mL of *water R* and cool in an ice bath. No precipitate is formed. Add 0.05 mL of a 4.9 g/L solution of *potassium dichromate R*. A violet colour develops which does not change to red.

E. It gives the reaction of acetyl (*2.3.1*). Heat over a naked flame.

TESTS

Related substances. Liquid chromatography (*2.2.29*). *Prepare the solutions immediately before use.*

Test solution. Dissolve 0.200 g of the substance to be examined in 2.5 mL of *methanol R* containing 4.6 g/L of a 400 g/L solution of *tetrabutylammonium hydroxide R* and dilute to 10.0 mL with a mixture of equal volumes of a 17.9 g/L solution of *disodium hydrogen phosphate dodecahydrate R* and of a 7.8 g/L solution of *sodium dihydrogen phosphate R*.

Reference solution (a). Dilute 1.0 mL of the test solution to 50.0 mL with the mobile phase. Dilute 5.0 mL of this solution to 100.0 mL with the mobile phase.

Reference solution (b). Dilute 1.0 mL of reference solution (a) to 10.0 mL with the mobile phase.

Reference solution (c). Dissolve 5.0 mg of *4-aminophenol R*, 5 mg of *paracetamol CRS* and 5.0 mg of *chloroacetanilide R* in *methanol R* and dilute to 20.0 mL with the same solvent. Dilute 1.0 mL to 250.0 mL with the mobile phase.

Reference solution (d). Dissolve 20.0 mg of *4-nitrophenol R* in *methanol R* and dilute to 50.0 mL with the same solvent. Dilute 1.0 mL to 20.0 mL with the mobile phase.

Column:

– *size*: l = 0.25 m, Ø = 4.6 mm,

– *stationary phase*: octylsilyl silica gel for chromatography R (5 µm),

– *temperature*: 35 °C.

Mobile phase: mix 375 volumes of a 17.9 g/L solution of *disodium hydrogen phosphate dodecahydrate R*, 375 volumes of a 7.8 g/L solution of *sodium dihydrogen phosphate R* and 250 volumes of *methanol R* containing 4.6 g/L of a 400 g/L solution of *tetrabutylammonium hydroxide R*.

Flow rate: 1.5 mL/min.

Detection: spectrophotometer at 245 nm.

Injection: 20 µL.

Run time: 12 times the retention time of paracetamol.

Relative retentions with reference to paracetamol (retention time = about 4 min): impurity K = about 0.8; impurity F = about 3; impurity J = about 7.

System suitability: reference solution (c):

– *resolution*: minimum 4.0 between the peaks due to impurity K and to paracetamol,

– *signal-to-noise ratio*: minimum 50 for the peak due to impurity J.

Limits:

– *impurity J*: not more than 0.2 times the area of the corresponding peak in the chromatogram obtained with reference solution (c) (10 ppm),

– *impurity K*: not more than the area of the corresponding peak in the chromatogram obtained with reference solution (c) (50 ppm),

– *impurity F*: not more than half the area of the corresponding peak in the chromatogram obtained with reference solution (d) (0.05 per cent),

– *any other impurity*: not more than half the area of the principal peak in the chromatogram obtained with reference solution (a) (0.05 per cent),

– *total of other impurities*: not more than the area of the principal peak in the chromatogram obtained with reference solution (a) (0.1 per cent),

– *disregard limit* for the calculation of the total of other impurities: the area of the principal peak in the chromatogram obtained with reference solution (b) (0.01 per cent).

Loss on drying (*2.2.32*): maximum 0.5 per cent, determined on 1.000 g by drying in an oven at 105 °C.

Figure 18.5 *Ph. Eur. monograph for paracetamol. Source: Reproduced from Ph. Eur. 2017 (Ninth Edition) with permission.*

Sulfated ash (*2.4.14*): maximum 0.1 per cent, determined on 1.0 g.

ASSAY

Dissolve 0.300 g in a mixture of 10 mL of *water R* and 30 mL of *dilute sulfuric acid R*. Boil under a reflux condenser for 1 h, cool and dilute to 100.0 mL with *water R*. To 20.0 mL of the solution add 40 mL of *water R*, 40 g of ice, 15 mL of *dilute hydrochloric acid R* and 0.1 mL of *ferroin R*. Titrate with *0.1 M cerium sulfate* until a greenish-yellow colour is obtained. Carry out a blank titration.

1 mL of *0.1 M cerium sulfate* is equivalent to 7.56 mg of $C_8H_9NO_2$.

STORAGE

Protected from light.

IMPURITIES

A. *N*-(2-hydroxyphenyl)acetamide,

B. *N*-(4-hydroxyphenyl)propanamide,

C. *N*-(3-chloro-4-hydroxyphenyl)acetamide,

D. *N*-phenylacetamide,

E. 1-(4-hydroxyphenyl)ethanone,

F. 4-nitrophenol,

G. 1-(4-hydroxyphenyl)ethanone oxime,

H. 4-(acetylamino)phenyl acetate,

I. 1-(2-hydroxyphenyl)ethanone,

J. *N*-(4-chlorophenyl)acetamide (chloroacetanilide),

K. 4-aminophenol.

01/2008:1034

PARAFFIN, HARD

Paraffinum solidum

DEFINITION

A purified mixture of solid saturated hydrocarbons generally obtained from petroleum. It may contain a suitable antioxidant.

CHARACTERS

Appearance: colourless or white or almost white mass; the melted substance is free from fluorescence in daylight.

Solubility: practically insoluble in water, freely soluble in methylene chloride, practically insoluble in ethanol (96 per cent).

IDENTIFICATION

First identification: A, C.

Second identification: B, C.

A. Infrared absorption spectrophotometry (*2.2.24*).

 Comparison: hard paraffin CRS.

 Preparation: place about 2 mg on a sodium chloride plate, heat in an oven at 100 °C for 10 min, spread the melted substance with another sodium chloride plate and remove one of the plates.

B. Acidity or alkalinity (see Tests).

C. Melting point (*2.2.16*): 50 °C to 61 °C.

TESTS

Acidity or alkalinity. To 15 g add 30 mL of boiling *water R* and shake vigorously for 1 min. Allow to cool and to separate. To 10 mL of the aqueous layer add 0.1 mL of *phenolphthalein solution R*. The solution is colourless. Not more than 1.0 mL of *0.01 M sodium hydroxide* is required to change the colour of the indicator to red. To a further 10 mL of the aqueous layer add 0.1 mL of *methyl red solution R*. The solution is yellow. Not more than 0.5 mL of *0.01 M hydrochloric acid* is required to change the colour of the indicator to red.

Polycyclic aromatic hydrocarbons. *Use reagents for ultraviolet absorption spectrophotometry*. Dissolve 0.50 g in 25 mL of *heptane R* and place in a 125 mL separating funnel with unlubricated ground-glass parts (stopper, stopcock). Add 5.0 mL of *dimethyl sulfoxide R*. Shake vigorously for 1 min and allow to stand until 2 clear layers are formed. Transfer the lower layer to a 2nd separating funnel, add 2 mL of *heptane R* and shake the mixture vigorously. Allow to stand until 2 clear layers are formed. Separate the lower layer and measure its absorbance (*2.2.25*) between 265 nm and 420 nm using as the compensation liquid the clear lower layer obtained by vigorously shaking 5.0 mL of *dimethyl sulfoxide R* with 25 mL of *heptane R* for 1 min. Prepare a 7.0 mg/L reference solution of *naphthalene R* in *dimethyl sulfoxide R* and measure the absorbance of this solution at the absorption maximum at 278 nm using *dimethyl sulfoxide R* as the compensation liquid.

Figure 18.5 (*Continued*)

addition to this, the *Latin name* ('Paracetamolum') is included. The *chemical structure*, the *chemical formula* ($C_8H_9NO_2$ for paracetamol), and the *relative molar mass* ($M_r = 151.2$ for paracetamol) are also at the top of the monograph.

The following section is termed DEFINITION and states the requirement for content (from 99.0% to 101.0% for paracetamol). The limits for the content defines the purity. The content is expressed in per cent % (*w/w*). Results obtained from ASSAY (quantitative measurement of purity) at the end of the monograph have to be within these limits for the ingredient to be released for pharmaceutical production. The upper limit is above 100% and is due to the standard deviation of the assay. When titration methods are used for the assay, the limits are often $100.0\% \pm 1.0\%$. For assays based on liquid chromatography (LC) and UV spectrophotometry, limits are typically $100.0\% \pm 2.0\%$ and $100.0\% \pm 3.0\%$, respectively. The purity requirements do not refer to the trade name, but rather to the *systematic chemical name* for unequivocal definition. Thus, paracetamol consists of 99.0–101.0% *N*-(4-hydroxyphenyl)acetamide, which is the systematic chemical name for the substance. Furthermore, the assay result has to be corrected for any content of moisture (and other volatile impurities), as the content is defined with reference to the dried substance. As discussed later, moisture (and other volatile impurities) is determined by the test *Loss on drying*, and for paracetamol the limit is 0.5%.

The section CHARACTERS contains qualitative information on appearance and solubility. This section can also contain information on polymorphism and other physical characters. The colour and physical state of the substance is described under *Appearance*. In addition, information on moisture sensitivity can be given. If a substance is hygroscopic it has to be stored in an airtight container. Appearance is for a first-hand visual inspection of the pharmaceutical ingredient. Most pharmaceutical ingredients are colourless compounds (white powders), and if such ingredients have a slight colour, it may indicate the presence of impurities. Usually the solubility of the pharmaceutical ingredient is given for a variety of solvents such as water, dilute acid, dilute base, acetone, and ethanol. The solubility is expressed using qualitative terms (such as *very soluble*, *soluble*, and *slightly soluble*), which are defined in Table 18.4.

The section IDENTIFICATION provides the procedures for identification of the pharmaceutical ingredient. It is of vital importance to verify that the content of a given container holding a pharmaceutical ingredient is in accordance with the labelling and to ensure that

Table 18.4   Definition of solubility (Ph. Eur.)

Description	Solubility (g/mL at $20 \pm 5\,°C$)
Very soluble	>1.0
Freely soluble	0.1–1.0
Soluble	0.03–0.1
Sparingly soluble	0.01–0.03
Slightly soluble	0.001–0.01
Very slightly soluble	0.0001–0.001
Practically insoluble	<0.0001

the correct ingredients are used for a given pharmaceutical preparation. Identity tests are not intended to provide a complete chemical identification (this would typically require NMR spectroscopy), but rather a confirmation of the identity of the pharmaceutical ingredient at an 'acceptable' level of certainty. The primary technique for identification is infrared (IR) spectrophotometry, but several other techniques and procedures are also in use. Identification procedures may be divided into *First identification* and *Second identification*. The procedures under *First identification* are the primary ones, and every production batch of a pharmaceutical ingredient should be tested according to these. Testing according to *Second identification* is an option, provided that the pharmaceutical ingredient has been tested according to *First identification* by the manufacturer or supplier.

The next section is TESTS. The test procedures are intended to check that pharmaceutical ingredients do not contain significant contamination of other chemical compounds such as related substances and residual solvents. Several of the tests are relatively simple and do not intend to control for all possible contaminants. Thus, the tests performed have been chosen to cover the known major impurities for the particular pharmaceutical ingredient. This means that if other impurities not covered by the tests are or can be present, *good pharmaceutical practices* should be followed and the pharmaceutical ingredient should not be released for production before the 'new' impurities have been examined. Section 18.5 describes the different techniques and procedures used under TESTS.

The ASSAY section describes the experimental procedure and equations for the quantitative determination, to measure the purity of the pharmaceutical ingredient. Results from the assay are compared with the requirement for content under DEFINITION. If the result is in compliance with DEFINITION, and the pharmaceutical ingredient also fulfils IDENTIFICATION and TESTS, the chemical quality is satisfactory for pharmaceutical production. Assays are often performed by titration, but titration is of limited specificity. Thus, related substances with a similar structure as the pharmaceutical ingredient can interfere during titration, but such substances are checked by the test for *Related substances*. Thus, TESTS and ASSAY are complementary and the combined efforts ensure that the pharmaceutical ingredient is of high purity. Techniques and procedures for assay are discussed in Section 18.6.

Monographs often contain a STORAGE section with recommendations for storage. If a pharmaceutical ingredient is hydroscopic, storage in an airtight container is prescribed. Some pharmaceutical ingredients are sensitive to light, and these should be stored protected from light. At the end of the monographs, there is often an IMPURITIES section. This section contains a list of systematic names and chemical structures of impurities typically found in the pharmaceutical ingredient. The list can be divided into two parts: the first part includes specified impurities with limits established by medicinal authorities. Such impurities are termed *Qualified impurities* as they have been tested to be of low toxicity in clinical trials. Qualified impurities are typically accepted in concentration levels of 0.1–0.5% in the pharmaceutical ingredient. The IMPURITIES section also covers *Other detectable impurities*. These are impurities with a known structure that can be detected in the given test system (typically LC). They are normally limited to less than 0.1%. Being very toxic, impurities are treated on an individual basis and their limits can be very low. One example is the impurity methyl phenyl-tetrahydro-pyridine in pethidine, where the limit is 0.1 ppm.

Acceptance criteria: 90.0%–110.0%

SPECIFIC TESTS
- **PH** ⟨791⟩: 4.5–5.8
- **BACTERIAL ENDOTOXINS TEST** ⟨85⟩: NMT 4.5 USP Endo-
  toxin Units/mg of acepromazine maleate
- **STERILITY TESTS** ⟨71⟩: It meets the requirements when
  tested as directed for *Test for Sterility of the Product to Be
  Examined, Membrane Filtration.*
- **OTHER REQUIREMENTS:** It meets the requirements in *Injec-
  tions and Implanted Drug Products* ⟨1⟩.

ADDITIONAL REQUIREMENTS
- **PACKAGING AND STORAGE:** Preserve in tight, light-resistant,
  single-dose or multiple-dose containers as described in
  Packaging and Storage Requirements ⟨659⟩, *Injection Pack-
  aging.* Store at controlled room temperature.
- **LABELING:** Label it to indicate that it is for veterinary use
  only.
- **USP REFERENCE STANDARDS** ⟨11⟩
  USP Acepromazine Maleate RS
  USP Endotoxin RS

Acepromazine Maleate Tablets

DEFINITION
Acepromazine Maleate Tablets contain NLT 90.0% and NMT
110.0% of the labeled amount of acepromazine maleate
($C_{19}H_{22}N_2OS \cdot C_4H_4O_4$).
Throughout the following procedures, protect samples, the
USP Reference Standard, and solutions containing them,
by conducting the procedures without delay, under subdued light, or using low-actinic glassware.

IDENTIFICATION
- **A. INFRARED ABSORPTION** ⟨197K⟩
  Sample: To a quantity of powdered Tablets, equivalent
  to 20 mg of acepromazine maleate, add 2 mL of water
  and 3 mL of 2 N sodium hydroxide, and extract with
  two 5-mL portions of cyclohexane. Combine the cyclohexane extracts, and evaporate to dryness under vacuum, using gentle heat if necessary.
  Acceptance criteria: Meet the requirements
- **B.** The retention time of the major peak of the *Sample
  solution* corresponds to that of the *Standard solution,* as
  obtained in the *Assay.*

ASSAY
- **PROCEDURE**
  Buffer: Add 6 mL of triethylamine to 700 mL of water,
  and adjust with phosphoric acid to a pH of 2.5.
  Mobile phase: Acetonitrile and *Buffer* (300:700)
  Standard stock solution: 1 mg/mL of USP
  Acepromazine Maleate RS in 0.05 N hydrochloric acid
  Standard solution: 0.1 mg/mL of USP Acepromazine
  Maleate RS in water from *Standard stock solution*
  Sample stock solution: Transfer NLT 10 Tablets to a
  200-mL volumetric flask, add 100 mL of 0.05 N hydro-
  chloric acid, and sonicate for 10 min. Shake by me-
  chanical means for 30 min, and dilute with 0.05 N hy-
  drochloric acid to volume.
  Sample solution: Nominally 0.1 mg/mL of
  Acepromazine Maleate in water from *Sample stock solu-
  tion.* Pass a portion of this solution through a filter of
  0.5-μm or finer pore size.
  Chromatographic system
  (See *Chromatography* ⟨621⟩, *System Suitability.*)

Mode: LC
Detector: UV 280 nm
Column: 4-mm × 15-cm; 5-μm packing L7
Flow rate: 1 mL/min
Injection volume: 10 μL
System suitability
Sample: *Standard solution*
Suitability requirements
  Column efficiency: NLT 1500 theoretical plates
  Tailing factor: NMT 2.5
  Relative standard deviation: NMT 2.0%
Analysis
Samples: *Standard solution* and *Sample solution*
Calculate the percentage of acepromazine maleate
($C_{19}H_{22}N_2OS \cdot C_4H_4O_4$) in the portion of Tablets taken:

$$\text{Result} = (r_U/r_S) \times (C_S/C_U) \times 100$$

r_U = peak area from the *Sample solution*
r_S = peak area from the *Standard solution*
C_S = concentration of USP Acepromazine Maleate
  RS in the *Standard solution* (mg/mL)
C_U = nominal concentration of the *Sample solution*
  (mg/mL)
Acceptance criteria: 90.0%–110.0%

ADDITIONAL REQUIREMENTS
- **PACKAGING AND STORAGE:** Preserve in tight, light-resistant
  containers, and store at controlled room temperature.
- **LABELING:** Label the Tablets to indicate that they are for
  veterinary use only.
- **USP REFERENCE STANDARDS** ⟨11⟩
  USP Acepromazine Maleate RS

Acetaminophen

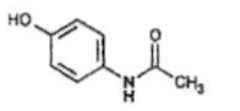

$C_8H_9NO_2$ 151.16
Acetamide, *N*-(4-hydroxyphenyl)-;
4'-Hydroxyacetanilide [103-90-2].

DEFINITION
Acetaminophen contains NLT 98.0% and NMT 102.0% of
acetaminophen ($C_8H_9NO_2$), calculated on the dried basis.

IDENTIFICATION
- **A. INFRARED ABSORPTION** ⟨197K⟩
- **B.** The retention time of the major peak of the *Sample
  solution* corresponds to that of the *Standard solution,* as
  obtained in the *Assay.*

ASSAY
- **PROCEDURE**
  Use low-actinic glassware for preparation of the *Sample
  solution.*
  Solution A: 1.7 g/L of monobasic potassium phosphate
  and 1.8 g/L of dibasic sodium phosphate, anhydrous
  Solution B: Methanol
  Mobile phase: See *Table 1.*

Table 1

Time (min)	Solution A (%)	Solution B (%)
0.0	99	1
3.0	99	1
7.0	19	81

Figure 18.6   *USP monograph for acetaminophen (paracetamol). Source: Reproduced from
USP 2017 with permission.*

Table 1 *(Continued)*

Time (min)	Solution A (%)	Solution B (%)
7.1	99	1
10.0	99	1

Standard solution: 0.1 mg/mL of USP Acetaminophen RS in methanol
Sample solution: 0.1 mg/mL of Acetaminophen in methanol
Chromatographic system
 (See *Chromatography* ⟨621⟩, *System Suitability*.)
 Mode: LC
 Detector: UV 230 nm
 Column: 4.6-mm × 10-cm; 3.5-μm packing L7
 Column temperature: 35°
 Flow rate: 1.0 mL/min
 Injection volume: 5 μL
System suitability
 Sample: *Standard solution*
 Suitability requirements
  Tailing factor: NMT 2.0
  Relative standard deviation: NMT 1.0%
Analysis
 Samples: *Standard solution* and *Sample solution*
 Calculate the percentage of acetaminophen ($C_8H_9NO_2$) in the portion of Acetaminophen taken:

$$\text{Result} = (r_U/r_S) \times (C_S/C_U) \times 100$$

r_U	= peak response from the *Sample solution*
r_S	= peak response from the *Standard solution*
C_S	= concentration of USP Acetaminophen RS in the *Standard solution* (mg/mL)
C_U	= concentration of Acetaminophen in the *Sample solution* (mg/mL)

 Acceptance criteria: 98.0%–102.0% on the dried basis

IMPURITIES
• **RESIDUE ON IGNITION** ⟨281⟩: NMT 0.1%

Delete the following:

•• **HEAVY METALS,** *Method II* ⟨231⟩: NMT 10 ppm● *(Official 1-Jan-2018)*
• **LIMIT OF FREE 4-AMINOPHENOL**
 Solution A, Solution B, Mobile phase, and **Chromatographic system:** Proceed as directed in the *Assay*.
 Standard solution: 1.25 μg/mL of USP 4-Aminophenol RS in methanol
 Sample solution: 25 mg/mL of Acetaminophen in methanol
 System suitability
  Sample: *Standard solution*
  [NOTE—The relative retention times for 4-aminophenol and acetaminophen are 0.6 and 1.0, respectively.]
  Suitability requirements
   Relative standard deviation: NMT 5.0%
 Analysis
  Samples: *Standard solution* and *Sample solution*
  Calculate the percentage of 4-aminophenol in the portion of Acetaminophen taken:

$$\text{Result} = (r_U/r_S) \times (C_S/C_U) \times 100$$

r_U	= peak response of 4-aminophenol from the *Sample solution*
r_S	= peak response from the *Standard solution*
C_S	= concentration of USP 4-Aminophenol RS in the *Standard solution* (μg/mL)
C_U	= concentration of Acetaminophen in the *Sample solution* (μg/mL)

 Acceptance criteria: NMT 0.005%
• **ORGANIC IMPURITIES**
 Solution A: Methanol, water, glacial acetic acid (50:950:1)
 Solution B: Methanol, water, glacial acetic acid (500:500:1)
 Mobile phase: See *Table 2*.

Table 2

Time (min)	Solution A (%)	Solution B (%)
0	82	18
8	82	18
53	0	100
58	0	100
59	82	18
73	82	18

 Diluent: Methanol
 System suitability solution: 20 μg/mL of USP Acetaminophen RS and 80 μg/mL each of USP Acetaminophen Related Compound B RS and USP Acetaminophen Related Compound C RS in *Diluent*
 Standard solution: 1.25 μg/mL of USP Acetaminophen Related Compound D RS and 0.25 μg/mL of USP Acetaminophen Related Compound J RS in *Diluent*
 Sample solution: 25 mg/mL of Acetaminophen in *Diluent*
 Chromatographic system
  (See *Chromatography* ⟨621⟩, *System Suitability*.)
  Mode: LC
  Detector: UV 254 nm
  Column: 4.6-mm × 25-cm; 5-μm packing L7
  Flow rate: 0.9 mL/min
  Column temperature: 40°
  Injection volume: 5 μL
 System suitability
  Samples: *System suitability solution* and *Standard solution*
  [NOTE—See *Table 3* for relative retention time values.]
  Suitability requirements
   Tailing factor: NMT 2.0 for acetaminophen related compound D, *Standard solution*
   Resolution: NLT 2.0 between acetaminophen and acetaminophen related compound B; NLT 1.5 between acetaminophen related compound B and acetaminophen related compound C, *System suitability solution*
   Relative standard deviation: NMT 5.0% for acetaminophen related compound D, *Standard solution*
 Analysis
  Samples: *Standard solution* and *Sample solution*
  Calculate the percentage of acetaminophen related compound J in the portion of Acetaminophen taken:

$$\text{Result} = (r_U/r_S) \times (C_S/C_U) \times 100$$

r_U	= peak response of acetaminophen related compound J from the *Sample solution*
r_S	= peak response of acetaminophen related compound J from the *Standard solution*
C_S	= concentration of USP Acetaminophen Related Compound J RS in the *Standard solution* (μg/mL)
C_U	= concentration of Acetaminophen in the *Sample solution* (μg/mL)

  Calculate the percentage of acetaminophen related compounds B, C, and D and any unspecified impurity in the portion of Acetaminophen taken:

$$\text{Result} = (r_U/r_S) \times (C_S/C_U) \times (1/F) \times 100$$

Figure 18.6   (Continued)

r_U = peak response of each specified or unspecified impurity from the *Sample solution*

r_S = peak response of acetaminophen related compound D from the *Standard solution*

C_S = concentration of USP Acetaminophen Related Compound D RS in the *Standard solution* (μg/mL)

C_U = concentration of Acetaminophen in the *Sample solution* (μg/mL)

F = relative response factor for each impurity shown in *Table 3*

Acceptance criteria: See *Table 3*. [NOTE—The relative retention times and relative response factors in *Table 3* (where applicable) are calculated relative to those of acetaminophen related compound D.]

Table 3

Name	Relative Retention Time	Relative Response Factor	Acceptance Criteria, NMT (%)
Acetaminophen	0.43	—	—
Acetaminophen related compound B[a]	0.67	1.2	0.05
Acetaminophen related compound C[b]	0.71	0.38	0.05
Acetaminophen related compound D[c]	1.0	1.0	0.05
Acetaminophen related compound J[d]	·1.73	—	0.001
Individual unspecified impurity	—	1.0	0.05
Total impurities	—	—	0.1

[a] N-(4-Hydroxyphenyl)propanamide.
[b] N-(2-Hydroxyphenyl)acetamide.
[c] N-Phenylacetamide.
[d] N-(4-Chlorophenyl)acetamide (*p*-chloroacetanilide).

SPECIFIC TESTS
• **LOSS ON DRYING** ⟨731⟩
  Analysis: Dry at 105° to constant weight.
  Acceptance criteria: NMT 0.5%

ADDITIONAL REQUIREMENTS
• **PACKAGING AND STORAGE:** Preserve in tight, light-resistant containers, and store at room temperature. Protect from moisture and heat.
• **USP REFERENCE STANDARDS** ⟨11⟩
  USP Acetaminophen RS
  USP Acetaminophen Related Compound B RS
  N-(4-Hydroxyphenyl)propanamide.
  C₉H₁₁NO₂   165.19
  USP Acetaminophen Related Compound C RS
  N-(2-Hydroxyphenyl)acetamide.
  C₈H₉NO₂   151.16
  USP Acetaminophen Related Compound D RS
  N-Phenylacetamide.
  C₈H₉NO   135.17
  USP Acetaminophen Related Compound J RS
  N-(4-Chlorophenyl)acetamide (*p*-chloroacetanilide).
  C₈H₈ClNO   169.61
  USP 4-Aminophenol RS
  C₆H₇NO   109.13

Acetaminophen Capsules

DEFINITION
Acetaminophen Capsules contain NLT 90.0% and NMT 110.0% of the labeled amount of acetaminophen (C₈H₉NO₂).

IDENTIFICATION
• **A.** The retention time of the major peak of the *Sample solution* corresponds to that of the *Standard solution*, as obtained in the *Assay*.
• **B. THIN-LAYER CHROMATOGRAPHIC IDENTIFICATION TEST** ⟨201⟩
  Sample solution: 1 mg/mL of acetaminophen prepared as follows. Triturate from contents of the Capsules in methanol. Filter, and use the clear filtrate.
  Chromatographic system
    Developing solvent system: Methylene chloride and methanol (4:1)
  Acceptance criteria: Meet the requirements

ASSAY
• **PROCEDURE**
  Mobile phase: Methanol and water (1:3)
  Standard solution: 0.01 mg/mL of USP Acetaminophen RS in *Mobile phase*
  Sample stock solution: Weigh the contents of NLT 20 Capsules, and calculate the average weight of the contents of each Capsule. Mix the combined contents of the Capsules, and transfer a portion, equivalent to 100 mg of acetaminophen, to a 200-mL volumetric flask. Add 100 mL of *Mobile phase*, shake by mechanical means for 10 min, and dilute with *Mobile phase* to volume. Transfer 5.0 mL of this solution to a 250-mL volumetric flask, and dilute with *Mobile phase* to volume. Pass a portion of this solution through a filter of 0.5-μm or finer pore size, discarding the first 10 mL of the filtrate.
  Sample solution: Nominally 0.01 mg/mL of acetaminophen from the *Sample stock solution* in *Mobile phase*. Pass a portion of this solution through a filter of 0.5-μm or finer pore size, discarding the first 10 mL of the filtrate.
  Chromatographic system
  (See *Chromatography* ⟨621⟩, *System Suitability*.)
    Mode: LC
    Detector: UV 243 nm
    Column: 3.9-mm × 30-cm; packing L1
    Flow rate: 1.5 mL/min
    Injection volume: 10 μL
  System suitability
    Sample: *Standard solution*
    Suitability requirements
      Column efficiency: NLT 1000 theoretical plates
      Tailing factor: NMT 2
      Relative standard deviation: NMT 2.0%
  Analysis
    Samples: *Standard solution* and *Sample solution*
    Calculate the percentage of the labeled amount of acetaminophen (C₈H₉NO₂) in the portion of Capsules taken:

$$\text{Result} = (r_U/r_S) \times (C_S/C_U) \times 100$$

r_U = peak response from the *Sample solution*
r_S = peak response from the *Standard solution*
C_S = concentration of USP Acetaminophen RS in the *Standard solution* (mg/mL)
C_U = nominal concentration of acetaminophen in the *Sample solution* (mg/mL)

Figure 18.6   (Continued)

USP has individual monographs for pharmaceutical ingredients similar to Ph. Eur. Figure 18.6 presents the USP monograph for paracetamol (acetaminophen) as an example. The monographs in USP have the following general structure:

• Definition
• Identification
• Assay
• Impurities

Correct the area of the peak due to methyl ricinoleate, by multiplying by a factor R calculated using the following expression:

$$\frac{m_1 \times A_2}{A_1 \times m_2}$$

m_1 = mass of methyl ricinoleate in the reference solution;

m_2 = mass of methyl stearate in the reference solution;

A_1 = area of the peak due to methyl ricinoleate in the chromatogram obtained with the reference solution;

A_2 = area of the peak due to methyl stearate in the chromatogram obtained with the reference solution.

Composition of the fatty-acid fraction of the oil:

– *palmitic acid*: maximum 2.0 per cent;
– *stearic acid*: maximum 2.5 per cent;
– *oleic acid*: 2.5 per cent to 6.0 per cent;
– *linoleic acid*: 2.5 per cent to 7.0 per cent;
– *linolenic acid*: maximum 1.0 per cent;
– *eicosenoic acid*: maximum 1.0 per cent;
– *ricinoleic acid*: 85.0 per cent to 92.0 per cent;
– *any other fatty acid*: maximum 1.0 per cent.

Water (*2.5.32*): maximum 0.3 per cent, or maximum 0.2 per cent if intended for use in the manufacture of parenteral preparations, determined on 1.00 g.

STORAGE

In an airtight, well-filled container, protected from light.

LABELLING

The label states, where applicable, that the substance is suitable for use in the manufacture of parenteral preparations.

 01/2015:0051

CASTOR OIL, VIRGIN

Ricini oleum virginale

DEFINITION

Fatty oil obtained by cold expression from the seeds of *Ricinus communis* L. A suitable antioxidant may be added.

PRODUCTION

During the expression step, the temperature of the oil must not exceed 50 °C.

CHARACTERS

Appearance: clear at 40 °C, slightly yellow, viscous, hygroscopic liquid.

Solubility: slightly soluble in light petroleum, miscible with ethanol (96 per cent) and with glacial acetic acid.

Relative density: about 0.958.

Refractive index: about 1.479.

IDENTIFICATION

First identification: B, C.

Second identification: A, B.

A. A mixture of 2 mL of the substance to be examined and 8 mL of *ethanol (96 per cent) R* is clear (*2.2.1*).

B. Specific absorbance (see Tests).

C. Composition of fatty acids (see Tests).

TESTS

Optical rotation (*2.2.7*): + 3.5° to + 6.0°.

Specific absorbance (*2.2.25*): maximum 0.7, determined at the absorption maximum at 270 nm.

To 1.00 g add *ethanol (96 per cent) R* and dilute to 100.0 mL with the same solvent.

Acid value (*2.5.1*): maximum 1.5.

Dissolve 5.00 g in 25 mL of the prescribed mixture of solvents.

Hydroxyl value (*2.5.3, Method A*): minimum 160.

Peroxide value (*2.5.5, Method A*): maximum 10.0.

Unsaponifiable matter (*2.5.7*): maximum 0.8 per cent, determined on 5.0 g.

Composition of fatty acids. Gas chromatography (*2.4.22*) with the following modifications.

Use the mixture of calibrating substances in Table 2.4.22.-3.

Test solution. Introduce 75 mg of the substance to be examined into a 10 mL centrifuge tube with a screw cap. Dissolve in 2 mL of *1,1-dimethylethyl methyl ether R1* with shaking and heat gently (50-60 °C). Add, while still warm, 1 mL of a 12 g/L solution of *sodium R* in *anhydrous methanol R*, prepared with the necessary precautions, and mix vigorously for at least 5 min. Add 5 mL of *distilled water R* and mix vigorously for about 30 s. Centrifuge for 15 min at 1500 g. Use the upper layer.

Reference solution. Dissolve 50 mg of *methyl ricinoleate CRS* and 50 mg of *methyl stearate CRS* in 10.0 mL of *1,1-dimethylethyl methyl ether R1*.

Column:

– *material*: fused silica;
– *size*: l = 30 m, Ø = 0.25 mm;
– *stationary phase*: macrogol 20 000 R (film thickness 0.25 μm).

Carrier gas: helium for chromatography R.

Flow rate: 0.9 mL/min.

Split ratio: 1:100.

Temperature:

	Time (min)	Temperature (°C)
Column	0 - 55	215
Injection port		250
Detector		250

Detection: flame ionisation.

Injection: 1 μL.

Calculate the percentage content of each fatty acid by the normalisation procedure.

Correct the area of the peak due to methyl ricinoleate, by multiplying by a factor R calculated using the following expression:

$$\frac{m_1 \times A_2}{A_1 \times m_2}$$

m_1 = mass of methyl ricinoleate in the reference solution;

m_2 = mass of methyl stearate in the reference solution;

A_1 = area of the peak due to methyl ricinoleate in the chromatogram obtained with the reference solution;

A_2 = area of the peak due to methyl stearate in the chromatogram obtained with the reference solution.

Figure 18.7 *Ph. Eur. monograph for Castor oil, virgin. Source: Reproduced from Ph. Eur. 2017 (Ninth Edition) with permission.*

Composition of the fatty-acid fraction of the oil:
- *palmitic acid*: maximum 2.0 per cent;
- *stearic acid*: maximum 2.5 per cent;
- *oleic acid* : 2.5 per cent to 6.0 per cent;
- *linoleic acid* : 2.5 per cent to 7.0 per cent;
- *linolenic acid* : maximum 1.0 per cent;
- *eicosenoic acid* : maximum 1.0 per cent;
- *ricinoleic acid* : 85.0 per cent to 92.0 per cent;
- *any other fatty acid*: maximum 1.0 per cent.

Water (*2.5.32*): maximum 0.3 per cent, determined on 1.00 g.

STORAGE

In an airtight, well-filled container, protected from light.

01/2017:0986

CEFACLOR

Cefaclorum

$C_{15}H_{14}ClN_3O_4S,H_2O$ M_r 385.8
[70356-03-5]

DEFINITION

(6R,7R)-7-[[(2R)-2-Amino-2-phenylacetyl]amino]-3-chloro-8-oxo-5-thia-1-azabicyclo[4.2.0]oct-2-ene-2-carboxylic acid monohydrate.

Semi-synthetic product derived from a fermentation product.

Content: 96.0 per cent to 102.0 per cent of $C_{15}H_{14}ClN_3O_4S$ (anhydrous substance).

CHARACTERS

Appearance: white or slightly yellow powder.

Solubility: slightly soluble in water, practically insoluble in methanol and in methylene chloride.

IDENTIFICATION

Infrared absorption spectrophotometry (*2.2.24*).

Comparison: cefaclor CRS.

TESTS

pH (*2.2.3*): 3.0 to 4.5.

Suspend 0.250 g in *carbon dioxide-free water R* and dilute to 10 mL with the same solvent.

Specific optical rotation (*2.2.7*): + 101 to + 111 (anhydrous substance).

Dissolve 0.250 g in a 10 g/L solution of *hydrochloric acid R* and dilute to 25.0 mL with the same solution.

Related substances. Liquid chromatography (*2.2.29*).

Test solution. Dissolve 50.0 mg of the substance to be examined in 10.0 mL of a 2.7 g/L solution of *sodium dihydrogen phosphate R* adjusted to pH 2.5 with *phosphoric acid R*.

Reference solution (a). Dissolve 2.5 mg of cefaclor CRS and 5.0 mg of *delta-3-cefaclor CRS* (impurity D) in 100.0 mL of a 2.7 g/L solution of *sodium dihydrogen phosphate R* adjusted to pH 2.5 with *phosphoric acid R*.

Reference solution (b). Dilute 1.0 mL of the test solution to 100.0 mL with a 2.7 g/L solution of *sodium dihydrogen phosphate R* adjusted to pH 2.5 with *phosphoric acid R*.

Column:
- *size*: l = 0.25 m, Ø = 4.6 mm;
- *stationary phase*: end-capped octadecylsilyl silica gel for chromatography R (5 μm).

Mobile phase:
- *mobile phase A*: 7.8 g/L solution of *sodium dihydrogen phosphate R* adjusted to pH 4.0 with *phosphoric acid R*;
- *mobile phase B*: mix 450 mL of *acetonitrile R* with 550 mL of mobile phase A;

Time (min)	Mobile phase A (per cent V/V)	Mobile phase B (per cent V/V)
0 - 30	95 → 75	5 → 25
30 - 45	75 → 0	25 → 100
45 - 55	0	100

Flow rate: 1.0 mL/min.

Detection: spectrophotometer at 220 nm.

Injection: 20 μL.

System suitability: reference solution (a):
- *resolution*: minimum 2 between the peaks due to cefaclor and impurity D; if necessary, adjust the acetonitrile content in the mobile phase;
- *symmetry factor*: maximum 1.2 for the peak due to cefaclor; if necessary, adjust the acetonitrile content in the mobile phase.

Limits:
- *any impurity*: for each impurity, not more than 0.5 times the area of the principal peak in the chromatogram obtained with reference solution (b) (0.5 per cent);
- *total*: not more than twice the area of the principal peak in the chromatogram obtained with reference solution (b) (2 per cent);
- *disregard limit*: 0.1 times the area of the principal peak in the chromatogram obtained with reference solution (b) (0.1 per cent).

Water (*2.5.12*): 3.0 per cent to 6.5 per cent, determined on 0.200 g.

ASSAY

Liquid chromatography (*2.2.29*).

Test solution. Dissolve 15.0 mg of the substance to be examined in the mobile phase and dilute to 50.0 mL with the mobile phase.

Reference solution (a). Dissolve 15.0 mg of cefaclor CRS in the mobile phase and dilute to 50.0 mL with the mobile phase.

Reference solution (b). Dissolve 3.0 mg of cefaclor CRS and 3.0 mg of *delta-3-cefaclor CRS* (impurity D) in the mobile phase and dilute to 10.0 mL with the mobile phase.

Column:
- *size*: l = 0.25 m, Ø = 4.6 mm;
- *stationary phase*: octadecylsilyl silica gel for chromatography R (5 μm).

Mobile phase: add 220 mL of *methanol R* to a mixture of 780 mL of *water R*, 10 mL of *triethylamine R* and 1 g of *sodium pentanesulfonate R*, then adjust to pH 2.5 with *phosphoric acid R*.

Flow rate: 1.5 mL/min.

Detection: spectrophotometer at 265 nm.

Injection: 20 μL.

System suitability:
- *resolution*: minimum 2.5 between the peaks due to cefaclor and impurity D in the chromatogram obtained with reference solution (b); if necessary, adjust the concentration of methanol in the mobile phase;

Figure 18.7   (Continued)

- Specific tests
- Additional requirements

DEFINITION, IDENTIFICATION, and ASSAY are very similar to Ph. Eur. in terms of purpose. The two sections IMPURITIES and SPECIFIC TESTS are similar to the section TESTS in Ph. Eur. and prescribe procedures to assess the purity and quality of the pharmaceutical ingredient. The methods in use for identification, assay, impurities, and specific tests are very similar to those used in Ph. Eur. The section ADDITIONAL REQUIRE-MENTS lists precautions related to packaging and storage of the pharmaceutical ingredient, and also list USP reference standards (chemical reference substances, CRSs) required for the monograph.

The discussion and examples above were focused on monographs for pure chemical ingredients. Monographs for organic multi-chemical ingredients are generally organized in the same way. Such monographs do not contain ASSAY, but it is customary to quantify one or more components in the mixture. Methods and procedures for this are given in the TESTS section. Procedures used under IDENTIFICATION and TESTS are not only chemical methods, and often different physical and biological methods are used in addition. These methods are, however, outside the scope of this book. An example of a Ph. Eur. monograph for a multi-chemical ingredient, castor oil (excipient), is shown in Figure 18.7.

18.3 Impurities in Pharmaceutical Ingredients

The following section discusses the different types of impurities found in pharmaceutical ingredients in more detail.

18.3.1 Impurities in Pure Chemical Ingredients

Pure chemical ingredients can be divided into *organic* and *inorganic* compounds. Examples of the former are paracetamol and ibuprofen, while examples of the latter are sodium chloride and ferrous sulfate. The API in a pharmaceutical preparation is, with few exceptions, an organic compound. Excipients are both organic and inorganic compounds. Generally, the purity definitions set by Ph. Eur. and USP are strict. As shown in Table 18.5, however, the purity definition can be different for different compounds in Ph. Eur. and USP. The upper limit of content is set based on the precision of the quantitation method used for the assay. The lower limit of content reflects the precision of the assay as well as the content of accepted impurities. The purity definition is given with reference to the *dried substance*. This implies that the assay should be performed with a dried substance, or the assay results should be corrected for the actual content of water/moisture. Due to the strict purity definitions, typically 99.0–99.5% of the pharmaceutical ingredient is pure substance and impurities are limited to less than 0.5–1.0%.

Impurities constitute a trace amount of the pharmaceutical ingredient, but they are strictly controlled. The reason for this is that even small amounts of impurities can influence the quality, stability, and efficacy of the pharmaceutical ingredient. Table 18.6 shows the major types of impurities found in organic and inorganic pure chemical ingredients. Impurity testing of pharmaceutical ingredients is described in Section 18.5 and Table 18.6 refers to the individual test for impurities described in this chapter.

Table 18.5 Examples of purity definitions given in Ph. Eur. and USP for APIs

Ingredient	Ph. Eur.		USP	
	Content (%)	Assay	Content (%)	Assay
Omeprazole	99.0–101.0	Titration	98.0–102.0	HPLC
Chlorpromazine hydrochloride	99.0–101.0	Titration	98.0–101.0	Titration
Simvastatin	97.0–102.0	HPLC	98.0–102.0	HPLC
Furosemide	98.5–101.0	Titration	98.0–102.0	HPLC
Hydrocortisone	97.0–103.0	UV	97.0–102.0	HPLC
Ibuprofen	98.5–101.0	Titration	97.0–103.0	HPLC
Paracetamol	99.0–101.0	Titration	98.0–102.0	HPLC
Riboflavin	97.0–103.0	UV	98.0–102.0	Fluorescence
Budesonide	97.5–102.0	HPLC	98.0–102.0	HPLC

Table 18.6 Overview of common types of impurities in organic and inorganic pure chemical ingredients

API type	Impurity	Origin of impurity	Test to be performed	Described in section
Organic				
	Minor ions	Synthesis	pH/Acidity/ Alkalinity	18.5.3
	Related substances (organic impurities)	Synthesis/degradation	Related substances	18.5.4
	Organic solvent	Synthesis	Residual solvents	18.5.5
	Inorganic impurities	Synthesis/production equipment	Foreign anions/ Sulfated ash	18.5.6 18.5.7
	Metals/catalysts	Production equipment	Elemental impurities	18.5.8
	Water	Synthesis/absorption from air	Loss on drying Water	18.5.9 18.5.10
Inorganic				
	Inorganic impurities	Synthesis/production equipment	Foreign anions	18.5.6
	Metals/catalysts	Synthesis/production equipment	Elemental impurities	18.5.8
	Water	Synthesis/absorption from air	Loss on drying Water	18.5.9 18.5.10

Pharmaceutical ingredients that are organic compounds often contain impurities having a related structure. These impurities are termed *related substances* in Ph. Eur. and *organic impurities* in USP, and can be traces of excess starting materials from the synthesis, by-products from the synthesis, or degradation products formed during storage.

Related substances have to be controlled in pharmaceutical ingredients for the following reasons:

- They may have a pharmacological or toxic effect
- They can influence the effect of the pharmaceutical preparation
- They can influence the stability of the pharmaceutical preparation

Ph. Eur. requires that the amount of related substances should not exceed a certain percentage of the API. If the maximum daily dose is larger than 2 g per day, impurities above 0.05% (*w/w*) have to be *identified* and *qualified*. Qualification implies that the toxicity of the impurity should be clarified. For APIs with a daily dose less than 2 g, the limit for related substances is typically at 0.1% (*w/w*) level. Similar requirements are found in USP.

In addition to limits for related substances, Ph. Eur. and USP set strict requirements for the content of *water* in pharmaceutical ingredients. Normally, the content of the compound is given based on the *dried substance* when the test for loss on drying has been performed, or is based on an *anhydrous substance* if the test for *water* has been performed. Thus the assay has to be corrected for the result obtained in the test for *loss on drying* or the test for *water* before the result can be compared with the DEFINITION. For most pharmaceutical ingredients, the limit for water is set to 0.5% (*w/w*). For pharmaceutical ingredients containing crystal water, the purity definition is adjusted accordingly. The water content has to be controlled as it may:

- Reduce the stability of the pharmaceutical ingredient (chemical reactions are accelerated in the presence of moisture)
- Inaccurate dosing of the final drug if the water concentration is not taken into consideration (a significant part of the API consists of water)

Residual solvents can be found in pharmaceutical ingredients. Some organic solvents (e.g. benzene and 1,2-dichloroethane) are very toxic and should not be present in pharmaceutical ingredients, while other solvents like ethanol are less toxic. In Ph. Eur. the solvents are classified according to toxicity. *Class 1* is carcinogenic and are highly toxic solvents that should not be used in the production of pharmaceutical ingredients. *Class 2* is toxic solvents with strict requirements to maximum content, while the solvents in *Class 3* are less toxic with a typical limit at 0.5% (*w/w*).

Traces of metals can be found in pharmaceutical ingredients as a result of contamination from industrial equipment, due to use as catalysts, or can be found in ingredients derived from vegetable or animal products (e.g. Hg in cod liver oil). Heavy metals and catalysts are hazardous and can compromise chemical stability. The test for *elemental impurities* in Ph. Eur. has recently been implemented and this test is harmonized with the test in USP by the common International Council for Harmonization (ICH) Guideline for Elemental Impurities Q3D (1) to provide a global policy for limitation of elemental impurities in drug products.

The total content of *inorganic material* in pure chemical ingredients is normally limited to 0.1% (*w/w*) determined by *sulfated ash*. In addition, some ingredients are tested for the presence of specific foreign inorganic anions such as *chloride, sulfate,* or *nitrate.* This is done because strong inorganic acids and bases are often used in the syntheses, and residues can be found in pharmaceutical ingredients. Tests for *pH* or *acidity* and *alkalinity* are also used to verify protolytic residues from synthesis. The limits for the presence of foreign ions vary depending on daily doses, and typically vary between 5 and 250 ppm (equivalent to 0.0005–0.025%).

The quality requirements discussed to this point are from Ph. Eur. The requirements set by USP are very similar. The discussion above was restricted to organic pure chemical ingredients. The requirements for inorganic pure chemical ingredients are essentially the same.

18.3.2   Impurities in Organic Multi-Chemical Ingredients

Organic multi-chemical ingredients are tested for a more diverse range of contaminants. The reason is that this group consists of very different ingredients and that they can be manufactured in very different ways. Thus, as examples, microcrystalline cellulose and carbomers are organic polymers, cetostearyl alcohol and hard fat are waxes, polysorbate 80 and cetrimide are solubilizers, and castor oil and cod liver oil are fatty oils. These mixtures often have to be characterized using other tests than those used for pure chemical ingredients. However, the principle of what is allowed to be present in the ingredients is basically the same as discussed for pure chemical ingredients. Limits are stricter for contaminants of high toxicity and for ingredients with a high expected daily intake.

18.4   Identification of Pharmaceutical Ingredients

In Ph. Eur., two sets of identification procedures termed *First identification* and *Second identification* are customary. This is exemplified in Figure 18.8. The procedures under *First identification* are the primary ones, and every production batch of a pharmaceutical ingredient should be tested accordingly. Testing according to *Second identification* is an option for a given production batch, provided that this has been tested according to *First identification* by the manufacturer or supplier. First identification normally includes testing by IR spectrophotometry. For laboratories where such instrumentation is not available, the procedures under *Second identification* can be used. In new and revised monographs, it is now usual only to have one set of identification procedures including IR spectrophotometry. This is similar to USP, where most monographs have a set of two or three primary identification procedures.

For identification of pure chemical ingredients, several analytical techniques are used. The most important ones are summarized in Table 18.7 and discussed in the following. In addition, a variety of colour and precipitation reactions are used. The latter tests are limited in use and are therefore not discussed further in this text.

18.4.1   IR Spectrophotometry

IR absorption spectrophotometry is used extensively for identification both in Ph. Eur. and USP. The fundamentals of IR spectrophotometry are presented in Chapter 8. Using

IDENTIFICATION

First identification: B, D.

Second identification: A, C, D.

A. Ultraviolet and visible absorption spectrophotometry (2.2.25).

Test solution. Dissolve 50.0 mg in water R and dilute to 50.0 mL with the same solvent. Dilute 5.0 mL of the solution to 50.0 mL with water R.

Spectral range: 230-350 nm.

Absorption maximum: at 279 nm.

Specific absorbance at the absorption maximum: 64 to 72.

B. Infrared absorption spectrophotometry (2.2.24).

Comparison: mianserin hydrochloride CRS.

If the spectra obtained in the solid state show differences, dissolve the substance to be examined and the reference substance separately in methanol R, evaporate to dryness and record new spectra using the residues.

C. Thin-layer chromatography (2.2.27).

Test solution. Dissolve 10 mg of the substance to be examined in methylene chloride R and dilute to 5 mL with the same solvent.

Reference solution (a). Dissolve 10 mg of mianserin hydrochloride CRS in methylene chloride R and dilute to 5 mL with the same solvent.

Reference solution (b). Dissolve 10 mg of mianserin hydrochloride CRS and 10 mg of cyproheptadine hydrochloride CRS in methylene chloride R and dilute to 5 mL with the same solvent.

Plate: TLC silica gel GF_{254} plate R.

Mobile phase: diethylamine R, ether R, cyclohexane R 5:20:75 (*v/v/v*).

Application: 2 μL.

Development: over 2/3 of the plate.

Detection: examine in ultraviolet light at 254 nm.

System suitability: reference solution (b):
- the chromatogram shows 2 clearly separated principal spots.

Results: the principal spot in the chromatogram obtained with the test solution is similar in position and size to the principal spot in the chromatogram obtained with reference solution (a).

D. It gives reaction (a) of chlorides (2.3.1).

Figure 18.8 *First and second identification for mianserin hydrochloride (Ph. Eur.)*

an IR spectrophotometer, a spectrum of the pharmaceutical ingredient is recorded (*sample spectrum*) and compared with a spectrum of *CRS* recorded with the same IR spectrophotometer. Alternatively, the spectrum of the pharmaceutical ingredient can be compared with a published *reference spectrum*. In the latter case, recording the sample spectrum should be performed under the same experimental conditions as reported for the reference spectrum.

The sample spectrum can be obtained directly on the pharmaceutical ingredient in the solid state and can be recorded using the attenuated total reflectance (ATR) technique. For

Table 18.7   Overview of the main procedures used for the identification of pure chemical ingredients

IR spectrophotometry
UV-Vis spectrophotometry
Melting point
Polarimetry
LC
TLC colour reactions

the pharmaceutical ingredient to be identified unambiguously, the sample spectrum and the CRS spectrum (or reference spectrum) has to be identical. This implies that transmission minima (absorption maxima) in the sample spectrum correspond in position and relative size to those in the CRS spectrum (or reference spectrum). One example from USP for identification of ibuprofen is shown in Box 18.1.

Box 18.1   Identification of ibuprofen by IR spectrophotometry according to USP

STRUCTURE

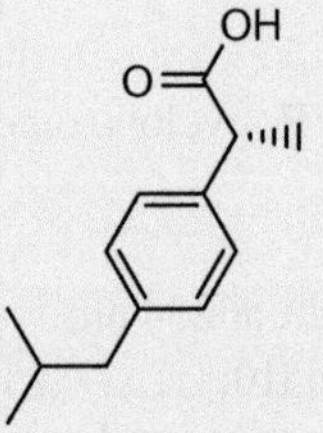

PHARMACOPEIA TEXT – INDIVIDUAL MONOGRAPH FOR IBUPROFEN

IDENTIFICATION
A. IR Absorption – Do Not Dry Specimens

PHARMACOPEIA TEXT – GENERAL INFORMATION ABOUT IR SPECTROPHOTOMETRY

Record the spectra of the test specimen and the corresponding USP Reference Standard over the range from about 2.6 mm to 15 mm (3800/cm to 650/cm) unless otherwise specified in the individual monograph. The IR absorption spectrum of the preparation of the test specimen, previously dried under conditions specified for the corresponding Reference Standard unless otherwise specified, or unless the Reference Standard is to be used without drying, exhibits maxima only at the same wavelengths as that of a similar preparation of the corresponding USP Reference Standard. The ATR technique can be used to record the spectra of the test specimen and the corresponding USP Reference Standard.

PROCEDURE STEP-BY-STEP

Spectra are recorded using ATR. First, a small portion of sample is placed in the ATR source and the sample spectrum is recorded. Second, the ATR source is cleaned, a small portion of ibuprofen USP Reference Standard is placed in the source and the CRS spectrum is recorded.

EXAMPLE

The spectra for a sample of ibuprofen API (test specimen) and CRS of ibuprofen (USP Reference Standard) are recorded by ATR, and are shown below.

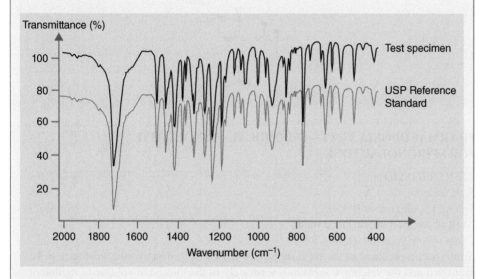

As seen from the spectra, the test specimen (sample) exhibits maxima only at the same wavelengths as that of a similar preparation of the corresponding USP Reference Standard. The sample is therefore in accordance with the pharmacopeia requirement.

DISCUSSION

When the information from the ibuprofen monograph and from the general text is combined, the entire procedure and the experimental conditions are specified. ATR is convenient because spectra can be recorded directly on the ibuprofen powder. Cleaning the ATR source according to the manufacturer's in-between sample and the USP Reference Standard is important to avoid cross-contamination of the source.

A number of pharmaceutical ingredients can show *polymorphism* (different crystal structure). For such ingredients, the appearance of the spectrum can depend on the crystal structure. In these cases, it can be necessary to recrystallize the substance in a suitable solvent, and the CRS has to be treated similarly. Alternatively, the spectra of pharmaceutical ingredient and CRS can be recorded in solution. In solution, the crystal structure is destroyed and the spectra are identical. An example of IR spectrophotometry in solution is shown in Box 18.2 for spironolactone.

Box 18.2   Identification of spironolactone by infrared absorption spectrophotometry according to Ph. Eur.

STRUCTURE

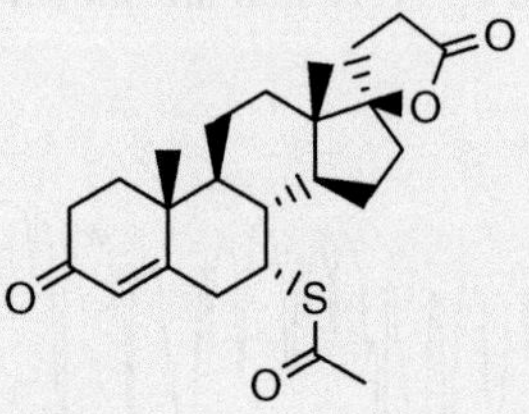

PHARMACOPOEIA TEXT – INDIVIDUAL MONOGRAPH FOR SPIRONOLACTONE

IDENTIFICATION
First identification: A.
Second identification: B, C.
A. IR absorption spectrophotometry
Comparison: spironolactone CRS
If the spectra obtained in the solid state show differences, dissolve the substance to be examined and the reference substance separately in the minimum volume of methanol R, evaporate to dryness and record new spectra using the residues.

PROCEDURE STEP-BY-STEP

Spectra are recorded using ATR similar to Box 18.1. A small portion of ingredient is placed in the ATR source and the sample spectrum is recorded. The ATR source is cleaned, a small portion of spironolactone CRS is placed in the source, and the CRS spectrum is recorded. This procedure is repeated after recrystallization if required.

EXAMPLE

As seen from the spectra below, the sample spectrum (A) and the CRS spectrum (B) are not similar due to differences in crystal form. After recrystallization, the sample spectrum (C) and the CRS spectrum (D) are similar. The sample is therefore in accordance with the pharmacopoeia requirement.

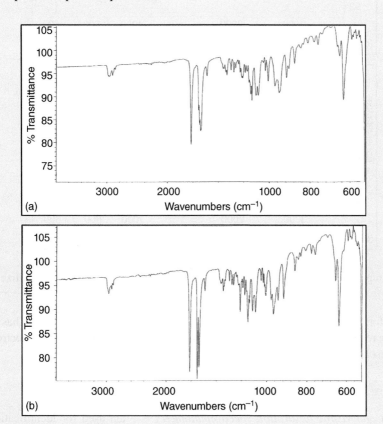

(A) Sample spectrum before recrystallization; (B) CRS spectrum before recrystallization

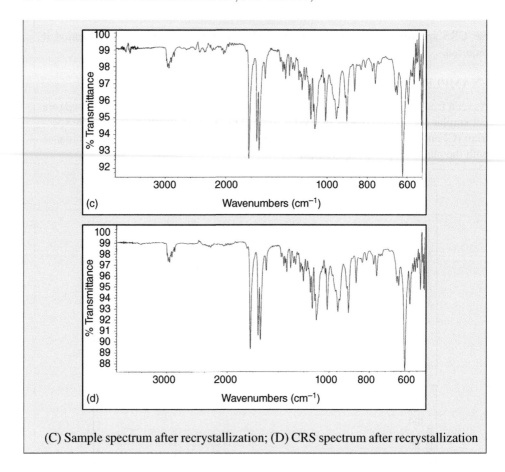

(c)

Wavenumbers (cm⁻¹)

(d)

Wavenumbers (cm⁻¹)

(C) Sample spectrum after recrystallization; (D) CRS spectrum after recrystallization

The performance of IR absorption spectrophotometers used for identification has to be checked on regular basis. As discussed in Chapter 8, this implies control of the spectrophotometric resolution and the wave-number scale.

18.4.2   UV-Vis Spectrophotometry

Ultraviolet and visible absorption spectrophotometry is often used for identification of pharmaceutical ingredients both in Ph. Eur. and USP. The fundamentals of UV-Vis spectrophotometry are presented in Chapter 7. The *sample spectrum* is recorded for the pharmaceutical ingredient dissolved in an ultraviolet-transparent solvent, and absorption maxima are observed. In some procedures, the sample spectrum is compared with a spectrum of *CRS* recorded under identical conditions on the same spectrophotometer. More frequently, however, the absorption maxima are provided in the monograph and no CRS spectrum is required.

The requirements for identification are met if the UV absorption spectra of the test solution exhibit maxima and minima at specified wavelengths and if absorptivities and/or absorbance ratios are within specified limits. UV spectrophotometry for identification of mianserin hydrochloride according to Ph. Eur. is shown in Box 18.3.

Box 18.3   Identification of mianserin hydrochloride by ultraviolet absorption spectrophotometry according to Ph. Eur.

STRUCTURE (FIGURE SHOWS FREE BASE)

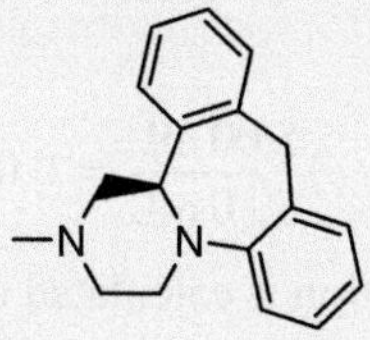

PHARMACOPOEIA TEXT – INDIVIDUAL MONOGRAPH FOR MIANSERIN HYDROCHLORIDE

IDENTIFICATION
First identification: B, D.
Second identification: A, C, D.
A. Ultraviolet and visible absorption spectrophotometry
Test solution: Dissolve 50.0 mg in water R and dilute to 50.0 mL with the same solvent. Dilute 5.0 mL of the solution to 50.0 mL with water R.
Spectral range: 230–350 nm.
Absorption maximum: at 279 nm.
Specific absorbance at the absorption maximum: 64–72.

PHARMACOPOEIA TEXT – GENERAL INFORMATION ABOUT UV-VIS SPECTROPHOTOMETRY

Unless otherwise prescribed, measure the absorbance at the prescribed wavelength using a path length of 1 cm. Unless otherwise prescribed, the measurements are carried out with reference to the same solvent or the same mixture of solvents.

PROCEDURE STEP-BY-STEP

About 50 mg of ingredient is weighed accurately, dissolved in purified water, and diluted to 50.0 mL with purified water in a volumetric flask. From this solution, 5.00 mL is pipetted and diluted with purified water to 50.0 mL in a new volumetric flask. The spectrum is recorded for this solution in the range 230–350 nm. An absorption maximum should be observed at 279 nm and the specific absorbance A (1 cm, 1%) calculated at this wavelength should be in the range of 64–72.

EXAMPLE

0.0503 g (50.3 mg) of the ingredient is weighed, dissolved, and diluted as specified in the monograph. The absorbance (A) is measured to be 0.658 at 279 nm with a light path of 1 cm, and 279 nm is observed as the absorption maximum. The specific absorbance at 279 nm is calculated in the following way.

The concentration of the ingredient after the first dilution (50.3 mg dissolved in 50.0 ml) is

$$C_1 = \frac{50.3 \text{ mg}}{50.00 \text{ mL}} = 1.01 \text{ mg/mL} = 0.101\%(w/v)$$

The concentration of the ingredient in the second dilution used for the measurement (5.00 mL diluted to 50.0 mL) is

$$C_2 = 0.101\%(w/v)\frac{5.00 \text{ mL}}{50.0 \text{ mL}} = 0.0101\%(w/v)$$

Using Beer's law the specific absorption is calculated to be

$$a = \frac{A}{bc} = \frac{0.658}{1 \text{ cm} \times 0.0101\%} = 65.4$$

Both the measured absorbance maximum wavelength and the specific absorbance are within the requirement given for mianserin hydrochloride, and identification is positive.

DISCUSSION

The experimental procedure is detailed in the monograph. Because the letter R is behind water in the text, *water* is a reagent listed under REAGENTS. Water R is purified water, which is defined in the monograph WATER, PURIFIED: *Purified water is prepared by distillation, by ion exchange, or by reversed osmosis.* The general description of ultraviolet and visible absorption spectrophotometry is given in Section 2.2.25.

The reason for weighing about 50.0 mg of ingredient for the test is to reduce the uncertainty of weighing as this uncertainty increases with decreasing amounts. Stating 50.0 mg rather than 50 mg is because this is a semi-quantitative test where the specific absorption is to be determined, and therefore high accuracy is needed. The sample must be diluted to give an absorbance in the range 0.2–0.8, and the dilution is done in two steps to avoid the use of too large volumetric flasks. In this case the concentration of the final test solution is relatively high (0.1 mg/mL) which is necessary as the two isolated benzene rings are the only UV-absorbing chromophores in the molecule. The UV spectrum of mianserin is shown below. The absorbance maximum at 279 nm can be seen.

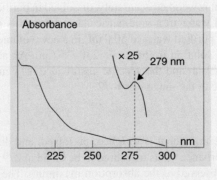

The solvent used for dissolution of the pharmaceutical ingredient is prescribed in the monograph. The solvents in use are ultraviolet-transparent, which implies that absorbance due to the solvent can be neglected. Such solvents include water, aqueous phosphate buffers, dilute hydrochloric acid, and dilute sodium hydroxide. The prescribed solvent should be used, as a change of solvent can change the absorption maxima. Before performing identification using UV-Vis spectrophotometry, it is important that both absorbance and wavelength readings of the spectrophotometer are controlled, as discussed in Chapter 7. Also, prior to recording spectra, a blind sample (pure solvent in the cuvette) should be recorded to eliminate non-specific absorbance (absorbance not originating from the pharmaceutical ingredient).

18.4.3   Thin-Layer Chromatography

Thin-layer chromatography (TLC) is another technique used for identification of pharmaceutical ingredients. The fundamentals of TLC are presented in Chapter 11. Generally, TLC is applied to the sample and the corresponding CRS, dissolved separately, and spotted in parallel lanes on the TLC plate. Identification is positive if the R_f value, size, and colour of the sample spot correspond to the CRS spot. An example of identification based on TLC is shown in Box 18.4 for metrifonate.

Box 18.4   Identification of metrifonate by TLC according to USP

STRUCTURE

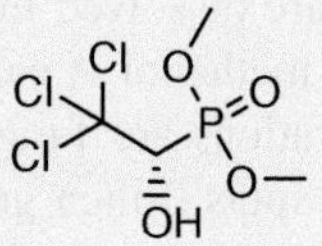

PHARMACOPEIA TEXT – INDIVIDUAL MONOGRAPH FOR METRIFONATE

IDENTIFICATION
A: IR Absorption
B: Thin-Layer Chromatographic Identification Test
Test solution: Dissolve 10 mg of metrifonate in methanol and dilute with methanol to 10.0 mL.
Developing solvent system: a mixture of toluene, dioxane, and glacial acetic acid (70 : 25 : 5).
Procedure: Proceed as directed in the chapter. After allowing the plate to air-dry, spray the plate with a 5% solution of 4-(*p*-nitrobenzyl)pyridine in acetone, and heat at 120 °C for 15 minutes. Before the plate cools, spray it with a 10% solution of tetraethylene-pentamine in acetone, and immediately examine the plate: the principal spot in the

chromatogram obtained for the *Test solution* corresponds in R_f value, size, and blue colour to that in the chromatogram obtained from the standard solution.

PHARMACOPEIA TEXT – GENERAL INFORMATION ABOUT THIN-LAYER CHROMATOGRAPHY

Prepare a test solution as directed in the individual monograph. On a line parallel to and about 2 cm from the edge of a suitable thin-layer chromatographic plate, coated with a 0.25 mm layer of chromatographic silica gel mixture (see Chromatography) apply 10 μL of this solution and 10 μL of a standard solution from the USP Reference Standard for the drug substance being identified, in the same solvent and at the same concentration as the test solution, unless otherwise directed in the individual monograph. Allow the spots to dry and develop the chromatogram in a solvent system consisting of a mixture of chloroform, methanol, and water (180 : 15 : 1) unless otherwise directed in the individual monograph, until the solvent front has moved three-quarters of the length of the plate. Remove the plate from the developing chamber, mark the solvent front, and allow the solvent to evaporate. Unless otherwise directed in the individual monograph, locate the spots on the plate by examination under short-wavelength UV light. The R_f value of the principal spot obtained from the test solution corresponds to that obtained from the Standard solution.

DISCUSSION

The procedure is obtained by combining the information in the metrifonate monograph with the information from the general text on the thin-layer chromatographic identification test. The ingredient and CRS are dissolved in methanol because this is a good solvent for metrifonate, and because methanol is easy to evaporate after spotting the solutions on the TLC plate. Concentrations are 1 mg/mL, and this high concentration is used for clear visualization of the spots. Silica gel is used as the stationary phase (normal phase chromatography), and a mixture of toluene, dioxane, and water has been optimized as the mobile phase for this particular ingredient. Spraying with 4-(*p*-nitrobenzyl)pyridine in acetone, heating, and spraying with a 10% solution of tetraethylenepentamine is to perform a chemical reaction where the colourless spots of metrifonate are converted to a product with blue colour. This helps the examination of the chromatogram.

In the pharmacopeia TLC is gradually being replaced by LC and IR spectrophotometry.

18.4.4 Melting Point

Identification tests based on determination of the melting point using a melting point apparatus are used both in Ph. Eur. (termed *Melting point*) and USP (termed *Melting Range or Temperature*). A melting point apparatus consists of a metal heating block with one or more compartments for capillary tubes, or of a suitable glass vessel containing a liquid bath (e.g. water, liquid paraffin, or silicone oil; see Figure 18.9) and equipped with a suitable means of heating and stirring. The apparatus is equipped with a temperature sensor or a suitable

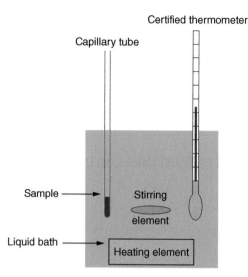

Figure 18.9 *Schematic view of an apparatus for melting point determination*

certified thermometer allowing readings at least to the nearest 0.1 °C. A small amount of sample is inserted in a capillary tube of glass sealed in one end. The capillary with the sample is then placed in the melting point apparatus and the temperature is increased typically by 10 °C/min to 5 °C below the expected melting point; then the temperature is increased at 1 °C/min until the entire sample has melted. Melting of the sample is observed, and the melting point is read as the temperature where the last remnant of sample has melted. The melting point can also be measured automatically by means of a beam of light and a detector, coupled to a computer. The measured value is compared with the melting point requirement given in the individual monograph. The melting point should be determined on dried material. If moisture is present it can affect the observed melting point. A melting point test is illustrated in Box 18.5 for paracetamol. The melting point apparatus has to be calibrated regularly using melting point reference substances.

18.4.5 Optical Rotation

Optical rotation of *plane-polarized light* is used to test compounds with one or more chiral centres. A number of pharmaceutical ingredients are *chiral compounds*, which means that they contain one or more carbon atoms bound to four different ligands or atoms, as shown in Figure 18.10. A carbon atom bound to four different ligands is termed a *chiral centre*, and the spacious configuration of the four ligands make two *enantiomers* possible, being mirror images of one another, similar to the right and left hands.

The two enantiomers of the same chemical substance have identical physicochemical properties such as melting point, boiling point, density, colour, pK_a, and log P, but they can have very different pharmacological effects. A special feature of enantiomeric compounds is that they can rotate the polarization plane of plane-polarized light. The electromagnetic radiation of the plane-polarized light shows oscillations in two directions in one plane only (Figure 18.11).

Box 18.5   Melting point for paracetamol according to Ph. Eur.

STRUCTURE

PHARMACOPOEIA TEXT – INDIVIDUAL MONOGRAPH FOR PARACETAMOL

IDENTIFICATION
First identification: A, C.
Second identification: A, B, D, E.
A. Melting point: 168–172 °C.

DISCUSSION

The equipment for the melting point determination is presented in Figure 18.9. The melting point of paracetamol should be within the range 168–172 °C.

Some enantiomers rotate plane-polarized light to the right, which is highlighted by placing (+) before the name of the compound (or D for *dextrorotatory*). The mirror image rotates plane-polarized light the same angle to the left, which is marked with (−) or L in front of the name (L for *levorotatory*). If a compound contains an equal amount of (+) and (−) enantiomers, it is termed a racemic mixture or a *racemate*, and as the two enantiomers rotate the plane-polarized light equally to the right and left, respectively, no net rotation is observed. Since optical rotation is the only physical property that is different for enantiomers, it is often used for testing chiral pharmaceutical ingredients.

Optical rotation is measured with a *polarimeter*, and the principle of this instrument is shown in Figure 18.11. The light source in the instrument is a sodium lamp that emits electromagnetic radiation at 589.3 nm. Light passes through a fixed Nicol prism (polarizer) that plane-polarizes the light. The plane-polarized light passes through a tube containing the sample solution (*polarimeter tube*). If the compound is a liquid, it can be filled directly into the polarimeter tube and if the compound is a solid it has to be dissolved before filling the polarimeter tube. At the other end of the polarimeter tube there is a second Nicol

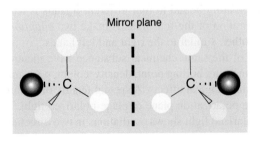

Figure 18.10   Visualization of the chirality concept

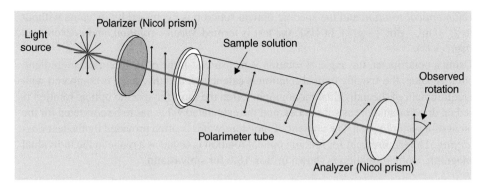

Figure 18.11   *Principle for measurement of optical rotation*

prism (analyser), which can be rotated. The operator observes the light intensity through this prism. If there is no solution in the polarimeter tube (or if there is no optically active substance in the solution), a maximum of light will pass through when the polarization planes of the two prisms are aligned. When the Nicol prism at the observer is rotated, the observed light intensity is reduced because the polarization planes are out of alignment. At 90° rotation, the operator will not observe any light transmission at all. When an optically active sample is introduced into the polarimeter tube, the plane of the polarized light is rotated and the operator observes a reduction in light intensity. The operator must rotate the prism (analyser) either right or left to re-establish maximum light transmission. The angle can be read accurately on a scale divided into degrees, and this angle is termed the *angle of rotation*.

The angle of rotation depends on: (i) the chemical structure of the pharmaceutical ingredient (sample), (ii) the length of the polarimeter tube, (iii) the concentration of the sample, (iv) the solvent used for dissolution of the sample, and (v) the temperature of the sample. For identification purposes it is convenient to eliminate variability due to (ii), (iii), (iv), and (v), and only operate with a parameter that depends on the structure of the compound. Therefore the optical rotation is calculated as the *specific optical rotation* $[\alpha]_D^{20}$ defined as follows in Ph. Eur. for solutions of solids and undiluted liquids, respectively:

$$[\alpha]_D^{20} = \frac{1000\alpha}{lc} \text{ (for solutions of solids)} \tag{18.1}$$

$$[\alpha]_D^{20} = \frac{\alpha}{l\rho_{20}} \text{ (for neat liquids)} \tag{18.2}$$

where α is the measured angle of rotation expressed in degrees (°), l is the length of the polarimeter tube (in decimetres, dm), c is the concentration of the compound (in g/L) and ρ_{20} is the density of the liquid at 20 °C (in g/cm³). D indicates that the measurement is done using a sodium lamp at 589.3 nm, as this is termed the *D line*.

Pharmaceutical ingredients that are liquids at room temperature are usually measured directly without dilution, while solids must be dissolved before they can be examined by optical rotation. Pharmaceutical ingredients are typically dissolved either in water, dilute HCl, dilute NaOH, methanol, ethanol, acetonitrile, or methylene chloride, and the concentration is typically in the range 10–100 mg/mL. The specific optical rotation can be measured at other wavelengths (λ) and other temperatures (t) but this should then be stated as $[\alpha]_\lambda^t$. Optical rotation is used both in Ph. Eur. and USP. In the former, the test is termed

Specific optical rotation and the specific optical rotation is expressed by its value without units $((°)\cdot mL \cdot dm^{-1} \cdot g^{-1})$. In USP, the test is termed *Specific rotation*, and is expressed in degrees (°).

Using a polarimeter, the angle of rotation is measured for the pharmaceutical ingredient. From this value, the specific optical rotation is calculated, and this value is compared with the requirement of the individual monograph. Often the limit for specific optical rotation is based on dried or anhydrous substances, and the calculated value has to be corrected for the content of moisture/water. The content of moisture/water is often provided by the test *Loss on drying*. The requirement for specific optical rotation is stated as a range in the individual monograph. Optical rotation is shown in Box 18.6 for simvastatin.

Box 18.6   Optical rotation for simvastatin according to Ph. Eur.

STRUCTURE

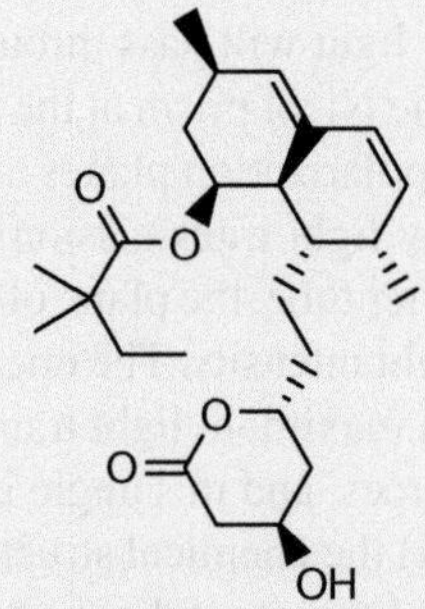

PHARMACOPOEIA TEXT – INDIVIDUAL MONOGRAPH FOR SIMVASTATIN

IDENTIFICATION
A. Specific optical rotation (see Tests).
B. IR absorption spectrophotometry.
Comparison: simvastatin CRS.
TESTS
Specific optical rotation (2.2.7): + 285 to + 300 (dried substance).
Dissolve 0.125 g in acetonitrile R and dilute to 25.0 mL with the same solvent.

PHARMACOPOEIA TEXT – GENERAL INFORMATION ABOUT OPTICAL ROTATION

The polarimeter must be capable of giving readings to the nearest 0.01°. The scale is usually checked by means of certified quartz plates. The linearity of the scale may be checked by means of sucrose solutions.

Method: Determine the zero of the polarimeter and the angle of rotation of polarized light at the wavelength of the D-line of sodium ($\lambda = 589.3$ nm) at 20 ± 0.5 °C, unless otherwise prescribed. Measurements may be carried out at other temperatures only where the monograph indicates the temperature correction to be made to the measured optical rotation. Determine the zero of the apparatus with the tube closed; for liquids the zero is determined with the tube empty and for solids filled with the prescribed solvent.

EXAMPLE

0.127 g of ingredient (not previously dried) is weighed accurately, transferred to a 25 mL (0.02500 L) volumetric flask, dissolved in acetonitrile, and diluted to volume. The angle of rotation is measured to +2.9° at 20 °C using a polarimeter tube of 20 cm. The ingredient is tested for Loss of drying (discussed later), which is determined to 0.3%. Basically, this is a measure of moisture, and the dried substance comprised 99.7% of the ingredient mass. Since the requirement for specific optical rotation is with reference to a dried substance, correction according to the loss on drying has to be included in the calculations. Thus, the concentration of the dried ingredient in the sample solution is

$$c = \frac{0.997 \times 0.127\,\text{g}}{0.02500\,\text{L}} = 5.06\,\text{g/L}$$

The specific optical rotation is calculated as follows:

$$[\alpha]_D^{20} = \frac{1000\alpha}{lc} = \frac{1000 \times 2.9}{2 \times 5.06} = +287 \text{ (degrees)}$$

This value for specific optical rotation is within the requirement of the monograph.

DISCUSSION

Simvastatin comprises several chiral centres, and is supplied optically active for use in pharmaceutical preparations. Identification of simvastatin is based on a combination of IR spectrophotometry and optical rotation. Basically, the substance is identified by IR spectrophotometry, but this technique cannot differentiate between enantiomers because they have identical spectra. Thus, identification by optical rotation is to confirm the correct enantiomeric form.

Polarimetry is the simplest technique available to differentiate between enantiomers. It is very often used to test pharmaceutical ingredients having a chiral center. The optical rotation both serves as an identification test and a test for enantiomeric purity. Thus, a value for the specific optical rotation in compliance with the monograph requirement supports the fact that the pharmaceutical ingredient is delivered as the correct enantiomer, and that this enantiomer is not seriously contaminated with other enantiomers. Optical rotation cannot stand alone as an identification test, but is typically combined with IR spectrophotometry. In USP, optical rotation is performed under *Specific tests*, while it can be performed either as an identification test or a purity test in Ph. Eur.

18.4.6 Liquid Chromatography

LC is frequently used for identification of pharmaceutical ingredients, especially in USP. The fundamentals of LC are described in Chapter 12. A *Test solution* is prepared by dissolution of a small portion of the pharmaceutical ingredient, and a *Reference solution/Standard preparation* is prepared by dissolution of a corresponding CRS. Both solutions are injected and analysed in the LC system. The requirement for positive identification is that the principal peak in the chromatogram obtained with the test solution is similar in retention time and size to the principal peak in the chromatogram obtained with the reference solution/standard preparation. Identification based on LC is shown in Box 18.7 for calcitriol.

Box 18.7   Identification of calcitriol by LC according to USP

STRUCTURE

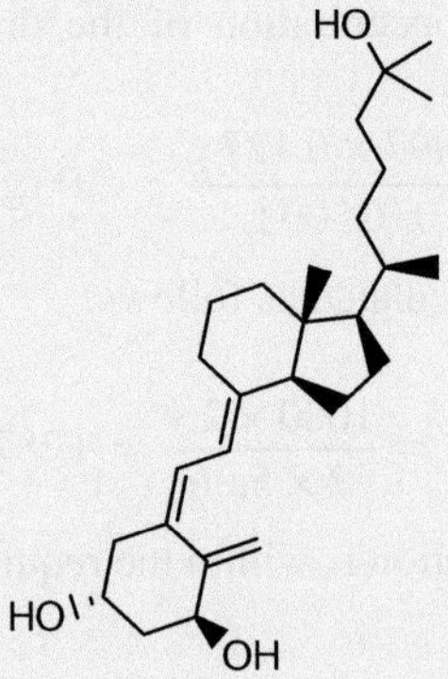

PHARMACOPOEIA TEXT – INDIVIDUAL MONOGRAPH FOR CALCITRIOL

IDENTIFICATION

A: IR Absorption

B: The retention time of the major peak in the chromatogram of the Assay preparation corresponds to that in the chromatogram of the Standard preparation, as obtained in the Assay.

Mobile phase: Prepare a filtered and degassed mixture of acetonitrile and Tris buffer solution (55 : 45).

Standard preparation: Transfer an accurately weighed quantity of USP Calcitriol RS to a suitable volumetric flask, dissolve first in acetonitrile (without heating), using 55% of the final volume, then dilute with Tris buffer solution to volume, and mix to obtain a solution having a known concentration of about 100 µg/mL of calcitriol.

Assay preparation: Transfer an accurately weighed quantity of Calcitriol to a suitable volumetric flask, dissolve first in acetonitrile (without heating), using 55% of the final volume, then dilute with Tris buffer solution to volume, and mix to obtain a solution having a known concentration of about 100 µg/mL of calcitriol.

Chromatographic system: The liquid chromatograph is equipped with a 230 nm detector and a 4.6 mm × 25 cm column that contains 5 μm packing L7 (C$_8$ column). The flow rate is about 1 mL/min. The column temperature is maintained at 40 °C.
Procedure: Separately inject equal volumes (about 50 μL) of the Standard preparation and the Assay preparation into the chromatograph and record the chromatograms.

EXAMPLE

Chromatograms of Standard preparation and Assay preparation are illustrated below. Retention times and peak size are similar in the two chromatograms and the requirements for identification are fulfilled.

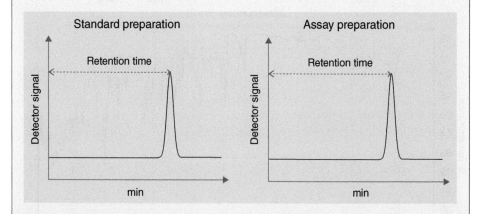

DISCUSSION

Calcitriol is a relatively hydrophobic compound, and therefore a C$_8$ column is chosen in order not to obtain too strong a retention. Isocratic elution is used because no sophisticated separation is required.

18.4.7 Chloride and Sulfate Identification

As mentioned previously, IR spectrophotometry is very powerful for the identity testing of pharmaceutical ingredients. However, a weakness of IR spectrophotometry is that it is less suitable for identification of inorganic ions. This is very relevant for a variety of pharmaceutical ingredients because they are supplied in salt form as hydrochlorides, sulfates, or phosphates – to mention a few examples. Normally, an organic base and the corresponding hydrochloride and sulfate salts show differences in the IR absorption spectra. This is illustrated in Figure 18.12 for morphine, morphine hydrochloride, and morphine sulfate. Although most of the fine structures in the spectra originate from the 'organic part' of the pharmaceutical ingredient, the different salts can be distinguished by

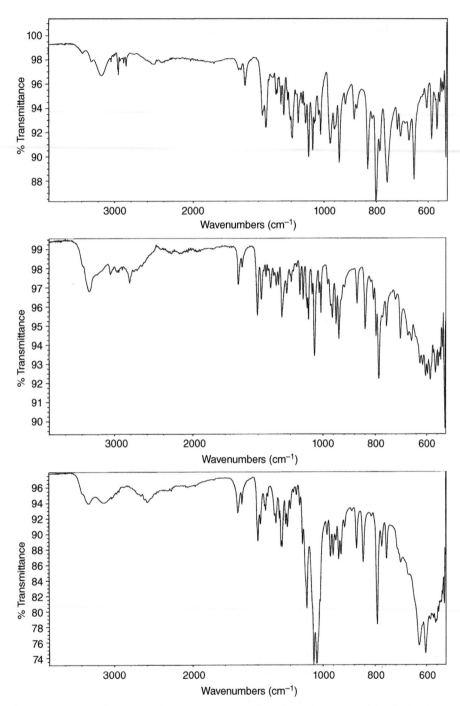

Figure 18.12 *Infrared absorption spectra of morphine base (top), morphine hydrochloride (middle), and morphine sulfate (bottom)*

IR spectrophotometry. In order to avoid mistakes, however, the anions are identified by a separate procedure. Upon dissolution of hydrochlorides, chloride ions will be present in the solution. Chloride ions can be identified by adding silver nitrate to the solution of the pharmaceutical ingredient, due to formation of a white precipitate of silver chloride (the solubility product, K_{sp} (AgCl) is 1.8×10^{-10} M^2) according to the following reaction:

$$Cl^-(aq) + Ag^+(aq) \rightarrow AgCl(s) \tag{18.3}$$

To verify that the precipitate is silver chloride, it is isolated and washed with water. Then the precipitate is suspended in water and concentrated ammonia is added. The silver chloride precipitate will dissolve easily. This is due to the complex formation of silver with ammonia:

$$Ag^+(aq) + 2NH_3(aq) \rightarrow Ag(NH_3)_2{}^+(aq) \tag{18.4}$$

Bromides and iodides also precipitate with silver ions (similar to the reaction in Eq. (18.3), but the precipitate dissolves with difficulty for bromides and does not dissolve for iodides when ammonia is added.

In a similar manner, sulfate is identified in pharmaceutical ingredients by precipitation with barium ions according to the following reaction:

$$SO_4{}^{2-}(aq) + Ba^{2+}(aq) \rightarrow BaSO_4 (s) \tag{18.5}$$

Barium sulfate is almost insoluble in aqueous solution (solubility product, K_{sp} (BaSO$_4$) is 1.1×10^{-10} M^2) and a white precipitate is observed when a solution of the pharmaceutical ingredient is mixed with a solution of barium chloride. An example on identification of chloride is shown in Box 18.8 for chlorcyclizine hydrochloride.

Box 18.8 Identification of chloride in chlorcyclizine hydrochloride according to Ph. Eur.

STRUCTURE (FIGURE SHOWS FREE BASE)

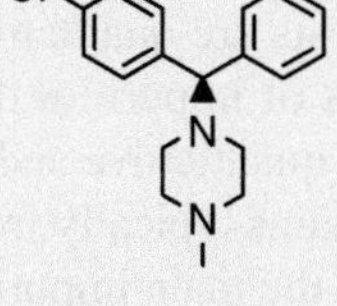

PHARMACOPOEIA TEXT – INDIVIDUAL MONOGRAPH FOR CHLORCYCLIZINE HYDROCHLORIDE

IDENTIFICATION
First identification: B, D.
Second identification: A, C, D.

B. Examine by IR absorption spectrophotometry comparing with the spectrum obtained with chlorcyclizine hydrochloride CRS. Examine the substances prepared as discs.
D. It gives reaction (a) of chlorides.

PHARMACOPOEIA TEXT – GENERAL INFORMATION ABOUT IDENTIFICATION REACTION OF CHLORIDES

(a) Dissolve in 2 mL of water R a quantity of the substance to be examined equivalent to about 2 mg of chloride (Cl^-) or use 2 mL of the prescribed solution. Acidify with dilute nitric acid R and add 0.4 mL of silver nitrate solution R1. Shake and allow to stand. A curdled, white precipitate is formed. Centrifuge and wash the precipitate with three quantities, each of 1 mL, of water R. Carry out this operation rapidly in subdued light, disregarding the fact that the supernatant solution may not become perfectly clear. Suspend the precipitate in 2 mL of water R and add 1.5 mL of ammonia R. The precipitate dissolves easily with the possible exception of a few large particles that dissolve slowly.

DISCUSSION

IR spectrophotometry of the ingredient is performed according to the same procedure as discussed in Box 18.1. To further support the fact that the ingredient is supplied as hydrochloride, identification is complemented by a test for chloride. The latter prescribe the use of an ingredient equivalent to 2 mg of Cl^- (~19 mg of the substance having a molar mass of 337.3 g/mol) dissolved in 2 mL of water. This solution is acidified with dilute nitric acid and 0.4 mL of silver nitrate solution R1 is added.

18.5   Impurity Testing of Pharmaceutical Ingredients (Pure Chemical Ingredients)

As discussed in Section 18.2, individual monographs prescribe a set of test procedures to check for certain potential impurities. Such procedures are found both in Ph. Eur. and USP. The purpose of these is to ensure that pharmaceutical ingredients do not contain organic or inorganic contaminants. Most purity tests are qualitative methods or limit tests, and they are intended to control that the content of impurities does not exceed the limit set in the monograph. In addition, some tests are quantitative and are used to quantify the content of a given impurity. Each monograph contains typically from two to six complementary tests. These are selected to identify and limit the main impurities that are known for the particular pharmaceutical ingredient, especially qualified organic impurities (related substances), volatile contaminants (solvents and water), and inorganic impurities. Such a limited range of purity tests will not detect all possible impurities. Extensive testing above this level is not desirable, as this could initiate further purification of pharmaceutical ingredients beyond what is medically relevant. The major methods used for impurity testing in Ph. Eur. are summarized in Table 18.8. In the discussion below, attention is focused on pure chemical ingredients. Most methods are also used in USP, although sometimes they are named differently.

Table 18.8   *Overview of major methods for impurity testing in Ph. Eur. and USP*

Method	Impurities tested for
Appearance of solution	Particles, insoluble and coloured impurities
Absorbance, colour of solution	Impurities with strong UV or VIS absorbance
pH, acidity or alkalinity	Protolytic impurities
Related substances, organic impurities	Organic impurities
Residual solvents	Organic solvents
Inorganic anions	Unfamiliar anions
Heavy metals	Heavy metals
Sulfated ash	Inorganic material
Loss on drying	Volatile substances
Water	Water

18.5.1   Appearance of Solution

The test *Appearance of solution* investigates the colour and the solubility of the pharmaceutical ingredient. The pharmaceutical ingredient is dissolved in a specified solvent, which can be water, dilute HCl, methanol, ethanol, dichloromethane, or dimethylformamide. This solution is termed *Solution S*. Solution S is a very concentrated solution, with a content of pharmaceutical ingredient at the 10–200 mg/mL level. Solution S is filled in a test-tube of glass, with a flat base and an internal diameter of 15–25 mm. First, *clarity* or *opalescence* of Solution S is compared with a *reference*. The reference is prepared according to the monograph and filled into a second test-tube. Clarity and opalescence are observed against a black background, as illustrated in Figure 18.13. If Solution S is required to be *clear* it must have clarity similar to pure water, and pure water is prescribed as the reference solution. If some *opalescence* is allowed, Solution S is compared to a reference suspension

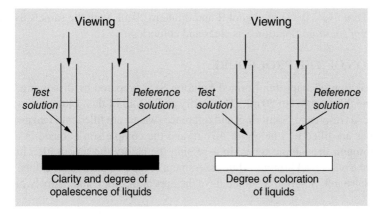

Figure 18.13   *Appearance of solution*

prepared from a mixture of hydrazine sulfate and hexamethylenetetramine. In this case, the opalescence of Solution S should not be more pronounced than that of the reference suspension.

Second, the *colour* of Solution S is compared with a *reference*. Colour is observed against a white background, as shown in Figure 18.13. The reference is prepared according to the monograph and filled into a second test-tube. If Solution S is required to be *colourless*, it has the appearance of water. If some colour is allowed, Solution S is compared to a coloured reference solution prepared according to the monograph. In such cases, the test is valid provided that Solution S is not more intensively coloured than the reference solution. Many pharmaceutical ingredients are colourless, while some degradation products can be coloured. Coloured degradation products may be discovered by Appearance of the solution, provided that they are present at levels above the visual detection limit. Box 18.9 provides an example of Appearance of solution for ibuprofen.

Box 18.9   Appearance of solution for ibuprofen according to Ph. Eur.

STRUCTURE

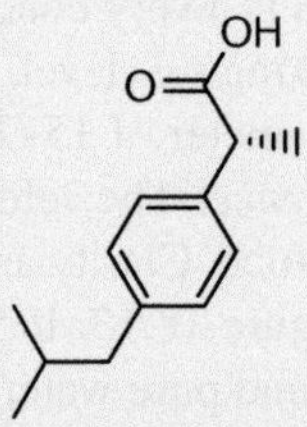

PHARMACOPOEIA TEXT – INDIVIDUAL MONOGRAPH FOR IBUPROFEN

TESTS
Solution S: Dissolve 2.0 g in methanol R and dilute to 20 mL with the same solvent.
Appearance of solution: Solution S is clear and colourless.

DISCUSSION OF THE PROCEDURE:

A sample solution of ibuprofen (termed Solution S) is prepared by dissolving 2.0 g in methanol and diluting this to 20 mL. The clarity and colour of this solution is evaluated against water as reference. Solution S and reference (water) are filled into two test-tubes. The solutions are filled to a height of 40 mm and the comparison is made in daylight vertically through the sample tubes. In assessing the clarity the observation has to be made against a black background, while assessment of the colour should be done against a white background. Solution S should have the appearance of water, namely clear and colourless.

In recent monographs for pharmaceutical ingredients not intended for parenteral use, the test for Appearance of solution is not prescribed. This is primarily due to the fact that the test for related substances, based on LC, is supposed to be able to detect any significant coloured impurity and normally pharmaceutical ingredients do not contain insoluble impurities.

In USP, there is a similar test termed *Completeness of Solution*. This is a test for clarity, and the pharmaceutical ingredient is dissolved in a prescribed solvent (test solution) and compared with pure solvent as the reference. The test is valid if the test solution is not less clear than an equal volume of the solvent contained in a similar vessel and examined similarly.

18.5.2 Absorbance

The test for *Absorbance* is prescribed as a purity test in several monographs in Ph. Eur. The same test is also found in USP termed *Color of Solution*. A concentrated solution of the pharmaceutical ingredient (10–100 mg/mL) is prepared in dilute HCl, methanol, ethanol, dichloromethane, or dimethylformamide. The absorbance of this solution is measured using a UV-Vis spectrophotometer. The measurement is performed at a wavelength where one or more known impurities have strong absorbance and where the absorbance from the pharmaceutical ingredient itself is insignificant. The measured value for absorbance is compared with a limit value as prescribed in the individual monograph. The test is sensitive for detection of impurities with ultraviolet or visible light absorption characteristics very different from the pharmaceutical ingredient. One example is shown in Box 18.10 for esomeprazole magnesium, where the same test (but named differently) is prescribed both in Ph. Eur. and USP.

18.5.3 pH and Acidity or Alkalinity

Measurement of pH is a common test in Ph. Eur. and USP. The purpose of this test is to control the level of protolytic impurities, which are impurities having acid or base properties that are different from the pharmaceutical ingredient. pH is measured directly in an aqueous solution of the pharmaceutical ingredient using a pH meter. Water is used for dissolution of the pharmaceutical ingredient. Because of the buffering capacity of dissolved carbon dioxide, which can affect pH measurement, purified and recently boiled water has to be used. Such water is termed *carbon dioxide-free water*. The measured pH value should be in compliance with the requirement of the individual monograph. One example of pH measurement is shown in Box 18.11 for esmolol hydrochloride.

In Ph. Eur., protolytic impurities are often tested by *acidity* or *alkalinity*. This is an alternative method to pH and is a semi-quantitative titration. The pharmaceutical ingredient is dissolved in carbon dioxide-free water and an acid–base colour indicator is added. Then a small amount of base is added, and a colour change should be observed due to increased pH. Subsequently, a small amount of acid is added to lower the pH again, and a new colour change should be observed. The base and acid are added in small quantities just sufficiently to cause the colour shifts. If the pharmaceutical ingredient contains significant amounts of protolytic impurities, the colour shifts are suppressed or disturbed. If the pharmaceutical ingredient provides a certain buffer capacity in water, pH measurements are preferred. A test for acidity or alkalinity is illustrated in Box 18.12.

Box 18.10   Absorbance according to Ph. Eur. and color of solution according to USP for esomeprazole magnesium

STRUCTURE (FIGURE SHOWS FREE ANHYDROUS ACID)

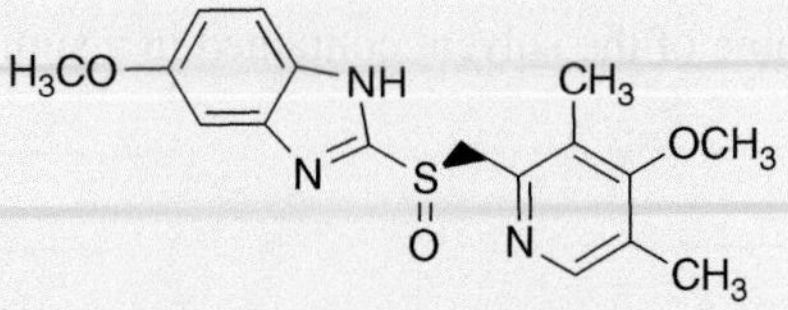

PH. EUR. TEXT – INDIVIDUAL MONOGRAPH FOR ESOMEPRAZOLE MAGNESIUM

TESTS

Absorbance: maximum 0.20 at 440 nm.

Dissolve 0.500 g in methanol R and dilute to 25.0 mL with the same solvent. Filter through a membrane filter (nominal pore size 0.45 µm).

USP TEXT – INDIVIDUAL MONOGRAPH FOR ESOMEPRAZOLE MAGNESIUM

COLOR OF SOLUTION

Sample solution: 20 mg/mL of Esomeprazole Magnesium in methanol, filtered.

Analysis: Determine the absorbance of this solution at 440 nm, in 1 cm cells, using methanol as the blank.

Acceptance criteria: not more than (NMT) 0.2.

DISCUSSION

The procedure is exactly the same in the two pharmacopeias, but the descriptions are somewhat different. The abbreviation NMT in USP means 'not more than'. The pharmaceutical ingredient is dissolved in methanol because this is a good solvent for esomeprazole and because methanol has no absorbance itself at 440 nm. The solution being tested is highly concentrated, and therefore it is possible to detect even very low levels of impurity F and G at 440 nm where esomeprazole does not absorb, but where impurities F and G show significant absorbance (see the figure below). To avoid false high absorbance due to particles in the sample solution, it is filtered through a 0.45 µm filter.

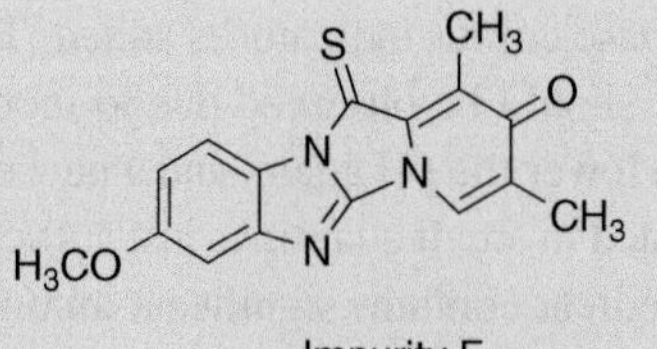

Impurity F

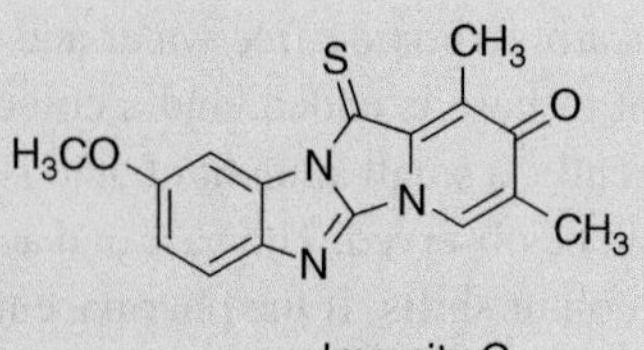

Impurity G

Box 18.11   pH according to USP for esmolol hydrochloride

STRUCTURE (FIGURE SHOWS FREE BASE)

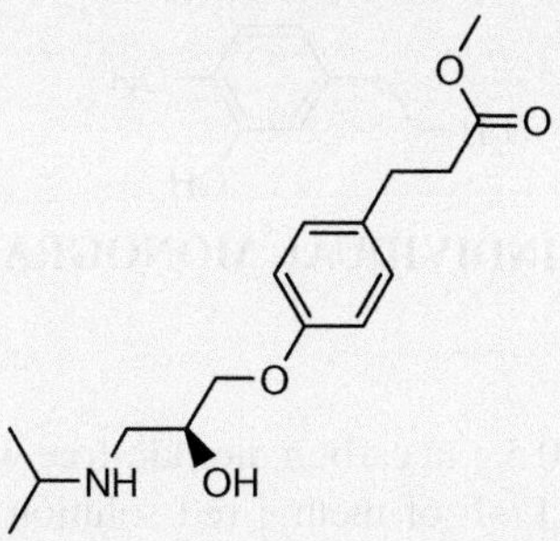

PHARMACOPEIA TEXT – INDIVIDUAL MONOGRAPH FOR ESMOLOL HYDROCHLORIDE

SPECIFIC TESTS
pH: 3.0–5.0

PHARMACOPEIA TEXT – GENERAL INFORMATION ABOUT PH

All test samples should be prepared using Purified Water, unless otherwise specified in the monograph. All test measurements should use manual or automated Nernst temperature compensation.

Prepare the test material according to requirements in the monograph or according to specific procedures. If the pH of the test sample is sensitive to ambient carbon dioxide, then use Purified Water that has been recently boiled, and subsequently stored in a container designed to minimize ingress of carbon dioxide.

Rinse the pH sensor with water, then with a few portions of the test material.

Immerse the pH sensor into the test material and record the pH value and temperature.

In all pH measurements, allow sufficient time for stabilization of the temperature and pH measurement.

Diagnostic functions such as glass or reference electrode resistance measurement may be available to determine equipment deficiencies. Refer to the electrode supplier for diagnostic tools to assure proper electrode function.

DISCUSSION

The pharmaceutical ingredient is dissolved in carbon dioxide-free water and pH is measured directly in the solution. The value should be in the range 3.0–5.0 for the test to be valid. Calibration of the pH meter before use is mandatory and this is performed according to the manufacturer.

Box 18.12 Acidity or alkalinity according to Ph. Eur. for dopamine hydrochloride

STRUCTURE (FIGURE SHOWS FREE BASE)

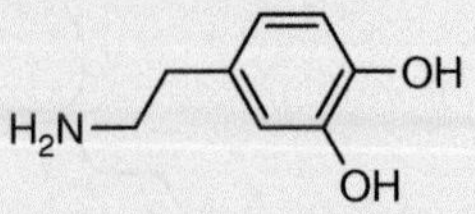

PHARMACOPEIA TEXT – INDIVIDUAL MONOGRAPH FOR DOPAMINE HYDROCHLORIDE

TESTS

Acidity or alkalinity: Dissolve 0.5 g in carbon dioxide-free water R and dilute to 10 mL with the same solvent. Add 0.1 mL of methyl red solution R and 0.75 mL of 0.01 M sodium hydroxide. The solution is yellow. Add 1.5 mL of 0.01 M hydrochloric acid. The solution is red.

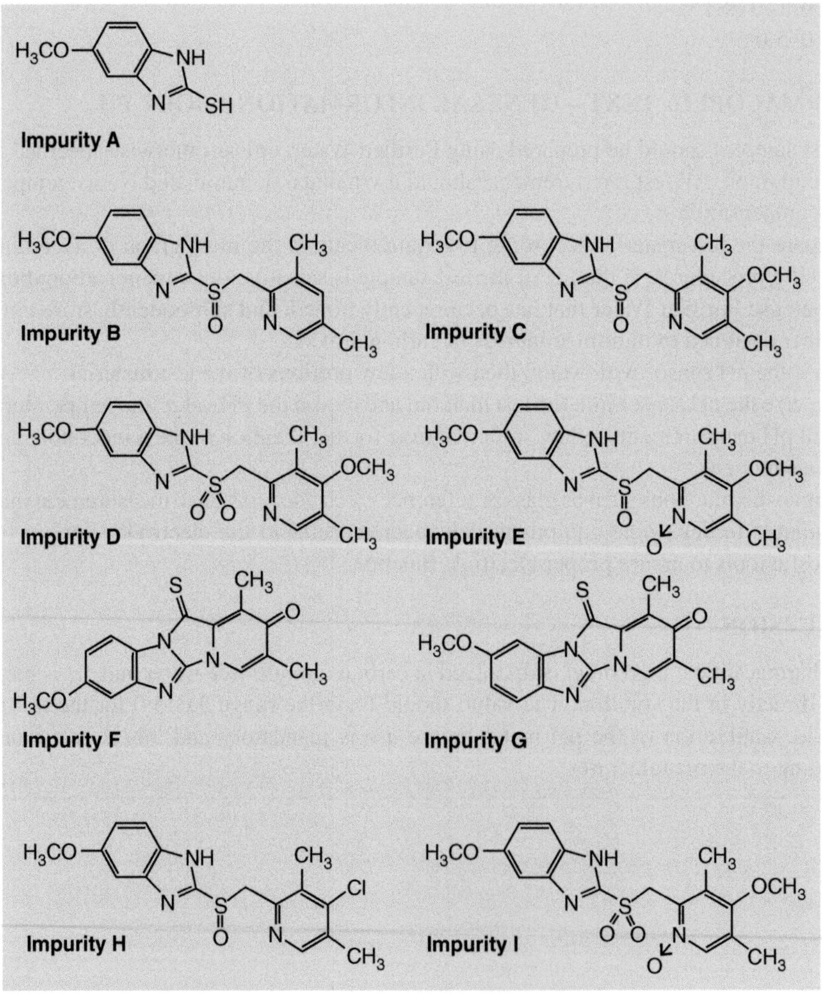

Figure 18.14   Related substances for omeprazole

18.5.4   Related Substances

In Ph. Eur., most monographs for organic compounds of the type of pure chemical ingredients prescribe a test for *Related substances*. This is similar in USP, where the test is termed *Organic Impurities*. This is a very important test for assessment of purity, where the level of organic impurities is investigated. Related substances are traces of starting materials from the synthesis, by-products from the synthesis, or degradation products formed during storage. The variety of related substances is exemplified in Figure 18.14 for omeprazole. A small sample of the pharmaceutical ingredient is dissolved in mobile phase (test solution). The concentration is relatively high in order to enable detection of impurities at a very low level (corresponding to less than 0.05% of the main compound). The test solution is analysed by LC with UV detection, and related substances are observed as peaks in the chromatogram at retention times different from the main peak. Box 18.13 exemplifies a test for related substances for omeprazole.

Box 18.13   Related substances according to Ph. Eur. for omeprazole

STRUCTURE

PHARMACOPEIA TEXT – INDIVIDUAL MONOGRAPH FOR OMEPRAZOLE

TESTS
Related substances: LC. Prepare the solutions immediately before use.
Test solution: Dissolve 3 mg of the substance to be examined in the mobile phase and dilute to 25.0 mL with the mobile phase.
Reference solution (a): Dissolve 1 mg of omeprazole CRS and 1 mg of omeprazole impurity D CRS in the mobile phase and dilute to 10.0 mL with the mobile phase.
Reference solution (b): Dilute 1.0 mL of the test solution to 100.0 mL with the mobile phase. Dilute 1.0 mL of this solution to 10.0 mL with the mobile phase.
Reference solution (c): Dissolve 3 mg of omeprazole for peak identification CRS (containing impurity E) in the mobile phase and dilute to 20.0 mL with the mobile phase.
Column:
• size: $l = 0.125$ m, $Ø = 4.6$ mm;
• stationary phase: octylsilyl silica gel for chromatography R (5 μm).

Mobile phase: mix 27 volumes of acetonitrile R and 73 volumes of a 1.4 g/L solution of disodium hydrogen phosphate dodecahydrate R previously adjusted to pH 7.6 with phosphoric acid R.
Flow rate: 1 mL/min.
Detection: spectrophotometer at 280 nm.
Injection: 40 μL.
Run time: five times the retention time of omeprazole.
   Identification of impurities: use the chromatogram obtained with reference solution (a) to identify the peak due to impurity D; use the chromatogram supplied with

omeprazole for peak identification CRS and the chromatogram obtained with reference solution (c) to identify the peak due to impurity E.

Relative retention with reference to omeprazole (retention time = about 9 minutes): impurity E = about 0.6; impurity D = about 0.8.

System suitability: reference solution (a):

- resolution: minimum 3.0 between the peaks due to impurity D and omeprazole; if necessary, adjust the pH of the aqueous part of the mobile phase or the concentration of acetonitrile R; an increase in the pH will improve the resolution.

Limits:

- impurities D, E: for each impurity, NMT 1.5 times the area of the principal peak in the chromatogram obtained with reference solution (b) (0.15%);
- unspecified impurities: for each impurity, NMT the area of the principal peak in the chromatogram obtained with reference solution (b) (0.10%);
- total: NMT five times the area of the principal peak in the chromatogram obtained with reference solution (b) (0.5%);
- disregard limit: 0.5 times the area of the principal peak in the chromatogram obtained with reference solution (b) (0.05%).

DISCUSSION

The test solution and reference solutions are dissolved in the mobile phase as this is a suitable solvent for the compounds and because the mobile phase is a suitable medium for injection into the LC system. The column is a C_8 column, which means that the separation principle is RP chromatography. The C_8 stationary phase together with a mobile phase containing 27% of acetonitrile provides suitable retention of the analytes. The test and reference solutions are made up in the mobile phase and as omeprazole has limited stability, the solutions have to be prepared immediately before use. The pH of the mobile phase is adjusted to 7.6. Omeprazole is a weak acid (pK_a approx. 7.9) and for all ionizable substances it is important to have pH control in the mobile phase to ensure stable retention times. Normally substances are separated at a pH significantly different from the pK_a value. In this example, this is not the case since a phosphate buffer at pH 7.6 is selected. This is because the related substances are better separated at pH 7.6 and because omeprazole is unstable in acidic solution. A system suitability test has to be performed prior to analysis of the sample. If the system suitability requirement of a minimum resolution of 3.0 between impurity D and omeprazole cannot be met, advice is given on how to obtain this by modification of the mobile phase. A decrease in the acetonitrile concentration will increase the retention time but it is not known if this will provide the sufficient separation. An increase in pH is reported to improve the resolution, which is probably the way to go if optimization is needed. A standard flow rate of 1 mL/min is used and a run time of five times the retention of omeprazole is needed to elute all possible impurities. A run time is always prescribed when an isocratic high performance liquid chromatography (HPLC) system is used in order to inform the user about the time needed to elute all possible impurities from the column. The injection volume (40 μL) is above the usual volume, but it is still acceptable with a column of

4.6 mm internal diameter. Omeprazole and impurities are detected at 280 nm, where all relevant compounds have relatively high UV absorption.

The resolution, R_S, between omeprazole and impurity D in Reference solution (a) should be at least three. This is to ensure that the HPLC system has sufficient separation capability to separate the related substances. Reference solution (c) is used for the identification of impurity E.

Related substances are observed in the chromatogram for the Test solution in the example below. Thus, Impurity D, Impurity E, and an unknown impurity are detected. Reference solution (b) is a 1 : 1000 (= 0.1%) dilution of the Test solution, and the peak area for omeprazole in this chromatogram serves as reference. First, all impurity peaks are evaluated against the disregard limit and all impurity peaks below this limit are not considered. The peak areas for the impurities in the Test solution are compared against the peak area for omeprazole in Reference solution (b). Impurities D and E should not exceed 1.5 times the peak in Reference solution (b), and this corresponds to a 0.15% level when assuming that the impurities are similar in absorbance to that of omeprazole. Unknown impurities and the sum of impurities are treated in a similar way.

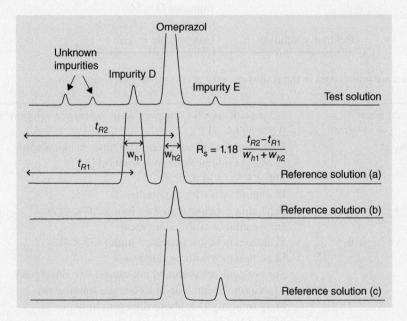

EXAMPLE

System Suitability

The following retention times and peak widths are measured for Reference solution (a):

| Impurity D: | $t_{R1} = 7.657$ min | $w_{h1} = 0.21$ min |
| Omeprazole: | $t_{R2} = 9.226$ min | $w_{h2} = 0.25$ min |

The data are interpreted in the following way according to System suitability requirements in the monograph, and the resolution is calculated as follows:

$$R_S = 1.18 \times \frac{t_{R2} - t_{R1}}{W_{h1} + W_{h2}} = 1.18 \times \frac{9.226 - 7.657}{0.21 + 0.25} = 4.0$$

The calculated resolution is in compliance with the requirements and the LC system can be used for the test for related substances.

Limits

The following peak areas are measured for the Test solution and Reference solution (b):

Test solution:	Unknown impurity = 302
	Impurity E = 207
	Impurity D = 566
	Omeprazole = 825 076
Reference solution (b):	Omeprazole = 833

The data are interpreted in the following way according to the monograph:

Disregard limit:	$0.5 \times$ peak area for omeprazole in Reference solution (b)
	$0.5 \times 833 = 417$
Impurity D:	Peak area should not exceed 1.5 times peak area for omeprazole in Reference solution (b)
	$566 < 1.5 \times 833 = 1250$
	In compliance with requirement
Impurity E:	Peak area is below disregard limit ($207 < 417$)
	In compliance with requirement
Unknown impurity:	Peak area is below disregard limit ($302 < 417$)
	In compliance with requirement
Total:	Sum of peak areas should not exceed five times peak area for omeprazole in Reference solution (b)
	Only Impurity D is above disregard limit
	$566 < 5 \times 833 = 4165$
	In compliance with requirement

In conclusion, the experimental data for related substances were in compliance with the requirements as stated in the monograph.

The test for related substances generally does not allow unknown related substances to be present in the raw material in quantities exceeding 0.10% (*w/w*). Impurities present in concentrations above 0.10% are to be identified, and if the concentration exceeds 0.15% the impurity also has to be qualified, which means its toxicity has to be known. In most cases these limits represent sufficient demands seen from a medical perspective. Impurities below 0.10% (*w/w*) will result in a very low daily intake when using the finished product. For drug substances given in higher doses over longer periods of time, lower limits for the impurities are valid. In some cases, however, it is known that a specific impurity can have a significant toxic effect, and for such contaminants much stricter limits will apply. In pethidine hydrochloride used for parenteral administration as an example, a limit for methyl-phenyl-tetrahydropyridine is as low as 0.1 ppm (one part in 10 millions). If the test for related substances is not adequate to detect such low levels of a toxic impurity, a separate LC test is prescribed that has been optimized for the toxic impurity. In some monographs, two different LC methods are therefore prescribed in order to enable separation of all relevant related substances.

18.5.5   Residual Solvents

Pharmaceutical ingredients may contain traces of *residual solvents* originating from synthesis or purification processes. The content of residual solvents must be very low to prevent their intake by patients. In Ph. Eur., residual solvents are divided into three different classes (*Classes 1, 2, and 3*). Examples of different solvents are given in Table 18.9. *Class 1* solvents are carcinogenic and highly toxic and these should not be used in the production of pharmaceutical ingredients. *Class 2* solvents are toxic solvents with strict requirements to maximum content, while the solvents in *Class 3* are less toxic with a typical limit at 0.5% (*w/w*).

Table 18.9   Examples of solvents in Classes 1, 2, and 3 according to Ph. Eur.

Class 1 solvents (should be avoided)	Class 2 solvents	Class 3 solvents
Benzene	Acetonitrile	Acetic acid
Carbon tetrachloride	Chloroform	Acetone
1,2-Dichloroethane	Cyclohexane	1-Butanol
1,1-Dichloroethene	Dichloromethane	Butylacetate
1,1,1-Trichloroethane	*N*,*N*-Dimethylacetamide	Ethanol
	N,*N*-dimethylformamide	Ethylacetate
	1,4-Dioxane	Diethyl ether
	Ethyleneglycol	1-Propanol
	Formamide	2-Propanol
	n-Hexane	
	Methanol	
	Pyridine	
	Tetrahydrofurane	
	Toluene	

Ph. Eur. prescribes that the solvents in Class 1 should not be present. The content of Class 2 solvents permitted in pharmaceutical ingredients is calculated on the basis of the *permitted daily exposure* (PDE) value:

$$Concentration\,(ppm) = \frac{1000 \times PDE}{Dose} \qquad (18.6)$$

PDE is in mg/day and dose is given in g/day. The PDE as well as the concentration limit for each Class 2 solvent is given in Ph. Eur. Class 3 solvents are typically limited to a maximum of 0.5% (*w/w*) determined by *Loss on drying*.

The control of residual solvents is mainly performed using gas chromatography (GC) with static headspace sampling and injection. A small amount of the pharmaceutical ingredient is dissolved in a suitable solvent such as water or *N,N*-dimethylformamide, and placed in an air-tight glass vial. The vial is heated for a prescribed period of time, residual solvent in the sample partly evaporates into the headspace, and equilibrium is established. At equilibrium, the amount of residual solvent in the headspace is proportional to the original amount of residual solvent in the pharmaceutical ingredient. A small sample of the headspace is sampled with a syringe and injected into the GC system. Different residual solvents are separated on the column and detected by a flame ionization detector. The residual solvents are identified based on their retention times and are quantitated using calibration standards of solvents prepared similarly to the sample or using standard addition. Similar tests are found in USP, and an example is given in Box 18.14 for Olmesartan Medoxomil. This is intended for traces of acetone and is termed *Limit of Acetone*.

Box 18.14   Limit of acetone according to USP for olmesartan medoxomil

STRUCTURES (OLMESARTAN MEDOXOMIL AND ACETONE)

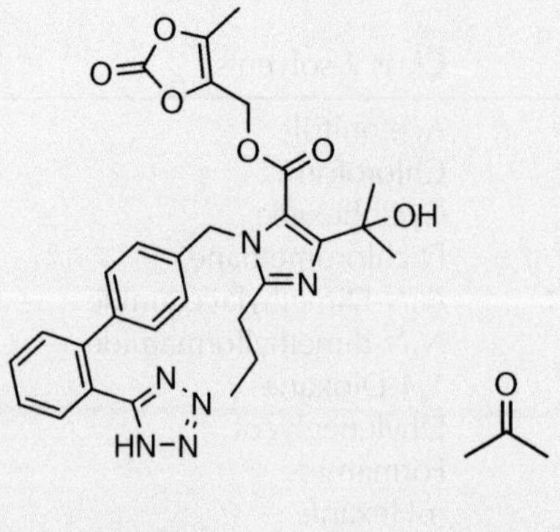

PHARMACOPEIA TEXT – INDIVIDUAL MONOGRAPH FOR OLMESARTAN MEDOXOMIL

SPECIFIC TESTS

Limit of Acetone (If Present)

Internal standard solution: 1% solution of 1-butanol in dimethylsulfoxide

Standard solution: 0.37 μL/mL of acetone and 2 μL/mL of 1-butanol from Internal standard solution in dimethylsulfoxide
Sample solution: 25 mg/mL of olmesartan medoxomil and 2 μL/mL of 1-butanol from Internal standard solution in dimethylsulfoxide

Chromatographic System

(See Chromatography, System Suitability.)
Mode: GC
Detector: Flame ionization
Column: 30 m × 0.53 mm column bonded with a 1 μm film of phase G14 (PEG)
Column temperature: 50 °C (5 minutes) and then 10 °C/min to 180 °C (5 minutes)
Injection port temperature: 200 °C
Detector temperature: 200 °C
Autosampler temperature: 80 °C
Carrier gas: Helium
Flow rate: 4 mL/min
Injection size: 1 mL
Split ratio: 5 : 1

System Suitability

Sample: standard solution
Suitability requirements
Resolution: NLT 60 between the acetone and 1-butanol peaks
Relative standard deviation: NMT 5.0% for the peak area ratio of acetone and 1-butanol

ANALYSIS

Samples: Standard solution and Sample solution. Calculate the percentage of acetone in the portion of olmesartan medoxomil taken:

$$\text{Result} = \frac{r_u}{r_s} \times \frac{C_s}{C_u} \times 100$$

r_u = peak response of acetone from the Sample solution
r_s = peak response of acetone from the Standard solution
C_s = concentration of acetone in the Standard solution (mg/mL)
C_u = concentration of olmesartan medoxomil in the Sample solution (mg/mL)
Acceptance criteria: NMT 0.6%

EXAMPLE

The Sample solution is prepared by weighing 0.2564 g of pharmaceutical ingredient on analytical balance, followed by transfer to a 10 mL volumetric flask, addition of 2.00 mL Internal standard solution and dimethylsulfoxide to volume. The Standard solution is 0.37 μL/mL of acetone and 2 μL/mL of 1-butanol from the Internal standard solution in dimethylsulfoxide. The Sample solution and Standard solution are analysed by the prescribed GC system, providing the following results:

Sample solution:	Peak area acetone = 1667
	Peak area 1-butanol = 9887
	Peak response (ratio) $r_u = \dfrac{1667}{9887} = 0.169$
Standard solution:	Peak area acetone = 13 552
	Peak area 1-butanol = 9693
	Peak response (ratio) $r_s = \dfrac{13\,552}{9693} = 1.40$

The result can be calculated directly from the equation in the monograph, but understanding this equation is important. Thus, peak areas measured for acetone are divided by the corresponding peak areas for 1-butanol as the internal standard, providing peak response values. The purpose of this is to correct for small variations in the injected volume. Based on the Standard solution, a measured peak area response of r_s corresponds to a concentration of acetone of C_s. A measured peak area of r_u corresponds to the following concentration of acetone in the sample:

$$\frac{r_u}{r_s} \times C_s$$

The percentage of acetone is calculated as the acetone ($r_u/r_s \times C_s$) to the olmesartan medoxomil (C_u) concentration ratio multiplied by 100%, providing the equation given in the monograph:

$$\frac{r_u}{r_s} \times \frac{C_s}{C_u} \times 100\%$$

The content of acetone in the example is calculated as follows. First, C_u is calculated as follows:

$$C_u = \frac{0.2564 \text{ g} \times 1000 \text{ mg/g}}{10.00 \text{ mL}} = 25.64 \text{ mg/mL}$$

Then C_s is calculated taking $0.7845 \text{ g/mL} = 0.7845 \text{ mg/µL}$ as the density of acetone:

$$C_s = 0.37 \text{ µL/mL} \sim 0.37 \text{ µL/mL} \times 0.7845 \text{ mg/µL} = 0.29 \text{ mg/mL}$$

$$\text{Result} = \frac{0.169}{1.40} \times \frac{0.29 \text{ mg/mL}}{25.64 \text{ mg/mL}} \times 100 = 0.14\%$$

The content of acetone is compliant with the requirement of the monograph.

DISCUSSION

Helium is used as the carrier gas. The column used is a capillary column with standard length and inner diameter. The stationary phase is PEG, and due to polarity this stationary phase is ideal for GC of the polar solvents acetone and 1-butanol. The film thickness of the stationary phase is relatively large, to give the volatile components sufficient retention. The GC analysis is performed in the temperature programmed mode using a short program up to 180 °C. An FID detector is used because this is relatively simple to operate and detects both solvents (it is a universal detector). Injection is by static headspace, and the sample solution is heated to 80 °C in a closed vessel to obtain equilibrium between the residual solvent present in the headspace and in the sample. A 1 mL sample is collected from the headspace and is injected into the gas chromatograph by split injection.

18.5.6 Foreign Anions

Some monographs specify tests for the presence of specific foreign anions. The anions are typically chloride and sulfate, and such tests are found both in Ph. Eur. (termed *Chlorides* and *Sulfates*) and in USP (often termed *Limit of chloride* and *Limit of sulfate*). Acids are often used during synthesis of pharmaceutical ingredients and traces of these can remain in the material. In the test for chloride, the pharmaceutical ingredient is dissolved, the solution is acidified with diluted nitric acid, and silver nitrate solution is added (the final solution is termed *test solution*). If traces of chloride are present in the pharmaceutical ingredient, the test solution becomes opalescent due to precipitation of traces of silver chloride:

$$Cl^-(aq) + Ag^+(aq) \rightarrow AgCl(s) \tag{18.7}$$

Similarly, testing for trace amounts of sulfates in pharmaceutical ingredients can be carried out by addition of barium chloride solution to a solution of the pharmaceutical ingredient (*test solution*). If sulfate is present, the test solution becomes opalescent due to trace precipitation of barium sulfate:

$$SO_4^{2-}(aq) + Ba^{2+}(aq) \rightarrow BaSO_4(s) \tag{18.8}$$

Both in the test for chloride and for sulfate, *standards* are prepared for comparison of opalescence. The standard contains an exact amount of chloride or sulfate, corresponding to the limit as defined in the monograph. The test is valid if any opalescence in the test solution is not more intense than that in the standard. The test solution and standard are observed against a black background. An example of a monograph where tests for chlorides and sulfates are included is shown in Box 18.15.

Box 18.15   Test for foreign chlorides and sulfates in furosemide according to Ph. Eur.

STRUCTURE

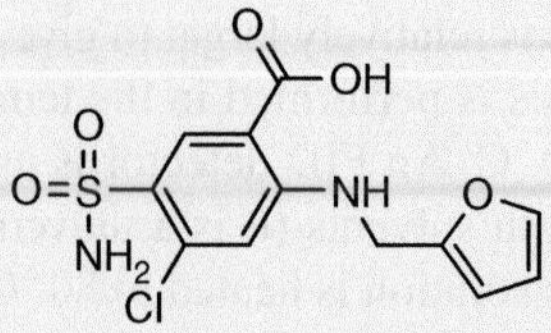

PHARMACOPOEIA TEXT – INDIVIDUAL MONOGRAPH FOR FUROSEMIDE

TESTS
Chlorides: maximum 200 ppm.
To 0.5 g add a mixture of 0.2 mL of nitric acid R and 30 mL of water R and shake for five minutes. Allow to stand for 15 minutes and filter.
Sulfates: maximum 300 ppm.
To 1.0 g add a mixture of 0.2 mL of acetic acid R and 30 mL of distilled water R and shake for five minutes. Allow to stand for 15 minutes and filter.

PHARMACOPOEIA TEXT – GENERAL INFORMATION ABOUT CHLORIDES AND SULFATES

Chlorides

To 15 mL of the prescribed solution add 1 mL of dilute nitric acid R and pour the mixture as a single addition into a test-tube containing 1 mL of silver nitrate solution R2. Prepare a standard in the same manner using 10 mL of chloride standard solution (5 ppm Cl) R and 5 mL of water R. Examine the tubes laterally against a black background. After standing for five minutes protected from light, any opalescence in the test solution is not more intense than that in the standard.

Sulfates

All solutions used for this test must be prepared with distilled water R. Add 3 mL of a 250 g/L solution of barium chloride R to 4.5 mL of sulfate standard solution (10 ppm SO_4) R1. Shake and allow to stand for one minute. To 2.5 mL of this suspension add 15 mL of the prescribed solution and 0.5 mL of acetic acid R. Prepare a standard in the same manner using 15 mL of sulfate standard solution (10 ppm SO_4) R instead of the prescribed solution. After five minutes, any opalescence in the test solution is not more intense than that in the standard.

DISCUSSION

Chlorides

Furosemide is practically insoluble in acidic aqueous solution, while chloride impurities are soluble. Therefore the solution has to be filtered to obtain a clear solution. To 15 ml of this filtrate is added 1 mL of dilute nitric acid (125 g/L), and the total solution is then poured into a test-tube containing 1 mL of silver nitrate solution (17 g/L). A standard is prepared in the same manner using 10 mL of chloride standard solution (5 ppm Cl) R and 5 mL of water. This mixture is poured into a test-tube containing 1 mL of silver nitrate solution R (17 g/L). Both tubes are observed vertically against a black background.

Sulfates

Similar to the test for chlorides, the sample solution is filtered. Then 15 mL of the filtrate is collected, acidified, and added to 2.5 mL of barium chloride solution (250 g/L) prepared in the following way: 3 mL of a 250 g/L solution of barium chloride R is added to 4.5 mL of sulfate standard solution (10 ppm SO_4) R1. A standard is prepared in the same manner using 15 mL of sulfate standard solution (10 ppm SO_4) R instead of the solution to be examined. To obtain a repeatable precipitation of barium sulfate a small amount of sulfate has to be present in advance in both solutions. The opalescence in the solutions is compared after five minutes.

CALCULATION OF CHLORIDE LIMIT

In the test for chloride a maximum limit of 200 ppm is set. This means that chloride ions in amounts up to 200 ppm are allowed in the pharmaceutical ingredient. The standard corresponds to 200 ppm according to the following calculation.

The content of chloride ions in the standard is

$$\frac{10\,mL}{10\,mL + 5\,mL + 1\,mL} \times 5\,ppm = 3.125\,ppm = 3.125\,\mu g/mL$$

Thus NMT 3.125 µg of chloride ions per mL is allowed in the test solution:
3.125 µg/mL in (15 mL + 1 mL) corresponds to 16 mL × 3.125 µg/mL = 50 µg

The test solution contains half of the amount of the pharmaceutical ingredient that was weighed (15 mL of 30 mL were used for the test solution). This means that 50 µg is allowed in 0.25 g of pharmaceutical ingredient, which corresponds to 50 µg/0.25 g = 200 ppm.

18.5.7 Sulfated Ash

Frequently, monographs in Ph. Eur. prescribe a test for *Sulfated ash*. This test is also found in USP and is termed *Residue on Ignition*. This test is used for determining the content of inorganic impurities in organic substances. Sulfated ash is performed by placing a small amount of pharmaceutical ingredient in a crucible. The raw material is moistened with sulfuric acid and the crucible is heated. During this process, all organic material is combusted and disappears primarily as CO_2 and H_2O, while inorganic cations remain in the crucible as sulfates (sulfated ash). The exact mass of both the crucible and the raw material is determined before and after the procedure. The mass of the sulfated ash should generally not exceed 0.1% (*w/w*) of the raw material. The test Residue on Ignition is exemplified in Box 18.16.

Box 18.16   Residue on ignition for acetaminophen according to USP

STRUCTURE

PHARMACOPEIA TEXT – INDIVIDUAL MONOGRAPH FOR ACETAMINOPHEN

IMPURITIES
RESIDUE ON IGNITION: NMT 0.1%

PHARMACOPEIA TEXT – GENERAL INFORMATION ABOUT RESIDUE ON IGNITION

Ignite a suitable crucible (for example silica, platinum, quartz, or porcelain) at $600 \pm 50°$ for 30 minutes, cool the crucible in a desiccator (silica gel or other suitable desiccant), and weigh it accurately. Weigh accurately 1–2 g of the substance, or the amount specified in the individual monograph, in the crucible. Moisten the sample with a small amount (usually 1 mL) of sulfuric acid, then heat gently at a temperature as low as practicable until the sample is thoroughly charred. Cool; then, unless otherwise directed in the individual monograph, moisten the residue with a small amount (usually 1 mL) of sulfuric acid; heat gently until white fumes no longer evolve; and ignite at 600 ± 50 °C, unless another temperature is specified in the individual monograph, until the residue is completely incinerated. Ensure that flames are not produced at any time during the procedure. Cool the crucible in a desiccator (silica gel or other suitable desiccant), weigh accurately, and calculate the percentage of residue.

EXAMPLE

1.0660 g of pharmaceutical ingredient is transferred to a crucible that weighs 23.5664 g. After the ashing procedure the mass of the crucible is found to be 23.5671 g. The residue on ignition corresponds to

$$\text{Mass of residue} = 23.5671 \text{ g} - 23.5664 \text{ g} = 0.0007 \text{ g}$$

$$\%\text{Residue on Ignition} = \frac{0.0007 \text{ g}}{1.0660 \text{ g}} \times 100\% = 0.07\%$$

The test is valid.

18.5.8 Elemental Impurities

In addition to anionic impurities, some monographs include tests for traces of foreign inorganic cations termed *elemental impurities*, which are inorganic traces of (metal) ions. The trace impurities can originate from many sources. They can be intentionally added inorganic reagents or catalysts, or be contaminants originating from interaction with manufacturing equipment or container closure systems. The presence of elemental impurities may therefore indicate a lack of control with the manufacturing and handling processes. Traces of metals can have catalytic effects, which may compromise the stability of pharmaceutical ingredients. As the elemental impurities provide no beneficial effect to the patient and may be toxic, it is a regulatory requirement that levels are controlled.

Previously, the test for heavy metal impurities was based on addition of sulfide ions to a solution of the pharmaceutical ingredient. The presence of heavy metal impurities was observed based on a change of colour due to the formation of sulfides. Determination of *Elemental impurities* is now described in a guideline from *The International Council for Harmonisation of Technical Requirements for Pharmaceuticals for Human Use* (ICH), named the ICH Q3D guideline for elemental impurities in medicinal products. Ph. Eur. as well as USP applies this guideline. The guideline prescribes a PDE according to the route of administration for elements of toxicological concern. The guideline contains established PDEs for 24 elements, which are grouped into four groups (1, 2A, 2B, and 3) according to their toxicity and likelihood of presence in the pharmaceutical preparation, as shown in Table 18.10.

The purpose of the risk assessment strategy is to focus the control on elemental impurities that pose a risk. If the risk assessment of the drug product reveals that a given element is unlikely to appear or the element is non-toxic in the intended route of administration, the element can be excluded from the analytical testing. This could be the case for Class 2B elements if they are not intentionally added (e.g. as catalyst).

Potential sources of elemental impurities are divided into five main categories, water, equipment container closure system, manufacturing equipment, drug substance, and excipients, as illustrated in Figure 18.15. The final risk assessment combines the risks of all potential sources of elemental impurities and evaluates the presence of these impurities by comparing the observed level to the concentration limit established on the basis of the

Table 18.10 Classification of element to be considered in the risk assessment according to ICH Q3D

Class	Elements	Elements
1	As, Cd, Hg, Pb	Human toxicants that have limited or no use in the drug product
2A	Co, V, Ni	Route-dependent toxicants that have a high probability occurrence in the drug product
2B	Ag, Au, Ir, Os, Pd, Pt, Rh, Ru, Se, Tl	Route-dependent toxicants with a reduced probability of occurrence in the drug product
3	Ba, Cr, Cu, Li, Mo, Sb, Sn	Relatively low toxicity for the oral route of administration but may be toxic by parenteral or inhalation route of administration
Other	Al, B, Ca, Fe, K, Mg, Mn, Na, W, Zn	Low inherent toxicity

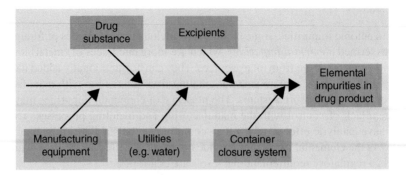

Figure 18.15 Potential sources of elemental impurities

PDE. Thus, it is essential that pharmaceutical manufacturers have a thorough knowledge and understanding of the manufacturing processes.

Chapter 2.4.20 in Ph. Eur. is a general description of the determination of elemental impurities. As substances and limits of impurities can vary, any sample preparation method and analytical method may be used if appropriate system suitability tests have demonstrated that it fulfils the requirements. Validation requirements include specificity, range, accuracy, repeatability, intermediate precision, and limits of detection and quantitation.

The suitable analytical methods for determination of elemental impurities are mainly based on atomic absorption spectrometry (AAS), atomic emission spectrometry (AES), inductively coupled plasma atomic emission spectrometry (ICP-AES), or inductively coupled plasma mass spectrometry (ICP-MS). The major advantages of these methods are that they are element specific and very sensitive, allowing for quantitation of elements at very low concentration levels. As an example, the test for zinc in human insulin is given in Box 18.17.

Box 18.17   Test for foreign zinc in human insulin according to Ph. Eur.

TESTS

PHARMACOPOEIA TEXT – INDIVIDUAL MONOGRAPH FOR HUMAN INSULIN

Zinc: maximum 1.0% (dried substance).

Atomic Absorption Spectrometry

Test solution: Dissolve 50.0 mg of the substance to be examined in 0.01 M hydrochloric acid and dilute to 25.0 mL with the same acid. Dilute if necessary to a suitable concentration (for example, 0.4–1.6 µg of Zn per millilitre) with 0.01 M hydrochloric acid.
Reference solutions: Use solutions containing 0.40, 0.80, 1.00, 1.20, and 1.60 µg of Zn per millilitre, freshly prepared by diluting zinc standard solution (5 mg/mL of Zn) R with 0.01 M hydrochloric acid.

Source: zinc hollow cathode lamp.
Wavelength: 213.9 nm.
Atomization device: air–acetylene flame of suitable composition (for example, 11 L of air and 2 L of acetylene per minute).

EXAMPLE

48.9 mg of a substance is weighed on an analytical balance and diluted to 25.0 mL with 0.01 M hydrochloric acid as the Test solution. The Test solution and the Reference solutions are measured by AAS, providing the following results:

Sample	Concentration (μg/mL)	Measured absorbance
Test solution		0.3166
Reference solutions	0.40	0.1265
	0.80	0.2621
	1.00	0.3036
	1.20	0.3589
	1.60	0.5099

Based on measurement of the reference solutions, the following equation was established between the measured signal (y) and the concentration of Zn in solution (x):

$$y = 0.3118x + 0.0005$$

From this equation, the concentration of Zn in the Test solution is calculated:

$$0.3166 = 0.3118x + 0.0005$$

$$x = 1.0\,\mu\text{g/mL}$$

The volume of test solution was 25.0 mL and in this solution the following amount of Zn was present:

$$1.0\,\mu\text{g/mL} \times 25.0\,\text{mL} = 25\,\mu\text{g}$$

This amount of Zn was present in 48.9 mg of pharmaceutical ingredient, equivalent to

$$\frac{25\,\mu\text{g}}{1000\,\mu\text{g/mg} \times 48.9\,\text{mg}} \times 100\%(w/w) = 0.052\%(w/w)$$

The test is valid.

DISCUSSION

AAS is used as a specific test for Zn. Zn is measured with a Zn hollow cathode lamp at a wavelength of 213.9 nm. This wavelength is specific for Zn. The amount of Zn is quantified, based on calibration of the instrument with reference solutions with a known concentration of Zn.

18.5.9 Loss on Drying

The test *Loss on drying* is very common both in Ph. Eur. and USP. Loss on drying determines the total amount of volatile matter of any kind present in the pharmaceutical ingredient. An exact amount of the substance is dried either in an oven at 105 °C (or any other temperature specified in the monograph), in a desiccator containing P_2O_5, or under vacuum at a given temperature. After drying, the sample is weighed again and the loss can be calculated.

Box 18.18   Loss on drying for paracetamol according to Ph. Eur.

STRUCTURE

PHARMACOPOEIA TEXT – INDIVIDUAL MONOGRAPH FOR PARACETAMOL

TESTS
Loss on drying: maximum 0.5%, determined on 1.000 g by drying in an oven at 105 °C.

PHARMACOPOEIA TEXT – GENERAL INFORMATION ABOUT LOSS ON DRYING

Method: Place the prescribed quantity of the substance to be examined in a weighing bottle previously dried under the conditions prescribed for the substance to be examined. Dry the substance to constant mass or for the prescribed time by one of the following procedures. Where the drying temperature is indicated by a single value rather than a range, drying is carried out at the prescribed temperature of ± 2 °C. The drying is carried out in an oven within the temperature range prescribed in the monograph; instrument qualification is carried out according to established quality system procedures, for example using a suitable certified reference material (amoxicillin trihydrate for performance verification CRS may be used).

EXAMPLE

1.1023 g of pharmaceutical ingredient is transferred to a weighing glass that weighs 18.2361 g. After drying to a constant weight, the total mass (weighing glass + dry pharmaceutical ingredient) was 19.3361 g. The loss on drying is calculated to be

$$\text{Loss on drying} = 18.2361 \text{ g} + 1.1023 \text{ g} - 19.3361 \text{ g} = 0.0023 \text{ g}$$

$$\%\text{Loss on drying} = \frac{0.0023 \text{ g}}{1.1023 \text{ g}} \times 100\% = 0.21\% \ (w/w)$$

The test is valid.

Volatile compounds such as water and residual solvents evaporate during drying, and the mass loss determined is equal to the sum of these components. Thus, loss on drying determines the total amount of volatile components. If a more specific test for residual solvents is necessary, GC is used as previously described under Residual solvents (Section 18.5.5). In some monographs, a more specific test for water is prescribed. This test is based on titration (Section 18.5.10). An example illustrating Loss on drying is shown in Box 18.18.

18.5.10 Water

Many pharmaceutical ingredients are hydrates or contain adsorbed water. As a result, the determination of water is important in demonstrating compliance with the monograph. In Ph. Eur. the test is termed *Water* and in USP the term *Water Determination* is used. Determination of water is based on the Karl–Fischer redox titration, and the fundamentals were discussed in Section 5.5. A small amount of pharmaceutical ingredient is dissolved in anhydrous methanol (or a similar type of solvent) and is titrated with the Karl–Fischer reagent. Stoichiometry is not exact and therefore standardization is performed by titration of an exact amount of water. The Karl–Fischer titration is not a limit test but a quantitative method that accurately and specifically determines the water in pharmaceutical ingredients. Box 18.19 shows an example of the determination of water.

Box 18.19   Determination of water in ephedrine according to Ph. Eur.

STRUCTURE

PHARMACOPEIA TEXT – INDIVIDUAL MONOGRAPH FOR EPHEDRINE

Water: NMT 0.5%, determined on 2.000 g by the semi-micro determination of water.

PHARMACOPOEIA TEXT – GENERAL INFORMATION ABOUT WATER

Water. Semi-micro determination

Standardization: To the titration vessel, add methanol R, dried if necessary, or the solvent recommended by the supplier of the titrant. Where applicable for the apparatus used, eliminate residual water from the measurement cell or carry out a pre-titration. Introduce a suitable amount of water in an appropriate form (water R or a certified reference material) and carry out the titration, stirring for the necessary time. The water equivalent is not less than 80% of that indicated by the supplier. Standardize the titrant before the first use and at suitable intervals thereafter.

Unless otherwise prescribed, use Method A.

Method A. Introduce into the titration vessel methanol R, or the solvent indicated in the monograph or recommended by the supplier of the titrant. Where applicable for the apparatus used, eliminate residual water from the measurement cell or carry out a pre-titration. Introduce the substance to be examined rapidly and carry out the titration, stirring for the necessary extraction time.

EXAMPLE

First, anhydrous methanol (such as 100 mL) is filled into the titration vessel. Next, 20 μL (20 mg) of pure water is added to the titration vessel and titrated with 5.13 mL of Karl–Fischer reagent. This is done for standardization purposes and provides the relationship between the amount of water and consumption of Karl–Fischer reagent. Then 2.1016 g of pharmaceutical ingredient is dissolved in the titration vessel, and titrated with 0.72 mL of Karl–Fischer reagent.

Based on the standardization, 5.13 mL of Karl–Fischer reagent is equivalent to 20 mg of water. Thus, 0.72 mL of Karl–Fischer reagent is equivalent to

$$\frac{0.72 \text{ mL}}{5.13 \text{ mL}} \times 20 \text{ mg} = 2.8 \text{ mg water}$$

2.8 mg of water in 2.1016 g of pharmaceutical ingredient is equivalent to

$$\frac{2.8 \text{ mg}}{2.1016 \text{ g} \times 1000 \text{ mg/g}} \times 100\% = 0.13\%(w/w)$$

The test is valid.

18.6 Identification and Impurity Testing of Organic Multi-Chemical Ingredients

Organic multi-chemical ingredients can roughly be divided into polysaccharides, synthetic polymers, fatty oils, and hydrocarbons. With few exceptions, all these pharmaceutical ingredients are used as excipients in pharmaceutical preparations.

Identification and purity testing are also needed for this group of substances. This involves a number of test procedures of which some are different from those described above. This section focuses on the most important tests, some of which are used for both identification and for purity testing. For organic multi-chemical ingredients, no distinction is normally made between first and second identifications, and for some compounds, no identification procedures are given at all. The tests for appearance of solution, melting point, and acidity/alkalinity/pH are the same as described previously. The principle of the tests for the freezing point and boiling point do not differ significantly from the test for the melting point.

18.6.1   Oxidizing Substances

Using the test for *Oxidizing substances* it is verified that the pharmaceutical ingredient does not contain significant amounts of compounds that can oxidize potassium iodide according to the following reaction:

$$2I^- + \text{oxidizing impurity} \rightarrow I_2 + \text{reduced impurity} \tag{18.9}$$

An excess of potassium iodide is added to a known quantity of the pharmaceutical ingredient. A fraction of the iodide ions will react with the oxidizing impurity and form iodine, I_2. Iodine can then be determined by redox titration with sodium thiosulfate and starch as the indicator according to the following reaction:

$$I_2 + 2S_2O_3^{2-} \rightarrow 2I^- + S_4O_6^{2-} \tag{18.10}$$

Although the test is used for purity testing, it is performed as a quantitative measurement. The test is important because the presence of oxidizing compounds can reduce the stability of pharmaceutical preparations. Occasionally there is also a test for *reducing substances*.

18.6.2   Acid Value

The purpose of measuring *Acid value* is to determine the total amount of free acids in the pharmaceutical ingredient. The acid value is defined as the amount of potassium hydroxide that must be added to neutralize the free acids in 1 g of pharmaceutical ingredient. The procedure is to weigh m grams of pharmaceutical ingredient and dissolve it in a mixture of equal amounts of ethanol (96%) and light petroleum (boiling range 100–120 °C). This solution is titrated with n mL of 0.1 M KOH or 0.1 M NaOH to the colour changes from colourless to pink with phenolphthalein as the indicator. The acid value (I_A) is calculated in the following way:

I_A = mg of KOH required to neutralize the free acids present in 1 g of the substance:

$$I_A = \frac{5.610n}{m} \tag{18.11}$$

Here m is the mass (g) of pharmaceutical ingredient and n is the volume (mL) of 0.1 M KOH or 0.1 M NaOH. One mole of KOH corresponds to 56.1 g. Observe that the acid value is expressed as x mg KOH even if 0.1 M NaOH is used for titration. The acid value is often determined in fatty oils and synthetic polymers to ensure that pharmaceutical ingredients do not contain too high a level of unreacted free acids. For fatty oils a high acid value indicates the presence of large amounts of free fatty acids. An upper limit for the acid value is therefore given in each individual monograph.

18.6.3   Hydroxyl Value

The *Hydroxyl value* evaluates the degree of polymerization. This is particularly relevant for synthetic polymers, where the number of free hydroxy groups depends on the degree of polymerization. The hydroxyl value (I_{OH}) is defined as the number of milligrams of potassium hydroxide required to neutralize the acid liberated by acylation of 1 g of the substance. Acylation is performed using acetic acid anhydride:

$$R-CH_2OH + (CH_3CO)_2O \rightarrow R-CH_2OCOCH_3 + CH_3COOH \qquad (18.12)$$

For each mole of OH groups in the raw material, 1 mol of acetic acid is produced in the reaction. The exact amount of acetic acid is determined by titration with 0.5 M alcoholic potassium hydroxide solution (n_1 mL) until a colour change occurs with phenolphthalein from colourless to pink is observed. A blank test is carried out under the same conditions (n_2 mL of 0.5 M alcoholic potassium hydroxide solution is consumed). I_A is the acid value determined as described above:

$$I_{OH} = \frac{28.05(n_2 - n_1)}{m} + I_A \qquad (18.13)$$

Any free acid is of course also neutralized by potassium hydroxide and is included in the n_1 volume, which is subtracted from the blank. Therefore the acid value has to be added in the calculation. The requirement to the hydroxyl value is given in the individual monograph.

18.6.4 Iodine Value

The *Iodine value* (I_I) is another test performed for a variety of pharmaceutical ingredients such as synthetic polymers and hydrocarbons. The iodine value expresses the degree of unsaturation in 100 g of pharmaceutical ingredient, that is the extent to which various compounds in the material contain carbon–carbon double bonds. The iodine value is defined as the number of grams of iodine that can be consumed by (substituted into) 100 g of pharmaceutical ingredient. An accurately weighed quantity of pharmaceutical ingredient (m grams) is dissolved in chloroform and a known quantity of IBr solution. The following reaction occurs between the double bonds in the raw material and IBr (IBr is added to the C—C double bonds):

$$-C = C- + IBr \rightarrow -CI-CBr- \qquad (18.14)$$

The surplus of IBr reacts with potassium iodide according to the following reaction:

$$IBr + I^- \rightarrow I_2 + Br^- \qquad (18.15)$$

The iodine formed by this reaction is titrated with 0.1 M sodium thiosulfate according to the following reaction (n_1 mL of thiosulfate solution is consumed):

$$I_2 + 2S_2O_3^{2-} \rightarrow 2I^- + S_4O_6^{2-} \qquad (18.16)$$

A blank test is carried out under the same conditions (n_2 mL of thiosulfate solution consumed).

The amount of IBr that has reacted with the raw material can now be calculated using the difference between the volume of the blank and the volume corresponding to the amount of unreacted IBr:

$$I_I = \frac{1.269(n_2 - n_1)}{m} \qquad (18.17)$$

The Iodine value gives an indication of the amount of double bonds contained in the pharmaceutical ingredient. The iodine value test is often included in monographs for oils and fats, and in monographs for oleic and stearic acid among others. The requirement for the Iodine value is given in the individual monograph.

18.6.5 Peroxide Value

Peroxides are undesirable in large quantities in pharmaceutical ingredients because they behave as oxidants and therefore can affect the stability of the pharmaceutical ingredient and of pharmaceutical preparations. *Peroxide value* is a test used to determine the amount of peroxides. To a known quantity of pharmaceutical ingredient, a surplus of potassium iodide is added and the following redox reaction takes place:

$$R—OOH + 2I^- + 2H^+ \rightarrow R—OH + I_2 + H_2O \quad (18.18)$$

The amount of iodine formed is subsequently determined by titration with 0.01 M sodium thiosulfate according to the following reaction:

$$I_2 + 2S_2O_3^{2-} \rightarrow 2I^- + S_4O_6^{2-} \quad (18.19)$$

Starch is used as a colour indicator to determine the titration endpoint. The peroxide value is expressed as the number of milli-equivalents of active oxygen in 1000 g of the pharmaceutical ingredient, and the number is usually located within the range 1–100. The requirement for the peroxide value varies from monograph to monograph and is set as an upper limit. The peroxide value is often prescribed in monographs for synthetic polymers and fatty oils.

18.6.6 Saponification Value

Determination of *Saponification value* is typically prescribed for synthetic polymers and fatty oils. The saponification value (I_S) is defined as the number of milligrams of potassium hydroxide needed to neutralize free acids and to saponify (hydrolyse) esters present in 1 g of pharmaceutical ingredient. The procedure is to dissolve a known quantity (m gram) of the pharmaceutical ingredient in 25.0 mL of alcoholic potassium hydroxide. Potassium hydroxide has to be in excess. The mixture is refluxed for 30 minutes unless otherwise prescribed. The following reactions occur:

$$R_1—COOH + OH^- \rightarrow R_1—COO^- + H_2O \text{ (neutralization of free acids)} \quad (18.20)$$

$$R_2—COOR_3 + OH^- \rightarrow R_2—COO^- + R_3—OH \text{ (hydrolysis of esters)} \quad (18.21)$$

The amount of potassium hydroxide is added in excess and the remaining potassium hydroxide is titrated with 0.5 M hydrochloric acid (n_1 mL) using phenolphthalein as the indicator. A blank test is carried out under the same conditions (n_2 mL of 0.5 M hydrochloric acid consumed). From these titrations, the amount of reacted potassium hydroxide is calculated and the saponification value can be determined:

$$I_S = \frac{28.05(n_2 - n_1)}{m} \quad (18.22)$$

The requirement for the saponification value is usually given as an interval in the individual monographs as exemplified by the value for hard fat, which should be in the range 210–260. Hard fat is a mixture of tri-, di-, and mono-glycerides.

18.6.7   Unsaponifiable Matter

Determination of *Unsaponifiable matter* is a test that is used to characterize fatty oils that are composed of triglycerides. The procedure involves saponification of a known amount (*m* gram) of the pharmaceutical ingredient in the first place:

$$\text{Triglyceride} + \text{KOH} \rightarrow \text{glycerol} + \text{potassium salts of fatty acids} \qquad (18.23)$$

After the saponification the reaction mixture is diluted with water and extracted with peroxide-free diethyl ether in a separating funnel. Glycerol and the potassium salts of the fatty acids will remain in the water phase while lipophilic constituents that could not be hydrolysed will be extracted to the ether phase. The ether phase is purified by extraction with potassium hydroxide solution and water to neutrality and then evaporated to dryness. The mass of the residue (*a* gram) is determined and the content of unsaponifiable matter is calculated:

$$\text{Unsaponifiable matter} = \frac{100a}{m}\% \qquad (18.24)$$

Lipophilic organic compounds (e.g. hydrocarbons), which were not originally on the ester form and have not been saponified, will be extracted into diethyl ether, and the amount is typically NMT 1–2% (*w/w*) of the pharmaceutical ingredient.

18.6.8   Other Tests

For fatty oils and essential oils, which are based on mixtures of different triglycerides or terpenes, respectively, control of the composition using GC is often prescribed. For fatty oils comprising triglycerides, the triglycerides are hydrolysed and the free fatty acids are transformed into methyl esters. The methyl esters are analysed by GC in order to give the composition of fatty acids. The main components are determined quantitatively by *normalization* by comparison with a standard mixture of fatty acid methyl esters as the reference standard.

When analysing essential oils, a small amount of the oil is injected directly without derivatization into the GC system, and the individual components are separated. It is usual to first inject a solution of a standard that contains one or more of the main components of the oil and then inject a volume of the oil. For a positive identification the characteristic (main) peaks in the chromatogram of the oil should be similar in retention time to peaks of the reference solution. An example of the use of GC for the identification of an essential oil is shown in Box 18.20.

Box 18.20   Chromatographic profile for peppermint oil according to Ph. Eur.

PHARMACOPEIA TEXT – INDIVIDUAL MONOGRAPH FOR PEPPERMINT OIL

Chromatographic profile: GC: use the normalization procedure.

Test solution: Mix 0.20 mL of the substance to be examined with heptane R and dilute to 10.0 mL with the same solvent.

Reference solution (a): Dissolve 10 µL of limonene R, 20 µL of cineole R, 40 µL of menthone R, 10 µL of menthofuran R, 10 µL of (+)-isomenthone R, 40 µL of menthyl acetate R, 20 µL of isopulegol R, 60 mg of menthol R, 20 µL of pulegone R, 10 µL of piperitone R, and 10 µL of carvone R in heptane R and dilute to 10.0 mL with the same solvent.

Reference solution (b): Dissolve 5 µL of isopulegol R in heptane R and dilute to 10.0 mL with the same solvent. Dilute 0.1 mL of the solution to 5.0 mL with heptane R.

Column:
- material: fused silica;
- size: $l = 60$ m, $\emptyset = 0.25$ mm;
- stationary phase: macrogol 20 000 R (film thickness 0.25 µm).

Carrier gas: helium for chromatography R.
Flow rate: 1.5 mL/min.
Split ratio: 1 : 50.
Temperature:
- Column: 60 °C (10 minutes), then 2 °C/min to 180 °C (5 minutes)
- Injection port: 200 °C
- Detector: 220 °C

Detection: flame ionization.
Injection: 1 µL.
Elution order: order indicated in the composition of reference solution (a); record the retention times of these substances.

   Identification of peaks: using the retention times determined from the chromatogram obtained with reference solution (a), locate the components of reference solution (a) in the chromatogram obtained with the test solution.

   System suitability: reference solution (a):
- resolution: minimum 1.5 between the peaks due to limonene and 1,8-cineole; minimum 1.5 between the peaks due to piperitone and carvone.

   Determine the percentage content of each of the following components. The limits are within the following ranges:

- limonene: 1.0% to 3.5%;
- 1,8-cineole: 3.5% to 8.0%;
- menthone: 14.0% to 32.0%;
- menthofuran: 1.0% to 8.0%;
- isomenthone: 1.5% to 10.0%;
- menthyl acetate: 2.8% to 10.0%;
- isopulegol: maximum 0.2%;
- menthol: 30.0% to 55.0%;
- pulegone: maximum 3.0%;
- carvone: maximum 1.0%;
- disregard limit: the area of the principal peak in the chromatogram obtained with reference solution (b) (0.05%).

The ratio of 1,8-cineole content to limonene content is minimum 2.

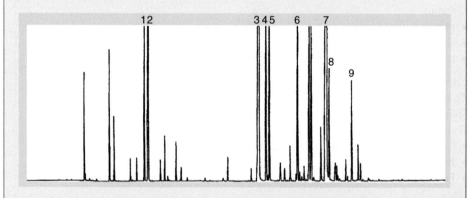

Peaks 1 = limonene, 2 = cineol, 3 = menthone, 4 = menthofuran, 5 = isomenthone, 6 = menthyl acetate, 7 = menthol, 8 = pulegone, 9 = carvone.

DISCUSSION

Helium is used as a carrier gas because it provides rapid analysis compared to nitrogen and because it is safer to use than hydrogen. Split injection is used because a sample volume of 1 µL is too large for a column with an internal diameter of only 0.25 mm. With a split ratio of 1 : 50, only 2% of the sample is introduced into the column, thus ensuring that the column is not overloaded with subsequent loss of chromatographic resolution. The temperature in the injector is set at 200 °C to ensure that all test components are quickly evaporated and introduced into the column. The column used is a relatively long capillary column, and the length is selected to provide high separation efficiency. The stationary phase is polar and separates the analytes according to both volatility and polarity. The GC analysis is performed using temperature programming, which means that the temperature is changed during the analysis. Initially the temperature is kept at 60 °C for 10 minutes. The most volatile components will be eluted through the column, while less volatile analytes have a much higher affinity to the stationary phase at the entrance of the column. Then the temperature is increased gradually. The temperature increase is low (2 °C/min), which provides high resolution. At the end the temperature is kept for five minutes at 180 °C to ensure that less volatile sample components are eluted from the column before the system cools down to its initial temperature for the next analysis. Flame ionization detection is used, as this detector has a large linear range and responds to all components. In this case it is not necessary to use a more selective detector. The temperature of the detector is kept at 220 °C to ensure that no sample components condense in the detector.

18.7   Assay of Pharmaceutical Ingredients

Monographs for pharmaceutical ingredients of the type of pure chemical ingredient prescribe an *assay* for quantitative determination by which the content or purity is determined. The unit for purity is percentage (*w/w*). The definition of content is very strict, and is often in the range 99.0–101.0% (Table 18.5). The definition of content is, however, also dependent on the precision of the method of analysis used for the assay. Thus, titrations can normally be performed with high interlaboratory precision, and three times the standard deviation is within 1%. Therefore, the definition can be limited to 99.0–101.0%. The interlaboratory precision of an assay performed by LC is less and therefore the definition is often 98.0–102.0% when the assay is based on LC. When using UV-Vis spectrophotometry the definition is even wider and is typically 97.0–103.0%. The lower limit of the definition may sometimes be even lower to allow for some impurities to be present. The upper limit is above 100% due to analytical uncertainty as well as the lack of specificity of titration and UV-Vis spectrophotometry. The most important assay techniques are summarized in Table 18.11 and are discussed in the following.

18.7.1   Aqueous Acid–Base Titration

Acid–base titrations in aqueous solution are frequently used in Ph. Eur. and USP. The fundamentals of aqueous acid–base titrations are discussed in Chapter 5. An example of a quantitative determination of an acidic pharmaceutical ingredient by aqueous acid–base titration is shown in Box 18.21. Similarly, an example of quantitative determination of a

Table 18.11   Overview of the main techniques for assay of pharmaceutical ingredients

Technique	Specificity	Typical range for content (%)
Aqueous acid–base titration	Low	99.0–101.0
Non-aqueous acid–base titration	Low	99.0–101.0
Redox titration	Low	99.0–101.0
Liquid chromatography	High	98.0–102.0
Ultraviolet and visible spectrophotometry	Low	97.0–103.0

hydrochloride of a basic pharmaceutical ingredient using aqueous acid–base titration is shown in Box 18.22.

Box 18.21   Assay of omeprazole by aqueous acid–base titration according to Ph. Eur.

PHARMACOPEIA TEXT – INDIVIDUAL MONOGRAPH FOR OMEPRAZOLE

ASSAY
Dissolve 0.250 g in a mixture of 10 mL of water R and 40 mL of ethanol (96%) R. Titrate with 0.1 M sodium hydroxide, determining the endpoint potentiometrically. Then 1 mL of 0.1 M sodium hydroxide is equivalent to 34.54 mg of $C_{17}H_{19}N_3O_3S$.

DISCUSSION

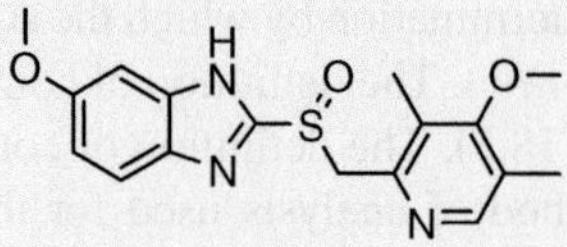

Content 99.0–101.0%
   Omeprazole is a weak acid, and to make sure that titration is 100% the substance is titrated with a strong base (OH^-). Omeprazole reacts in the mole ratio 1 : 1 with OH^-. Omeprazole is very slightly soluble in water, and therefore the sample is dissolved in a mixture of ethanol and water.

EXAMPLE

A 0.2506 g sample is dissolved according to the procedure and titrated with 7.20 mL of 0.1008 M NaOH. The molar mass of omeprazole is 345.4 g/mol. The purity of the sample is calculated as follows:
   The number of mol of base (OH^-) consumed

$$= \frac{7.20 \text{ mL}}{1000 \text{ mL/L}} \times 0.1008 \text{ M} = 7.26 \times 10^{-4} \text{mol}$$

The number of mol of omeprazole

$$= 7.26 \times 10^{-4} \text{mol}$$

Omeprazole in g

$$= 7.26 \times 10^{-4} \text{mol} \times 345.4 \text{ g/mol} = 0.251 \text{ g}$$

$$\text{Content} = \frac{0.251 \text{ g}}{0.2506 \text{ g}} \times 100\% = 100.0\% (w/w)$$

The assay is valid.

Box 18.22   Assay of amitriptyline hydrochloride by aqueous acid–base titration according to Ph. Eur.

PHARMACOPEIA TEXT – INDIVIDUAL MONOGRAPH FOR AMITRIPTYLINE HYDROCHLORIDE

ASSAY

Dissolve 0.250 g in 30 mL of ethanol (96%) R. Titrate with 0.1 M sodium hydroxide, determining the endpoint potentiometrically.

Then 1 mL of 0.1 M sodium hydroxide is equivalent to 31.39 mg of $C_{20}H_{24}ClN$.

DISCUSSION (STRUCTURE SHOWS FREE BASE)

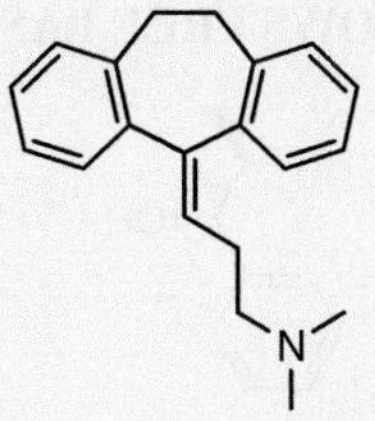

Content 99.0–101.0%

Amitriptyline hydrochloride is a weak acid, and to make sure that titration is 100%, the substance is titrated with a strong base (OH^-). Amitriptyline hydrochloride reacts in the mol ratio 1 : 1 with OH^-. Amitriptyline hydrochloride is freely soluble in water, but still the sample is dissolved in ethanol. This is because the reaction product, namely amitriptyline as a free base, is soluble in ethanol but not in water. In other words, to make sure precipitation does not occur during titration, the solvent is ethanol. However, the titration is still aqueous because 96% ethanol contains water and because the titrant is aqueous.

The calculation of purity from titration data is identical to Box 18.21.

For pharmaceutical ingredients produced as hydrochlorides, traces of hydrochloric acid may be adsorbed and the pharmaceutical ingredient is delivered with an excess of hydrochloric acid. The reverse may also occur; the pharmaceutical ingredient may not be fully protonated due to insufficient hydrochloric acid, and small amounts of the ingredient are delivered as a free base. In both cases, assays based on acid–base titration will be inaccurate. To avoid such errors, it is common to add a minor amount of hydrochloric acid prior to titration. The sample solution will then contain both the fully protonated drug substance and strong hydrochloric acid, and this mixture is then titrated with an aqueous solution of a strong base. This is performed by potentiometric titration and a titration curve is recorded. The curve will show two inflexion points. The first inflexion point corresponds to hydrochloric acid (strong acid) and the second corresponds to the protonated ingredient. Thus, the volume between the first and the second inflexion point corresponds to the amount of protonated base and is used for the calculation of content. An example of this procedure is shown in Box 18.23.

Box 18.23   Assay of ephedrine hydrochloride by aqueous acid–base titration according to Ph. Eur.

PHARMACOPEIA TEXT – INDIVIDUAL MONOGRAPH FOR EPHEDRINE HYDROCHLORIDE

ASSAY

Dissolve 0.150 g in 50 mL of ethanol (96%) R and add 5.0 mL of 0.01 M hydrochloric acid. Carry out a potentiometric titration, using 0.1 M sodium hydroxide. Read the volume added between the two points of inflexion.

1 mL of 0.1 M sodium hydroxide is equivalent to 20.17 mg of $C_{10}H_{16}ClNO$.

DISCUSSION (STRUCTURE SHOWS FREE BASE)

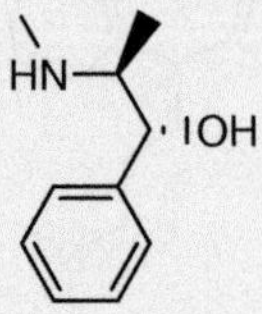

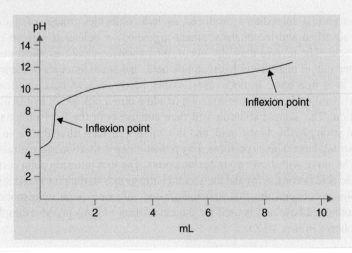

Content 99.0–101.0%

Ephedrine hydrochloride is a weak acid, and to make sure that titration is 100% the substance is titrated with a strong base (OH⁻). Ephedrine hydrochloride reacts in the mole ratio 1 : 1 with OH⁻. To correct for potential excess or insufficiency of hydrochloric acid, hydrochloric acid is added in excess to the sample before titration.

EXAMPLE

0.1498 g of sample is dissolved and added to 5.00 mL 0.00990 M HCl. The solution is titrated using 0.52 mL of 0.0956 M NaOH to the first inflexion point and 8.30 mL to the second inflexion point. The molar mass of ephedrine hydrochloride is 201.7 g/mol. The titration curve is illustrated below.

The content is calculated as follows:

The number of mol of base (OH^-) consumed between the two inflexion points

$$= \frac{8.30\,mL - 0.52\,mL}{1000\,mL/L} \times 0.0956\,M = 7.44 \times 10^{-4}\,mol$$

The number of mol of ephedrine hydrochloride

$$= 7.44 \times 10^{-4}\,mol$$

Ephedrine hydrochloride in g

$$= 7.44 \times 10^{-4}\,mol \times 201.7\,g/mol = 0.150\,g$$

$$Content = \frac{0.150\,g}{0.1498\,g} \times 100\% = 100.1\%(w/w)$$

The assay is valid.

Pharmaceutical ingredients that are free bases (B) often have limited solubility in water. For such substances, an exact amount of hydrochloric acid (H_3O^+) can be added to the sample in excess. The substance becomes fully protonated (BH^+) and an excess of H_3O^+ is determined by titration with a strong base $(OH^-$, back titration). The amount of compound in the sample solution can then be determined as the difference between the added strong acid and the excess of strong acid determined by titration. An example of this is shown in Box 18.24.

Box 18.24 Assay of ephedrine by aqueous acid–base titration according to Ph. Eur.

PHARMACOPEIA TEXT – INDIVIDUAL MONOGRAPH FOR EPHEDRINE

ASSAY
Dissolve 0.200 g in 5 mL of alcohol R and add 20.0 mL of 0.1 M hydrochloric acid. Using 0.05 mL of methyl red solution R as indicator, titrate with 0.1 M sodium hydroxide until a yellow colour is obtained. 1 mL of 0.1 M hydrochloric acid is equivalent to 16.52 mg of $C_{10}H_{15}NO$.

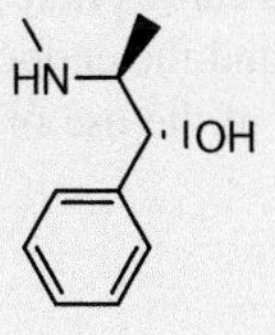

Content 99.0–101.0%

EXAMPLE

0.2099 g of sample is dissolved in 5 mL of ethanol and added to 20.00 mL of 0.1024 M HCl. This solution is titrated with 7.75 mL of 0.0998 M NaOH. The molar mass of ephedrine is 165.2 g/mol.

The content is calculated as follows:

The number of mol of base (OH^-) consumed

$$= \frac{7.75 \text{ mL}}{1000 \text{ mL/L}} \times 0.0998 \text{ M} = 7.73 \times 10^{-4} \text{ mol}$$

The number of mol of HCl in excess

$$= 7.73 \times 10^{-4} \text{ mol}$$

The total number of mol of HCl added

$$= \frac{20.00 \text{ mL}}{1000 \text{ mL/L}} \times 0.1024 \text{ M} = 2.048 \times 10^{-3} \text{mol}$$

The number of mol of HCl consumed = moles of ephedrine

$$= 2.048 \times 10^{-3} \text{ mol} - 7.73 \times 10^{-4} \text{ mol} = 1.28 \times 10^{-3} \text{ mol}$$

Ephedrine in g $= 1.28 \times 10^{-3} \text{ mol} \times 165.2 \text{ g/mol} = 0.211 \text{ g}$

$$\text{Content} = \frac{0.211 \text{ g}}{0.2099 \text{ g}} \times 100\% = 100.3\%(w/w)$$

The assay is valid.

18.7.2   Non-Aqueous Acid–Base Titration

Non-aqueous acid–base titration is another type of titration often used for assay of pharmaceutical ingredients. As discussed previously (Section 5.4), this method is either used when the substance is a very weak acid ($pK_a > 8$) or a very weak base ($pK_b > 8$), or when the substance is poorly soluble in aqueous solution or in polar organic solvents. Non-aqueous titrations are found both in Ph. Eur. and USP, and are primarily prescribed for bases, and for hydrochlorides, sulfates, phosphates, and carboxylates of bases. The titrant $CH_3COOH_2^+$ is gradually added to the sample solution and the endpoint is determined potentiometrically or by using an indicator. Two examples of the use of non-aqueous acid–base titration are shown in Boxes 18.25 and 18.26.

Box 18.25   Assay of metronidazole benzoate by non-aqueous acid–base titration according to USP

PHARMACOPEIA TEXT – INDIVIDUAL MONOGRAPH FOR METRONIDAZOLE BENZOATE

ASSAY
PROCEDURE

Sample solution	Dissolve by stirring 250 mg of metronidazole benzoate in 50.0 mL of glacial acetic acid
Titrimetric system	See Titrimetry
Mode	Direct titration
Titrant	0.1 M perchloric acid
Endpoint detection	Potentiometric
Analysis	Titrate the Sample solution with Titrant. Perform a blank determination and make any necessary correction. Each mL of 0.1 M perchloric acid is equivalent to 27.53 mg of metronidazole benzoate ($C_{13}H_{13}N_3O_4$)
Acceptance criteria	98.5–101.0% on the dried basis

DISCUSSION

Metronidazole benzoate is a weak base ($pK_a = 2.44$) and therefore titration in aqueous solution is not feasible. Under non-aqueous titration, metronidazole reacts with the titrant in the mol ratio 1 : 1.

EXAMPLE

0.2506 g of the sample is dissolved according to the procedure and titrated with 9.01 mL of 0.1004 M perchloric acid. The molar mass of metronidazole benzoate is 275.26 g/mol.

The content is calculated as follows:
The number of mol of perchloric acid consumed

$$= \frac{9.01 \text{ mL}}{1000 \text{ mL/L}} \times 0.1004 \text{ M} = 9.05 \times 10^{-4} \text{ mol}$$

The number of mol of metronidazole benzoate

$$= 9.05 \times 10^{-4} \text{ mol}$$

Metronidazole benzoate in g

$$= 9.05 \times 10^{-4} \text{ mol} \times 275.26 \text{ g/mol} = 0.249 \text{ g}$$

$$\text{Content} = \frac{0.249 \text{ g}}{0.2506 \text{ g}} \times 100\% = 99.4\%(w/w)$$

The assay is valid.

Box 18.26   Assay of lidocaine by non-aqueous acid–base titration according to Ph. Eur.

PHARMACOPEIA TEXT – INDIVIDUAL MONOGRAPH FOR LIDOCAINE

ASSAY
To 0.200 g add 50 mL of anhydrous acetic acid R and stir until dissolution is complete. Titrate with 0.1 M perchloric acid, determining the endpoint potentiometrically.
   1 mL of 0.1 M perchloric acid is equivalent to 23.43 mg of $C_{14}H_{22}N_2O$.

DISCUSSION

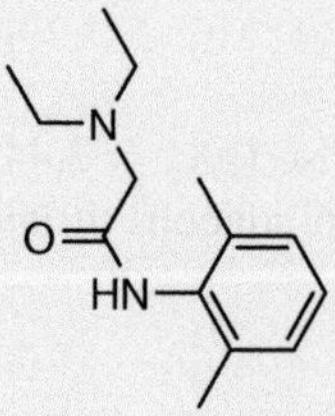

Content 99.0–101.0%
   Under non-aqueous titration, lidocaine reacts with the titrant in the mol ratio 1 : 1.

EXAMPLE

0.2008 g of sample is dissolved according to the procedure and titrated with 8.81 mL of 0.0968 M perchloric acid. The molar mass of lidocaine is 234.3 g/mol. The titration curve is illustrated below.

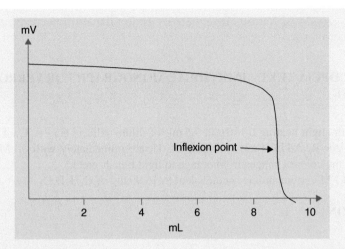

The content is calculated as follows:
The number of mol of perchloric acid consumed

$$= \frac{8.81 \text{ mL}}{1000 \text{ mL/L}} \times 0.0968 \text{ M} = 8.53 \times 10^{-4} \text{ mol}$$

The number of mol of lidocaine

$$= 8.53 \times 10^{-4} \text{mol}$$

Lidocaine in g

$$= 8.53 \times 10^{-4} \text{mol} \times 234.3 \text{ g/mol} = 0.200 \text{ g}$$

$$\text{Content} = \frac{0.200 \text{ g}}{0.2008 \text{ g}} \times 100\% = 99.5\%(w/w)$$

The assay is valid.

18.7.3 Redox Titrations

Redox titrations are also used for assay in Ph. Eur. and USP. The fundamentals of redox titration are discussed in Chapter 5. To a solution of the pharmaceutical ingredient, the titrant is added gradually until all the substance has reacted (the endpoint). The endpoint of the titration is detected either with a redox colour indicator or by potentiometric detection. Box 18.27 shows an example of a quantitative determination by means of a redox titration.

Box 18.27 **Assay of ferrous fumarate by redox titration according to Ph. Eur.**

PHARMACOPEIA TEXT – INDIVIDUAL MONOGRAPH FOR FERROUS FUMARATE

ASSAY

Dissolve with slight heating 0.150 g in 7.5 mL of dilute sulfuric acid R. Cool and add 25 mL of water R. Add 0.1 mL of ferroin R. Titrate immediately with 0.1 M cerium sulfate until the colour changes from orange to light bluish-green.

1 mL of 0.1 M cerium sulfate is equivalent to 16.99 mg of $C_4H_2FeO_4$.

DISCUSSION

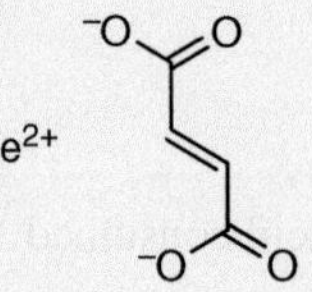

Content 93.0–101.0%

Titration is performed according to the following reaction:

$$Fe^{2+} + Ce^{4+} \rightarrow Fe^{3+} + Ce^{3+}$$

Immediate titration is important in order to avoid instantaneous oxidation of Fe^{2+}.

EXAMPLE

0.1487 g of the sample is dissolved according to the procedure and titrated with 8.81 mL of 0.0996 M cerium sulfate. The molar mass of ferrous fumarate is 169.9 g/mol.

The content is calculated as follows:

The number of mol of cerium sulfate consumed

$$= \frac{8.81 \text{ mL}}{1000 \text{ mL/L}} \times 0.0996 \text{ M} = 8.77 \times 10^{-4} \text{ mol}$$

The number of mol of ferrous fumarate

$$= 8.77 \times 10^{-4} \text{mol}$$

Metronidazole benzoate in g

$$= 8.77 \times 10^{-4} \text{mol} \times 169.9 \text{ g/mol} = 0.149 \text{ g}$$

$$\text{Content} = \frac{0.149 \text{ g}}{0.1487 \text{ g}} \times 100\% = 100.3\% (w/w)$$

The assay is valid.

18.7.4 Liquid Chromatography

LC is used extensively for assays in USP, and is now being increasingly used for assays also in Ph. Eur. The principle of LC is described in Chapter 12. The LC system is calibrated using a standard solution of CRS, and one-point calibration is used. A solution of the pharmaceutical ingredient is analysed in the same LC system, and quantitation is based on a comparison of peak areas for the principal peak in the two chromatograms. An example of an assay based on LC is shown in Box 20.34. LC is much more specific than titration and UV-Vis spectrophotometry, and is therefore less prone to interferences.

Box 18.28   Assay of simvastatin by LC according to USP

STRUCTURE

PHARMACOPEIA TEXT – INDIVIDUAL MONOGRAPH FOR SIMVASTATIN

ASSAY

System suitability preparation: Dissolve accurately weighed quantities of USP Simvastatin RS and USP Lovastatin RS in Diluent, and dilute quantitatively, and stepwise if necessary, with Diluent to obtain a solution having known concentrations of about 1.5 mg per mL of USP Simvastatin RS and 0.015 mg per mL of USP Lovastatin RS.

Standard preparation: Dissolve an accurately weighed quantity of USP Simvastatin RS in Diluent to obtain a solution having a known concentration of about 1.5 mg/mL.

Assay preparation: Transfer about 75 mg of Simvastatin, accurately weighed, to a 50 mL volumetric flask, dissolve in and dilute with Diluent to volume, and mix.

Chromatographic system (see Chromatography): The liquid chromatograph is equipped with a 238 nm detector and a 4.6 × 33 mm column that contains packing L1 (C$_{18}$). The flow rate is about 3.0 mL/min (separation is by gradient elution, but the gradient conditions are not included in this example).

Chromatograph the System suitability preparation, and record the peak responses as directed for Procedure: the relative retention times are about 0.60 for lovastatin and 1.0 for simvastatin, and the resolution, R, between simvastatin and lovastatin is greater than 3. Chromatograph the Standard preparation, and record the peak responses as directed for Procedure: the relative standard deviation for replicate injections is NMT 1.0%.

Procedure: Separately inject equal volumes (about 5 μL) of the Standard preparation and the Assay preparation into the chromatograph, record the chromatograms, and measure the areas for the major peaks. Calculate the quantity, in mg, of $C_{25}H_{38}O_5$ in the portion of Simvastatin taken by the formula

$$VC\frac{r_u}{r_s}$$

in which V is the volume, in mL, of the Assay preparation; C is the concentration, in mg/mL, of USP Simvastatin RS in the Standard preparation; and r_u and r_s are the responses of the simvastatin peak obtained from the Assay preparation and the Standard preparation, respectively.

EXAMPLE

System suitability preparation contains 1.5 mg/mL of simvastatin and 0.015 mg/mL of lovastatin.

Standard preparation is prepared by dissolution and dilution of 74.6 mg of USP Simvastatin RS (accurately weighed) to 50.0 mL. The concentration (C) is calculated as follows:

$$C = \frac{74.6\ \text{mg}}{50.0\ \text{mL}} = 1.49\ \text{mg/mL}$$

Assay preparation is prepared by dissolution and dilution of 76.2 mg of pharmaceutical ingredient (accurately weighed) to 50.0 mL.

First, system suitability is checked and the test is valid. Resolution is higher than 3 and replicate injections are within 1.0%.

Then Standard preparation and Assay preparation are analysed with the following results:

Standard preparation:	Area of simvastatin peak = 10 788
Assay preparation:	Area of simvastatin peak = 10 875

The quantity (mg) of simvastatin in the Assay preparation is calculated as follows:

$$= VC\frac{r_u}{r_s} = 50.0\ \text{mL} \times 1.49\ \text{mg/mL} \times \frac{10\,875}{10\,788} = 75.2\ \text{mg}$$

The content is calculated as follows:

$$\text{Content} = \frac{75.2\ \text{mg}}{76.2\ \text{mg}} \times 100\% = 98.7\%(w/w)$$

The requirement is 98.0–102.0% calculated on the dried basis, and the assay is valid provided that Loss on drying is NMT 0.5%.

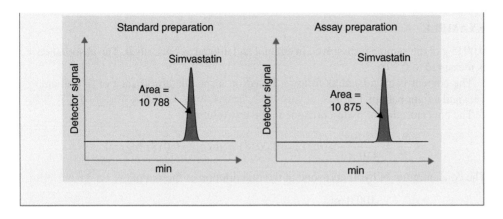

18.7.5   UV-Vis Spectrophotometry

UV-Vis spectrophotometry is used to some extent for assays both in Ph. Eur. and USP. The principle of UV-Vis spectrophotometry is described in Chapter 7. An exactly weighed amount of pharmaceutical ingredient is dissolved in a suitable solvent and the solution is placed in an UV-Vis spectrophotometer. The absorbance is directly proportional to the concentration of the pharmaceutical ingredient (Beer's law). Calibration of the UV-Vis spectrophotometer is normally not required, since the specific absorbance is given in the monograph. An example of assay based on UV-Vis spectrophotometry is shown in Box 18.29.

Box 18.29   Assay of hydrocortisone by UV spectrophotometry according to Ph. Eur.

STRUCTURE

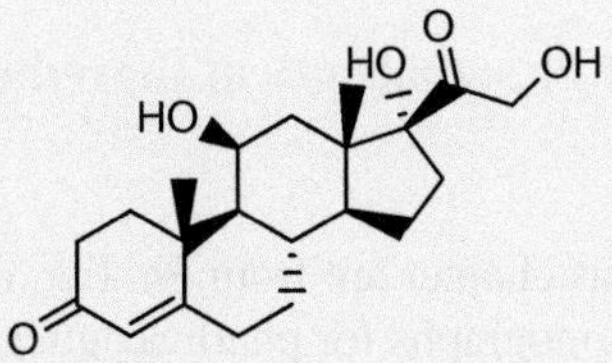

Content 97.0–103.0%

PHARMACOPEIA TEXT – INDIVIDUAL MONOGRAPH FOR HYDROCORTISONE

ASSAY
Dissolve 0.100 g in ethanol (96%) R and dilute to 100.0 mL with the same solvent. Dilute 2.0 mL of the solution to 100.0 mL with ethanol (96%) R. Measure the absorbance at the absorption maximum at 241.5 nm.
 Calculate the content of $C_{21}H_{30}O_5$ taking the specific absorbance to be 440.

EXAMPLE

0.0915 g of the sample is dissolved in ethanol and diluted as prescribed. The absorbance is measured to 0.798.

The content is calculated as follows, based on a specific absorbance of 440 (1 cm, 1%) and a light path of 1 cm.

The concentration of hydrocortisone in the test solution

$$= \frac{A}{ab} = \frac{0.798}{440 \times 1} \%(w/v) = 0.00181\%(w/v) \sim 0.0181 \text{ mg/mL}$$

The concentration of hydrocortisone in the first dilution of the sample

$$= \frac{100.0 \text{ mL}}{2.00 \text{ mL}} \times 0.0181 \text{ mg/mL} = 0.9050 \text{ mg/mL}$$

Amount of hydrocortisone in the first dilution of the sample

$$= 0.9050 \text{ mg/mL} \times 100.0 \text{ mL} = 90.50 \text{ mg} = 0.0905 \text{ g}$$

$$\text{Content} = \frac{0.0905 \text{ g}}{0.0915 \text{ g}} \times 100\% = 98.9\%(w/w)$$

The assay is valid.

COMMENTS

Ethanol is used as solvent because hydrocortisone dissolves easily in this solvent, and because it has no absorbance at 241.5 nm. The weighed amount of sample is 100 mg to ensure that uncertainty is kept to a minimum. The sample is diluted to obtain an absorbance within the range of 0.2–0.8 where the highest precision is obtained.

18.8   Chemical Analysis of Pharmaceutical Ingredients Not Included in Pharmacopoeias

All the examples discussed in this chapter are from Ph. Eur. or USP. Both pharmacopeias comprise a large collection of monographs for pharmaceutical ingredients, including both APIs and excipients. Newer drug substances are often protected by patents, and they are not included in Ph. Eur. or USP. The reason for this is that the substances are only used by one manufacturer, namely, the company that holds the patent on the compound. It is of little relevance to publish monographs for such drug substances until the patent expires. The manufacturer using such a substance has to develop a monograph for the substance, and this will be submitted to the medicinal authorities as part of the documentation for marketing authorization. These monographs are designed in the same way as those found in Ph. Eur. and USP. When the patent expires, it is common for the monograph to be included in the official pharmacopoeias, because other manufacturers will appear on the market with the same substance. Both Ph. Eur. and USP are revised and updated regularly. This involves inclusion of new substances, and also improvement of the test procedures in existing monographs.

19

Chemical Analysis of Pharmaceutical Preparations

19.1 Chemical Analysis of Pharmaceutical Preparations

Quality control of the pharmaceutical preparation is an important part of the entire pharmaceutical production. Chemical analysis of pharmaceutical ingredients is important to ensure high quality, which in turn is of the highest importance for the safety and efficacy of pharmaceutical preparations. The quality of all ingredients is tested prior to production, while the finished product (pharmaceutical preparation) is subjected to further quality control after production. This control can be divided into three categories, according to the nature of the experimental methods in use:

- Microbiological testing
- Pharmaceutical testing
- Chemical testing (chemical analysis)

Introduction to Pharmaceutical Analytical Chemistry, Second Edition.
Stig Pedersen-Bjergaard, Bente Gammelgaard and Trine Grønhaug Halvorsen.
© 2019 John Wiley & Sons Ltd. Published 2019 by John Wiley & Sons Ltd.

The *microbiological tests* are used to check the pharmaceutical preparation with respect to different bacteria and related microorganisms. As microbiological tests rely on the concepts of microbiology, these tests are beyond the scope of this textbook. However, it is important to emphasize that microbiological tests are important for sterile as well as non-sterile preparations, and microbiology is therefore an important subject in the pharmaceutical curriculum. *Pharmaceutical testing* checks the quality of the dosage form itself. Pharmaceutical testing is based on physical methods and is discussed in textbooks in pharmaceutics (outside the scope of this textbook).

Chemical tests (chemical analysis) are intended to check the quality of the pharmaceutical preparation in terms of the *chemical composition*. Chemical analysis of pharmaceutical preparations relies to a large extent on spectroscopic and chromatographic techniques.

Chemical analysis of pharmaceutical preparations is performed by quality control laboratories within the pharmaceutical industry prior to release of every single production batch. Analytical methods for chemical analysis of pharmaceutical preparations are also used during development of new drugs and preparations to test their stability (*stability indicating methods*). Such work is typically performed by research laboratories within the pharmaceutical industry or by external analytical services laboratories. Occasionally, medicinal agencies test production batches released to market. Therefore, chemical analysis of pharmaceutical preparations is also performed outside the pharmaceutical industry by *Official Medicines Control Laboratories* (*OMCLs*).

19.2   Monographs and Chemical Analysis

In Chapter 18, it was emphasized that chemical analysis of pharmaceutical ingredients is performed in compliance with the individual monographs in the European Pharmacopoeia (Ph. Eur.) when intended for the European market or in compliance with the United States Pharmacopoeia (USP) when intended for the United States market. Similar requirements hold for chemical analysis of pharmaceutical preparations. Previously, Ph. Eur. did not comprise individual monographs for final pharmaceutical preparations, but recently such monographs started emerging in Ph. Eur. and more are expected to come. Ph. Eur. mainly contains *general monographs for the dosage forms*, and chemical testing of products has to be in compliance with these general monographs. Examples of dosage forms are capsules, parenteral preparations, and tablets.

The general monographs in Ph. Eur. for dosage forms are composed of the following sections:

- DEFINITION
- PRODUCTION
- TESTS

The Ph. Eur. monograph for capsules is shown in Box 19.1. The DEFINITION section defines the dosage form and provides a general description of production, including an overview of the excipients typically used in the dosage form. Furthermore, the Definition section states the requirements for the material of the containers intended for the pharmaceutical preparation. All pharmaceutical preparations have to comply with the requirements stated in the sections *Materials used for the manufacture of containers* and the section *Containers* to make sure that the pharmaceutical preparation is not deteriorated by sunlight, moisture, air, contamination, leakage, or mechanical stress, or by adsorption of ingredients to the container wall. The *Materials used for the manufacture of containers* section contains monographs for the materials used for containers, like polyvinyl chloride (PVC), polyethylene (PE), and polypropylene (PP). These monographs are structured in much the same way as the monographs for pharmaceutical ingredients, and include sections about definition, production, identification, and tests. Containers and materials used for containers are outside the scope of this textbook.

Box 19.1   Ph. Eur. monograph for capsules

CAPSULES, CAPSULAE

The requirements of this monograph do not necessarily apply to preparations that are presented as capsules intended for use other than by oral administration. Requirements for such preparations may be found, where appropriate, in other general monographs, for example Rectal preparations (1145) and Vaginal preparations (1164).

DEFINITION

Capsules are solid preparations with hard or soft shells of various shapes and capacities, usually containing a single dose of active substance(s). They are intended for oral administration.

The capsule shells are made of gelatin or other substances, the consistency of which may be adjusted by the addition of substances such as glycerol or sorbitol. Excipients such as surface-active agents, opaque fillers, antimicrobial preservatives, sweeteners, colouring matter authorized by the competent authority and flavouring substances may be added. The capsules may bear surface markings.

The contents of capsules may be solid, liquid or of a more paste-like consistency. They consists of one or more active substances with or without excipients such as solvents, diluents, lubricants and disintegrating agents. The contents do not cause deterioration of the shell. The shell, however, is attacked by the digestive fluids and the contents are released.

Where applicable, containers for capsules comply with the requirements of *Materials used for the manufacture of containers* (3.1 and subsections) and *Containers* (3.2 and subsections).

Several categories of capsules may be distinguished:

- Hard capsules
- Soft capsules
- Gastro-resistant capsules
- Modified-release capsules
- Cachets

PRODUCTION

In the manufacture, packaging, storage, and distribution of capsules, suitable measures are taken to ensure their microbial quality; recommendations on this aspect are provided in the text on *Microbiological quality of pharmaceutical preparations (5.1.4)*.

TESTS

Uniformity of dosage units. Capsules comply with the test for uniformity of dosage units (*2.9.40*) or, where justified and authorized, with the tests for uniformity of content and/or uniformity of mass, shown below. Herbal drugs and herbal drug preparations present in the dosage form are not subject to the provisions of this paragraph.

Uniformity of content (2.9.6). Unless otherwise prescribed or justified and authorized, capsules with a content of active substance less than 2 mg or less than 2% of the fill mass comply with test B for uniformity of content of single-dose preparations. If the preparation has more than one active substance, the requirement applies only to those ingredients that correspond to the above conditions.

Uniformity of mass (2.9.5). Capsules comply with the test for uniformity of mass of single-dose preparations. If the test for uniformity of content is prescribed for all the active substances, the test for uniformity of mass is not required.

Dissolution. A suitable test may be carried out to demonstrate the appropriate release of the active substance(s), for example one of the tests described in *Dissolution test for solid dosage forms (2.9.3)*. Where a dissolution test is prescribed, a disintegration test may not be required.

STORAGE

Store at a temperature not exceeding 30 °C.

The following section is PRODUCTION. This section describes precautions related to the production of the specific dosage form. In addition, the Production section provides the requirements for different tests to make sure that the production process has been successful. The tests can be pharmaceutical tests to ensure the technical quality of the preparation or microbiological tests to ensure the microbiological quality of the preparation. For sterile preparations, the production also has to comply with the requirements stated in the section *Methods of preparation of sterile products*. For some preparations, antimicrobial preservatives are added, and in those cases, the preparation has to be tested according to the section *Efficacy of microbial preservation*, to make sure that the efficiency of the antimicrobial preservatives is sufficient.

The final section is TESTS. The purpose of the tests is mainly to ensure the pharmaceutical quality of the preparation. The prescribed tests are quite different from dosage form to dosage form. For tablets and capsules, the tests are typically intended to check that the dosage form is disintegrated and the active ingredient is released appropriately. Also, for tablets and capsules, tests are performed to check that the active pharmaceutical ingredient (API) is equally distributed between the individual dosage units. For parenteral preparations, tests are performed to check that the product is not contaminated with particulate matter and that the product is sterile.

In contrast to Ph. Eur., both the British Pharmacopoeia (BP) and USP have individual monographs for final pharmaceutical preparations. An example of such an individual monograph is shown in Box 19.2 for paracetamol tablets in BP.

Box 19.2   BP monograph for paracetamol tablets

PARACETAMOL TABLETS

Action and Use

Analgesic; antipyretic.

DEFINITION

Paracetamol Tablets contain Paracetamol.

The tablets comply with the requirements stated under Tablets and with the following requirements.

Content of Paracetamol, $C_8H_9NO_2$
95.0–105.0% of the stated amount.

IDENTIFICATION

Extract a quantity of the powdered tablets containing 0.5 g of Paracetamol with 20 mL of *acetone*, filter, evaporate the filtrate to dryness, and dry at 105 °C. The residue complies with the following tests.

A. The *infrared absorption spectrum*, Appendix II A, is concordant with the *reference spectrum* of Paracetamol (*RS 258*).
B. Boil 0.1 g with 1 mL of *hydrochloric acid* for three minutes, add 10 mL of *water* and cool; no precipitate is produced. Add 0.05 mL of 0.0167 M *potassium dichromate*; a violet colour is produced slowly that does not turn red.
C. *Melting point*, about 169 °C, Appendix V A.

TESTS

Dissolution

Comply with the requirements for the Monographs of the BP in the *dissolution test for tablets and capsules*, Appendix XII B1, using Apparatus 2. Use as the medium 900 mL of *phosphate buffer* pH 5.8 and rotate the paddle at 50 rpm. Withdraw a sample of 20 mL of the medium and filter. Dilute the filtrate with 0.1 M *sodium hydroxide* to give a solution expected to contain about 0.00075% (*w/v*) of Paracetamol. Measure the *absorbance* of this solution, Appendix II B, at the maximum at 257 nm using 0.1 M *sodium hydroxide* in the reference cell. Calculate the total content of paracetamol,

$C_8H_9NO_2$, in the medium, taking 715 as the value of A (1%, 1 cm) at the maximum at 257 nm.

Related Substances

Carry out the method for *liquid chromatography* (LC), Appendix III D, using the following solutions. Prepare the solutions immediately before use and protect from light. For solution (1), disperse a quantity of powdered tablets containing 0.2 g of Paracetamol in 8 mL of the mobile phase with the aid of ultrasound, add sufficient mobile phase to produce 10 mL, mix well, and filter. For solution (2), dilute 1 volume of solution (1) to 20 volumes with the mobile phase and dilute 1 volume of this solution to 20 volumes with the mobile phase. Solution (3) contains 0.002% (*w/v*) each of *4-aminophenol* and *paracetamol* CRS in the mobile phase. Solution (4) contains 0.00002% (*w/v*) of *4'-chloroacetanilide* in the mobile phase.

The chromatographic procedure may be carried out using (i) a stainless steel column (25 cm × 4.6 mm) packed with *octylsilyl silica gel for chromatography* (5 μm) (Zorbax Rx C_8 is suitable), (ii) as the mobile phase with a flow rate of 1.5 mL/min, at a temperature of 35 °C, a mixture of 250 volumes of *methanol* containing 1.15 g of a 40% (*v/v*) solution of tetrabutylammonium hydroxide with 375 volumes of 0.05 M *disodium hydrogen orthophosphate* and 375 volumes of 0.05 M *sodium dihydrogen orthophosphate*, and (iii) a detection wavelength of 245 nm.

The test is not valid unless, in the chromatogram obtained for solution (3), the *resolution factor* between the two principal peaks is at least 4.0.

Inject solution (1) and allow the chromatography to proceed for 12 times the retention time of the principal peak. In the chromatogram obtained with solution (1) the area of any peak corresponding to 4-aminophenol is not greater than the area of the corresponding peak in solution (3) (0.1%), the area of any peak corresponding to 4'-chloroacetanilide is not greater than the area of the principal peak in solution (4) (10 ppm), and no other impurity is greater than the area of the principal peak obtained with solution (2) (0.25%).

ASSAY

Weigh and powder 20 tablets. Add a quantity of the powder containing 0.15 g of Paracetamol to 50 mL of 0.1 M *sodium hydroxide*, dilute with 100 mL of *water*, shake for 15 minutes, and add sufficient *water* to produce 200 mL. Mix, filter, and dilute 10 mL of the filtrate to 100 mL with *water*. Add 10 mL of the resulting solution to 10 mL of 0.1 M *sodium hydroxide*, dilute to 100 mL with water and measure the *absorbance* of the resulting solution at the maximum at 257 nm, Appendix II B. Calculate the content of $C_8H_9 NO_2$ taking 715 as the value of A (1%, 1 cm) at the maximum at 257 nm.

STORAGE

Paracetamol Tablets should be protected from light.

The first part of the monograph gives a *Definition* of the pharmaceutical preparation. In the case of paracetamol tablets (Box 19.2), it is stated that paracetamol tablets contain paracetamol as the API, and that the tablets should comply with the requirements defined in the general monograph for tablets. The tolerance for the content of API is stated. In this case, the content should be within the range 95.0–105.0% of the stated amount.

The *Identification* section describes procedures for identification of the API in the pharmaceutical preparation. In the case of paracetamol tablets, the tablets are first powdered, the powder is extracted with acetone, and acetone is evaporated to dryness. The residue, which is the API, is either tested by IR spectrophotometry, by a colour test, or by the melting point for positive identification.

The next section is *Tests*, which describes test methods prescribed for the pharmaceutical preparation. In this case, *Dissolution* describes how to check the release of paracetamol from the tablets according to *Dissolution test for tablets and capsules* based on a spectrophotometric assay. The test for dissolution is intended to check that the API is released appropriately. Most often, the test section includes a test for *Related substances*, where the pharmaceutical preparation is tested for traces of degradation products or other trace level substances related to the API. For paracetamol tablets, the test is based on the LC of tablet powder dispersed in the mobile phase and subsequently filtered. The Test section may also contain test procedures as *Uniformity of content*, which is intended for checking that the content of the API only varies within given limits.

For liquid preparations, like injections and oral solutions, the *Tests* section may, among others, include tests for clarity, pH, light absorption, and refractive index. These tests are intended to check that the pharmaceutical preparation has the correct pH value and that there is no particulate matter in the product.

The final section is *Assay*. This section describes the procedure for quantitative analysis of the API in the pharmaceutical preparation. For paracetamol tablets, the assay is based on UV spectrophotometry.

Procedures adopted from BP and USP are used as examples in this chapter to illustrate and discuss in detail chemical analysis of pharmaceutical preparations. The examples are identical or similar to the methods used by the pharmaceutical industry. However, this discussion is not complete, as the pharmaceutical industry may use additional tests in their quality control.

19.3   Identification of the API

Identification tests are intended to identify the API in the pharmaceutical preparation. In some cases, particular excipients are included in the identification tests. This may for instance be the case for antimicrobial preservatives. The major analytical techniques used for identification testing are IR spectrophotometry, LC, and UV spectrophotometry. In older pharmacopoeias, thin-layer chromatography (TLC) was also used. In addition, colour reactions may be prescribed, but these are not further discussed as they are less in use in modern pharmaceutical analysis.

19.3.1 Identification by IR Spectrophotometry

IR spectrophotometry is frequently in use, and the fundamentals of this technique are discussed in Chapter 8. IR spectrophotometry is a specific method for identification, and an IR spectrum exactly matching a similar spectrum of chemical reference substance (CRS), provides a highly reliable identification. However, since excipients may interfere in the IR spectrum, the API must be isolated from the excipients before identification. Thus, sample preparation and extraction procedures are required depending on the type and complexity of the preparation. In common, the outcome of the procedures is a dry residue of the API, which is examined by IR spectrophotometry. A few such examples are discussed in Boxes 19.3 to 19.5 for a solid, a liquid, and a semi-solid preparation, respectively.

Box 19.3   Identification of aspirin (acetylsalicylic acid) in aspirin tablets (solid preparation) by IR spectrophotometry

CHEMICAL PROPERTIES OF THE API

Molar mass: 180.2 g/mol
Acid/base: acidic
pK_a: 3.5
$\log P$: 1.40

PROCEDURE (ADOPTED FROM USP)

Shake a quantity of finely powdered Tablets, equivalent to about 500 mg of aspirin, with 10 mL of alcohol for several minutes. Centrifuge the mixture. Pour off the clear supernatant and evaporate to dryness. Dry the residue in vacuum at 60 °C for one hour. The infrared absorption spectrum of the dried residue is concordant with the reference spectrum of aspirin.

DISCUSSION

The tablet powder is prepared by solid–liquid extraction (SLE). During extraction with ethanol, aspirin is isolated from the tablet excipients, as the latter are almost insoluble in ethanol, whereas aspirin is soluble in ethanol. The tablet excipients are then removed by centrifugation. Since IR spectrophotometry is preferably carried out in the solid state, the extraction solvent is evaporated to dryness to give a residue of pure aspirin. This residue is dried under vacuum at 60 °C to remove residues of ethanol, which can interfere during the IR spectrophotometry and an IR spectrum is obtained on the residue. The IR spectrum of the residue is then compared with a reference spectrum of aspirin (based on aspirin CRS, see below), and if they are identical, aspirin is positively identified in the tablets.

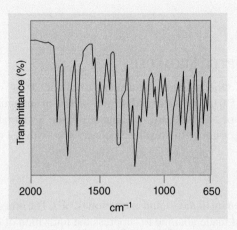

IR spectrum of aspirin

CALCULATIONS

Tablets containing 500 mg of aspirin are analysed. As an amount of tablet powder equivalent to 500 mg of aspirin is prescribed for the test, tablet powder corresponding to the average mass of a single tablet is used. Ten tablets are collected and the total mass is 5.6322 g. The average mass of each tablet (containing 500 mg of aspirin as the specified content) is

$$\frac{5.6322\,\text{g}}{10\,\text{tablets}} = 0.56\,\text{g/tablet}$$

Thus, 0.56 g of tablet powder should be used for the identification. Since this is not a quantitative method, exact weighing of the tablet powder is not critical. Note that each tablet contains 500 mg of aspirin and 60 mg of excipients.

Box 19.4   Identification of fluoxetine in fluoxetine hydrochloride oral solution (liquid preparation) by IR spectrophotometry

CHEMICAL PROPERTIES OF THE API'S FREE BASE

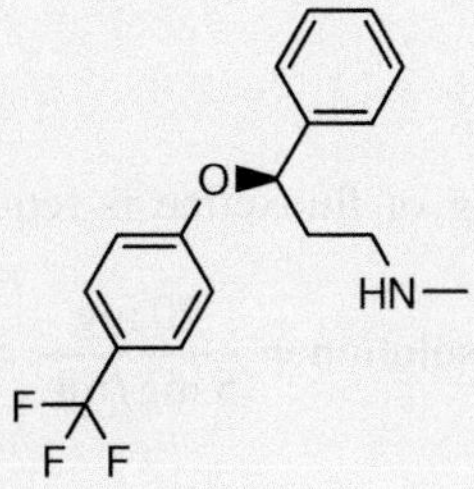

Molar mass: 309.3 g/mol
Acid/base: basic
pK_a: 9.8
$\log P$: 4.2

PROCEDURE (ADOPTED FROM USP)

Transfer a volume of Oral solution, equivalent to about 20 mg of fluoxetine, to a separatory funnel, add 5.0 mL of water and 0.5 mL of 1 M of sodium hydroxide, extract with 5 mL of chloroform, and discard the aqueous layer. Evaporate to dryness. The infrared absorption spectrum of the residue is concordant with the reference spectrum of fluoxetine.

DISCUSSION

Oral solution is prepared by liquid–liquid extraction (LLE). The oral solution, which has a pH of about 6, is first made alkaline to suppress the ionization of the basic API, and then this neutral substance is extracted into chloroform by LLE. The chloroform extract is filtered to remove any particulate matter, and the filtrate is evaporated to dryness to remove the solvent. The residue is the API, from which an IR spectrum can be obtained. The IR spectrum from this is compared with a reference spectrum of fluoxetine (based on fluoxetine CRS, see below), and if they are identical, fluoxetine is positively identified in the oral solution.

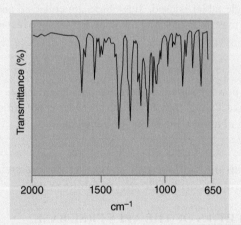

IR spectrum for fluoxetine

CALCULATIONS

A volume corresponding to 20 mg of fluoxetine is required and the oral solution is 5 mg/mL:

$$\text{mL of oral solution} = \frac{20\,\text{mg}}{5\,\text{mg/mL}} = 4\,\text{mL}$$

Thus, 4 mL of oral solution should be used for the identification. Since this is not a quantitative method, collection of the oral solution by an exact transfer pipette is not critical.

Box 19.5   Identification of mupirocin in mupirocin calcium nasal ointment (semi-solid preparation) by IR-spectrophotometry

CHEMICAL PROPERTIES OF THE API'S FREE ANHYDROUS ACID

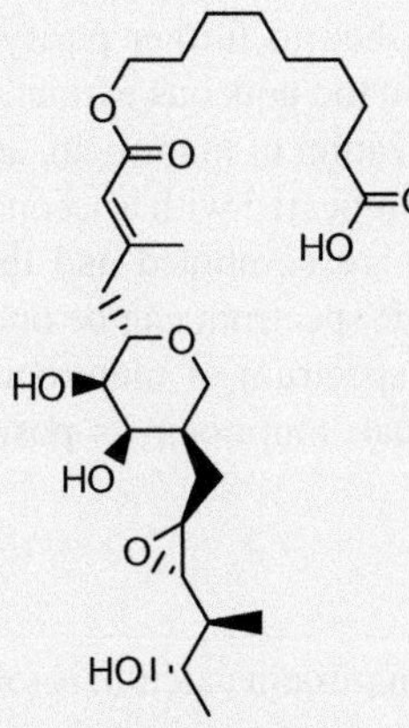

Molar mass: 500.6 g/mol
Acid/base: acidic
pK_a: 4.8
log P: 2.4

PROCEDURE (ADOPTED FROM BP)

Disperse a quantity of the nasal ointment containing the equivalent of 100 mg of mupirocin calcium in a mixture of 20 mL of dichloromethane and 4 mL of 0.2 M sodium borate buffer pH 8.4, with shaking, and centrifuge at 3000 rpm until separate layers are obtained. Remove and retain the upper aqueous layer and re-extract the lower layer with a further 4 mL of 0.2 M sodium borate buffer of pH 8.4. Combine the aqueous extracts, add 10 mL of dichloromethane to the combined extracts and centrifuge at 3000 rpm. Remove the upper aqueous layer and adjust the pH to about 4 with 1 M hydrochloric acid (a slight cloudiness may be observed), add 10 mL of dichloromethane and centrifuge. Separate and retain the lower layer; add a further 10 mL dichloromethane to the upper layer and centrifuge. Combine the dichloromethane extracts and evaporate to dryness at room temperature under a current of nitrogen; a jelly-like residue is obtained. The infrared absorption spectrum of the dried residue is concordant with the reference spectrum of mupirocin.

DISCUSSION

Ointment is a complex formulation where the API is distributed in a lipid vehicle. Thus, the formulation is not soluble in water. To disperse the vehicle and to get access to the API, dichloromethane is added to the sample. Borate buffer is then added, resulting in a two-phase LLE system. The fatty ointment vehicle remains in the dichloromethane phase, whereas the acidic drug substance (mupirocin) is extracted into the slightly basic borate buffer because mupirocin is charged. In this way, mupirocin is isolated from the bulk formulation. Extraction with borate buffer is accomplished two times to ensure an efficient extraction of mupirocin. The two portions of borate buffer (containing mupirocin) are combined, and extracted with a new portion of dichloromethane. This is accomplished to further purify the aqueous phase, to make sure that no excipients are present in the aqueous extract. Finally, the aqueous phase is acidified, which suppresses the ionization of mupirocin, and mupirocin is extracted into dichloromethane. This extraction is repeated with a second portion of dichloromethane. The two dichloromethane extracts are combined and the solvent is evaporated. The residue is the API, from which an IR spectrum can be obtained. The IR spectrum from this is compared with a reference spectrum of mupirocin (based on mupirocin CRS, see below), and if they are identical, mupirocin is positively identified in the nasal ointment.

CALCULATIONS

Ointment containing 2% (*w/w*) of mupirocin calcium is to be analysed, where 2% (*w/w*) corresponds to 20 mg/g ointment.

100 mg of mupirocin calcium is equivalent to the following amount of ointment:

$$\frac{100\,mg}{20\,mg/g} = 5\,g$$

Thus, 5 g of ointment should be used for the identification.

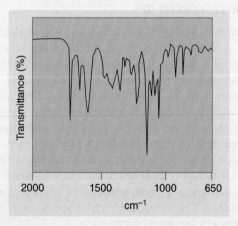

IR spectrum for mupirocin

19.3.2   Identification by Liquid Chromatography

LC with UV detection is often used for identification. The fundamentals of LC are discussed in Chapter 12. Liquid preparations can be injected more or less directly into the LC system, whereas solid or semi-solid preparations have to be dissolved in the mobile phase before injection. The retention time of the API is compared with the retention time for a standard solution (reference solution) prepared from a CRS of the API. If the retention times are equal, the API is positively identified. The major advantage of LC is that dissolved excipients do not have to be removed from the sample, as long as the LC system can separate them from the API. This means that dissolution of the preparation is the only action that has to be taken before analysis. Additionally, the same LC analysis may also be used for the assay (quantitative determination) of the API. Examples on the use of LC for identification are discussed in Boxes 19.6 to 19.8, for a solid, a liquid, and a semi-solid preparation, respectively.

Box 19.6   Identification of fluoxetine in fluoxetine hydrochloride capsules (solid preparation) by LC

CHEMICAL PROPERTIES OF THE API'S FREE BASE

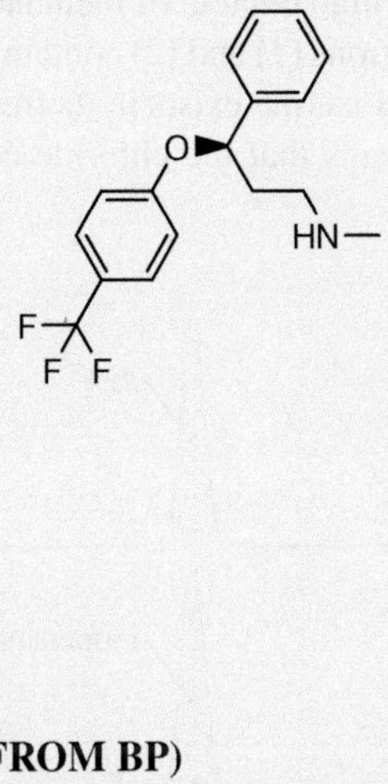

Molar mass: 309.3 g/mol
Acid/base: basic
pK_a: 9.8
log P: 4.2

PROCEDURE (ADOPTED FROM BP)

For solution (1) dissolve a quantity of the mixed contents of 20 capsules containing the equivalent of 20 mg of fluoxetine in a sufficient quantity of the mobile phase to produce 200 mL, filter and use the filtrate, discarding the first 2 mL. Solution (2) contains 0.011% (*w/v*) of fluoxetine hydrochloride (CRS) in the mobile phase. The chromatogram obtained with solution (1) shows a peak with the same retention time as the principal peak in the chromatogram obtained with solution (2).

LC Conditions
Column. C_8 (octyl silyl silica)

Mobile phase. 33 volumes of a solution containing 0.3% (*w/v*) of glacial acid and 0.64% (*w/v*) of sodium pentane sulfonate, adjusted to pH 5.0 with 5 M sodium hydroxide and 67 volumes of methanol

Detection. UV at 227 nm

Discussion

Capsule content corresponding to 20 mg of fluoxetine is added to the mobile phase, and the API is dissolved in the mobile phase. Most excipients are not soluble in the mobile phase and are removed by filtration. Since 200 mL of the sample solution is produced, but only a very small portion of this is injected in high performance liquid chromatography (HPLC), the first 2 mL of the filtrate can be discarded. This is done to saturate the filter with the sample solution, to ensure that no compositional changes occur during filtration of the sample. The retention time of the API in solution (1) is compared with the retention time for fluoxetine (CRS) in solution (2), and if they are equal the API is identified as fluoxetine. The chromatographic principle is ion-pair chromatography, where pentane sulfonate is used as the ion-pairing reagent. With acidic conditions in the mobile phase, fluoxetine is positively charged and forms ion pairs with pentane sulfonate ions. The ion pairs of the relative non-polar drug fluoxetine result in very strong retention, but this is balanced by a high content of methanol in the mobile phase (strong mobile phase). Note that both solutions (1) and (2) contain fluoxetine hydrochloride, but as soon as this salt is dissolved, fluoxetine exists in its free protonated form in solution and in the mobile phase. This means that the chloride counter ions do not affect the retention of fluoxetine.

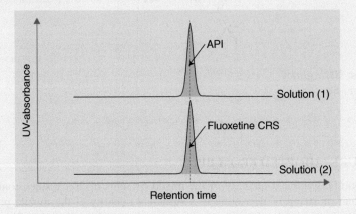

Calculations

Capsules containing fluoxetine hydrochloride, each equivalent to 20 mg of fluoxetine, are to be analysed. 20 capsules are collected and the total mass is 3.7520 g. The average content of each capsule (containing 20 mg of fluoxetine) is

$$\frac{3.7520\,\text{g}}{20\,\text{capsules}} = 0.19\,\text{g/capsule}$$

Capsule content corresponding to 20 mg of fluoxetine is equivalent to one capsule, which corresponds to 0.19 g of capsule powder. Thus, 0.19 g of capsule powder should be used for the identification.

Box 19.7   Identification of droperidol in droperidol injection by LC

CHEMICAL PROPERTIES OF THE API

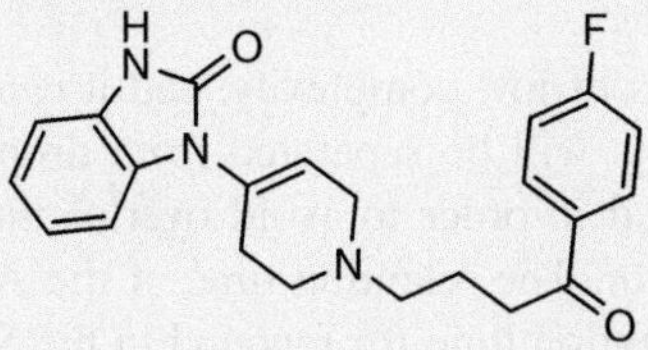

Molar mass: 379.4 g/mol
Acid/base: basic
pK_a: 7.6
$\log P$: 3.5

Procedure (Adopted from USP)

Assay preparation. Transfer an accurately measured volume of Injection, equivalent to about 5 mg of droperidol, to a 100-mL volumetric flask, dilute with *Mobile phase* to volume, and mix.

Standard preparation. Transfer 5.00 mL of *Standard stock preparation* to a 100-mL volumetric flask, dilute with *Mobile phase* to volume, and mix.

Standard stock preparation. Dissolve an accurately weighed quantity of *USP Droperidol RS* in methanol to obtain a solution having a known concentration of about 1 mg/mL.

Chromatographic system. The LC is equipped with a 280-nm detector and a 4.6 mm × 25 cm column that contains 10 μm of packing L1 (C_{18}). The flow rate is about 1 mL/min.

Mobile phase. Prepare a filtered and degassed mixture of methanol, water, and borate buffer (700 : 280 : 20).

Procedure. Separately inject equal volumes (about 10 μL) of the *Standard preparation* and the *Assay preparation* into the chromatograph and record the chromatograms. The retention time of the major peak in the chromatogram of the *Assay preparation* corresponds to that in the chromatogram of the *Standard preparation*.

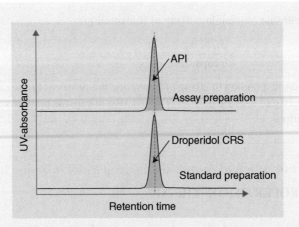

Discussion

In this case, the preparation is of low complexity, and it can be more or less injected directly in LC. Any excipients will be separated from droperidol in the LC system. The injection has to be diluted in order to avoid overloading of the LC column and overloading of the UV detector. The retention time of the API in the *Assay preparation* is compared with the retention time for fentanyl in the *Standard preparation*, and if they are equal the API is identified as droperidol. The chromatographic principle is reversed-phase chromatography.

Calculations

Injection containing 2.5 mg/mL of droperidol is to be analysed. For the *Assay preparation*, Injection corresponding to 5 mg of droperidol is required, which corresponds to a 2.00 mL Injection.

Box 19.8   Identification of Beclomethasone Dipropionate in Beclomethasone Dipropionate Ointment (Semi-Solid Preparation) by LC

CHEMICAL PROPERTIES OF THE API

Molar mass: 408.9 g/mol
Acid/base: neutral
log *P*: 2.1

PROCEDURE (ADOPTED FROM BP)

For solution (1), dilute 10 mL of methanol (80%) containing 0.01% (*w/v*) of beclomethasone dipropionate to 50 mL with the same solvent. For solution (2), disperse a quantity
of the ointment containing 1 mg of anhydrous beclomethasone dipropionate in 25 mL
of hot 2,2,4-trimethylpentane, cool and extract the mixture with successive quantities of
20, 10, and 10 mL of methanol (80%), filtering each extract in turn through a small plug
of absorbent cotton previously washed with methanol (80%). Combine the filtrates and
dilute to 50 mL with methanol. The chromatogram obtained with solution (2) shows a
peak with the same retention time as the peak due to beclomethasone dipropionate in
the chromatogram obtained with solution (1).

HPLC Conditions

Column. C_{18} (octadecyl silyl silica)
Mobile phase. 70 volumes of methanol and 30 volumes of water
Column temperature. 60 °C
Detection. UV at 238 nm

DISCUSSION

First the ointment sample is dispersed in hot 2,2,4-trimethylpentane to disintegrate the
semi-solid formulation and to give access to the API. Beclomethasone dipropionate is
extracted from this very non-polar phase and into 80% methanol in water by LLE. To
ensure an efficient extraction, this is repeated three times with successive portions of
80% methanol. The extracts are filtered to remove any particulate matter and combined
prior to the LC analysis. The retention time of the API in solution (2) is compared with
the retention time for beclomethasone dipropionate (CRS) in solution (1), and if they
are equal the API is identified as beclomethasone dipropionate. The chromatographic
principle is reversed-phase (RP) chromatography. Beclomethasone dipropionate is a relatively non-polar substance, and to avoid too long a retention, the content of methanol
in the mobile phase is relatively high (strong mobile phase).

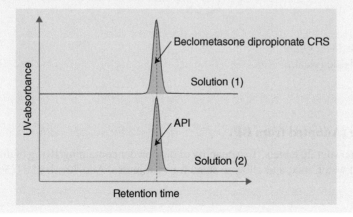

CALCULATIONS

Ointment containing 0.1% (*w/w*) beclomethasone dipropionate is to be analysed. 0.1% corresponds to 1 mg/g. Thus, 1.0 g ointment should be used for the identification.

19.3.3 Identification by UV-Vis Spectrophotometry

UV-Vis spectrophotometry can also be used for the identification of APIs. The fundamentals of UV-Vis spectrophotometry are discussed in Chapter 7. Typically, a UV spectrum from the API is recorded, and one or two absorption maxima should appear at specified wavelengths for positive identification. In some cases, the absorbance at the absorption maximum is required to be within a specified range for positive identification. Excipients may interfere in the UV measurement, and in such cases the API has to be isolated from the excipients by a sample preparation. Generally the procedures result in an aqueous solution of the active ingredient or a solution in a polar organic solvent such as methanol or ethanol. Aqueous solutions and solutions in methanol or ethanol are UV transparent and thereby suitable for UV measurements. Examples on the use of UV spectrophotometry for identification are discussed in Boxes 19.9 to 19.11, for a solid, a liquid, and a semi-solid preparation, respectively.

Box 19.9   Identification of Diazepam in Diazepam Tablets (Solid Preparation) by UV Spectrophotometry

CHEMICAL PROPERTIES OF THE API

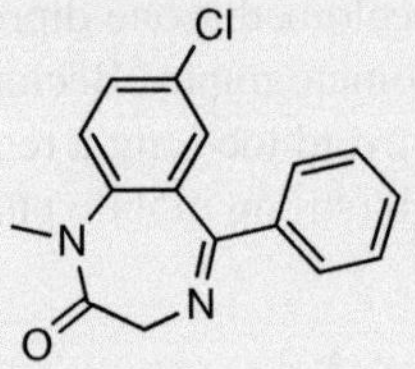

Molar mass: 284.7 g/mol
Acid/base: basic (weak)
pK_a: 2.9
$\log P$: 3.1

Procedure (Adopted from BP)

Weigh and powder 20 tablets. To a quantity of the powder containing 10 mg of diazepam add 5 mL of water, mix, and allow to stand for 15 minutes. Add 70 mL of a 0.5% (*w/v*)

---

solution of sulfuric acid in methanol, shake for 15 minutes, add sufficient of the methanolic sulfuric acid to produce 100 mL, and filter. Dilute 10 mL of the filtrate to 50 mL with the same solvent and measure the absorbance of the resulting solution in the range 230–350 nm. The solution shows two maxima, at 242 and 284 nm.

Discussion

The tablets are powdered to access the total amount of API (diazepam). Twenty tablets are selected to represent the sample, but only a small fraction of the resulting powder, corresponding to 10 mg of diazepam, is required for the identification. The tablet powder is suspended in pure water, and acidified methanol is added to dissolve the diazepam. Diazepam is soluble in methanol, but the solubility is improved by acidifying methanol with sulfuric acid. Most excipients from the tablets are not dissolved in acidic methanol and are removed by the subsequent filtration. The methanolic solution containing the API is further diluted in order not to overload the UV spectrophotometer. This final solution is filled into a quartz cuvette, and the UV spectrum is recorded between 230 and 350 nm. Absorbance maxima should be observed at 242 and 284 nm for positive identification of diazepam.

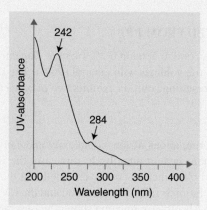

UV spectrum of diazepam

Calculations

Tablets containing 5 mg of diazepam are to be analysed. A quantity of 10 mg is prescribed for identification. This amount corresponds to two tablets. Twenty tablets are collected and the total mass is 2.6656 g. The average mass of each tablet (containing 5 mg of diazepam as the specified content) is

$$\frac{2.6656 \text{ g}}{20 \text{ tablets}} = 0.13 \text{ g/tablet}$$

Thus, 0.26 g of tablet powder is used for identification.

Box 19.10   Identification of Flupentixol Decanoate in Flupentixol Decanoate Injection (Liquid Preparation) by UV Spectrophotometry

CHEMICAL PROPERTIES OF THE API

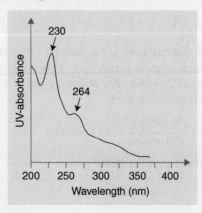

Molar mass: 588.8 g/mol
Acid/base: basic
pK_a: 7.8
log P: 8.8

PROCEDURE (ADOPTED FROM BP)

Dilute a volume of the injection to contain 0.2% (*w/v*) of flupentixol decanoate. Dilute 1 volume of this solution to 2 volumes with ethanol (96%). The light absorption in the range 205–300 nm of the resulting solution exhibits two maxima at 230 and 264 nm.

DISCUSSION

Injections are normally formulations of low complexity and can often be analysed by UV spectrophotometry more or less directly. In this case, the injection is diluted to ensure that the UV spectrophotometer is not overloaded and the absorbance is in the range 0.2–0.8. The final solution is filled into a cuvette and the UV spectrum is recorded between 205 and 300 nm. Absorbance maxima should be observed at 230 and 264 nm for positive identification of flupentixol decanoate.

UV spectrum of flupentixol decanoate

CALCULATIONS

An injection containing 20 mg/mL of flupentixol decanoate is to be analysed. 20 mg/mL corresponds to 2% (*w/v*). Injection containing 0.2% (*w/v*) (or 2 mg/mL) of flupentixol decanoate is equivalent to the original injection diluted by a factor of 10. Thus, the injection should be diluted 10 times with water prior to analysis. This can be accomplished by dilution of 1 mL of injection to 10 mL. This solution is further diluted by adding 2 volumes of ethanol to 1 volume of diluted solution.

Box 19.11   Identification of Miconazole in Miconazole Nitrate Cream (Semi-Solid Preparation) by UV Spectrophotometry

CHEMICAL PROPERTIES OF THE API'S FREE BASE

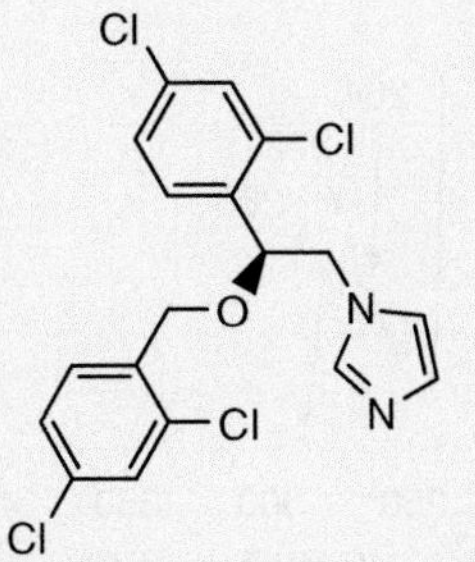

Molar mass: 416.1 g/mol
Acid/base: basic
pK_a: 6.8
log *P*: 6.1

PROCEDURE (ADOPTED FROM BP)

Mix a quantity containing 40 mg of miconazole nitrate with 20 mL of a mixture of 1 volume of 1 M sulfuric acid and 4 volumes of methanol and shake with two 50-mL quantities of hexane, discarding the organic layers. Make the aqueous phase alkaline with 2 M ammonia and extract with two 40-mL quantities of chloroform. Combine the chloroform extracts, shake with 5 g of anhydrous sodium sulfate, filter, and dilute the filtrate to 100 mL with chloroform. Evaporate 50 mL to dryness and dissolve the residue in 50 mL of a mixture of 1 volume of 0.1 M hydrochloric acid and 9 volumes of methanol. The light absorption of the resulting solution in the range 230–350 nm exhibits maxima at 264, 272, and 280 nm.

DISCUSSION

The cream sample is mixed in an LLE system, consisting of a non-polar phase (hexane) and a polar phase (methanol/sulfuric acid). The hydrophobic excipients are extracted into hexane and discarded, whereas the API (basic substance) is dissolved in the polar phase. Polar excipients will also remain in the polar phase. However, the polar phase is made alkaline in a subsequent step; the basic substance miconazole becomes uncharged and is extracted into chloroform. The chloroform extract is dried with anhydrous sodium sulfate to remove traces of water, and the extract is evaporated to remove chloroform. Chloroform is not desired as a solvent in the final UV measurement because chloroform gives strong UV absorption itself. Finally, the residue is dissolved in a mixture of hydrochloric acid and methanol; this final solution is filled into a cuvette and the UV spectrum is recorded to be between 230 and 350 nm. Absorbance maxima should be observed at 264, 272, and 280 nm for positive identification of miconazole. The mixture of methanol and hydrochloric acid is UV transparent above 205 nm, and does not contribute to absorption.

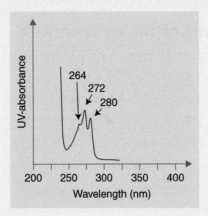

UV spectrum of miconazole.

CALCULATIONS

Cream containing 2% (*w/w*) miconazole nitrate (20 mg/g) is to be analysed.

40 mg of miconazole nitrate is equivalent to 2.0 g of cream. Thus, 2.0 g of cream should be used for the identification.

19.4   Assay of the Active Pharmaceutical Ingredient

The assay is intended to check that the content of an API is within a certain tolerance relative to the specified content, that is, to check that the pharmaceutical preparation contains the correct amount of drug substance. Assay is therefore a quantitative analysis. Typically, the analytical result from the assay should be within a ±5.0% tolerance relative to the specified content, but the tolerance may vary from product to product. The ±5.0% is the requirements of the authorities and often manufactures have stricter limits. In some cases, particular excipients may also be included in the assay. The major analytical techniques used for quantitation are LC, UV spectrophotometry, and titration, but also gas chromatography (GC) is used in some cases.

19.4.1 Assays Based on Liquid Chromatography

LC is the dominating technique for assays of APIs in pharmaceutical preparations. The major advantage of LC is that soluble excipients do not have to be removed from the sample, as long as the LC system can separate them from the API. Liquid preparations can be injected directly or after dilution, whereas solid and semi-solid preparations have to be dissolved in mobile phase before LC. The LC system has to be calibrated with a reference solution of the API, containing an exact concentration of CRS. With this calibration, the relationship between peak area and concentration is established. Calibration is normally performed using one point calibration. Based on this, the peak area measured for the pharmaceutical preparation can be converted to the exact content of API. Some examples on the use of LC assays are discussed in Boxes 19.12 to 19.14.

Box 19.12   Assay of Omeprazole in Gastro-Resistant Omeprazole Tablets (Solid Preparation) by LC

CHEMICAL PROPERTIES OF THE API

Molar mass: 345.4 g/mol
Acid/base: zwitterionic (acid and base)
pK_a: 4.8 (basic), 9.3 (acidic)
log P: 2.4

PROCEDURE (ADOPTED FROM BP)

Weigh and powder 20 tablets. Carry out the method for LC, using the following solutions:

(1) Disperse a quantity of the powdered tablets containing 24 mg of omeprazole in 150 mL of the mobile phase, mix with the aid of ultrasound for 30 minutes, dilute with sufficient mobile phase to produce 200 mL, mix, filter, and further dilute 10 volumes to 100 volumes.
(2) 0.0012% (*w/v*) of omeprazole CRS in the mobile phase.
(3) Mix 10 mg of each of omeprazole CRS and omeprazole impurity D CRS in the mobile phase and dilute to 100 mL with the same solvent.

HPLC Conditions

Column. C$_8$ (octyl silyl silica)
Mobile phase. 27 volumes of acetonitrile and 73 volumes of 0.14% (*w/v*) solution of disodium hydrogen orthophosphate that has been previously adjusted to pH 7.6 with orthophosphoric acid
Detection. UV at 280 nm

System Suitability

The test is not valid unless, in the chromatogram obtained with solution (3), the resolution factor between the peaks due to impurity D and omeprazole is higher than 3.0.

Determination of Content

Calculate the content of $C_{17}H_{19}N_3O_3S$ in the tablets from the chromatograms obtained using the declared content of $C_{17}H_{19}N_3O_3S$ in omeprazole CRS.

DISCUSSION

Twenty tablets are used for the analysis to give a representative sample. The tablets are first powdered to give access to the API. Powder corresponding to 24 mg omeprazole is required for the analysis. The required tablet powder is weighed accurately on an analytical balance and is then dispersed in the mobile phase to an exact volume of 200.0 mL. This causes the API to be dissolved. The majority of excipients are not dissolved in the mobile phase, and those are removed by filtration. The filtrate is further diluted utilizing a 10 mL transfer pipette and a 100 mL volumetric flask; the purpose of this is to avoid overloading the LC system. The diluted filtrate is then injected into the LC instrument and the peak area for omeprazole is measured. This peak area is compared with the peak area of 0.0012% (w/v) of omeprazole CRS in the mobile phase to calculate the content (one point calibration). The declared content of omeprazole in the CRS is used to calculate the exact concentration in the reference solution. The chromatographic principle is RP chromatography.

CALCULATIONS

Tablets containing 10 mg omeprazole (specified content) is to be analysed. Twenty tablets are collected and the total mass is 10.0665 g. The average mass of a tablet is

$$\frac{10.0665 \text{ g}}{20 \text{ tablets}} = 0.5033 \text{ g/tablet}$$

Thus, 24 mg of omeprazole corresponds to

$$\frac{24 \text{ mg}}{10 \text{ mg}} \times 0.5033 \text{ g} = 1.2 \text{ g}$$

Thus, approximately 1.2 g of tablet powder should be used for the assay. The tablet powder has to be weighed accurately on an analytical balance, in this case because it is a quantitative measurement.

As an example, 1.2097 g of tablet powder is analysed by the procedure, resulting in a peak area of 109 766 for the API (see the chromatogram below). The reference solution (2) is prepared from omeprazole CRS of reported purity 99.9%. Then 0.1213 g (121.3 mg) of omeprazole CRS is dissolved in the mobile phase and the volume is adjusted to 100.0 mL; 10.00 ml of this solution is diluted to 100.0 mL with the mobile phase and 10.00 mL of this solution is diluted further to 100.0 mL with the mobile phase. The peak area of solution (2) for omeprazole is 108 871.

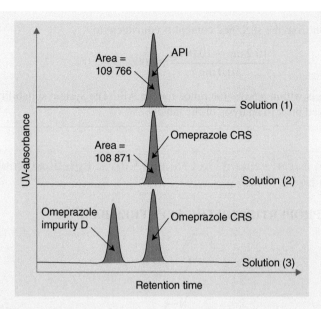

The content of omeprazole per tablet is calculated as shown below.

The concentration of omeprazole in reference solution (2) is calculated based on the reported purity and taking the dilutions into consideration (from 10.00 to 100.0 mL and thereafter from 10.00 to a new 100.0 mL):

$$0.999 \times \frac{121.3\,\text{mg}}{100.0\,\text{mL}} \times \frac{10.00\,\text{mL}}{100.0\,\text{mL}} \times \frac{10.00\,\text{mL}}{100.0\,\text{mL}} = 0.0121\,\text{mg/mL}$$

This means that 0.0121 mg/mL of omeprazole results in a peak area of 108 871. The peak area for the final sample solution was measured to 109 766. The concentration of omeprazole in the final solution (1), which is slightly higher than in the reference solution (2), is calculated as follows:

$$\frac{109\,766}{108\,871} \times 0.0121\,\text{mg/mL} = 0.0122\,\text{mg/mL}$$

The original amount of omeprazole in the sample is calculated from the concentration in the final solution (1), taking the dilution into consideration (from 10.00 to 100.0 mL) and the total volume of the original volumetric flask (200.0 mL):

$$0.0122\,\text{mg/mL} \times \frac{100.0\,\text{mL}}{10.00\,\text{mL}} \times 200.0\,\text{mL} = 24.4\,\text{mg}$$

This amount of omeprazole is present in 1.2097 g of tablet powder. The average mass of a tablet is

$$\frac{10.0665\,\text{g}}{20\,\text{tablets}} = 0.5033\,\text{g/tablet}$$

Therefore, the content of omeprazole in a tablet of average mass is

$$\frac{24.4\,\text{mg} \times 0.5033\,\text{g/tablet}}{1.2097\,\text{g}} = 10.2\,\text{mg/tablet}$$

The deviation from the specified content is equivalent to

$$\frac{10.2\,mg - 10.0\,mg}{10.0\,mg} \times 100\% = +2.0\%$$

This content is within $\pm 5.0\%$ tolerance for the API. The system suitability test also has to be checked prior to analysis of the tablets.

Box 19.13   Assay of Fentanyl in Fentanyl Citrate Injection (Liquid Preparation) by LC

CHEMICAL PROPERTIES OF THE API'S FREE BASE

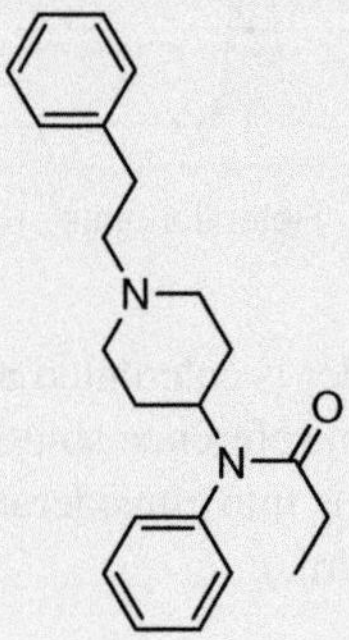

Molar mass: 336.5 g/mol
Acid/base: basic
pK_a: 8.8
log P: 3.8

PROCEDURE (ADOPTED FROM USP)

Mobile phase. Prepare a filtered and degassed mixture containing four volumes of ammonium acetate solution (1 in 100) and six volumes of a mixture of methanol, acetonitrile, and glacial acetic acid (400 : 200 : 0.6). Adjust this solution to a pH of 6.6 ± 0.1 by the dropwise addition of glacial acetic acid, and make adjustments if necessary to obtain a retention time of about five minutes for the fentanyl peak.

Standard preparation. Dissolve an accurately weighed quantity of *USP Fentanyl Citrate RS* in water, and quantitatively dilute with water to obtain a solution having a known concentration of about 80 µg/mL.

Assay preparation. If necessary, dilute the *Injection* with water so that each millilitre contains the equivalent of about 50 µg of fentanyl.

Chromatographic system. The LC is equipped with a 230-nm detector and a 4.6 mm × 25 cm column that contains packing L1 (C_{18}). The flow rate is about 2 mL/min. Chromatograph the *Standard preparation* and record the peak response: the tailing factor for the fentanyl peak is not more than 2.0 and the relative standard deviation for replicate injections is not more than 2.0%.

Procedure. Separately inject equal volumes (about 25 µL) of the *Standard preparation* and the *Assay preparation* into the chromatograph, record the chromatograms, and measure the responses for the major peaks. Calculate the quantity, in micrograms, of fentanyl ($C_{22}H_{28}N_3O$) in each millilitre of the Injection taken by the formula:

$$\frac{336.48}{528.59} \times CD \frac{r_u}{r_s}$$

in which 336.48 and 528.59 are the molar masses (g/mol) of fentanyl and fentanyl citrate, respectively; C is the concentration, in micrograms per millilitre, of *USP Fentanyl Citrate RS* in the *Standard preparation*; D is the dilution factor used to obtain the *Assay preparation*; and r_u and r_s are the peak responses for the fentanyl peak obtained from the *Assay preparation* and the *Standard preparation*, respectively.

DISCUSSION

The sample is diluted (*Assay preparation*) if the content of fentanyl is above 50 µg/mL to avoid overloading of the LC system. The *Assay preparation* is injected directly without any further sample preparation and the peak area for fentanyl is measured (r_u). This peak area is compared with the peak area of fentanyl for the *Standard preparation* (one point calibration). Note that the Standard preparation is prepared with fentanyl citrate, while the injection content is with respect to fentanyl. Therefore, mass correction by 336.48/528.59 has to be performed, where 336.48 and 528.59 are the molar masses (g/mol) of fentanyl and fentanyl citrate, respectively. The chromatographic principle is RP chromatography.

CALCULATIONS

Injection sample containing fentanyl citrate, corresponding to 50 µg/mL of fentanyl is to be analysed. The sample can be injected directly into the LC system according to the procedure. As an example, direct injection of the sample (*Assay preparation*) provides a principal peak with an area of 99 753 (r_u). The *Standard preparation* is prepared from *USP Fentanyl Citrate RS* (CRS) of a reported purity of 99.7%. A portion of 0.0810 g (81.0 mg) of *USP Fentanyl Citrate RS* is weighed on an analytical balance and dissolved in water and the volume is adjusted to 100.0 mL; 10.00 mL of this solution is diluted to 100.0 mL with water. The peak area for fentanyl in the *Standard preparation* (r_s) is 100 763.

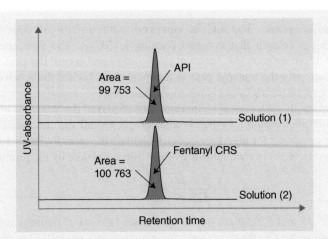

The content of fentanyl per millilitre in the injection is calculated as shown below.

The dilution factor (D) of the *Assay preparation* is 1 since the injection is not diluted prior to analysis. C is calculated as follows:

$$0.997 \times \frac{81.0 \text{ mg}}{100.0 \text{ mL}} \times \frac{10.00 \text{ mL}}{100.0 \text{ mL}} = 0.0808 \text{ mg/mL} = 80.8 \text{ µg/mL (fentanyl citrate)}$$

The concentration of fentanyl in the sample is calculated as follows:

$$\frac{336.48}{528.59} \times CD\frac{r_u}{r_s} = \frac{336.48}{528.59} \times 80.8 \text{ µg/mL} \times 1 \times \frac{99\,753}{100\,763} = 50.9 \text{ µg/mL (fentanyl)}$$

The deviation from the specified content is equivalent to

$$\frac{50.9 \text{ µg/mL} - 50.0 \text{ µg/mL}}{50.0 \text{ µg/mL}} \times 100\% = +1.8\%$$

This content is within ±5.0% tolerance for the API.

Box 19.14 Assay of Hydrocortisone in Hydrocortisone Ointment (Semi-Solid Preparation) by LC

CHEMICAL PROPERTIES OF THE API

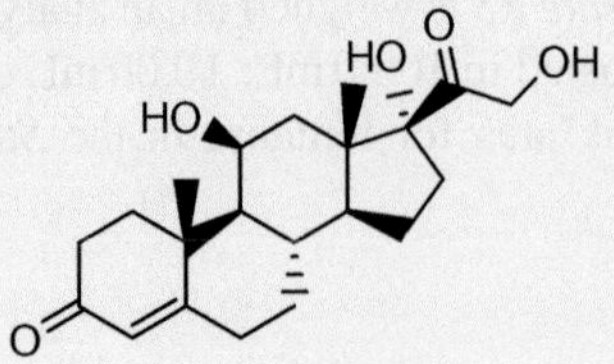

Molar mass: 362.5 g/mol
Acid/base: neutral
log P: 1.2

PROCEDURE (MODIFIED FROM BP)

Carry out the method for LC using the following solutions (for ointments containing more than 0.5% (*w/w*) of hydrocortisone). For solution (1), dissolve 25 mg of hydrocortisone CRS in 45 mL of methanol, add 5 mL of a 0.5% (*w/v*) solution of betamethasone (internal standard) in methanol and add sufficient water to produce 100 mL. For solution (2), disperse a quantity of ointment containing 25 mg of hydrocortisone in 100 mL of hot hexane, cool, and extract with 20 mL of a solution prepared by mixing 3 volumes of methanol with 1 volume of a 15% (*w/v*) solution of sodium chloride. Repeat the extraction using a further two 10 mL quantities of the methanolic sodium chloride solution. To the combined extracts add 5 mL of a 0.5% (*w/v*) solution of betamethasone (internal standard) in methanol and sufficient water to produce 100 mL, mix, and filter through a glass microfiber filter (Whatman GF/C is suitable).

HPLC Conditions

Column. C_{18} (octadecyl silyl silica)
Mobile phase. methanol mixed with water 50 : 50 (*v/v*)
Detection. UV at 240 nm

Determination of Content

Calculate the content of $C_{21}H_{30}O_5$ in the preparation being examined using the declared content of $C_{21}H_{30}O_5$ in hydrocortisone CRS.

DISCUSSION

First the ointment is dispersed in hot hexane. Hot hexane disintegrates the semi-solid preparation and the total amount of API becomes accessible. This hexane dispersion is then extracted (LLE) with a mixture of methanol and water. This mixture forms a two-phase system with hexane. The methanol/water phase contains large amounts of sodium chloride in order to increase the polarity of the extracting phase. Hydrocortisone is a relatively hydrophilic compound and is extracted into the methanol/water phase. This extraction is repeated with two more portions of methanol/water to ensure a quantitative transfer of hydrocortisone from the ointment. The three extracts are combined and a small amount of betamethasone is added as the internal standard. Hydrocortisone is measured relative to the internal standard, to improve the precision and the accuracy of the method (Internal standard method). The combined extracts are diluted to 100.0 mL in a volumetric flask and a small portion of this is filtered prior to LC. Filtration is done to remove any particulate matter that might plug the LC column. The peak area ratio (hydrocortisone/betamethasone) for solution (2) is compared with the peak area ratio of reference solution (1) to calculate the content of API. The declared content of hydrocortisone in the CRS is used to calculate the exact concentration in the reference solution. The chromatographic principle is RP chromatography.

CALCULATIONS

Ointment containing 1% (*w/w*) hydrocortisone (10 mg/g) is to be analysed. Internal standard is prepared by dissolution of 0.1249 g of betamethasone in 25.0 mL of methanol.

Solution (1) is prepared as specified by dissolution of 25.2 mg of hydrocortisone CRS with a specified content of 99.8% (*w/w*). The concentration of hydrocortisone in solution (1) is calculated based on the reported purity:

$$0.998 \times \frac{25.2 \text{ mg}}{100.0 \text{ mL}} = 0.251 \text{ mg/mL}$$

As an example, 2.4896 g of ointment is prepared as specified for solution (2). For solution (1), the peak areas measured for hydrocortisone and for the internal standard are 87 635 and 56 933, respectively. For solution (2), the peak areas measured for hydrocortisone and for the internal standard were 86 701 and 57 255, respectively.

The content of hydrocortisone in the ointment is calculated as follows. For solution (1), which is a reference solution, a concentration of 0.251 mg/mL provided a peak area ratio of

$$\frac{87\,635}{56\,933} = 1.539$$

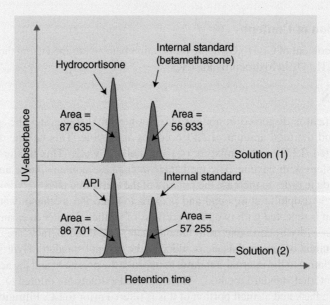

For solution (2), which is the sample solution, the following peak area ratio was obtained:

$$\frac{86\,701}{57\,255} = 1.514$$

This means that the concentration in the sample solution was slightly lower than in the reference solution. The concentration in the sample solution is calculated as follows:

$$\frac{1.514}{1.539} \times 0.251 \text{ mg/mL} = 0.247 \text{ mg/mL}$$

The volume of the final solution was 100.0 mL and the total amount of hydrocortisone in this solution was

$$0.247 \text{ mg/mL} \times 100.0 \text{ mL} = 24.7 \text{ mg}$$

This amount of hydrocortisone was in 2.4896 g (2489.6 mg) of ointment and the content is calculated as follows:

$$\frac{24.7 \text{ mg}}{2489.6 \text{ mg}} \times 100\% = 0.994\%(w/w)$$

The deviation from the specified content is equivalent to

$$\frac{0.994\% - 1.00\%}{1.00\%} \times 100\% = -0.62\%$$

This content is within $\pm 5.0\%$ tolerance for the API.

19.4.2 Assays Based on UV Spectrophotometry

UV spectrophotometry can be used for assays, especially for tablets and injections. The fundamentals of UV spectrophotometry are discussed in Chapter 7. UV spectrophotometry is always performed on accurately prepared solutions of the API, where interfering (UV absorbing) excipients have been removed prior to the measurement. The latter is important to make sure that the measured absorbance only arises from the API. Dissolution of the API and removal of UV absorbing excipients demands sample preparation. Examples on the use of UV spectrophotometry for assays are discussed in Boxes 19.15 and 19.16.

Box 19.15   Assay of Paracetamol in Paracetamol Tablets (Solid Preparation) by UV Spectrophotometry

CHEMICAL PROPERTIES OF THE API

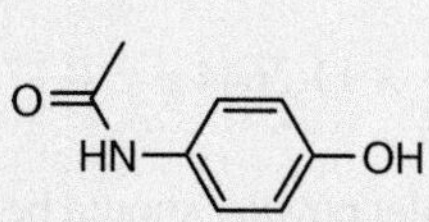

Molar mass: 151.2 g/mol
Acid/base: acidic (weak)
pK_a: 9.5
log P: 0.87

PROCEDURE (ADOPTED FROM BP)

Weigh and powder 20 tablets. Add a quantity of the powder containing 0.15 g of paracetamol to 50 mL of 0.1 M sodium hydroxide, dilute with 100 mL of water, shake for 15 minutes and add sufficient water to produce 200 mL. Mix, filter, and dilute 10 mL of the filtrate to 100 mL with water. Add 10 mL of the resulting solution to 10 mL of 0.1 M sodium hydroxide and dilute to 100 mL with water, and measure the absorbance of the resulting solution at the maximum at 257 nm. Calculate the content of $C_8H_9NO_2$ taking 715 as the value of A (1%, 1 cm) at the maximum at 257 nm.

DISCUSSION

The assay is based on powder from 20 tablets to establish a mean value for the content of API. The tablets are powdered to ensure accessibility to the total amount of API. Only a small portion of the powder is required for the quantitative analysis, namely powder corresponding to 0.15 g of paracetamol. The required tablet powder is weighed accurately on an analytical balance and mixed with a dilute solution of NaOH. Paracetamol is weekly acidic, and is therefore more soluble in water under alkaline conditions. The mixture is shaken for 15 minutes to dissolve the total amount of API, whereas most tablet excipients are not dissolved. The later are removed by filtration to avoid any interference from excipients during UV measurement. The filtrate is further diluted before the absorbance measurement to make sure that the absorbance is in the range 0.2–0.8. Dilution is accomplished in two steps, utilizing 10-mL transfer pipettes (10.00 mL) and a 100-mL volumetric flask (100.0 mL) to ensure high accuracy. The absorbance is measured at the absorption maximum for paracetamol, and the content of API is calculated as shown below. In this method, the specific absorbance, A (1%, 1 cm), for paracetamol in dilute NaOH at 257 nm with a 1 cm cuvette is 715, and this value is used on Beer's law to establish the relationship between measured absorbance and concentration.

CALCULATIONS

Tablets containing 500 mg (0.500 g) paracetamol (specified content) is to be analysed. As an example, 20 tablets are collected and the total mass of these 20 tablets is 11.2644 g. The total amount of paracetamol in 20 tablets according to the specification is

$$20 \text{ tablets} \times 0.500 \text{ g/tablet} = 10.0 \text{ g}$$

Tablet powder corresponding to 0.15 g of paracetamol is equivalent to

$$\frac{0.15 \text{ g}}{10.0 \text{ g}} \times 11.2644 \text{ g} = 0.17 \text{ g}$$

Thus, approximately 0.17 g of tablet powder should be used for the assay. This amount has to be weighed accurately on an analytical balance because this is a quantitative measurement, and the exact amount has to be known for the calculations.

As an example, 0.1703 g of tablet powder is analysed by the procedure, resulting in an absorbance of 0.539 with a 1 cm light path. The content of paracetamol per tablet is calculated as shown below.

The concentration of paracetamol in the final solution measured by UV spectrophotometry is calculated according to Beer's law:

$$c = \frac{A}{ab} = \frac{0.539}{1 \times 715}\%(w/v) = 7.54 \times 10^{-4}\%\,(w/v) \sim 7.54 \times 10^{-3}\,\text{mg/mL}$$

The concentration of paracetamol in the initial solution is calculated from the final solution taking the dilutions into consideration (from 10.00 to 100.0 mL and thereafter from 10.00 mL to a new 100.0 mL):

$$7.54 \times 10^{-3}\,\text{mg/mL} \times \frac{100.0\,\text{mL}}{10.00\,\text{mL}} \times \frac{100.0\,\text{mL}}{10.00\,\text{mL}} = 0.754\,\text{mg/mL}$$

The total amount of paracetamol in the initial solution is calculated from the concentration and the volume (200.0 mL):

$$0.754\,\text{mg/mL} \times 200.0\,\text{mL} = 151\,\text{mg}$$

This amount of paracetamol was present in 0.1703 g of tablet powder. The average mass of a tablet is

$$\frac{11.2644\,\text{g}}{20\,\text{tablets}} = 0.5632\,\text{g/tablet}$$

Therefore, the content of paracetamol in a tablet of average mass is

$$\frac{151\,\text{mg} \times 0.5632\,\text{g/tablet}}{0.1703\,\text{g}} = 499\,\text{mg/tablet}$$

The deviation from the specified content is equivalent to

$$\frac{499\,\text{mg} - 500\,\text{mg}}{500\,\text{mg}} \times 100\% = -0.27\%$$

This content is within ±5.0% tolerance for the API and the result from the assay is acceptable.

Box 19.16   Assay of Doxapram in Doxapram Hydrochloride Injection (Liquid Preparation) by UV Spectrophotometry

CHEMICAL PROPERTIES OF THE API'S FREE ANHYDROUS BASE

Molar mass: 378.5 g/mol
Acid/base: basic
pK_a: 7.2
log *P*: 3.2

PROCEDURE (ADOPTED FROM BP)

Dilute a volume containing 0.2 g of doxapram hydrochloride to 250 mL with water. Measure the absorbance of the resulting solution at the maximum at 258 nm. Calculate the content of $C_{24}H_{30}N_2O_2$, HCl, H_2O in the injection from the absorbance of a 0.08% (w/v) solution of doxapram hydrochloride CRS using the declared content of $C_{24}H_{30}N_2O_2$, HCl, H_2O in doxapram hydrochloride CRS.

DISCUSSION

Because the injection contains no other species absorbing at 258 nm, the only sample preparation required is dilution. Dilution is performed to make sure that the measured absorbance is within the range 0.2–0.8. Dilution is accomplished with a transfer pipette and a 250 mL volumetric flask (250.0 mL). A 0.08% (w/v) reference solution containing CRS of doxapram hydrochloride is used to calibrate the UV instrument, to establish the relationship between absorbance and concentration. The declared content of doxapram hydrochloride in the CRS is used to calculate the exact concentration in the reference solution.

CALCULATIONS

Injection containing 20 mg/mL of doxapram hydrochloride (specified content) is to be analysed. As an example, 10.00 mL of the injection is diluted to 250.0 mL with water, and the absorbance is measured to 0.766. A reference solution is prepared by dissolving 0.2016 g (201.6 mg) of doxapram hydrochloride CRS, with a reported content of 99.8%, in 250.0 mL of water, and the absorbance of this reference solution is measured to 0.753. The content of doxapram hydrochloride per millilitre is calculated as shown below.

The exact concentration of doxapram hydrochloride in the reference solution is calculated taking the reported purity into consideration:

$$0.998 \times \frac{201.6 \text{ mg}}{250.0 \text{ mL}} = 0.805 \text{ mg/mL}$$

This means that 0.805 mg/mL results in an absorbance of 0.753. The absorbance for the diluted injection was measured to 0.766. The concentration of doxapram hydrochloride in the diluted injection, which is slightly higher than in the reference solution, is calculated as follows:

$$\frac{0.766}{0.753} \times 0.805 \text{ mg/mL} = 0.819 \text{ mg/mL}$$

The concentration of doxapram hydrochloride in the original injection is calculated taking the dilution into consideration (from 10.00 to 250.0 mL):

$$\frac{250.0 \text{ mL}}{10.00 \text{ mL}} \times 0.819 \text{ mg/mL} = 20.5 \text{ mg/mL}$$

The deviation from the specified content is equivalent to

$$\frac{20.5 \text{ mg/mL} - 20.0 \text{ mg/mL}}{20.0 \text{ mg/mL}} \times 100\% = +2.5\%$$

This content is within $\pm 5.0\%$ tolerance for the API and the result from the assay is acceptable.

19.4.3 Assays Based on Titration

LC and UV spectrophotometry are by far the most popular techniques for assays. However, titrations are used to some extent. The principles of titration are discussed in Chapter 5. Both acid–base titrations and redox titrations are in use. Two examples on the use of titration assays are discussed in Boxes 19.17 and 19.18.

Box 19.17   Assay of Fe^{2+} in Ferrous Fumarate Tablets (Solid Preparation) by Titration

CHEMICAL PROPERTIES OF THE API

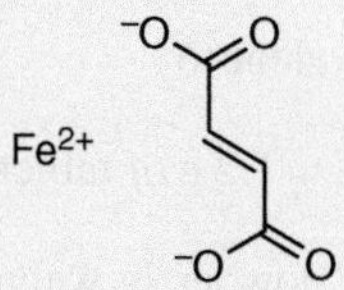

Molar mass: 169.9 g/mol

Procedure (Adopted from BP)

Weigh and powder 20 tablets. Dissolve a quantity of the powder containing 0.3 g of ferrous fumarate in 7.5 mL of 1 M sulfuric acid with gentle heating. Cool, add 25 mL of water, and titrate immediately with 0.1 M ammonium cerium (IV) sulfate VS using ferroin solution as indicator. Each 1.0 mL of 0.1 M ammonium cerium (IV) sulfate VS is equivalent to 5.585 mg of Fe (II).

Discussion

Twenty tablets are collected to have a representative sample and the tablets are powdered to make the API accessible to the assay. Only a minor fraction is required for titration, namely powder corresponding to 0.3 g of ferrous fumarate. This amount of powder is dissolved in sulfuric acid and heating is required to promote dissolution. Water is added to increase the volume of the sample solution and titration is performed immediately in order to avoid instantaneous oxidation of Fe^{2+} (to Fe^{3+}) by air. The titration is a redox titration according to the following reaction:

$$Fe^{2+} + Ce^{4+} \rightarrow Fe^{3+} + Ce^{3+}$$

Ferroin is a colour-indicator that detects the endpoint for the titration, the point where the total amount of Fe^{2+} from the sample has reacted with Ce^{4+}. The consumption of Ce^{4+} is measured and used to calculate the exact content of Fe^{2+}. The term VS behind ammonium cerium (IV) sulfate stands for volumetric solution and indicates that this is a solution for titration with an exactly known concentration.

Calculations

Tablets containing 50 mg (0.050 g) of Fe^{2+} are to be analysed. As an example, 20 tablets are collected and the total mass of tablets is 12.3673 g. The average mass per tablet is

$$\frac{12.3673 \text{ g}}{20 \text{ tablets}} = 0.6184 \text{ g/tablet}$$

Tablet powder corresponding to 0.3 g of ferrous fumarate (molar mass 169.9 g/mol) is equivalent to the following amount of Fe^{2+} (molar mass 55.85 g/mol):

$$0.3 \text{ g} \times \frac{55.85 \text{ g/mol}}{169.9 \text{ g/mol}} = 0.1 \text{ g Fe}$$

This corresponds to two tablets, equivalent to

$$2 \times 0.6184 \text{ g} = 1.2 \text{ g of tablet powder}$$

As an example, 1.2109 g of tablet powder is weighed on an analytical balance for the assay and titrated with 17.62 mL of 0.1006 M ammonium cerium (IV) sulfate. The content of Fe^{2+} per tablet is calculated as shown below.

The consumption of ammonium cerium (IV) sulfate is calculated from the titration volume and the molarity:

$$\frac{17.62 \text{ mL}}{1000 \text{ mL/L}} \times 0.1006 \text{ M} = 1.773 \times 10^{-3} \text{ mol}$$

This number of moles was equivalent to the number of moles of Fe^{2+} in 1.2109 g of tablet powder, which corresponded to the following mass found by multiplication with the molar mass of Fe^{2+} (55.85 g/mol):

$$1.773 \times 10^{-3} \text{ mol} \times 55.85 \text{ g/mol} = 0.09899 \text{ g} = 98.99 \text{ mg}$$

This amount of Fe^{2+} was present in 1.2109 g of tablet powder. The average mass of a tablet is

$$\frac{12.3673 \text{ g}}{20 \text{ tablets}} = 0.6184 \text{ g/tablet}$$

Therefore, the content of Fe^{2+} in a tablet of average mass is

$$\frac{98.99 \text{ mg} \times 0.6184 \text{ g/tablet}}{1.2109 \text{ g}} = 50.55 \text{ mg/tablet}$$

The deviation from the specified content is equivalent to

$$\frac{50.55 \text{ mg} - 50.0 \text{ mg}}{50.0 \text{ mg}} \times 100\% = +1.1\%$$

This content is within $\pm 5.0\%$ tolerance for the API.

Box 19.18   Assay of Diphenhydramine in Diphenhydramine Hydrochloride Oral Solution (Liquid Preparation) by Titration

CHEMICAL PROPERTIES OF THE API'S FREE BASE

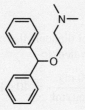

Molar mass: 255.3 g/mol
Acid/base: basic
pK_a: 8.9
log P: 3.7

Procedure (Adopted from BP)

Acidify a quantity containing 0.1 g of diphenhydramine hydrochloride with 2 M hydrochloric acid, shake with three 20-mL quantities of ether, discard the ether, make the aqueous solution alkaline with 5 M sodium hydroxide and extract with successive 15-mL quantities of ether until the extraction is complete. Wash the combined ether extracts with two 5-mL quantities of water, extract the combined washings with 15 mL of ether and evaporate the combined ether extracts to dryness. Dissolve the residue in 15 mL of 0.05 M sulfuric acid VS and titrate the excess of acid with 0.1 M sodium hydroxide VS using methyl red solution as the indicator. Each 1 mL of 0.05 M sulfuric acid VS is equivalent to 29.18 mg of $C_{17}H_{21}NO$, HCl.

Discussion

The oral solution (aqueous) is acidified to ensure high solubility of the basic API, followed by LLE with ether. In this extraction, important excipients are extracted into ether, whereas the API remains in the aqueous phase. Removal of excipients is important to avoid them interfering during the titration. Then the aqueous oral solution is made alkaline with sodium hydroxide to reduce the solubility of the API, and this is extracted into ether. The ether extracts are washed with water to remove further excipients, and the water washings are extracted with ether again to make sure that no diphenhydramine is lost during this quantitative procedure. Finally, the ether extracts are combined, and ether is evaporated. The residue, which is the API (diphenhydramine), is dissolved in an exact amount of sulfuric acid in excess. Sulfuric acid reacts with diphenhydramine by protonation:

$$2 \text{ Diphenhydramine} + H_2SO_4 \rightarrow 2 \text{ Diphenhydramine } H^+ + SO_4^{2-}$$

The excess amount of sulfuric acid, which has not reacted with the API, is then titrated with sodium hydroxide using methyl red as the indicator according to the following reaction:

$$H_2SO_4 + 2OH^- \rightarrow SO_4^{2-} + 2H_2O$$

Based on the difference between the total amount of sulfuric acid added and the excess titrated by sodium hydroxide, the amount of diphenhydramine can be calculated.

Calculations

Oral solution containing 2.5 mg/mL of diphenhydramine is to be analysed, where 0.1 g (100 mg) of diphenhydramine is equivalent to

$$\frac{100 \text{ mg}}{2.5 \text{ mg/mL}} = 40 \text{ mL}$$

As an example, 40.00 mL of oral solution is collected by transfer pipet and analysed according to the procedure. Note that oral solutions may be viscous, and therefore can be difficult to pipette accurately. In such cases, an exact amount of sample is weighed on analytical balance, and the exact volume is calculated from the mass and the density of the preparation. The residue is dissolved in 15.00 mL of 0.0493 M sulfuric acid, and excess sulfuric acid is titrated with 10.65 mL of 0.1026 M NaOH. The content of diphenhydramine is calculated as shown below.

The total amount of H_2SO_4 added to the residue is calculated from the volume and from the molarity:

$$\frac{15.00 \text{ mL}}{1000 \text{ mL/L}} \times 0.0493 \text{ M} = 7.40 \times 10^{-4} \text{ mol}$$

The amount of sodium hydroxide used during the final titration is calculated from the volume and from the molarity:

$$\frac{10.65 \text{ mL}}{1000 \text{ mL/L}} \times 0.1026 \text{ M} = 1.093 \times 10^{-3} \text{ mol}$$

This amount of sodium hydroxide reacted in the molar ratio 2 : 1 with excess sulfuric acid, and excess sulfuric acid is calculated as

$$\frac{1}{2} \times 1.093 \times 10^{-3} \text{ mol} = 5.464 \times 10^{-4} \text{ mol}$$

The number of moles of sulfuric acid reacted with diphenhydramine can be calculated as follows:

$$7.40 \times 10^{-4} \text{ mol} - 5.464 \times 10^{-4} \text{ mol} = 1.93 \times 10^{-4} \text{ mol}$$

Sulfuric acid reacted in the mol ratio 1 : 2 with diphenhydramine, and therefore the amount of diphenhydramine in the sample is calculated as follows:

$$2 \times 1.93 \times 10^{-4} \, \text{mol} = 3.86 \times 10^{-4} \, \text{mol}$$

This corresponded to the following mass of diphenhydramine found by multiplication with the molar mass of diphenhydramine (255.3 g/mol):

$$3.86 \times 10^{-4} \, \text{mol} \times 255.3 \, \text{g/mol} = 0.0986 \, \text{g} = 98.6 \, \text{mg}$$

This mass was present in 40.00 mL of oral solution, and the concentration in the oral solution is calculated as follows:

$$\frac{98.6 \, \text{mg}}{40.00 \, \text{mL}} = 2.46 \, \text{mg/mL}$$

The deviation from the specified content is equivalent to

$$\frac{2.46 \, \text{mg/mL} - 2.50 \, \text{mg/mL}}{2.50 \, \text{mg/mL}} \times 100\% = -1.4\%$$

This content is within $\pm 5.0\%$ tolerance for the API.

19.5   Chemical Tests for Pharmaceutical Preparations

In addition to identification and assay of the API, several tests are performed to check the quality of pharmaceutical preparations. This is only a selection of some of the common tests and not a complete overview.

19.5.1   Test for Related Substances

For many pharmaceutical preparations, a test for *related substances* is performed as a part of the chemical quality control. This test is intended to check that the content of impurities structurally related to the API (related substances) is at a very low level. This is important because related substances can affect the efficacy of the pharmaceutical preparation. A test for related substances can be performed as a part of the release control for a production batch, or can be performed on products stored for different periods of time and under different conditions to establish the shelf life for a pharmaceutical preparation. The latter is termed *stability testing* and is typically performed during pharmaceutical development. Tests for related substances are mainly performed by LC. The pharmaceutical preparation is typically dissolved or dispersed in the mobile phase to dissolve the API and related substances, and this solution is analysed after filtration by LC to detect related substances with retention times different from the API. Typically, any peak for related substances is compared with a reference solution serving as the limit. If any peak for related substances is greater than the peak for the reference, the level of related substances is exceeding the limit. An example of a test for related substances is shown in Box 19.19 for paracetamol tablets.

Box 19.19   Related Substances in Paracetamol Tablets (Solid Preparation) by LC

CHEMICAL PROPERTIES OF THE API

Molar mass: 151.2 g/mol
Acid/base: acidic (weak)
pK_a: 9.5
log P: 0.87

PROCEDURE (ADOPTED FROM BP)

Carry out the method for *LC* using the following solutions. Prepare the solutions immediately before use and protect from light. For solution (1), disperse a quantity of powdered tablets containing 0.2 g of paracetamol in 8 mL of the mobile phase with the aid of ultrasound, add sufficient mobile phase to produce 10 mL, mix well, and filter. For solution (2), dilute 1 volume of solution (1) to 20 volumes with the mobile phase, and dilute 1 volume of this solution to 20 volumes with the mobile phase. Solution (3) contains 0.002% (*w/v*) each of *4-aminophenol* and *paracetamol* CRS in the mobile phase. Solution (4) contains 0.00002% (*w/v*) of *4′-chloroacetanilide* in the mobile phase.

The chromatographic procedure may be carried out using: (i) a stainless steel column (25 cm × 4.6 mm) packed with *octylsilyl silica gel for chromatography* (5 μm; Zorbax Rx C_8 is suitable), (ii) as the mobile phase with a flow rate of 1.5 mL/min, at a temperature of 35 °C, a mixture of 250 volumes of *methanol* containing 1.15 g of a 40% (*v/v*) solution of tetrabutylammonium hydroxide with 375 volumes of 0.05 M *disodium hydrogen orthophosphate* and 375 volumes of 0.05 M *sodium dihydrogen orthophosphate*, and (iii) a detection wavelength of 245 nm.

The test is not valid unless, in the chromatogram obtained for solution (3), the *resolution factor* between the two principal peaks is at least 4.0.

Inject solution (1) and allow the chromatography to proceed for 12 times the retention time of the principal peak. In the chromatogram obtained with solution (1) the area of any peak corresponding to 4-aminophenol is not greater than the area of the corresponding peak in solution (3) (0.1%), the area of any peak corresponding to 4′-chloroacetanilide is not greater than the area of the principal peak in solution (4) (10 ppm), and no other impurity is greater than the area of the principal peak obtained with solution (2) (0.25%).

DISCUSSION

In solution (1), the tablet powder is dispersed in the mobile phase. Paracetamol and possible related substances are dissolved in the mobile phase, whereas the majority of excipients are not dissolved. The latter are therefore removed by filtration. Paracetamol is present at the 20 mg/mL level (2.0% (w/v)) in Solution (1). This is a very high concentration, but is required to be able to detect trace levels of related substances. Solution (1) is injected into the LC system and any peaks apart from the peak for paracetamol are observed (see the chromatogram below). Among the related substances, 4-aminophenol is of special interest. The retention time for 4-aminophenol is obtained from solution (3) (0.002% (w/v) of 4-aminophenol). If any peak is observed at this retention time for solution (1), 4-aminophenol is present in the tablets. In such a case, the area of the peak for 4-aminophenol in solution (1) should be less than the area for 4-aminophenol in solution (3). This corresponds to the following limit calculated relative to the amount of paracetamol in the tablets:

$$\frac{0.002\%(w/v)}{2.0\%(w/v)} \times 100\% = 0.1\%(w/w)$$

In addition, 4′-chloroacetanilide is of special interest. The retention time for this related substance is obtained from solution (4), and if 4′-chloroacetanilide is present in solution (1), the peak area should be less in solution (1) as compared to solution (4). This corresponds to the following limit calculated relative to the amount of paracetamol in the tablets:

$$\frac{0.00002\%(w/v)}{2.0\%(w/v)} \times 10^6 = 10\,\text{ppm}$$

Any other impurities should not be observed with peak areas exceeding the principal peak in solution (2). Solution (2) contains paracetamol at the following concentration level, taking the 1 : 20 dilution followed by 1 : 20 dilution from solution (1) into consideration:

$$2.0\%(w/v) \times \frac{1}{20} \times \frac{1}{20} = 0.0050\%(w/v)$$

Assuming that other impurities have the same specific absorbance at 245 nm as paracetamol, this corresponds to the following limit calculated relative to the amount of paracetamol in the tablets:

$$\frac{0.0050\%(w/v)}{2.0\%(w/v)} \times 100\% = 0.25\%(w/w)$$

A chromatogram for a test for related substances in paracetamol tablets are shown below.

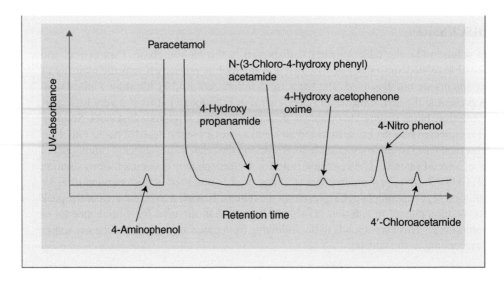

19.5.2 Uniformity of Content

Uniformity of content tests the variation of API from dosage unit to dosage unit. This test is typically used for pharmaceutical preparations containing 2 mg or less of API per dosage unit, or in other cases where dose variation is critical. In this test, typically 10 individual dosage units are collected randomly, such as 10 individual tablets, and each dosage unit is analysed individually to determine the amount of API in each unit. The results for the 10 individual dosage units should be within a certain specified range in order for the product to comply with specifications and requirements. The quantitative method used for *Uniformity of content* is often the same method as used for assay, the only difference being that the assay provides an average value for the content. Box 19.20 shows an example of the test *Uniformity of content* for phenindione.

Box 19.20 Uniformity of Content for Phenindione Tablets (Solid Preparation) by UV Spectrophotometry

CHEMICAL PROPERTIES OF THE API

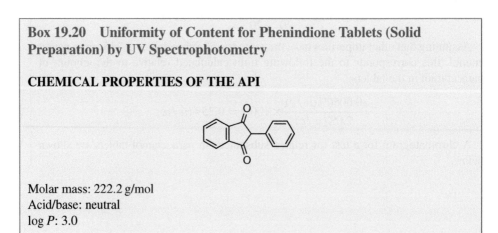

Molar mass: 222.2 g/mol
Acid/base: neutral
log *P*: 3.0

PROCEDURE (ADOPTED FROM BP)

Tablets containing 50 mg or less of Phenindione comply with the requirements stated under Tablets using the following method of analysis. Place one tablet in 50 mL of 0.1 M sodium hydroxide, dissolve the tablet completely by shaking gently, add a further 100 mL of 0.1 M sodium hydroxide, and shake for one hour. Dilute to 250 mL with 0.1 M sodium hydroxide, filter, and dilute a portion of the filtrate with sufficient 0.1 M sodium hydroxide to produce a solution containing 4 μg/mL of Phenindione. Measure the absorbance of the resulting solution at the maximum at 278 nm. Calculate the content of $C_{15}H_{10}O_2$ taking 1310 as the value of A (1%, 1 cm) at the maximum at 278 nm. Perform this procedure on 10 individual and randomly selected tablets.

Criteria

The preparation complies with the test if the individual content is between 85 and 115% of the average content. The preparation fails to comply with the test if more than one individual content are outside these limits or if one individual content is outside the limits of 75–125% of the average content.

DISCUSSION

This quantitative method is performed separately on 10 individual and randomly collected tablets, and the 10 results are checked according to the criteria. The quantitative analysis is by UV spectrophotometry.

CALCULATIONS

As an example, the following results (all in mg per tablet) are obtained for tablets specified to contain 10 mg phenindione (in increasing order of determined content):

10.1 mg	10.1 mg	10.3 mg	10.4 mg	10.4 mg
10.4 mg	10.8 mg	11.1 mg	11.2 mg	11.4 mg

The average content is calculated to be 10.62 mg per tablet; 85% and 115% of this value corresponds to

$$\frac{85\%}{100\%} \times 10.62 \text{ mg/tablet} = 9.03 \text{ mg/tablet}$$

$$\frac{115\%}{100\%} \times 10.62 \text{ mg/tablet} = 12.21 \text{ mg/tablet}$$

Clearly, all 10 tablets are within 85–115% of the average content, and the tested tablets comply with requirements.

19.5.3 Dissolution

Dissolution is a test intended for measuring the rate of release of API into solution from a pharmaceutical preparation. Dissolution testing is performed for tablet and capsule preparations. Information about the dissolution rate is important because it may have an important effect on the human absorption of the preparation. Basically, the dissolution test involves placing the tablet or capsule inside a stainless steel wire basket, which is rotated at a fixed speed while immersed in a dissolution medium. The steel wire basket and the dissolution medium are contained in a wide-mouthed cylindrical vessel. Samples of the dissolution medium are collected at specified times, filtered, and typically measured by UV spectrophotometry. In this way, the release of API to the dissolution medium can be measured versus time to check that the drug is released properly from the preparation.

20

Bioanalysis
Chemical Analysis of Pharmaceuticals
in Biological Fluids

20.1 Bioanalysis

Bioanalysis is a subdiscipline of pharmaceutical analysis, where drug substances and metabolites are measured in *biological fluids*. Typical biological fluids are *plasma*, *serum*, *whole blood*, *urine*, and *saliva*. The purpose of bioanalysis (bioanalytical methods) is either to quantitate drug substances in biological fluids or to screen for drug substances in biological fluids. Bioanalysis is mainly performed in the pharmaceutical industry

Introduction to Pharmaceutical Analytical Chemistry, Second Edition.
Stig Pedersen-Bjergaard, Bente Gammelgaard and Trine Grønhaug Halvorsen.
© 2019 John Wiley & Sons Ltd. Published 2019 by John Wiley & Sons Ltd.

(drug development), hospitals (therapeutic drug monitoring (TDM)), analytical services laboratories, forensic chemistry laboratories, and doping laboratories.

20.1.1 Drug Development

Drug discovery is the identification of drug candidates based on combinatorial chemistry, high throughput screening, and genomics. By *combinatorial chemistry* a large number of compounds are synthesized and these are tested for pharmacological activity and potency in high throughput screening systems, which simulate the interaction of the compounds with a specific biological receptor or target. Once a *lead compound* is found, a narrow range of similar compounds are synthesized and screened to improve the activity towards the specific target. Other studies investigate the *ADME* (absorption, distribution, metabolism, elimination) profile of drug candidates by analysing samples collected at different time points from tissue cultures (*in vitro* testing) and dosed laboratory animals (*in vivo* testing). In the *preclinical phase* (Figure 20.1), drug candidates are subjected to toxicity testing and further metabolism and pharmacological studies. Both *in vivo* and *in vitro* tests are conducted and various animal species are used to prove the pharmacokinetic profile of the candidate. Stability testing, research on dosage forms, and quality control are the following steps in the drug development.

The *clinical development* phase can begin when a regulatory body has evaluated the drug candidate to be efficient and safe in healthy volunteers. In *phase I* (Figure 20.1), the goal is to establish a safe and efficient dosage regimen and assess pharmacokinetics. Blood samples are collected and analysed from a small group of healthy volunteers (20–80). These data form the basis for developing controlled *phase II* studies with the purpose of demonstrating a positive benefit/risk balance in a larger group of patients (200–800), while monitoring efficacy and possible side effects. *Phase III* studies are large-scale efficacy studies with the focus on the effectiveness and safety of the drug candidate in a large group of patients. In most cases the drug candidate is compared with another drug used to treat the same condition. Phase III studies can last for two or three years or more and 3000–5000 patients can be involved. Carcinogenetic tests, toxicology tests, and metabolic studies in laboratory

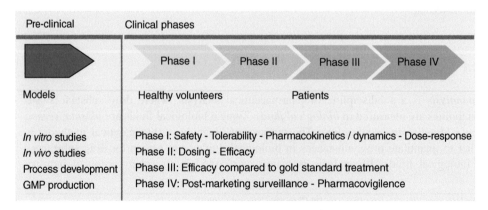

Figure 20.1 Schematic view of the drug development process

animals are conducted in parallel. *Phase IV* studies are studies conducted after a product launch to demonstrate long-term effects and examine possible drug–drug interactions. The time required from drugs discovery to product launch is up to 12 years.

The need for the pharmaceutical industry to develop new drugs faster and to reduce costs in the development process is a major driving force for inventions in new analytical instrumentation and new technologies. A combination of robotics, liquid handling workstations, and new formats for sample preparation now allows high-speed pharmaceutical analysis. New principles for chromatographic separation and new instrumentation, particularly in LC-MS, have greatly reduced time and costs in the identification of unknown substances and quantitation of known substances in biological fluids.

20.1.2   Therapeutic Drug Monitoring

TDM refers to the individualization of dosage by maintaining serum or plasma drug concentrations within a target range to optimize efficacy and to reduce the risk of adverse side effects. The target range of a drug is also termed *therapeutic range* or *therapeutic window*, and is the concentration range between the lowest drug concentration that has a positive effect and the concentration that gives more adverse effects than positive effects. Variability in the dose–response relationship between individual patients is due to pharmacokinetic and pharmacodynamic variability. *Pharmacodynamic variability* arises from variations in drug concentrations at the receptor and in variations in the drug–receptor interaction. *Pharmacokinetic variability* is due to variations in the relationship between the dose and plasma concentrations. Major sources of pharmacokinetic variability are age, physiology, disease, compliance, and genetic polymorphism of drug metabolism.

Most drugs are administered safely without monitoring plasma concentrations. However, in medication with drugs used prophylactically to maintain the absence of a condition (depressive or manic episodes, seizures, cardiac arrhythmias, organ rejection, asthma relapses), TDM may be advised to check that the patient is properly dosed. TDM is also used to avoid toxicity from drugs with a narrow therapeutic window. Examples are antiepileptic drugs, antidepressant drugs, digoxin, phenytoin, theophylline, cyclosporine, HIV protease inhibitors, and aminoglycoside antibiotics. The therapeutic ranges, in terms of surveillance of plasma concentrations of some typical drugs by TDM, are shown in Table 20.1.

Table 20.1   *Therapeutic range in plasma of common drugs subjected to therapeutic drug monitoring*

Drug	Therapeutic range	Drug	Therapeutic range
Amitriptyline	120–150 ng/mL	Nortriptyline	50–150 ng/mL
Carbamazepine	4–12 µg/mL	Phenobarbital	10–40 µg/mL
Desipramine	150–300 ng/mL	Phenytoin	10–20 µg/mL
Digoxine	0.8–2.0 ng/mL	Primidone	5–12 µg/mL
Disopyramide	2–5 µg/mL	Theophylline	10–20 µg/mL
Ethosuximide	40–100 µg/mL	Valproic acid	50–100 µg/mL

20.1.3   Forensic and Toxicological Analysis

Bioanalysis is also conducted in forensic chemistry and toxicology laboratories, as well as in doping laboratories. Biological samples are examined for the presence of drugs and drugs of abuse. In some cases, the identity of the analyte is known. Analysis of blood for alcohol or cannabinoids in connection with traffic incidents are examples of *targeted analysis*, where the identity of the analyte is known.

In most cases, the analytical work is based on screening of unknowns, which is the analytical process that takes place when no prior information about possible agents is available. The focus is on drugs and their metabolites, narcotics, and other substances that are toxicologically relevant. Serious cases that are of criminal relevance may include analysis of pharmaceuticals and addictive drugs that may impair human behaviour and detection of poisons and evaluation of their relevance in determining causes of death.

In the screening, the sample preparation is directed towards extraction of compounds with characteristic properties such as the selective extraction of basic drugs, neutral drugs, or acidic drugs. The extracts are analysed by LC with gradient elution or capillary GC with temperature programming. Due to the serious legal consequences of forensic cases, particular emphasis is placed on the quality and reliability of analytical results. Analyses by GC-MS or LC-MS are normally required to interpret analytical results for the court.

20.1.4   Doping Control Analysis

The World Anti-Doping Agency (WADA) was established in 1999 as an international agency to promote, coordinate, and monitor the fight against doping in sport. One of WADA's most significant achievements was the acceptance and implementation of the *World Anti-Doping Code*. The Code is the core document that provides the framework for anti-doping policies, rules, and regulations within sport organizations and among public authorities. The Code works in conjunction with five International Standards aimed at bringing harmonization among anti-doping organizations in various areas. The standards are:

- Prohibited list
- International Standard for Testing
- International Standard for Laboratories
- International Standard for Therapeutic Use Exemptions
- International Standard for the Protection of Privacy and Personal Information

The *prohibited list* is the standard that defines substances and methods prohibited to athletes at all times (both in competition and out of competition), substances prohibited in competition, and substances prohibited in particular sports. The prohibited list is updated regularly and falls into the categories shown in Box 20.1.

Box 20.1   Prohibited List in the World Anti-Doping Code

A. *Substances and methods prohibited at all times*:
   1. Anabolic agents:
      Anabolic androgenic steroids
      a. Exogenous anabolic androgenic steroids
      b. Endogenous anabolic androgenic steroids
   2. Other anabolic agents:
      S2 Peptide hormones, growth factors, and related substances
      S3 Beta-2 agonists
      S4 Hormone antagonists and modulators
      S5 Diuretics and other masking agents
   3. Prohibited methods:
      M1 Enhancement of oxygen transport
      M2 Chemical and physical manipulation
      M3 Gene doping
B. *Substances prohibited in competition*:
      S6 Stimulants
      S7 Narcotics
      S8 Cannabinoids
      S9 Glucocorticosteroids
C. *Substances prohibited in particular sports*:
      P1 Alcohol
      P2 Beta-blockers

The purpose of the international standards is to plan for effective in-competition and out-of-competition testing, and to maintain the integrity and identity of samples collected. They also ensure that the process of granting an athlete therapeutic use exemptions is harmonized and that all relevant parties adhere to the same set of privacy protections.

Only laboratories accredited by WADA take part in the testing to ensure that laboratories produce valid test results and that uniform and harmonized results are reported from all accredited laboratories. In addition, the document specifies the criteria that must be fulfilled by anti-doping laboratories to achieve and maintain their WADA accreditation. To confirm the presence of a performance-enhancing drug or method according to the prohibited list, mass spectrometry coupled to either GC or LC are the methods of choice. LC-MS and GC-MS are considered acceptable for both initial screening procedures and for confirmation procedures for a specific analyte. Updated information on prohibited substances and accredited laboratories can be found on the WADA web-page.

20.2 Biological Fluids

Blood plasma, serum, and urine samples are the preferred biological fluids for bioanalysis. Plasma and serum samples are preferred for exact measurement of drug exposure levels in humans and in test animals. Therefore, plasma and serum samples are used for TDM and for pharmacokinetic studies. Urine is the preferred matrix for screening of prohibited substances in doping analysis. An advantage of urine sampling is that samples are collected without intervention.

The composition of blood is shown in Figure 20.2. Blood is composed of *blood cells* suspended in a liquid termed *plasma*. The blood cells are *erythrocytes* (red blood cells), *leucocytes* (white blood cells), and *thrombocytes* (platelets). By volume, the red blood cells constitute about 46% of the blood volume for males and about 41% for females, plasma about 54–59%, and the white blood cells about 0.7%. The portion of blood volume occupied by red blood cells is measured as the *haematocrit value*. Plasma is an aqueous solution containing about 8% of proteins and trace amounts of other substances such as dissolved nutrients and waste products, electrolytes, lipoproteins, immunoglobulins, and blood clotting factors. Blood contains buffer components to maintain pH within a narrow range of 7.35–7.45.

Most drugs are bound to plasma proteins and exist in two forms, unbound or bound. Common proteins that bind drugs are serum albumin and glycoprotein. The protein binding is reversible and equilibrium exists between the bound and the unbound fraction:

$$\text{PROTEIN} + \text{DRUG} \rightleftharpoons \text{DRUG–PROTEIN} \qquad (20.1)$$

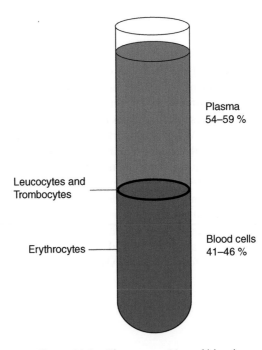

Plasma
54–59 %

Leucocytes and
Trombocytes

Erythrocytes

Blood cells
41–46 %

Figure 20.2   The composition of blood

Table 20.2 *Protein binding of common drugs*

Drug	Protein binding (%)	Drug	Protein binding (%)
Digoxin	20	Phenytoin	90
Captopril	30	Imipramine	93
Benzylpenicillin	50	Ibuprofen	98
Salicylic acid	80	Warfarin	99

The pharmacologic effect is exhibited by the *free fraction of drug*. The bound fraction acts as a reservoir or depot, from which the drug is slowly released to maintain equilibrium. Acidic and neutral drugs are primarily bound to *albumin*, while basic drugs are bound to acid *α-1 glycoprotein*. Table 20.2 shows the degree of protein binding of some common drugs. The protein binding is highly dependent on the physicochemical properties of the drug. Lipophilic drugs are more protein bound than hydrophilic drugs. The protein binding may also vary between persons, it may vary with age, and it is affected by disease and by other drugs that bind to the same protein. Sample preparation by protein precipitation (PP), liquid–liquid extraction (LLE), and solid-phase extraction (SPE) normally interrupt the drug–protein binding. Bioanalytical methods therefore determine the total concentration of drug in serum or plasma, and the result is unaffected by the protein binding.

When handling whole blood samples, care must be taken to prevent haemolysis. Haemolysis is the breakdown of red blood cells, which releases haemoglobin and other cell constituents into plasma, which changes colour from straw yellow to red. For this reason, blood cells should be removed as soon as possible to collect plasma or serum. Plasma or serum is preferred for analysis as these fluids are easier to handle than whole blood samples, and it is the drug concentration in serum/plasma that is clinically relevant. Therefore plasma is often the preferred matrix for pharmacokinetic studies and TDM, and therapeutic ranges are given as the serum/plasma concentration. To obtain *plasma*, the blood sample is collected in a clean tube containing an anticoagulant and plasma is collected after centrifugation (Figure 20.2). To obtain *serum*, the blood sample is stored until it is coagulated and serum is isolated from the coagulated blood. Serum corresponds to blood plasma with the fibrinogens removed.

The test tubes mostly used for collection of plasma or serum samples are *vacutainer tubes*. These are evacuated test tubes with a rubber cap. The vacutainer tubes are designed for *venipuncture*, and the blood is forced into the tubes by vacuum. The tubes may contain additional substances that preserve the blood for processing in the laboratory. *Heparin* is frequently added as an anticoagulant and is used for processing of plasma. Other anticoagulants are *potassium oxalate* and *EDTA* (ethylenediamine-tetraacetic acid). A *clot activator* or a gel may also be added to vacutainer tubes for rapid serum separation.

Urine is an aqueous solution of by-products of the body secreted by the kidneys. More than 95% of urine is water and the rest is a complex mixture of inorganic salts and organic compounds. Urine from healthy individuals is sterile. Some major constituents are *urea* 9.3 g/L, *chloride* 1.9 g/L, *sodium* 1.2 g/L, *potassium* 0.75 g/L, and *creatinine* 0.67 g/L. In addition to these substances, urine contains a range of substances that vary significantly, depending among others on the intake of food. The composition of urine therefore varies between different persons. The pH of urine is normally close to 7 but can vary between 4 and 8. Urinary pH effects the excretion of pharmaceuticals. Excretion of alkaline drugs is greatly reduced in alkaline urine because the solubility is reduced and pH values are normally checked in urine drug testing. Urine concentrations are highly dependent on the daily water intake and vary during the day. Thus, urine drug concentrations are measured simultaneously with the creatinine concentration. Creatinine is excreted at a relatively constant daily rate, and by reporting the result as the drug to creatinine concentration ratio, the samples are corrected for hydration.

A variety of drugs are metabolized by conjugation with glucuronic acid or sulfate and are excreted into urine as *glucuronides* and *sulfate conjugates*. These have a much higher water solubility than the parent drugs. In urine drug-testing, conjugates can be hydrolysed to release the parent drug and increase their concentration in the sample. For this purpose acid–base hydrolysis or enzymes that cleave the glycoside bond of a glucuronide, such as *β-glucuronidase*, are frequently used.

Normally, the concentrations of drugs and metabolites found in urine are higher than those found in plasma or serum. Urine is therefore a preferred matrix for screening of doping agents, illicit drugs, and poisons. Occasionally, *saliva* samples are used for bioanalysis. Saliva comprises 99.5% water, electrolytes, mucus, white blood cells, epithelial cells, glycoproteins, enzymes, and antimicrobial agents. Saliva can be used for screening purposes and is very convenient to use in terms of sampling.

20.3   Bioanalytical Methods – An Overview

Biological fluids are very complex and target analytes are often present at low concentrations. Therefore, LC-MS and GC-MS are the preferred techniques. Sample preparation is usually required prior to analysis to extract and isolate the analytes from the bulk sample and to make the sample injectable in the separation system. The various steps of a bioanalytical method are shown in Figure 20.3 and include sampling, sample preparation, separation (LC or GC), detection (MS), identification, and quantitation.

20.4   Sampling

Blood samples are preferably collected at fixed time points using vacutainer tubes and plasma or serum are separated as soon as possible. There are only minor differences between serum and plasma, and methods developed for plasma are usually applicable also for serum. Whole blood may be the sample of choice in forensic chemistry when processing of serum or plasma is not possible. Hemolysed whole blood samples are more complex and require

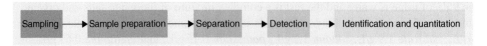

Figure 20.3   *The steps of a bioanalytical method*

extensive sample preparation. The blood volumes collected should be kept as small as possible and range from a few microlitres from laboratory animals and up to 5 mL from humans. When immediate analysis is impossible, serum and plasma samples are stored in a freezer at −18 or −80 °C prior to analysis. The stability of the analyte during storage and in the following freeze–thaw process must be documented. The collection and handling of blood and other biological fluids require great care because of the possible presence of transferable disease constituents. Gloves and protective eyeglasses must be worn at all times.

Urine is collected in clean sample cups and urine samples can be stored in a refrigerator (+5 °C) for a short period of time. Urine from healthy persons is sterile and is relatively free from protein. When properly handled and stored, the risk of bacterial degradation is greatly reduced. The time between sampling and analysis should be kept as short as possible to guarantee stability. If necessary, preservatives may be added to stabilize the analytes. Urine may contain compounds that can be irritating to eyes and skin. Gloves and protective eyeglasses are recommended when handling urine samples in the laboratory.

20.5 Sample Preparation

Sample preparation is crucial for successful bioanalysis. The major objectives of sample preparation are:

- Removal of major matrix components
- Enrichment of analytes to improve detection limits
- Transfer of analytes to a solution suitable for injection in LC-MS or GC-MS

For analysis of drugs in biological fluids and tissues, the analytes are separated from proteins in the sample matrix prior to injection. This is to avoid proteins precipitating inside the LC-MS or GC-MS instrument. It may also be desirable to isolate the analyte from other matrix components that may interfere with the analysis. One such example is *phospholipids*, which may cause *ion suppression* in the mass spectrometer.

Drug concentrations in biological fluids may be low and often it is necessary to pre-concentrate the drug substance prior to analysis. The analyte is then extracted and reconstituted in a small volume of solvent prior to injection into LC-MS or GC-MS. If a drug substance is extracted with 100% recovery from a 1.0 mL sample and reconstituted in 0.1 mL of solvent, the detection limit has been improved by a factor of 10.

The injected solution should be compatible with the LC-MS or GC-MS system, and exchange of solvent from the aqueous biological fluid to a solvent more suitable for injection is an important part of the sample preparation. A method that combines solvent exchange and analyte concentration is to evaporate a given volume of sample extract to dryness, followed by reconstitution in a smaller volume of solvent. For LC-MS analysis, the final solvent is typically the mobile phase, and for GC-MS analysis an organic solvent is used.

20.5.1 Protein Precipitation

Protein precipitation (PP) is often used for sample preparation in bioanalytical methods when samples contain significant amounts of proteins. This is the case with plasma and

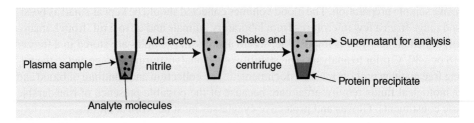

Figure 20.4 *Principle of protein precipitation*

serum samples. These samples contain about 8% (*w/w*) of protein. Urine samples do not contain large amounts of proteins, and they are not prepared by protein precipitation. The purpose of protein precipitation is to get rid of the proteins.

The principle of protein precipitation is illustrated in Figure 20.4. Common protein precipitation procedures are based on denaturation by water-soluble organic solvents such as acetonitrile or by acids. An excess of *precipitating reagent* is added to plasma or serum, the proteins are precipitated, and the supernatant is collected after centrifugation. When the concentration of analyte in the supernatant is above the quantitation limit and separation and detection is performed without interferences, no further concentration and clean-up steps are necessary, and the supernatant can be injected into an LC-MS system.

The typical protein precipitation procedure involves pipetting a certain volume of plasma or serum into a small centrifuge tube, typically 100–1000 µL, and mix the sample with acetonitrile. Acetonitrile is miscible with plasma and serum, and the presence of acetonitrile reduces the solubility of the proteins, resulting in precipitation. Acetonitrile acts as a *precipitant*. The mixture is shaken vigorously to facilitate the precipitation, and subsequently the mixture is centrifuged. After centrifugation, the clear supernatant is collected for direct injection in the LC-MS.

Protein precipitation can also be performed using other organic solvents, as illustrated in Table 20.3, among others being acetone and methanol. However, acetone and methanol are less efficient precipitants than acetonitrile. From Table 20.3 it is evident that the amount of precipitant plays an important role, as the protein removal increases with increasing volume

Table 20.3 *Efficiency of different approaches to protein precipitation*

Precipitant	Protein precipitation efficiency (%) per mL of plasma		
	0.2 mL precipitant	1.0 mL precipitant	2.0 mL precipitant
Acetonitrile	13.4	97.2	99.7
Acetone	1.5	96.2	99.4
Methanol	17.6	73.4	98.7
CCl_3COOH (10%)	99.7	99.5	99.8
$HClO_4$ (6%)	35.4	99.1	99.1
$CuSO_4$—Na_2WO_4	36.5	97.5	99.9
$ZnSO_4$—NaOH	41.1	94.2	99.3

of precipitant. Acids, like trichloroacetic acid (TCA) or perchloric acid, and metal ions can also be used as precipitants. This enables protein precipitation procedures to be carefully optimized.

Water-miscible organic solvents such as acetonitrile or methanol are the most popular precipitating reagents, due to their compatibility with common LC-MS mobile phases. The mild conditions are used to minimize the possibility for decomposition of labile drug substances, and normally the analyte is recovered close to 100% in the supernatant because most drugs are very soluble in the organic solvent–water mixture. A solvent–plasma ratio of $2:1$ has to be used to precipitate more than 98% of the proteins. The supernatant is often evaporated, and the sample is reconstituted in the LC-MS mobile phase prior to analysis.

As shown in Table 20.3, a 10% (*w/v*) solution of TCA is a highly efficient precipitation reagent and 0.2 mL of this precipitates more than 99% of the proteins in 1 mL of plasma. The supernatant is highly acidic in this case. Acidic drugs are poorly soluble in the acidic solvents and recovery in the supernatant can be low. For basic drugs, on the other hand, acidic conditions in the supernatant are in favour for high recovery. In some cases, drugs can partly precipitate with the proteins, but this reduces recovery.

As only proteins are removed by protein precipitation, the low molecular components of plasma or serum are still present in the supernatant. Nevertheless, protein precipitation is the preferred sample preparation technique, because of simplicity and because the technique is easily automated in the 96-well format. An example of a protein precipitation procedure is presented in Box 20.2.

Extraction methods such as LLE or SPE are preferred when sample clean-up, analyte enrichment, or solvent exchange is required.

20.5.2 Liquid–Liquid Extraction

The fundamentals of LLE are described in Chapter 16. In LLE, drug substances in their neutral form are extracted from the biological fluid and into a water immiscible organic solvent (organic phase). Subsequently, the organic phase is collected, the solvent is evaporated, and the residue is dissolved in a liquid compatible with LC-MS or GC-MS. For extraction of basic drugs, the biological fluid is made alkaline to neutralize the drug, and similarly for extraction of acidic drugs the biological fluid is acidified. When selecting a suitable solvent, volatility, polarity, selectivity, solubility, and density must be considered. Solvents that are harmful to health and to the environment should not be used. Evaporation of solvent is facilitated by a solvent with high volatility (low boiling point). Solvent polarity is important for the solubility of the analyte and for partitioning into the organic phase. Solvents with densities lower than water stay on top of the two-phase extraction system and are easily collected with a pipette.

Normally a single extraction is performed and the volume of solvent is adjusted to achieve satisfactory recovery. With proper selection of organic solvent and adjustment of pH, clean extracts are obtained with LLE. Inorganic salts, proteins, and hydrophilic matrix components remain in the aqueous biological fluid. LLE methods are more labour-intensive than protein precipitation. In spite of this, LLE is widely used due to the efficient clean-up, and because preconcentration can be achieved. An example of an LLE procedure is presented in Box 20.3.

Box 20.2   Protein Precipitation Procedure for Fluoxetine in Plasma

CHEMICAL PROPERTIES OF THE DRUG SUBSTANCE

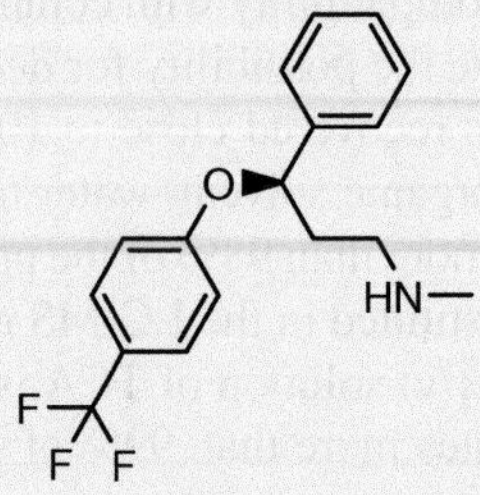

Molar mass:	309.33 g/mol
Acid/base:	base
pK_a:	9.80
$\log P$:	4.17
Therapeutic range:	160–500 ng/mL

PROCEDURE

400 μL of acetonitrile is pipetted to a centrifuge tube, followed by 100 μL of plasma sample and 20 μL of internal standard solution (100 ng/mL of fluoxetine-D6 (deuterated fluoxetine)). The centrifuge tube is vortexed and centrifuged for six minutes at 15 000 rotations per minute (rpm). The supernatant is analysed by LC-MS/MS.

DISCUSSION

The precipitation reagent, acetonitrile, is mixed with plasma approximately in the ratio 4 : 1. The high excess of acetonitrile provides very efficient removal of the proteins. The plasma volume is 100 μL and the internal standard is added to the sample to correct for variations in recovery and to improve the precision and accuracy. Thus, fluoxetine is measured relative to fluoxetine-D6 added to the sample. Vortexing ensures proper mixing. The precipitated proteins are centrifuged to the bottom of the tube and the clear supernatant is injected into the LC-MS/MS system. This procedure removes most of the proteins in the plasma sample. However, the low molar mass constituents of plasma are all present in the supernatant and sample clean-up is limited to protein removal.

Box 20.3   Liquid–Liquid Extraction Procedure for Amitriptyline in Serum

CHEMICAL PROPERTIES OF THE DRUG SUBSTANCE

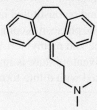

Molar mass:	277.41 g/mol
Acid/base:	base
pK_a:	9.76
log *P*:	4.81
Therapeutic range:	50–300 ng/mL

PROCEDURE

One millilitre of serum is pipetted into a centrifuge tube, followed by 50 μL of an internal standard solution (containing amitriptyline-d3), 500 μL of 0.75 M sodium bicarbonate/carbonate buffer (pH 10), and 5 mL of methyl tert-butyl ether – *n*-hexane 75 : 25 (*v/v*). The mixture is shaken for 30 seconds on a vortex mixer and then centrifuged at 4000 rpm for 10 minutes to separate the aqueous and the organic phases. The organic phase is transferred to a clean test-tube and evaporated to dryness under a stream of nitrogen at 40 °C. The residue is reconstituted in 100 μL of the mobile phase prior to analysis by LC-MS/MS.

DISCUSSION

The internal standard is deuterated amitriptyline, and this is added to lower measurement uncertainty. Amitriptyline is a hydrophobic base with a pK_a value of 9.76, and carbonate buffer pH 10 is added to neutralize the drug and to enable the uncharged drug to be extracted into the organic phase. The addition of *n*-hexane increases the hydrophobicity of the solvent mixture. Only neutral hydrophobic substances (including the analyte and the internal standard) are extracted into the organic phase. The polar and acidic constituents of serum remain in the biological fluid. With an excess of the organic phase

(5 mL), amitriptyline is extracted completely from serum. Mixing the liquids is facilitated by the vortex mixer and 30 seconds of mixing is sufficient. High-speed centrifugation for 10 minutes is necessary to separate the biological fluid and the organic phase into two separate layers. The solvent mixture has a low boiling point and is easily evaporated at 40 °C. Ten times enrichment is obtained because the analyte is extracted with 100% recovery from a 1 mL sample and reconstituted in 100 μL of mobile phase. When the extract is reconstituted in the mobile phase, it is compatible with the LC-MS/MS system. The original mixture of *n*-hexane and methyl tert-butyl ether is not well suited for injection in LC-MS, because this solvent mixture is immiscible with the mobile phase used in LC-MS/MS (acetonitrile mixed with dilute formic acid).

20.5.3   Solid-Phase Extraction

SPE is a more popular alternative than LLE. The fundamentals of SPE are discussed in Chapter 16. SPE is based on partition of the analyte between the aqueous biological fluid and a solid phase (sorbent). The amount of sorbent in the SPE column determines extraction capacity, and 1 mL SPE columns packed with 30–100 mg of sorbent are normally used for extraction of up to 1 mL of plasma or serum. The SPE columns are intended for single use and are discarded after extraction. Reversed-phase, ion exchange, and mixed mode sorbents are used for extraction of drug substances from biological fluids. Conditioning is performed by methanol, followed by washing with water or an aqueous buffer. Biological fluids are often diluted with an aqueous buffer prior to loading. Dilution decreases sample viscosity and the solution is sucked more efficiently through the sorbent bed. Sample pH can be adjusted to increase analyte retention. Matrix components not retained by chemical interactions remain in the aqueous phase and are diverted to waste. In reversed-phase SPE, hydrophilic matrix components and most of the proteins go to waste. The wash solvent removes contaminating matrix components while the analytes are retained in the sorbent bed. Analyte elution is normally carried out with methanol to suppress hydrophobic interactions. Pure methanol can often be used, but modification with acid or base, or use of other solvents can be required for substances with poor solubility in methanol, or for substances retained by secondary interactions. Solvent evaporation and sample reconstitution is often performed prior to injection into LC-MS, because neat methanol hampers the chromatography in reversed-phase LC-MS. SPE provides efficient sample clean-up from biological fluids, enables preconcentration, and can be automated in a 96-well configuration for high throughput applications. An example of an SPE procedure is presented in Box 20.4.

20.6   Separation and Detection

LC-MS is the standard instrumental technique used for bioanalysis of pharmaceuticals. UHPLC-MS methods are used increasingly because of high speed, reduced solvent consumption, and lower detection limits. Reversed phase separations utilizing C18 columns dominate. GC-MS is used for the most volatile drug substances. In GC-MS analysis, sample preparation procedures include derivatization to increase the volatility of target analytes, to improve the chromatographic separation, and to lower detection limits.

Box 20.4   Solid-Phase Extraction Procedure for Amitriptyline in Plasma

CHEMICAL PROPERTIES OF THE DRUG SUBSTANCE

See Box 20.2.

PROCEDURE

The SPE column is a 1-mL column packed with 100 mg of C18 sorbent. Vacuum-mediated elution is used and the flow rate is adjusted to 1 mL/min. The SPE columns are first conditioned with 2×1 mL of methanol followed by 1 mL of water. After conditioning, 0.5 mL of plasma is added to 100 μL of internal standard (amitriptyline-d3) solution and 0.5 mL of water, and is then applied to the SPE column. The SPE column is washed with 2×1 mL of water and with 2×1 mL of 25% methanol in water. The analyte is finally eluted with 0.5 mL of methanol containing 1% (*v/v*) perchloric acid. An aliquot of the eluate is injected into the LC-MS/MS system after 1 : 1 dilution with water.

DISCUSSION

The non-polar SPE sorbent is first conditioned with methanol. With an environment of methanol surrounding the C18 groups the sorbent becomes more open and available for interaction with the analyte. Excess methanol held in the bed volume of the SPE column is undesirable and is washed out with water. Amitriptyline, which is a hydrophobic base, is retained by hydrophobic interactions with the C18 chains. It can also bind to residual silanol groups by secondary interactions. At neutral pH, residual silanols are negatively charged, and positively charged compounds can be retained by ionic interaction. After the sample solution is drawn through the column by vacuum, the column is first washed twice with water to remove proteins and polar matrix components from plasma. Washing with 25% methanol in water removes compounds that are more polar and less retained than amitryptyline. Amitryptyline is eluted with methanol containing 1% perchloric acid that effectively suppresses primary and secondary interactions. Perchloric acid neutralizes residual silanol groups and suppresses ionic interactions, and methanol suppresses hydrophobic interactions. Since amtritriptyline is transferred from 1 mL plasma into 0.5 mL eluate and then diluted 1 : 1 with water, the analyte is not preconcentrated.

Mass spectrometry is used for detection, and in most cases the mass spectrometers are operated in selective reaction monitoring (SRM) mode. Mass spectrometry provides highly specific detection based on *m/z* values and provides mass spectral information which, when combined with retention data from the chromatographic separation, is used to identify unknown substances. Recording the intensities of masses that are characteristic and specific for target analytes is the basis of quantitation.

20.7   Quantitation

Bioanalysis is most often performed using internal standards and calibration based on calibration samples, to correct for variability in analyte recovery, sample volume evaporation, and volume injected into the analytical instrument. The preferred internal standard is an isotopically labelled analogue of the analyte. Isotopically labelled drug substances, however, are expensive and not always available; therefore compounds with similar extraction recoveries and similar LC-MS or GC-MS characteristics may be used. Isotopically labelled analytes are typically deuterated, containing 3–5 deuterium atoms. Other stable isotopes can also be used such as ^{13}C and ^{18}O.

Quantitation is based on standard curves set up after analysis of calibration samples. Calibration samples are prepared from drug-free biological fluid. Thus, calibration samples for plasma analysis are prepared in drug-free plasma and calibration samples for urine analysis are prepared in drug-free urine. Such samples are also termed *matrix matched standards*. Calibration samples containing known concentrations of the analyte are prepared from stock solutions of the chemical reference substance (CRS). The concentration range should be the same as the expected concentration range of the target analyte. Normally stock solutions are prepared in an aqueous solution or in a polar organic solvent such as methanol or acetonitrile. A small volume of the stock solution is then pipetted into a fixed volume of the drug-free biological fluid to prepare calibration samples of known concentrations.

Internal standards should preferably be added to real samples and to calibration samples prior to the sample preparation. A constant volume of the internal standard solution is pipetted to all real samples and calibration samples, to keep the concentration of the internal standard constant. Calibration samples and real samples are treated equally during sample preparation. The instrumental signal measured for the analyte may fluctuate with small variations in recovery, sample volume, and LC-MS injection volume, but the ratio between the instrumental signals for the analyte and the internal standard will remain unaffected. Box 20.5 illustrates calibration and quantitation based on the use of the internal standard with LC-MS, and a similar method by GC-MS is shown in Box 20.6.

Box 20.5   Quantitative Determination of Amitriptyline in Serum by LC-MS

CHEMICAL PROPERTIES OF THE DRUG SUBSTANCE

See Box 20.3.
   The procedure is based on SPE combined with UHPLC-MS/MS. The details are given below.

PROCEDURE – SAMPLE PREPARATION

Plasma samples are first prepared by SPE as described in Box 20.3: 1 mL of sample is extracted and dissolve in 1 mL of acidic methanol. The sample is analysed by the UHPLC-MS/MS system described below.

PROCEDURE – SEPARATION AND DETECTION

Analysis is by UHPLC-MS/MS using the following procedure:

Column:	UHPLC column (100 mm × 2.1 mm) packed with C18 particles of 1.7 μm
Column temperature:	65 °C
Mobile phase:	5 mM ammonium formate, pH 5.0
	B: Acetonitrile
Flow rate:	0.5 mL/min
Injection volume:	3 μL
Ionization:	Electrospray, positive mode
Selected reaction:	Amitriptyline m/z 278.1 → 233.0
Monitoring:	Amitriptyline-d3 (internal standard) m/z 281.3 → 236.0
Calibration range:	25–400 ng/mL
Internal standard:	100 ng/mL
Analysis time:	8 min

DISCUSSION

Reversed phase LC separations of basic drugs are usually performed with an acidic mobile phase. Mobile phases must be volatile in order to be compatible with the mass spectrometer. Therefore a volatile ammonium formate buffer is used to adjust the pH of the mobile phase to 5.0. Detection is performed by triple quadrupole mass spectrometry operated in a selective reaction monitoring mode. Amitriptyline and the internal standard have nominal masses (M) of 277 and 280, respectively, and the $(M + H)^+$ ions from electrospray ionization pass the first mass analyser (Q1), which is locked to m/z 278.1 and 281.3. In the second quadrupole, amitriptyline and amitriptyline-d3 are fragmented, and the third mass analyser (Q3) is locked to the selected fragments m/z 233 and 236, respectively. In this way, chromatograms are recorded for the m/z 278.1 → 233.0 and 281.3 → 236.0 mass transitions that are unique for amitriptyline and amitriptyline-d3, respectively, and provide highly specific detection.

QUANTITATION – EXAMPLE

Calibration samples containing 25, 50, 100, 150, 200, and 400 ng/mL of amitriptyline and 100 ng/mL of amitriptyline-d3 (internal standard) were prepared in drug-free plasma. The calibration samples were analysed according to the procedure above and the peak areas of amitriptyline and the internal standard are shown in the table below.

Concentration of amitriptyline (ng/mL)	Peak area of amitriptyline (m/z 278.1 → 233.0)	Peak area of internal standard (m/z 281.3 → 236.0)	Peak area ratio
25	64 710	253 764	0.255
50	126 880	257 363	0.493
100	249 100	255 487	0.975
150	378 166	258 134	1.465
200	498 079	256 874	1.939
400	956 135	253 886	3.766

The same amount of internal standard was added to all calibration samples and real sample, and the calibration curve is a plot of peak area ratio versus concentration.

The regression line $y = 0.009\,37x + 0.0378$ (where y is the peak area ratio and x is the concentration of amitriptyline (in ng/mL) used for the quantitation of amitriptyline in the real samples. In the chromatogram of a real sample, the peak areas for amitriptyline ($m/z\ 278.1 \rightarrow 233.0$) and the internal standard ($m/z\ 281.3 \rightarrow 236.0$) were 354 048 and 255 380, respectively.

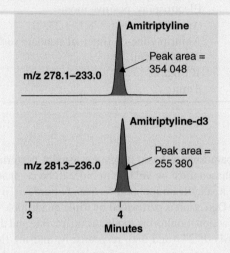

The peak area ratio for the real sample was

$$\frac{354\,048}{255\,380} = 1.386$$

The unknown concentration, x, was calculated from the regression line:

$$1.386 = 0.00937x + 0.0378$$

$$x = \frac{1.386 - 0.0378}{0.00937} = 144 \text{ ng/mL}$$

The concentration of amitriptyline in the sample was 144 ng/mL.

Box 20.6   Quantitative determination of valproic acid in serum by GC-MS

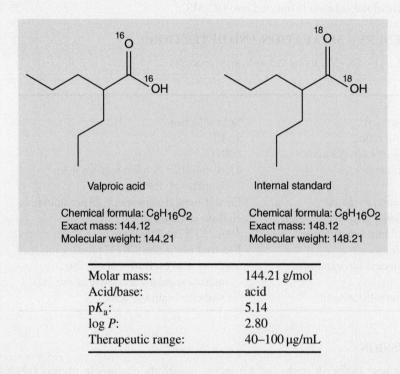

Valproic acid	Internal standard
Chemical formula: $C_8H_{16}O_2$	Chemical formula: $C_8H_{16}O_2$
Exact mass: 144.12	Exact mass: 148.12
Molecular weight: 144.21	Molecular weight: 148.21

Molar mass:	144.21 g/mol
Acid/base:	acid
pK_a:	5.14
log P:	2.80
Therapeutic range:	40–100 µg/mL

The procedure is based on LLE followed by derivatization and GC-MS. The details are given below.

PROCEDURE – SAMPLE PREPARATION

0.2 ml of plasma is added to 0.1 ml of a solution containing 100 µg of the internal standard. The mixture is added to 0.1 mL of 1 M HCl and 0.5 mL of hexane. After vortex mixing for one minute the phases are separated by centrifugation. Then 0.2 mL of the hexane phase is collected and derivatized with 0.05 mL of derivatizing reagent comprising 1% trimethylchlorosilane-pyridine 2:1 (*v/v*) dissolved in *N,O*-bis(trimethysilyl)-trifluoroacetamide (BSTFA) (heating for 30 minutes). One microlitre of the solution is injected into GC-MS.

PROCEDURE – SEPARATION AND DETECTION

Analysis is by GC-MS using the following procedure:

Injection:	Split injection
Split ratio:	1 : 10
Injector temperature:	260 °C
Column:	15-m capillary column with an inner diameter of 0.25 mm
Stationary phase:	Phenyl-methyl-siloxane, 0.25 μm thickness
Carrier gas:	Helium at 1.5 mL/min
Column temperature:	100–250 °C at 40 °C/min
Ionization:	Electron ionization, 70 eV, positive mode
Selected ion monitoring:	Valproic acid derivative at *m/z* 201, internal standard derivative at *m/z* 205
Internal standard:	^{18}O-labelled valproic acid

DISCUSSION

Valproic acid has a pK_a value of 4.8 and is negatively charged in plasma (pH 7.4). Addition of hydrochloric acid prior to LLE neutralizes the analyte and facilitates the extraction into hexane. After LLE, valproic acid is converted to the trimethylsilyl (TMS) derivative by the following reaction:

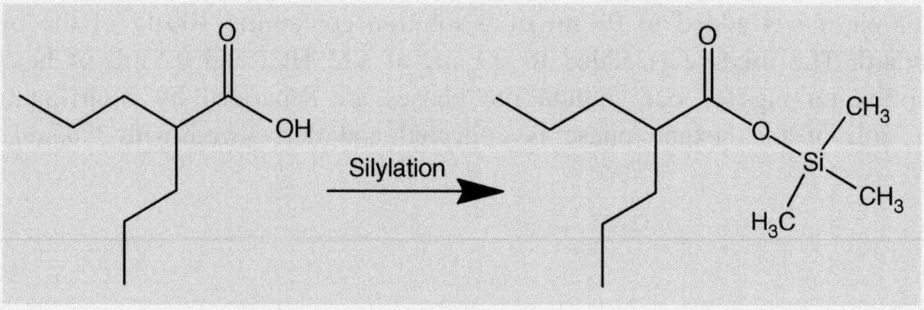

The derivatization increases analyte volatility. Samples are analysed by GC-MS with split injection, and about 10% (1 : 10) of the sample is introduced into the GC column. High temperature during injection ensures efficient transfer of the sample to the GC column. The GC column is a general non-polar capillary column with broad applicability. Because split injection is used, solvent focusing is not required, and the initial column temperature can be set to 100 °C, which is higher than the boiling point of the injection solvent (hexane, boiling point 68 °C). The temperature is increased rapidly (40 °C/min) to keep the analysis time short. The temperature program ends at 250 °C to make sure that matrix components less volatile than valproic acid are also eluted from the column, before the column is cooled for injection of a new sample. The mass spectrometer is operated in the selected ion monitoring mode, which is preferred for quantitation. Quantitation is based on the ratio of the signals at *m/z* 201 for valproic acid and at *m/z* 205 for the internal standard.

20.8   Screening

Screening is typically performed in cases related to drugs of abuse and doping in sport. Traditionally, urine has been the sample of choice for screening purposes. Urine sampling is non-invasive and large volumes are available. Drug substances are to a large degree found as metabolites in urine, and screening therefore includes both drug substances and their corresponding metabolites. Drug substances are often found as glucoronides or sulfates in urine, and therefore enzymatic hydrolysis using β-glucuronidase/arylsulfatase is often the first step in the procedure. Subsequently, extraction is usually performed to isolate alkaline substances, acidic substances, and neutral substances. Screening based on GC-MS is exemplified in Box 20.7.

Box 20.7   Screening for drugs of abuse in urine based on GC-MS

The procedure is based on (i) enzymatic hydrolysis, (ii) SPE, (iii) derivatization, and (iv) GC–MS. The mass spectrometer is operated in the full scan mode. Identification is based on retention time data and the full scan mass spectra.

PROCEDURE – SAMPLE PREPARATION

One millilitre of urine sample is mixed with 2 mL of 0.1 M acetate buffer, pH 4.8. The mixture is added to 25 μL of β-glucuronidase/arylsulfatase and enzymatically deconjugated for two hours at 50 °C, followed by cooling to room temperature. The SPE column is a 10-mL column packed with 130 mg of mixed mode sorbent with C_8 and SCX (strong cation exchanger). Vacuum-mediated elution is used and the flow rate is adjusted to 2 mL/min. The columns are conditioned with 3 mL of methanol followed by 3 mL of 0.25 M acetate buffer, pH 4.8. The hydrolysed urine sample is drawn through the column. The column is washed with 1 mL of water and 1 mL of 0.01 M hydrochloric acid. Before elution, the column is dried by vacuum. Elution at 1 mL/min is by (i) 1 mL of methanol and (ii) 1 mL of methanol/ammonia 98 : 2 (*v/v*). The two fractions are collected separately and are evaporated to dryness in a stream of nitrogen at 40 °C. The dry residues are derivatized with a mixture of *N,O*-bis(trimethysilyl)-trifluoroacetamide–trimethylchlorosilane (BSTFA-TMCS); 99 : 1 (*v/v*) at 70 °C for 30 minutes, and injected in GC-MS.

PROCEDURE – SEPARATION, DETECTION, AND QUANTITATION

GC-MS analysis:

Injection:	Splitless injection
Injection volume:	2 μL
Injector temperature:	240 °C
Column:	12-m capillary column with an inner diameter of 0.20 mm
Stationary phase:	Phenyl-methyl-siloxane, 0.25 μm thickness
Carrier gas:	Helium at 1.5 mL/min
Column temperature:	90 °C (1 min), 30 °C/min to 200 °C, 5 °C/min to 250 °C
Ionization:	Electron ionization, 70 eV, positive mode
Full scan mode:	m/z range 40–500

DISCUSSION

The first step is an enzymatic hydrolysis to convert glucuronide or sulfate metabolites to the original drug substance. The second step is a mixed mode SPE where the sorbent contains both hydrophobic (C8) and SCX functionalities, which makes it possible to retain urine constituents by hydrophobic interactions as well as by ionic interactions. The column is conditioned with methanol. The pH is adjusted to 4.8 with acetate buffer in the sample solution and in the conditioning step. A weakly acidic pH in the sample solution promotes the enzymatic hydrolysis, keeps basic compounds ionized, and reduces the ionization of acidic compounds. When the sample is loaded, basic compounds are primarily retained by ionic interactions to the negatively charged groups of SCX, while neutral and acidic compounds are primarily retained by hydrophobic interactions with C8 chains. Washing with water removes enzymes and salts. Washing with 0.01 M HCl regenerates the capacity of the cation exchanger, which may have lost capacity when the sample was loaded. The acidic pH maintains the ionization of basic substances. Traces of aqueous wash solutions are removed by vacuum prior to elution with an organic solvent. A fractionation is performed in which acidic/neutral components are eluted in the first methanol fraction, and basic compounds are eluted in the second methanol–ammonia fraction. Methanol elutes the neutral and acidic compounds that are retained by hydrophobic interaction. Basic compounds, which are retained by ionic interactions, are eluted with methanol–ammonia. Ammonia is added to neutralize the cationic retention mechanism and methanol–ammonia therefore suppresses all interaction between the sorbent and basic analytes. By evaporation to dryness, the sample can be reconstituted in an organic solvent optimal for the GC-MS analysis. Compounds are converted to TMS ethers by derivatization. The TMS derivatization is carried out to improve the chromatographic properties of polar analytes. With morphine as an example, the derivatization converts the polar —OH groups into less polar —OTMS groups, as shown in the figure below.

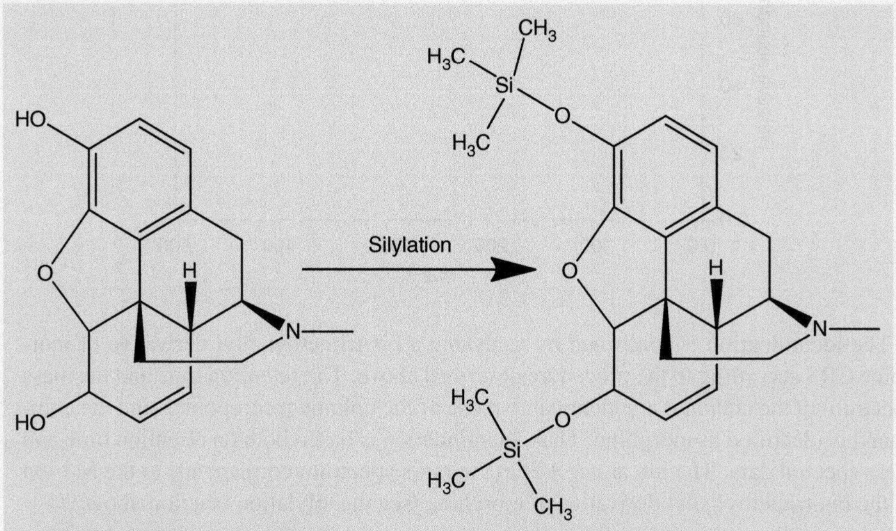

Chemical formula: $C_{17}H_{19}NO_3$
Exact mass: 285.14
Molecular Weight: 285.34

Chemical formula: $C_{23}H_{35}NO_3Si_2$
Exact mass: 429.22
Molecular Weight: 429.70

The GC-MS method is similar to the method discussed in Box 20.6, with three exceptions. First, injection is now performed in the splitless mode where the entire sample is introduced in the GC column, to lower detection limits. Second, the analysis is now performed in the temperature programmed mode. This is required because the screening method is intended for a broad range of substances of different volatilities. Third, the mass spectrometer is now operated in the full scan mode to collect full electron ionization mass spectra for identification. Thus, mass spectra for all the volatile and semi-volatile components in the sample are stored in a computer, and the mass spectrum corresponding to a given compound is available for interpretation.

IDENTIFICATION – EXAMPLE

A component elutes with a retention time of 12.51 minutes. The full scan mass spectrum of the unknown compound is searched against the mass spectra data base of the system, and the search suggests the unknown compound to be the bis-trimethyl silyl derivative of morphine (reference spectrum shown below).

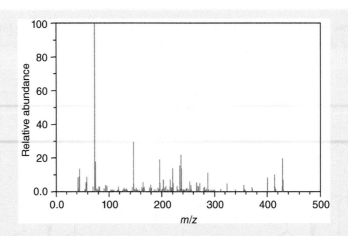

The identification is confirmed by analysing a bis-trimethyl silyl derivative of morphine CRS according to the procedure described above. The retention time and the mass spectrum of the standard are identical to those of the unknown compound, and the compound is identified as morphine. Thus, identification is based both on retention time and mass spectral data. The ion at *m/z* 429 in the mass spectrum corresponds to the M$^+$ ion of the bis-trimethyl silyl derivative of morphine (see the silylation reaction above).

The major advantage of GC-MS is that large database collections of electron ionization mass spectra (70 eV) are available, and these can be used to support identification. In addition, GC-MS provides high sensitivity and specificity. On the other hand, most drug substances are polar or non-volatile, and derivatization is required prior to GC-MS. Derivatization is not required in LC-MS and the chromatographic run times are normally shorter. Therefore, LC-MS is largely replacing GC-MS also for screening purposes. An example of the use of LC-MS (UHPLC-MS/MS) for screening for drugs of abuse is given in Box 20.8.

Box 20.8   Screening for a Specific Selection of Drugs of Abuse in Urine Based on UHPLC-MS/MS

The procedure is based on (i) enzymatic hydrolysis, (ii) SPE, and (iii) LC-MS. The mass spectrometer is operated in the selected reaction monitoring mode. Identification is based on retention data and specific mass transitions. The details are given below.

PROCEDURE – SAMPLE PREPARATION

One millilitre of urine sample is mixed with 2 mL of 0.1 M acetate buffer, pH 4.8. The mixture is added to 25 μL β-glucuronidase/arylsulfatase and enzymatically deconjugated for two hours at 50 °C, and is then cooled to room temperature. The SPE column is a 10-mL column packed with 130 mg of mixed mode sorbent with C_8 and SCX. Vacuum-mediated elution is used and the flow rate is adjusted to 2 mL/min. The columns are conditioned with 3 mL of methanol followed by 3 mL of 0.25 M acetate buffer, pH 4.8. The hydrolysed urine sample is drawn through the column. The column is washed with 1 mL of water and with 1 mL of 0.01 M hydrochloric acid. Before elution, the column is dried by vacuum. Elution at 1 mL/min is by (i) 1 mL of methanol and (ii) 1 mL of methanol/ammonia 98 : 2 (*v/v*). The two fractions are collected separately and evaporated to dryness in a stream of nitrogen at 40 °C. The alkaline extract from the mixed-mode SPE is evaporated at 40 °C using a flow of nitrogen gas and the residue is reconstituted in 200 µL of water–acetonitrile 95 : 5 (*v/v*).

PROCEDURE – SEPARATION AND DETECTION

Analysis is by UHPLC-MS/MS using the following procedure:

Column:	UHPLC column (100 mm × 2.1 mm) column packed with 1.7 µm C18 particles
Column temperature:	35 °C
Mobile phase:	Mobile phase A is 0.1% formic acid and mobile phase B is acetonitrile – 0.1% formic acid 9 : 1 (*v/v*)
Gradient:	0–1 min: 0% B
	1–4 min: B increases linearly up to 40%
	4–10 min: B increases up to 100%
	10–12 min: 100% B
Flow rate:	0.4 mL/min
Injection volume:	20 µL
Ionization:	Electrospray, positive mode
MS/MS:	A large number of mass transitions can be measured simultaneously, each transition representing a specific drug of abuse. Thus, the method can be set up to screen samples for many different drugs of abuse, including:
	Morphine, *m/z* 286 (protonated molecular ion) → *m/z* 201 (product ion)
	Codeine *m/z* 300 → *m/z* 165
	Heroin *m/z* 370 → *m/z* 165
	Cocaine *m/z* 304 → *m/z* 82
	Amphetamine *m/z* 136 → *m/z* 65
	Methamphetamine *m/z* 150 → *m/z* 65
Analysis time:	13 min

DISCUSSION OF THE PROCEDURE

Sample preparation is identical to Box 20.6. The UHPLC gradient starting at 0% B and ending with 100% B is optimized to separate common drugs of abuse. At the start no organic modifier is present in the mobile phase and hydrophobic compounds are completely focused on the column giving narrow peaks in the chromatographic separation. The run time of the UHPLC system is short, and new samples can be injected every 13 minutes. The precursor ion selected in the first quadrupole of the MS-MS instrument is the protonated molecular ion $(M + H)^+$. Morphine (as an example) has an exact mass of 285.14 and the precursor ion is at m/z 286. The precursor ion is fragmented in the second quadrupole through collision with neutral gas molecules. The characteristic product ion monitored in the third quadrupole is at m/z 201. The $286 \rightarrow 201$ mass transition chromatogram shows a peak with a retention time identical to morphine in cases of positive samples. The final confirmation is based on the analysis of the morphine standard. In a similar way, other drugs of abuse can be measured with the same method based on their specific mass transitions.

21

Chemical Analysis of Biopharmaceuticals

21.1   Biopharmaceuticals

The European Medicines Agency (EMA) has defined *Biological medicines* as 'a medicine whose active substance is made by a living organism'. Further it states that these are 'often

---

Introduction to Pharmaceutical Analytical Chemistry, Second Edition.
Stig Pedersen-Bjergaard, Bente Gammelgaard and Trine Grønhaug Halvorsen.
© 2019 John Wiley & Sons Ltd. Published 2019 by John Wiley & Sons Ltd.

produced by cutting-edge biotechnology'. The American Food and Drug Administration (FDA) uses the term *Biological products* or *Biologics* which 'include a wide range of products such as vaccines, blood and blood components, allergenics, somatic cells, gene therapy, tissues, and recombinant therapeutic proteins'. In addition, the FDA states that 'Biologics can be composed of sugars, proteins, or nucleic acids or complex combinations of these substances, or may be living entities such as cells and tissues. Biologics are isolated from a variety of natural sources – human, animal, or microorganism – and may be produced by biotechnology methods and other cutting-edge technologies'. Biological medicines can be used both for therapeutic purposes and for *in vivo* diagnostics. Biological medicines produced by biotechnology methods, i.e. recombinant deoxyribonucleic acid (rDNA) technology are often referred to as *biopharmaceuticals*. The group of biopharmaceuticals includes therapeutic peptides and proteins, antibodies, oligonucleotides and nucleic acid derivatives, and deoxyribonucleic acid (DNA) preparations. As most of the biological medicines in current clinical use contain active substances made of peptides and proteins, the focus of this chapter will be on these compounds including monoclonal antibodies. The word *biopharmaceutical* will in the following be referring to peptides, proteins, and monoclonal antibodies produced by rDNA technology. Therapeutic peptides and proteins span a wide size range from the simple peptides glucagon and teriparatide (both less than 5 kDa), to more complex proteins such as alteplase (68 kDa) and monoclonal antibodies (approximately 150 kDa). Despite the variation in size, the biopharmaceuticals have several properties in common as all peptides and proteins are built using the same building blocks.

21.1.1   Amino Acids, the Building Blocks of Biopharmaceuticals

Peptides and proteins are built from *amino acids*. There are 20 naturally occurring amino acids. An overview of the amino acids and their structure, pK_a, and isoelectric point, pI, can be found in Table 21.1. The amino acids are organic compounds that consist of an amino group, a carboxylic acid group, and a side chain. The side chain might be polar, non-polar, aromatic, acidic, or basic. Amino acids are *zwitterionic* due to the presence of both a basic and an acid group. The α-carbon, where the amino group, acid group, and side chain are bound, is chiral for all amino acids (except glycine). The amino acids natural occurring in protein molecules are the L-stereoisomers.

21.1.2   Structure of Proteins

In peptides and proteins the amino acids are linked by *peptide bonds* (amide bonds). A polypeptide is a linear chain of amino acid residues. Polypeptides of less than 20–30 amino acid residues are commonly termed *peptides* while larger chains are considered to be *proteins*. The protein structure can be divided into four levels (Figure 21.1). The *primary structure* is the linear sequence of the amino acids. The *secondary structure* is the local folding of the polypeptide chain into helixes and sheets, held together by hydrogen bonds. The *tertiary structure* is the three-dimensional folding pattern which occurs due to interaction of the amino acid side chains, and the *quaternary structure* is the combination of more than one amino acid chain into a complex structure.

   The amino acid residues in a protein may be chemically modified after synthesis, which is termed *post-translational modification*. In this process, certain amino acid residues can be

Table 21.1   Overview of the amino acids, abbreviations, and structures

Amino acid (three and one letter abbreviation)	pK_a (acid)	pK_a (base)[a]	pK_a (side chain)	Isoelectric point (pI)	Structure
Non-polar, aliphatic side chain					
Glycine (His, H)	2.34	9.58		5.96	
Alanine (Ala, A)	2.33	9.71		6.02	
Valine (Val, V)	2.27	9.52		5.90	
Leucine (Leu, L)	2.32	9.58		5.95	
Isoleucine (Ile, I)	2.26	9.60		5.93	
Proline (Pro, P)	1.95	10.47		6.21	
Aromatic side chain					
Phenylalanine (Phe, F)	2.18	9.09		5.64	
Tyrosine (Tyr, Y)[a]	2.24	9.04	10.10	5.64	
Tryptophan (Trp, W)	2.38	9.34		5.86	

(continued overleaf)

Table 21.1   (Continued)

Amino acid (three and one letter abbreviation)	pK_a (acid)	pK_a (base)[a]	pK_a (side chain)	Isoelectric point (pI)	Structure
Polar, uncharged side chain					
Serine (Ser, S)	2.13	9.05		5.59	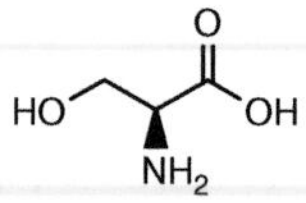
Threonine (Thr, T)	2.20	8.96		5.58	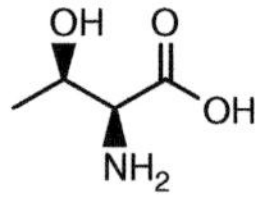
Cysteine (Cys, C)[a]	1.91	10.28	8.14	5.03	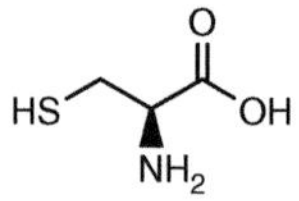
Methionine (Met, M)	2.16	9.08		5.62	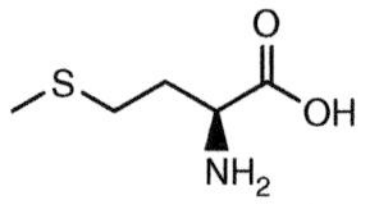
Asparagine (Asn, N)	2.16	8.76		5.46	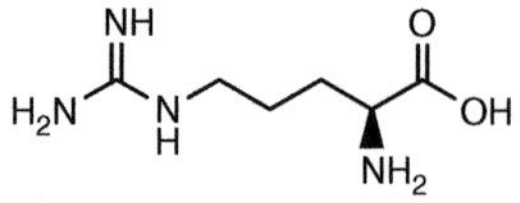
Glutamine (Gln, Q)	2.18	9.00		5.59	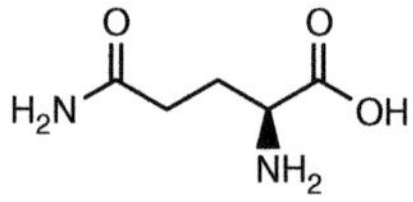
Negatively charged side chain					
Aspartic acid (Asp, D)	1.95	9.66	3.71	2.83	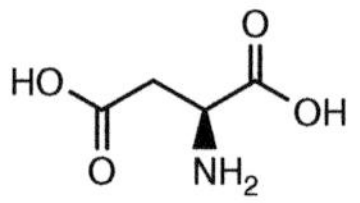
Glutamic acid (Glu, E)	2.16	9.58	4.15	3.16	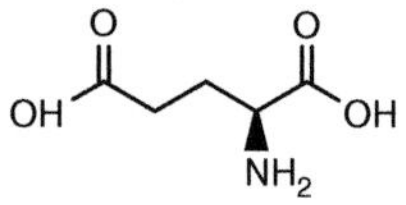
Positively charged side chain					
Lysine (Lys, K)	2.15	9.16	10.67[b]	10.07	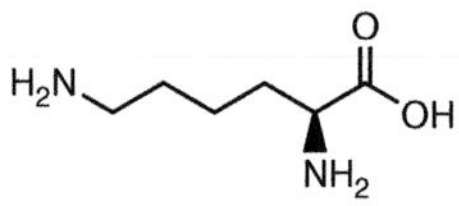
Arginine (Arg, R)	2.03	9.00	12.10[b]	10.55	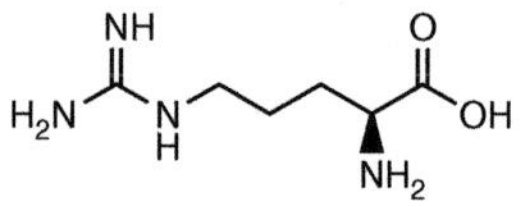
Histidine (His, H)	1.70	9.09	6.04[b]	7.57	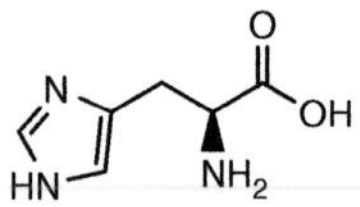

[a] Tyrosine and cysteine are normally grouped together with the aromatic and polar uncharged amino acids, respectively, although they have weakly acidic side chains.

[b] pK_a value refers to the pK_a value of the conjugated acid.

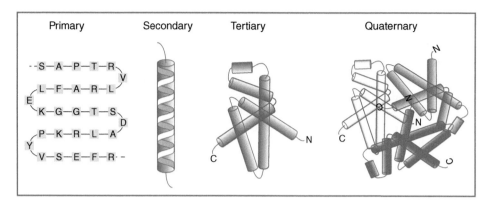

Figure 21.1 *Overview of the four levels of protein structure.*

modified by *methylation, phosphorylation,* or by addition of sugar groups (*glycosylation*). The physicochemical properties of a protein are dependent of the structure of the protein, including the post-translational modifications.

21.1.3 Glycosylation of Proteins and Peptides

The process of adding sugar groups to the protein backbone is termed *glycosylation.* This is one of the most common post-translational modifications. During glycosylation, oligosaccharide chains are covalently linked to amino acid side chains. A common name for these oligosaccharides is *glycans,* and glycans can be linked through binding to oxygen or nitrogen (O- and N-linked glycans, respectively). In *O-linked glycans* or O-glycans, the oligosaccharide is attached to the hydroxyl group of serine (S) or threonine (T). The oligosaccharides of O-glycans are generally short (consisting of 1–4 sugar residues). In *N-linked glycans* or N-glycans, the oligosaccharide is attached to the amide nitrogen of asparagine (N). Only asparagine, which is part of the amino acid sequence N—X—S or N—X—T, where X can be any amino acid except proline (P) or aspartic acid (D), will be glycosylated. N-glycans consist of a minimum of five sugar residues and can contain different amounts of sialic acid. They also exhibit different *glycoforms*. These glycoforms have the same protein backbone but differ in the sequences, locations, and number of bound oligosaccharides.

21.2 Biopharmaceuticals versus Small Molecule APIs

Biopharmaceuticals represent a class of medicinal products increasing in number. These products often represent the cutting-edge of biomedical research. Biopharmaceuticals differ in several aspects from the small molecule active pharmaceutical ingredients (APIs) derived from synthetic organic chemistry (and discussed in Chapters 1 to 20). The differences include, but are not limited to, the source of product and active ingredients, manufacturing procedure, formulation, identity, structure, and composition. While small molecule APIs are pure and well-defined chemicals, biopharmaceuticals are more complex mixtures that are not as easily identified or characterized in terms of chemical composition. The molecular weight of biopharmaceuticals is often two orders of magnitude higher than for small molecule APIs, and as they are composed of up to several hundred monomers the possibility of structural variation is great. It is also generally accepted that the structure–function relationship is not fully determined. Biopharmaceuticals tend to be heat sensitive, and the production is not as harsh as for conventional active ingredients. Due to this, biopharmaceuticals are more susceptible to microbial contamination. Since most biopharmaceuticals are administered as injections, sterility of the final product is important. This requires aseptic principles from the initial manufacturing steps.

21.3 Biopharmaceuticals and Pharmacopoeias

Biopharmaceuticals are regulated through a set of general and specific monographs in the European Pharmacopoeia (Ph. Eur.) and the United States Pharmacopoeia (USP). In the quality control of biopharmaceuticals, chemical analytical methods are only part of the methods used. For most biopharmaceuticals, biological testing is necessary, and for assay of active substances, the concentration or activity is often determined by potency or a biological assay.

For a biopharmaceutical where the patent time has elapsed, competing products are introduced to the market. These products are termed *biosimilar products* (*biosimilars*, EMA) or *follow-on biologics* (FDA), and are not generic products. Biosimilars are mimicking the activity and safety of the original biopharmaceutical but are not necessarily produced in the same way, and are hence not chemically identical as is the case with generic products. Box 21.1 shows examples of biosimilars.

Box 21.1   Biosimilar Products (Biosimilars/Follow-on Biologics)

Definition: Biopharmaceutical, which is similar in terms of quality, safety, and efficacy to an already licensed reference biopharmaceutical. The US-FDA uses the term *follow-on biologic*, which is highly similar to the reference product without clinically meaningful differences in safety, purity, and potency.

Examples of available biosimilars:

Recombinant non-glycosylated proteins	Insulin, human growth hormone, interferons
Recombinant glycosylated proteins	Erythropoietin, monoclonal antibodies, follicle stimulating hormone (FSH)
Recombinant peptides	Glucagon, parathyroid hormone

21.4   Production of Biopharmaceuticals

Protein-based drugs may be mass produced by using *rDNA technology* (see Box 21.2). The process is more complex than synthesis of small molecule APIs and several measures are taken to ensure that quality is embedded in the product during production. The *host–vector system* and the production process must be well-characterized and validated. After expression, the protein is extracted and purified to remove and/or inactivate contaminants derived from the host cell or culture medium. A variety of techniques is used for this such as extraction, precipitation, centrifugation, concentration, filtration, and/or chromatography. As can be seen in the pharmacopoeia monographs, biopharmaceuticals (active substances) are often freeze-dried powders or concentrated solutions.

During the production process related impurities including residual host cell or vector DNA and host-cell protein must be reduced to acceptable levels. These two components are of major interest for safety and tolerance of the product. The main reason for measuring *residual DNA* is to demonstrate the effectiveness of the purification process during manufacture. This is important as DNA from microbial sources may increase the risk of an immune response toward the product (increased *immunogenicity*), and it might be a possibility that residual DNA from mammalian cell lines add a risk to development of cancer (increased *tumorigenesis*). The most common methodology for rDNA quantitation is real-time polymerase chain reaction (rt-PCR). *Residual host cell protein* may also increase the immunogenicity of a final drug. This is independent of the expression/production system used. Generally, to test for immunogenicity of residual host proteins a broad spectrum screening assay is first applied and the results are confirmed by a more specific confirmatory assay. Box 21.3 shows the Ph. Eur. requirements for production of calcitonin (salmon) as an example of an active substance. For other biopharmaceuticals the production requirements are often more extensive.

Some of the therapeutic proteins produced by rDNA technology are *glycoproteins*. The glycosylation pattern is dependent on the host cell system used, and this pattern is important for both the efficacy and the safety of the product. Therefore, for glycoproteins, both careful selection of the host cell system and characterization of the glycan pattern during production is important.

Box 21.2   Brief Introduction to Recombinant DNA (rDNA) Technology

In rDNA technology, the gene of interest is introduced into a suitable *microorganism* or cell line. This is most often done by means of a *vector*. The gene is then expressed and translated into a protein. In production of biopharmaceuticals the gene is coding for the desired protein drug (i.e. human insulin).

Definitions:

Microorganism:	Bacteria such as *Escherichia coli* (*E. coli*)
Vector:	Plasmid, circular double-stranded DNA molecules that occur naturally and can replicate itself in, for example, bacteria
Host cell:	Microorganism or cell line
Host–vector system:	Stable complex of host cell and vector introducing the gene

Brief overview of the process:

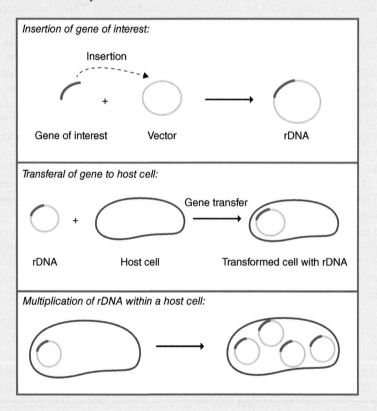

Insertion of gene of interest:

Insertion

Gene of interest + Vector → rDNA

Transferal of gene to host cell:

rDNA + Host cell → Gene transfer → Transformed cell with rDNA

Multiplication of rDNA within a host cell:

After multiplication of the rDNA, the cells containing the rDNA will be selected and allowed to grow. The protein will then be transcribed and can also be post-translationally modified (including phosphorylation, methylation, and glycosylation). Finally the protein is extracted and purified.

Box 21.3   Production Requirements for Recombinant Calcitonin (Salmon) by Ph. Eur.

STRUCTURE

H-Cys-Ser-Asn-Leu-Ser-Thr-Cys-Val-Leu-Gly-
Lys-Leu-Ser-Gln-Glu-Leu-His-Lys-Leu-Gln-
Thr-Tyr-Pro-Arg-Thr-Asn-Thr-Gly-Ser-Gly-
Thr-Pro-NH$_2$

$C_{990}H_{1528}N_{262}O_{300}S_7$ M_r 22 125

PHARMACOPOEIA TEXT – INDIVIDUAL MONOGRAPH FOR CALCITONIN (SALMON)

PRODUCTION

The following requirements apply only to calcitonin (salmon) produced by a method based on rDNA technology.

Prior to release the following tests are carried out on each batch of final bulk product unless exemption has been granted by the competent authority.

Host-cell-derived proteins. The limit is approved by the competent authority.
Host-cell- and vector-derived DNA. The limit is approved by the competent authority.

21.5   Identification Procedures for Biopharmaceuticals (Active Substance)

The identification procedures for biopharmaceuticals differ to a large extent from those of small molecule APIs. First of all, infrared (IR) spectroscopy is not commonly used. This is mainly due to the complexity of the molecules. In Ph. Eur. monographs only one set of identification tests is prescribed for biopharmaceuticals. All identification procedures must be positive in order to positively identify the active substance. Some of the identification tests are identical to those used for small molecule APIs, such as UV spectrophotometry and liquid chromatography (LC). In addition, several other tests are applied depending on the actual structure of the biopharmaceutical. An overview of common identification tests can be found in Table 21.2. The identification tests address molecular structure and composition

Table 21.2 *Overview of the main procedures used for identification of biopharmaceuticals (Ph. Eur.)*

Analytical technique	Specificity	Reference material[a] needed
Peptide mapping by reversed phase liquid chromatography with UV detection (RP-LC-UV)	Information about primary structure Evaluation of profile and sometimes retention time of selected peak(s)	Yes
Potency assay/biological assay	Potency/biological activity serves to identify	Yes
Liquid chromatography	Evaluation of retention time and size of principal peak	Yes
Capillary zone electrophoresis	Evaluation of migration time and size of peaks/number of peaks present	Yes
Polyacrylamide gel electrophoresis (PAGE) and immunoblotting	Evaluation of band position and intensity	Yes
Isoelectric focusing	Evaluation of band position and pI of principal peak(s) or band pattern	Yes
Size-exclusion chromatography (SEC)	Evaluation of retention time and sometimes also size of principal peak	Yes
Glycan analysis by anion exchange liquid chromatography w/fluorescence or ampherometric detection	Evaluation of profile and content of different sialylated forms	Yes
PAGE – reducing conditions	Evaluation of band position of principal peak	Yes
Amino acid analysis	Information about primary structure	—
N-terminal sequence assay	Information about primary structure	—

[a] Or representative batch of material.

in addition to other relevant physicochemical, biological, and immunochemical properties. The purity and biological activity can also be used for identification, and a comparison is done with well-defined reference material or a representative batch of the material.

21.5.1 Peptide Mapping – Information About Primary Structure

Peptide mapping is a powerful identification technique common for all biopharmaceuticals. The technique can be used to detect almost any single amino acid change, and will detect changes resulting from errors in the reading of complementary DNA during production or changes due to point mutations. Peptide mapping is a non-generic technique and a specific map is made for each unique protein. A fingerprint of the primary structure of the protein is obtained through a three-step procedure. First the protein is selectively cleaved to peptides, second the peptides are separated by chromatography, and finally the peptides are analysed and identified. The technique is comparative and therefore a reference standard or representative batch of the material must be available for confirmation. The test sample and the reference are treated in the same manner and analysed in parallel.

Peptide mapping is used to confirm the primary structure of the protein. As the protein is cleaved into peptides prior to analysis, information about the secondary, tertiary, or quaternary structure will not be obtained. Alterations in structure can be detected and information about process consistency and genetic stability is gained.

The first step in the procedure is selective cleavage of the protein. Cleavage of proteins into peptides is commonly termed *digestion* and can be performed either chemically or enzymatically using a protease. A list of different cleavage agents can be found in Table 21.3. The choice of treatment, including the kind of enzyme, requires knowledge about protein characteristics as the peptide map needs to have certain specificity. Complete digestion is more likely to occur using an enzyme compared to chemical cleavage. The goal is to produce enough fragments, but not too many because then specificity can be lost as several proteins might have the same profiles.

For some enzymes, the sample, the cleavage agent, and the protein need to be pretreated. Pretreatment of the sample prior to digestion is dependent on the size and the configuration

Table 21.3   *Examples of cleavage agents*

Type	Agent	Specificity	Optimum pH
Enzymatic digestion	Trypsin	C-terminal side of Arg and Lys	7–9
	Chymotrypsin	C-terminal side of hydrophobic residues[c]	7–9
	Lysyl endopeptidase (Lys-C)	C-terminal side of Lys	7–9
	Glutamyl endopeptidase (Glu-C)[a]	C-terminal side of Glu (and Asp)[d]	4 and 8[d]
	Peptidyl-Asp metalloendopeptidase (Asp-N endoproteinase)	N-terminal side of Asp	7–8
	Clostripain	C-terminal side of Arg	7.4–7.8
	Pepsin	Non-specific digest	1.8–2.5
Chemical digestion	Cyanogen bromide	C-terminal side of Met	1–2
	2-Nitro-5-thio-cyanobenzoic acid (NTCB)	N-terminal side of Cys	8–9
	O-Iodosobenzoic acid	C-terminal side of Trp and Tyr	60–80% acetic acid
	Dilute acid	Asp and Pro	70% formic acid
	BNPS-skatole[b]	Trp	0.1% TFA or 1% formic acid/acetic acid

[a] From *S. aureus* strain V8.
[b] BNPS-Skatole = 3-bromo-3-methyl-2-[(2-nitro-phenyl)thio]-3H-indole.
[c] For example Leu, Met, Ala, and aromatic amino acids.
[d] Cleavage of Asp is buffer and pH dependent and often slow.
Source: Adapted from Ph. Eur. and USP.

of the protein, and it may include protection of certain cleavage sites to reduce the amount of peptides produced. Pretreatment of the cleavage agent, on the other hand, might be necessary to ensure reproducibility of the peptide map. This may include a clean-up step and is especially important for enzymes. Finally, depending on the protein structure, pretreatment of the protein might be necessary: the cleavage step might be preceded by a denaturation step for *unfolding of the protein* and a reduction and alkylation step for *cleavage of disulfide bonds*. Typical pretreatment performed when using the cleavage agent trypsin is described in Box 21.4.

Box 21.4   Trypsin as Cleavage Agent

Pretreatment of sample: Lysine residues must be protected by citraconylation or maleylation if the protein has a molecular mass greater than 100 000 Da to avoid generation of too many peptides.

Pretreatment of cleavage agent: Trypsin is treated with tosyl-L-phenylalanin chloromethyl ketone to inactivate chymotrypsin.

Pretreatment of protein: Dependent of the protein structure and incudes denaturation of protein and/or reduction and alkylation of disulfide bonds.

Digestion: Trypsin digestion may introduce ambiguities in the peptide map due to side reactions occurring during digestion. Examples of this are non-specific cleavage, deamidation, disulfide isomerization, oxidation of methionine residues, and formation of pyroglutamic groups. In addition, peaks may be produced due to autohydrolysis of enzyme.

Optimum digestion conditions for trypsin:
pH 8 – temperature 37 °C – time 14–16 hours – amount 20:1 to 40:1 (protein: enzyme)

The digestion conditions are important and special attention is paid to the following parameters:

- pH
- temperature
- time
- amount of enzyme

Optimum pH is dependent on the cleavage agent (see Table 21.3). The pH should not denature the enzyme, it should be constant during cleavage, and it should not alter the integrity of the biopharmaceutical. An adequate temperature is typically 25–37 °C. The temperature should minimize side reactions or modifications of the biopharmaceutical and it should not denature the protein (or enzyme) during digestion. Digestion times vary from 2 to 30 hours. The optimal digestion time results in a reproducible map and avoids incomplete digestion. The digestion reaction is stopped by freezing or by addition of an acid that does not interfere with the map. Typically a protein-to-enzyme ratio between 20 : 1 and 200 : 1 is used. This is a compromise between the efficiency of the cleavage and the introduction of interferences in the chromatogram due to *auto-hydrolysis* of the enzyme. A more optimal digestion might be achieved by addition of the cleavage agent in two or more stages. A *blank digestion* (without target protein) should always be performed to evaluate the contribution of the enzyme to the chromatographic pattern. The enzyme solution is typically prepared by making a solution with non-optimal pH for the enzyme and then diluting the enzyme with the digestion buffer just before addition to the sample. Typical conditions for trypsin digestion are included in Box 21.4.

The second step in peptide mapping is separation. Several different techniques are used to separate the peptides after digestion: reversed-phase LC, ion-exchange chromatography (IEC), hydrophobic interaction chromatography (HIC), sodium dodecyl sulfate polyacrylamide gel electrophoresis (SDS-PAGE), capillary electrophoresis (CE), paper chromatography-high voltage (PCHV), and high voltage-paper electrophoresis (HVPE).

The most frequently prescribed technique in Ph. Eur. and USP is LC with UV detection. Porous silica particles modified with butyl (C_4), octyl (C_8), or octadecyl (C_{18}) are commonly used as the stationary phase. C_8 and C_{18} are more efficient for smaller peptides than C_4. The particle size of the silica particles varies between 3 and 10 µm and the pore size varies between 10 and 30 nm. Separation is typically performed with gradient elution using water and acetonitrile with 0.1% trifluoroacetic acid as the mobile phase. UV detection is performed at 214 or 215 nm due to the low absorbance maximum of most amino acids.

After chromatographic separation and UV detection the chromatogram of the test solution is compared to a chromatogram obtained after identical treatment of a reference standard. A complete match in retention time and peak intensity for all peaks is necessary. Identification of the peaks is normally performed when establishing the method during development of the product. An example demonstration of the power of peptide mapping using insulin as the model is described in Box 21.5.

Box 21.5   Identification of Insulin Analogues

Human insulin consists of two chains, A and B, connected by two disulfide bridges (for the structure, see below). Insulin is important for regulation of the blood glucose level and several variants of this hormone are available on the market. Some of these insulin variants are isolated from animals while others are produced by rDNA technology. Some of the insulin analogues differ from human insulin by only one amino acid (see the table below). However, by careful selection of the digestion and separation system it is possible to identify the different analogues by peptide mapping (including the animal-derived insulins).

STRUCTURE OF HUMAN INSULIN

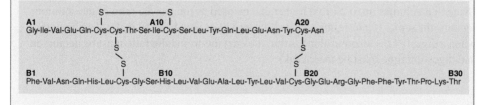

Amino acid differences between recombinant produced insulin analogues and human insulin

Insulin	Amino acid differences from human insulin[a]
Insulin lispro	B28 Lys, B29 Pro
Insulin aspart	B28 Asp
Insulin glargine	A21 Gly, two Arg added to C-terminus of B chain
Insulin glulisine	B3 Lys, B29 Glu

[a]The letter and the number indicates the chain and amino acid in human insulin that is modified.

In USP a common peptide mapping procedure is described for all the various insulins. The procedure is performed without a preceding denaturation and reduction and alkylation step, as described below.

PHARMACOPOEIA TEXT – EXCERPT FROM USP PHYSICOCHEMICAL PROCEDURES FOR INSULINS

Peptide Mapping Procedure

The following procedure is applicable for preparing peptide maps of Insulin Human, insulin analogues, and animal-derived insulins.

HEPES buffer	Dissolve 2.38 g of HEPES (*N*-2-hydroxyethylpiperazine-*N'*-2-ethanesulfonic acid) in about 90 mL of water in a 100-mL volumetric flask. Adjust with 5 M sodium hydroxide to a pH of 7.5, dilute with water to volume, and mix.
Sulfate buffer	2.0 M ammonium sulfate and 0.5 M sulfuric acid (1 : 1). Mix, and filter.

Insulin Digestion
The following procedure provides efficient cleavage of the glutamyl bonds of insulin. (Note that volumes up to 20-fold higher can be used as long as the ratio of the solutions remains the same. If interfering autolysis by-products are observed in the chromatogram when a digest of the enzyme alone is run, the enzyme-to-insulin ratio must be decreased and digestion time must be increased.)

Enzyme solution	Prepare a 1-mg/mL solution of *S. aureus* V8 protease in water (approximately 500 units/mg using casein as the substrate).
Sample solution	Prepare a 2.0-mg/mL solution of the insulin to be examined in 0.01 N hydrochloric acid. To a clean vial add 25 µL of the 2.0-mg/mL insulin solution, 100 µL of *HEPES buffer*, and 20 µL of *Enzyme solution* (final ratio is 25 : 100 : 20). Cap the vial and incubate at 25° for 6 hours. Stop the reaction by adding an equal volume of *Sulfate buffer*. Longer incubation times may be needed for analogues with poor solubility at pH 7.5.
Standard solution	Prepare at the same time and in the same manner a solution of the appropriate USP Insulin Reference Standard as directed in the *Sample solution*.

Peptide Fragment Determination
Determine the peptide fragments using the following peptide mapping procedure.

Solution A	Acetonitrile, water, and Sulfate buffer (100 : 700 : 200). Filter and degas.
Solution B	Acetonitrile, water, and Sulfate buffer (400 : 400 : 200). Filter and degas.
Mobile phase	See the table below.

Time (min)	Solution A (%)	Solution B (%)
0	95	5
3	95	5
30	41	59
35	20	80
40	95	5
50	95	5

Chromatographic system

Mode	LC
Detector	UV 214 nm
Column	4.6 mm × 10 cm; 5 µm packing L1 (octadecyl silane)
Column temperature	40 °C
Flow rate	1 mL/min
Injection volume	50–100 µL

System suitability

Sample	Standard solution

Suitability requirements

Chromatogram comparability	In the chromatogram obtained from the Standard solution, identify the peaks due to digest fragments I, II, III, and IV. The chromatogram of the Standard solution corresponds to that of the typical chromatogram provided with the appropriate USP Insulin Reference Standard.
Resolution	There should be complete separation of the peaks due to fragments II and III. The resolution is defined in the applicable insulin monograph.
Tailing factor	NMT 1.5 for digest fragments II and III

Analysis

Samples	Sample solution and Standard solution

Condition the Chromatographic system by running at initial conditions, $t = 0$ minutes, for at least 15 minutes.

Carry out a blank gradient program before injecting the digests. Separately inject equal volumes of the Standard solution and the Sample solution, and record the responses of each peak.

Acceptance criteria	The chromatographic profile of the Sample solution corresponds to that of the Standard solution.

DISCUSSION

The digestion of insulin is performed using HEPES buffer at pH 7.5. The choice of buffer pH corresponds well with the optimum pH of the digestion enzyme (pH 4–8). The enzyme used is *Staphylococcus aureus* V8. This protease is also called glutamyl endopeptidase (Glu-C). It cleaves the carboxy terminal side of Glu (and Asp). The enzyme is specific for Glu in ammonium bicarbonate and other non-phosphate containing buffers. It will hence cleave all glutamyl bonds in insulin without cleavage of the aspartyl bonds in the above described procedure. For most insulins, four fragments are produced. The fragments produced for human insulin are shown in the figure below:

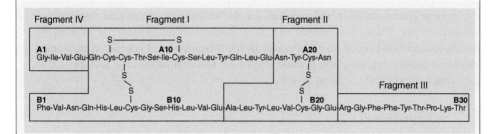

Digestion is performed at 25 °C for six hours. This corresponds well with the suggested digestion temperature and time for enzymatic digestion. For digestion, 25 μL of 2.0 mg/mL insulin solution is mixed with 20 μL of 1 mg/mL enzyme solution (and diluted with 100 μL of digestion buffer). This means that the mass ratio is 2.5 : 1. This is a relatively high ratio of protein-to-enzyme. The reaction is stopped by adding an equal volume of the sulfate buffer containing a mixture of 2.0 M ammonium sulfate and 0.5 M sulfuric acid (1 : 1).

In addition to digesting the sample and the standard solution, a digestion of the enzyme alone is performed to check for autolysis by-products. If interfering peaks are seen when analysing this sample, a reduced enzyme-to-protein ratio and longer digestion times should be applied.

Chromatographic separation of digested samples is done using a reversed phase system capable of separating the four fragments and giving a different pattern for the different insulins. In which fragment the different amino acid modifications can be seen is shown in the table below. As can be seen from the table, fragment IV is identical for all recombinant insulins, and all insulins have an amino acid modification in fragment III.

Overview of differences in the amino acid sequence and the digestion fragments that are affected by the modification. Differences with insulin human are shown in bold.

Insulin	Amino acid differences compared to insulin human[a]	
Insulin lispro	Fragment III	(B22) Arg-Gly-Phe-Phe-Tyr-Thr-**Lys**-**Pro**-Thr (B30)
Insulin aspart	Fragment III	(B22) Arg-Gly-Phe-Phe-Tyr-Thr-**Asp**-Lys-Thr (B30)
Insulin glargine	Fragment II	(A18) Asn-Tyr-Cys-**Gly** (A21)
	Fragment III	(B22) Arg-Gly-Phe-Phe-Tyr-Thr-Pro-Lys-Thr-**Arg-Arg** (B32)
Insulin glulisine	Fragment I	(B1) Phe-Val-**Lys**-Gln-His-Leu-Cys-Gly-Ser-His-Leu (B11)
	Fragment III	(B22) Arg-Gly-Phe-Phe-Tyr-Thr-Pro-**Glu**-Thr (B30)

[a]A and B denote the A and B chains of insulin, respectively; numbers denote the amino acid position in the chain.

Chromatograms showing the patterns for the different recombinant insulins are included below. These demonstrate that it is possible to distinguish the insulins by the above described procedure. However, in identification of a specific recombinant insulin by peptide mapping, the peptide map of the recombinant insulin is compared to a reference standard of the same insulin.

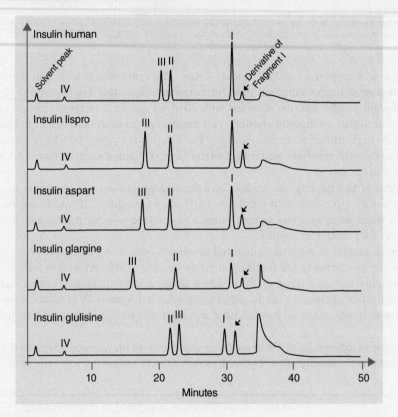

The figure shows the peptide map of five different recombinant insulins (including human insulin).

21.5.2 Separation Techniques

Also for biopharmaceuticals (active substances), separation techniques are frequently applied for identification. In addition to LC, several other techniques such as capillary electrophoresis, polyacrylamide gel electrophoresis (PAGE) with and without subsequent immunoblotting, isoelectric focusing, size-exclusion chromatography, and mass spectrometry are used for identification of biopharmaceuticals. The variety of techniques is due to the complexity of biopharmaceuticals, which might consist of different isoforms and subunits. Identification using a separation technique is performed by comparison of a *Test solution* containing the dissolved or diluted substance to be examined with a

Reference solution/Standard preparation containing a reference standard at the same concentration level. The requirement of a positive identification is that the principal peak (or the principle band) obtained with the test solution is similar in retention/migration time and size (or band position and band intensity) to what is obtained with the reference solution/standard preparation. An example using isoelectric focusing for identification is described in Box 21.6.

Box 21.6   Identification of Molgramostim in Molgramostim Concentrated Solution Using Isoelectric Focusing According to Ph. Eur.

Molgramostim is a recombinant form of human granulocyte-macrophage colony-stimulating factor.

Structure

APARSPAPST	QPWEHVNAIQ	EARRLLNLSR	DTAAEMNETV
EVISEMFDLQ	EPTCLQTRLE	LYKQGLRGSL	TKLKGPLTMM
ASHYKQHCPP	TPETSCATQI	ITFESFKENL	KDFLLVIPFD
CWEPVQE			

$C_{639}H_{1007}N_{171}O_{196}S_8$ M_r 14 477

PHARMACOPOEIA TEXT – INDIVIDUAL MONOGRAPH FOR MOLGRAMOSTIM CONCENTRATED SOLUTION

IDENTIFICATION

Test B. Isoelectric Focusing

Test solution. Dilute the preparation to be examined with water R to obtain a concentration of 0.25 mg/mL.

Reference solution (a). Dilute molgramostim chemical reference substance (CRS) with water R to obtain a concentration of 0.25 mg/mL.

Reference solution (b). Use an isoelectric point (pI) calibration solution, in the pI range of 2.5–6.5, prepared according to the manufacturer's instructions.

Focusing:
- pH gradient: 4.0–6.5,
- catholyte: 8.91 g/L (0.1 M) solution of 3-aminopropionic acid R,
- anolyte: 14.7 g/L (0.1 M) solution of glutamic acid R in a 50% *v/v* solution of dilute phosphoric acid R (0.5 M),
- application: 20 μL.

Detection. Immerse the gel in a 21.6 suitable volume of a solution containing 115 g/L of trichloroacetic acid R and 34.5 g/L of sulfosalicylic acid R and shake the container gently for 30 minutes. Transfer the gel to a mixture of 32 volumes of glacial acetic acid R, 100 volumes of ethanol R, and 268 volumes of water R (mixture A) and rinse for 5 minutes. Immerse the gel for 10 minutes in a staining solution prewarmed to 60 °C and prepared by adding acid blue 83 R at a concentration of 1.2 g/L to mixture A. Wash the gel in several containers with mixture A and keep the gel in this mixture until the background is clear (12–24 hours). After adequate destaining, soak the gel for 1 hour in a 10% *v/v* solution of glycerol R in mixture A.

System suitability:

- in the electropherogram obtained with reference solution (b), the relevant isoelectric point markers are distributed along the entire length of the gel,
- in the electropherogram obtained with reference solution (a), the pI of the principal band is 4.9 to 5.4.

Results. The principal band in the electropherogram obtained with the test solution corresponds in position to the principal band in the electropherogram obtained with reference solution (a). Plot the migration distances of the relevant pI markers versus their pI and determine the isoelectric points of the principal component of each of the test solution and reference solution (a). They do not differ by more than 0.2 pI units.

DISCUSSION

In isoelectric focusing, analytes are separated based their isoelectric point (pI). Small differences in an amino acid sequence can result in a shift in pI, i.e. if asparagine (neutral) is exchanged (or deamidated) to aspartic acid (acidic). In identification of molgramostim by isoelectric focusing a pH gradient of 4.0–6.5 is used. This pH gradient covers the calculated pI of molgramostim, which is 5.1. Since the pH gradient is rather narrow it will be possible to differentiate molgramostim from other substances with a close pI.

In addition to the test sample and the reference sample, a calibration solution comprising a mixture of proteins with different pI is analysed. The calibration standard used in the current setup consists of a set of proteins with pI in the range 2.5–6.5.

A system suitability test is performed to evaluate the separation power of the gel (as the distribution of proteins in the calibration standard across the gel) and the accuracy of the system (as the pI of the principle band in the reference solution). In the figure below the calibration solution consists of four markers with pI within the pH gradient (pI 4.5, 5.2, 5.3, and 6.0).

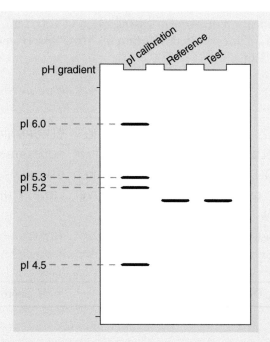

The identity is confirmed by comparing the position of the principle band in the electropherogram of the test solution with the principle band in the electropherogram of the reference solution. The comparison is based on calculation of the pI of both principle bands: a plot of the migration distance versus pI of the separated markers in the calibration solution is made, and from this plot the pI of the principle band in the test solution and the reference solution is calculated. The values are compared and the identification is positive if the pI does not differ by more than 0.2 pI units. For the test to be valid the calculated pI of the reference solution should be within 4.9–5.4 (described under the system suitability test above).

Some biopharmaceuticals consist of a mixture of isoforms or different subunits. For these substances, separation techniques are also used to identify the different subunits or isoforms and to measure their relative content. An example of this is shown in Box 21.7 for the identification of the isoform distribution of interferon beta-1a in Interferon beta-1a concentrated solution. Measurements are performed by mass spectrometry.

Box 21.7   Identification of Isoform Distribution of Interferon Beta-1a in Interferon Beta-1a Concentrated Solution by Mass Spectrometry According to Ph. Eur.

STRUCTURE

MSYNLLGFLQ	RSSNFQCQKL	LWQLNGRLEY	CLKDRMNFDI
PEEIKQLQQF	QKEDAALTIY	EMLQNIFAIF	RQDSSSTGWN*
ETIVENLLAN	VYHQINHLKT	VLEEKLEKED	FTRGKLMSSL
HLKRYYGRIL	HYLKAKEYSH	CAWTIVRVEI	LRNFYFINRL
TGYLRN			

*glycosylation site

$C_{908}H_{1406}N_{246}O_{252}S_7$ M_r approx. 22 500

PHARMACOPOEIA TEXT – INDIVIDUAL MONOGRAPH FOR INTERFERON

Beta-1a Concentrated Solution

IDENTIFICATION

   B. Isoform distribution. Mass spectrometry.

   Introduction of the sample: direct inflow of a desalted preparation to be examined or liquid chromatography–mass spectrometry (LC-MS) combination.

   Mode of ionization: electrospray.

   Signal acquisition: complete spectrum mode from 1100 to 2400.

   Calibration: use myoglobin in the m/z range of 600–2400; set the instrument within validated instrumental settings and analyse the sample; the deviation of the measured mass does not exceed 0.02% of the reported mass.

Interpretation of results: a typical spectrum consists of six major glycoforms (A–F), which differ in their degree of sialylation and/or antennarity type as shown in the table below.

MS peak	Glycoform[a]	Expected M_r	Sialylation level
A	2A2S1F	22 375	Disialylated
B	2A1S1F	22 084	Monosialylated
C	3A2S1F and/or 2A2S1F + 1 HexNacHex repeat	22 739	Disialylated
D	3A3S1F	23 031	Trisialylated
E	4A3S1F and/or 3A3S1F + 1 HexNacHex repeat	23 400	Trisialylated
F	2A0S1F	21 793	Non-sialylated

[a]2A = biantennary complex type oligosaccharide; 3A = triantennary complex type oligosaccharide; 4A = tetraantennary complex type oligosaccharide; 0S = non-sialylated; 1S = monosialylated; 2S = disialylated; 3S = trisialylated; 1F = fucosylated.

Results: the mass spectrum obtained with the preparation to be examined corresponds, with respect to the six major peaks, to the mass spectrum obtained with *interferon beta-1a CRS*.

DISCUSSION

Mass spectrometry is ideal for identification of isoforms, as isoforms with different masses can be seen as separate peaks in the mass spectrum. In the identification of interferon beta-1a, electrospray ionization (ESI) is used as the ion source. Electrospray generates multiple charged ions, and for each isoform a so-called *charge envelope* will be generated. Based on the masses of the isoforms (ranging from 21 793 to 23 400) and a scan range between 1100 and 2400 the charge envelope between +10 and +20 will be seen for most of the isoforms of interferon beta-1a. This means that for each isoform the signal will be distributed between 10 and 12 peaks in the mass range 1100–2400. A very complex spectrum with a unique pattern for interferon beta-1a with the desired glycoform distribution will be generated. Positive identification is based on comparison of the mass spectrum obtained from the test solution and the reference solution.

Analysis may be performed either by direct infusion of the test solution after desalting or through an LC-MS system. The desalting is important to remove non-volatile salts that are not compatible with the mass spectrometry (MS) system.

Prior to analysis of the sample, calibration is performed by analysis of myoglobin. From this, the mass accuracy is calculated to ensure optimum performance of the mass spectrometer. Other details are not stated and will be dependent of the type of mass spectrometer available.

A mass spectrum ($m/z = 600$–2400) of myoglobin (used for calibration) is shown below. A charge envelope from $+10$ to $+23$ can be seen.

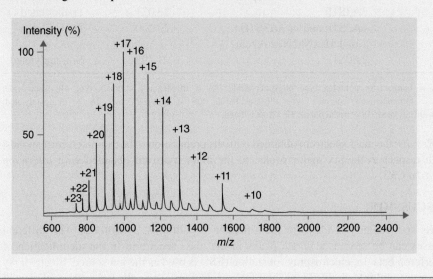

21.5.3 Glycan Analysis

For glycoprotein biopharmaceuticals, identification of sugar moieties is an important part of the identification procedure as different host systems will produce different glycosylation patterns. In Box 21.8, one of the glycan analysis methods used for identification of follitropin by Ph. Eur. is discussed.

Box 21.8   Glycan Analysis of Follitropin by Method A in Ph. Eur.

Follitropin is a recombinant version of FSH.

STRUCTURE

α-subunit

APDVQDCPEC TLQENPFFSQ PGAPILQCMG CCFSRAYPTP

LRSKKTMLVQ KNVTSESTCC VAKSYNRVTV MGGFKVENHT

ACHCSTCYYH KS

β-subunit

NSCELTNITI AIEKEECRFC ISINTTWCAG YCYTRDLVYK

DPARPKIQKT CTFKELVYET VRVPGCAHHA DSLYTYPVAT

QCHCGKCDSD STDCTVRGLG PSYCSFGEMK E

Glycosylation sites:

α-subunit: Asn-52, Asn-78

β-subunit: Asn-7, Asn-24

Disulfide sites:

α-subunit: 7–31, 10–60, 28–82, 32–84, 59–87

β-subunit: 3–51, 17–66, 20–104, 28–82, 32–84, 87–94

M_r approx. 30 000–40 000

PHARMACOPOEIA TEXT – INDIVIDUAL MONOGRAPH FOR FOLLITROPIN

IDENTIFICATION

E. Glycan analysis. Carry out either method A or method B.

METHOD A

PROTEIN DENATURATION

Test solution. Dissolve 500 μg of the substance to be examined in 60 μL of 0.05 M phosphate buffer solution pH 7.5 R. Add 6 μL of a 10 mg/mL solution of sodium dodecyl sulfate (SDS) R and 35 μL of a 1% *v/v* solution of 2-mercaptoethanol R. Mix using a vortex mixer, centrifuge, and incubate at 37 °C for 15 minutes.

Reference solution. Freeze-dry a sample of follitropin for peptide mapping and glycan analysis CRS that contains 500 μg of follitropin. Dissolve in 60 μL of 0.05 M phosphate buffer solution pH 7.5 R and continue as for the test solution.

SELECTIVE RELEASE OF THE GLYCANS

Test solution. To the test solution obtained in the previous step add 0.75 μL of octylphenyl-polyethylene glycol and mix using a vortex mixer. Add 25 mU of peptide N-glycosidase F R, mix using a vortex mixer, and centrifuge. Incubate at 37 °C for 24 hours. Remove the protein fraction using a suitable, validated procedure. The following method has been found to be appropriate. Add 600 μL of anhydrous ethanol R, previously cooled at −20 °C for 45 minutes. Mix using a vortex mixer and centrifuge. Precipitate the proteins at −20 °C for 15 minutes, then centrifuge at 10 600 g at 4 °C for 5 minutes. Transfer the supernatant to a separate tube and evaporate the ethanol for 15 minutes. Add 1 mL of particle-free water R and resume evaporating until the remaining volume is about 500–800 μL; then freeze-dry.

Label the liberated glycans contained in the sample with 2-aminobenzamide. The procedure employs a combination of reagents optimized and validated for the efficient labelling of glycans, and for the subsequent extraction and recovery of the labelled glycans from the reaction. Recover the sample in 1.5 mL of particle-free water R.

Reference solution. Prepare at the same time and in the same manner as for the test solution but using the reference solution obtained in the previous step.

CHROMATOGRAPHIC SEPARATION. Liquid chromatography
Column:
– size: $l = 0.075$ m, $\varnothing = 7.5$ mm;
– stationary phase: weak anion-exchange resin R (10 μm);
– temperature: 30 °C.

Mobile phase:
- mobile phase A: acetonitrile R;
- mobile phase B: 0.5 M ammonium acetate buffer solution pH 4.5 R; filter through a membrane filter (nominal pore size 0.22 µm);
- mobile phase C: particle-free water R;

Time (min)	Mobile phase A (per cent *v/v*)	Mobile phase B (per cent *v/v*)	Mobile phase C (per cent *v/v*)
0–5	20	0	80
5–21	20	0 → 4	80 → 76
21–61	20	4 → 25	76 → 55
61–62	20	25 → 50	55 → 30
62–71	20	50	30
71–72	20	50 → 0	30 → 80
72–117	20	0	80

Flow rate: 0.4 mL/min.
Detection: fluorimeter at 330 nm for excitation and at 420 nm for emission.
Injection: 50 µL.
System suitability: reference solution:
- the chromatogram obtained is qualitatively similar to the chromatogram supplied with follitropin for peptide mapping and glycan analysis CRS;
- by comparison with the chromatogram supplied with follitropin for peptide mapping and glycan analysis CRS, identify the peaks due to neutral, mono-, di-, tri- and tetra-sialylated forms; determine the area of each peak and express it as a percentage of the total; calculate the Z number using the following expression:

$$(A_0 \cdot 0) + (A_1 \cdot 1) + (A_2 \cdot 2) + (A_3 \cdot 3) + (A_4 \cdot 4)$$

A_0 = peak area percentage due to the neutral form; A_1 = peak area percentage due to the mono-sialylated form; A_2 = peak area percentage due to the di-sialylated form; A_3 = peak area percentage due to the tri-sialylated form; A_4 = peak area percentage due to the tetra-sialylated form.

The Z number obtained for the reference solution is in the range 177–233.

Examine the chromatogram obtained with the test solution and calculate the Z number as described above.

Result: Z = 177–233.

DISCUSSION

Step 1: *Protein denaturation.* It is recommended to denature glycoproteins prior to deglycosylation as the cleavage rate is increased by detergent and heat denaturation. The pH is kept around the physiological level (pH = 7.5) for optimum enzyme activity in the subsequent step. SDS is used to denature the protein, and the disulfide bonds in follitropin are reduced by addition of 2-mercaptoethanol; 2-mercaptoethanol (and other sulfhydryl reagents) can be used because it does not interfere with the enzyme activity in step 2. Denaturation and reduction are facilitated by an increased (physiological) temperature (37 °C).

Step 2: *Selective release of the glycans.* SDS (and other ionic detergents) inhibit the enzyme peptide N-glycosidase F (PNGase F), which is used to cleave the N-linked oligosaccharides from the protein backbone. Octophenyl-polyethylene glycol is added to counteract the SDS inhibition. Action of the mechanism is unknown but this effect is common for non-ionic detergents. Octophenyl-polyethylene glycol should be added in an approximately fivefold excess to SDS.

The active pH range for PNGase F activity is 6–10, with an optimum of 8.6. The reaction is commonly performed at 37 °C and for an optimized time range (optimized during method development). During the process the N-linked oligosaccharides are removed intact from the protein backbone, and the asparagine residue to which the saccharide was linked is transformed to aspartate.

Step 3: *Removal of the protein fraction.* According to the pharmacopoeia, the protein fraction can be removed using any suitable, validated method. The method described in the pharmacopoeia uses precipitation with ethanol. Ethanol is added in a more than five times excess to the deglycosylated sample. After precipitation, the sample is centrifuged and the protein-free supernatant is transferred to a new tube. Ethanol is removed by evaporation and the remaining water phase is freeze-dried. Ethanol is a well-known precipitation agent for proteins.

Step 4: *Labelling of glycans.* Prior to analysis, the glycans are fluorescence labelled with 2-aminobenzamide. A detailed procedure is not described in the pharmacopoeia, but it needs to be optimized and validated for the glycans in question. After labelling, the labelled glycans are extracted from the labelling solution.

Step 5: *Chromatographic separation and fluorescence detection.* The chromatographic separation is performed using a weak-anion exchanger and a buffer concentration gradient. The glycans will then elute in groups depending on the number of sialic acids. The detector is operating at an excitation wavelength of 330 nm and emission wavelength of 420, which are optimum wavelengths for 2-aminobenzamide.

Step 6: *Data analysis.* The area under the curve for the identified neutral, mono-, di-, tri, and tetra-sialylated forms are determined and the Z number is calculated. An example demonstrating calculation of the Z number is shown below the chromatogram.

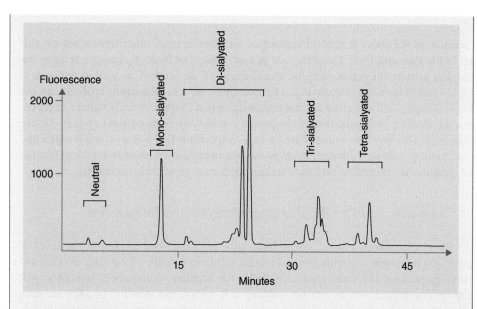

Chromatogram showing a typical glycan pattern of follitropin after deglycosylation and fluorescence labelling.

CALCULATION OF Z NUMBER

First the area under the curve is determined for all sialylated forms. Then the peak area percentage compared to the total peak is calculated. An example of typical peak areas is shown in the table below.

Sialylated form	Peak area	Peak area percentage compared to total
Neutral	943	1%
Mono-sialylated	13 803	15%
Di-sialylated	52 704	58%
Tri-sialylated	13 223	15%
Tetra-sialylated	9453	10%
Total	90 126	100%

The Z number can be calculated using the formula given in the pharmacopoeia:

$$Z = (A_0 \cdot 0) + (A_1 \cdot 1) + (A_2 \cdot 2) + (A_3 \cdot 3) + (A_4 \cdot 4)$$

$$Z = (1 \cdot 0) + (15 \cdot 1) + (58 \cdot 2) + (15 \cdot 3) + (10 \cdot 4) = 0 + 15 + 116 + 45 + 40 = 216$$

The Z number is within 177–233 and the glycan analysis of follitropin complies with Ph. Eur. requirements.

21.5.4 Other Techniques

In addition to the above described techniques for identification, other techniques are also used in Ph. Eur. and USP. These include potency assays or biological assays relating the biological activity (or potency) to the identification of the product, as well as amino acid analysis and N-terminal sequencing. The two latter are less frequently applied than the other techniques and therefore are not discussed in this chapter. A wide variety of procedures are used for determination of the *potency* or *activity* of biopharmaceuticals (active substances). This includes animal-based assays, cell culture-based assays, and receptor and ligand binding assays. Although such assays are important techniques for identification (and quantitative determination), they are outside the scope of the present textbook.

21.6 Impurity Tests for Biopharmaceuticals (Active Substances)

Also for biopharmaceuticals, a set of limit tests are prescribed to control the level of organic and inorganic contaminants and related substances. Some of the tests are identical to tests prescribed for small molecule APIs, for instance *appearance, solubility, pH, bacterial endotoxins, loss on drying, sulfated ash, water,* and *sterility.* In addition, tests similar to *related substances/organic impurities* are performed for biopharmaceuticals either by reversed-phase LC or by other techniques such as size exclusion chromatography (SEC), SDS-PAGE, and capillary zone electrophoresis (CZE). The latter techniques are used to reveal related organic impurities with different molecular mass, charge variants, or typical degradation products of biopharmaceuticals. An overview of the main test procedures for related substances/organic impurities is shown in Table 21.4. In addition, a test for determination of the *total amount of protein* using UV detection is included for many biopharmaceuticals to relate the biological activity to the amount of protein.

In the following, only tests special for biopharmaceuticals will be discussed. Tests also used for small molecule APIs have already been discussed (Chapter 18).

Table 21.4   Overview of tests revealing related organic impurities of biopharmaceuticals

Method for purity testing	Impurities to be controlled
Anion exchange liquid chromatography w/UV detection (SAX-LC-UV)	Deamidation of protein
Size exclusion chromatography (SEC)	Aggregation (dimers/oligomers) Single-chain content Monomer content
Sodium dodecyl sulphate polyacrylamide gel electrophoresis (SDS-PAGE)	Aggregation (dimers/oligomers) Impurities with differing molecular mass Free subunits
Reversed-phase liquid chromatography w/UV detection (RP-LC-UV)	Related organic impurities such as oxidized raw material
Capillary zone electrophoresis (CZE)	Charged variants; typically deamidation
Isoelectric focusing	Impurities with different charges; typically deamidation

21.6.1 Impurities with Differing Molecular Masses

SEC and SDS-PAGE are typically used for evaluation of the presence of impurities with differing molecular mass or to ensure a sufficient level of the monomeric biopharmaceutical. Impurities with differing molecular mass can either be *dimers* or other *oligomers* that are a result of aggregation of the protein, or free subunits or single chains that are a result of detachment of the protein complex. Both aggregation of monomers and detachment of a protein complex may affect the biological effect of the biopharmaceutical. It is therefore important to ensure that levels of these impurities are not too high. This is either done by direct determination of oligomers or free subunits, or by determination of the content of the monomeric protein complex. An example describing the determination of impurities with greater molecular masses for teriparatide is described in Box 21.9. This test is based on SEC.

Box 21.9   Determination of Impurities with Greater Molecular Masses for Teriparatide According to Ph. Eur.

Teriparatide is also called rhPTH1-34, as it is the recombinant version of the first 34 amino acids of recombinant human parathyroid hormone.

STRUCTURE

> H-Ser-Val-Ser-Glu-Ile-Gln-Leu-Met-His-Asn-
> Leu-Gly-Lys-His-Leu-Asn-Ser-Met-Glu-Arg-
> Val-Glu-Trp-Leu-Arg-Lys-Lys-Leu-Gln-Asp-
> Val-His-Asn-Phe-OH

$C_{181}H_{291}N_{55}O_{51}S_2$ M_r 4118

PHARMACOPOEIA TEXT – INDIVIDUAL MONOGRAPH FOR TERIPARATIDE

TESTS

Impurities with molecular masses greater than that of teriparatide. For size-exclusion chromatography, use the normalization procedure. Store the solutions at 2–8 °C and use them within 72 hours.

Test solution. Dissolve the substance to be examined in water R to obtain a concentration of 1 mg/mL.

Reference solution. Dissolve the contents of a vial of teriparatide CRS in water R to obtain a concentration of 1 mg/mL.

Blank solution. Water R.

Resolution solution. Incubate a vial of teriparatide CRS at 75 °C for 16–24 hours. After incubation, dissolve the contents of the vial in water R to obtain a concentration of 1 mg/mL of degraded teriparatide.

Column:

- size: $l = 0.30$ m, $\varnothing = 7.8$ mm;
- stationary phase: hydrophilic silica gel for chromatography R (5–10 μm) with a pore size of 12.5 nm, of a grade suitable for fractionation of globular proteins of relative molecular mass up to 150 000.

Mobile phase: add 1 mL of trifluoroacetic acid R to 750 mL of water R, mix with 250 mL of acetonitrile for chromatography R and degas.

Flow rate: 0.5 mL/min.
Detection: spectrophotometer at 214 nm.
Autosampler: set at 2–8 °C.
Injection: 20 μL.
Run time: 1.5 times the retention time of the teriparatide monomer.
Retention time: teriparatide monomer = about 17 minutes.
System suitability:

- the chromatogram obtained with the reference solution is similar to the chromatogram supplied with teriparatide CRS;
- resolution: minimum 2.0 between the peaks due to teriparatide dimer and monomer in the chromatogram obtained with the resolution solution.

Limit:

- sum of the peaks eluted before the principal peak: maximum 0.3%; disregard any peak with a retention time greater than that of the peak due to the teriparatide monomer.

DISCUSSION

Teriparatide is a small protein (peptide). Dimers and other aggregates will elute prior to the monomeric teriparatide. A chromatogram of teriparatide analysed by size-exclusion chromatography is shown in the figure below. By normalization, the sum area of all peaks (excluding peaks eluting later than the principal peak (= solvent peaks)) are set to 100%, and the percentage of each peak eluting before the principal peak is calculated as shown below.

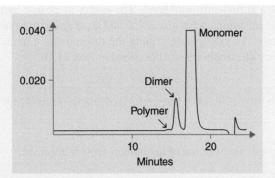

The figure shows the separation order of a teriparatide and its corresponding dimer and polymers by size-exclusion chromatography.

CALCULATIONS

The peak areas of teriparatide, its dimer, and polymer are as follows:

Analyte	Peak area	% of total peak area
Teriparatide (monomer)	287 572	99.82
Dimer	398	0.14
Polymer	127	0.04
Total area	288 097	100

The total % area of peaks eluting before the principal peak will then be 0.14% + 0.04% = 0.18%.

Since the total % area is less than the limit for the test (0.3%), the level of impurities with molecular masses greater than teriparatide complies with Ph. Eur.

21.6.2 Impurities with Different Charges/Charge Variants

Isoelectric focusing, CZE, and liquid chromatography with a strong anion exchange column are typically used for evaluation of impurities with different charges and charge variants. This test is important to keep the level of compounds/impurities with different charges or charge variants at a sufficiently low level. The origin of charge variants is typically *deamidation* of the neutral amino acids glutamine and asparagine to the negatively charged glutamate and aspartate. An example describing the determination of charged variants of somatropin by capillary electrophoresis is described in Box 21.10.

Box 21.10   Determination of Charged Variants of Somatropin According to Ph. Eur.

Somatropin is a recombinant version of the human growth hormone.

STRUCTURE

FPTIPLSRLF	DNAMLRAHRL	HQLAFDTYQE	FEEAYIPKEQ
KYSFLQNPQT	SLCFSESIPT	PSNREETQQK	SNLELLRISL
LLIQSWLEPV	QFLRSVFANS	LVYGASDSNV	YDLLKDLEEG
IQTLMGRLED	GSPRTGQIFK	QTYSKFDTNS	HNDDALLKNY
GLLYCFRKDM	DKVETFLRIV	QCRSVEGSCG	F

$C_{990}H_{1528}N_{262}O_{300}S_7$ M_r 22 125

PHARMACOPOEIA TEXT – INDIVIDUAL MONOGRAPH FOR SOMATROPIN

Charged variants. Capillary electrophoresis

   Test solution (a). Prepare a solution of the substance to be examined containing 1 mg/mL of somatropin.

   Test solution (b). Mix equal volumes of test solution (a) and the reference solution.

   Reference solution. Dissolve the contents of a vial of somatropin CRS in water R and dilute with the same solvent to obtain a concentration of 1 mg/mL.

   Capillary:

- material: uncoated fused silica;
- size: effective length = at least 70 cm, $\varnothing$ = 50 μm.

   Temperature: 30 °C.

   CZE buffer: 13.2 g/L solution of ammonium phosphate R adjusted to pH 6.0 with phosphoric acid R and filtered.

   Detection: spectrophotometer at 200 nm.

Set the autosampler to store the samples at 4 °C during analysis.

Preconditioning of the capillary: rinse with 1 M of sodium hydroxide for 20 minutes, with water R for 10 minutes and with the CZE buffer for 20 minutes.

Between-run rinsing: rinse with 0.1 M sodium hydroxide for two minutes and with the CZE buffer for six minutes.

Note: rinsing times may be adapted according to the length of the capillary and the equipment used.

Injection: test solution (a) and the reference solution, under pressure or vacuum, using the following sequence: sample injection for at least three seconds then CZE buffer injection for one second.

The injection time and pressure may be adapted in order to meet the system suitability criteria.

Migration: apply a field strength of 217 V/cm (20 kV for capillaries of 92 cm in total length) for 80 minutes, using the CZE buffer as the electrolyte in both buffer reservoirs.

Relative migration with reference to somatropin: deamidated forms = 1.02–1.11.

System suitability: reference solution:

– the electropherogram obtained is similar to the electropherogram of somatropin supplied with somatropin CRS; 2 peaks (I1, I2) eluting prior to the principal peak and at least two peaks (I3, I4) eluting after the principal peak are clearly visible.

Note: peak I2 corresponds to the cleaved form and peak I4 corresponds to the deamidated forms, eluting as a doublet.

Limits:

– deamidated forms: maximum 5.0%;
– any other impurity: for each impurity, maximum 2.0%;
– total: maximum 10.0%.

DISCUSSION

Somatropin contains nine asparagines and glutamines that may be deamidated to aspartate and glutamate. The amino acids that are most often deamidated are the asparagines in position 149 and 152. The deamidated forms can be separated from the none-deamidated somatropin by capillary electrophoresis as the pI of deamidated somatropin will be different from the pI of somatropin. At pH 6 (the pH used in the separation buffer) the deamidated amino acids are negatively charged and this will result in longer migration times in the applied separation system. The system is also able to separate a cleaved form (I_2 in the electropherogram) and an isoform (I_3 in the electropherogram: histidine in position 18 is exchanged with glutamate).

A representative electropherogram of somatropin (electropherogram of the reference solution):

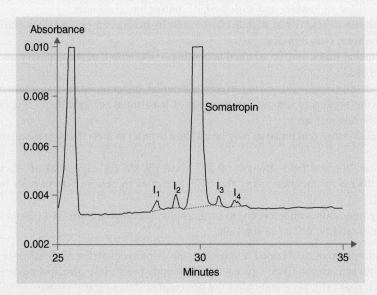

The capillary is rinsed with sodium hydroxide and re-equilibrated with separation buffer between each run. Ensure the removal of the remains of a prior sample and repeatable migration times.

After running the reference solution and the test solution, the areas of the impurities are compared to the area of somatropin and the percentage is calculated. Limits with respect to both the deamidated forms (denoted I_4 in the electropherogram), other impurities, and the total amount of impurities must be met. An example of how to calculate the amount of impurities is shown below. (The example is not based on the electropherogram above).

CALCULATIONS

Peak areas and percentage of the peak area relative to the peak area of somatropin is shown in the table.

Peak	Peak area	Percentage of peak area relative to peak area of somatropin (%)
I_1	124	1.0
I_2	158	1.2
Somatropin	12 753	100
I_3	197	1.5
I_4	236	1.9

Deamidated forms:

Peak I_4 is the deamidated form. The peak area of this peak constitutes 1.9% of the somatropin peak area and is within the limits (less than 5.0%).

Any other impurity:

The peak areas of the other impurities constitute 1.0%, 1.2%, and 1.5% of the somatropin peak area (I_1, I_2 and I_3, respectively), and all are within the limits (less than 2.0% each).

Total:

The total amount of impurities is found by adding the areas of each impurity:

$$\text{Total\%} = \%I_1 + \%I_2 + \%I_3 + \%I_4 = 1.0\% + 1.2\% + 1.5\% + 1.9\% = 5.6\%$$

The total level of impurities is also within the limits (less than 10%).

21.6.3 Total Protein/Protein Content

The *protein content/amount of total protein* is another typical test for biopharmaceuticals prescribed in Ph. Eur. and USP. The total protein content can be determined using several techniques, but mostly UV spectrophotometry is prescribed. In solution, proteins absorb ultraviolet light at a wavelength of 280 nm due to the presence of aromatic amino acids, mainly tyrosine and tryptophan, in the protein structure. If biological activity is part of the assay, this activity is often related to the protein content and the activity is given as units/mg of protein. An example of determination of the protein content as described in Ph. Eur. is discussed in Box 21.11 for erythropoietin.

Box 21.11   Determination of Protein Content of Erythropoietin in Erythropoietin Concentration Solution by UV Spectrophotometry According to Ph. Eur.

STRUCTURE

APPRLICDSR	VLERYLLEAK	EAENITTGCA	EHCSLNENIT
VPDTKVNFYA	WKRMEVGQQA	VEVWQGLALL	SEAVLRGQAL
LVNSSQPWEP	LQLHVDKAVS	GLRSLTTLLR	ALGAQKEAIS
PPDAASAAPL	RTITADTFRK	LFRVYSNFLR	GKLKLYTGEA
CRTGD			

PHARMACOPOEIA TEXT – INDIVIDUAL MONOGRAPH FOR ERYTHROPOIETIN CONCENTRATED SOLUTION

DEFINITION

Erythropoietin concentrated solution is a solution containing a family of closely related glycoproteins that are indistinguishable from the naturally occurring human erythropoietin (urinary erythropoietin) in terms of amino acid sequence (165 amino acids) and average glycosylation pattern, at a concentration of 0.5–10 mg/mL. It may also contain buffer salts and other excipients. It has a potency of not less than 100 000 IU/mg of active substance determined using the conditions described under Assay and in the test for protein.

TESTS

Protein (Method I): 80–120% of the stated concentration.

Test solution. Dilute the preparation to be examined in a 4 g/L solution of ammonium hydrogen carbonate R.

Record the absorbance spectrum between 250 nm and 400 nm. Measure the value at the absorbance maximum (276–280 nm), after correction for any light scattering, measured up to 400 nm. Calculate the concentration of erythropoietin taking the specific absorbance to be 7.43.

DISCUSSION

The protein content is calculated based on the absorbance from aromatic amino acids in the protein. For erythropoietin the concentrated solution is diluted with ammonium hydrogen carbonate. The general monograph for total protein states that the concentration of protein after dilution should be between 0.2 and 2 mg/mL. After dilution the absorbance spectrum is recorded from 250 to 400 nm and the absorbance maximum is determined. The absorbance is then measured at the absorbance maximum.

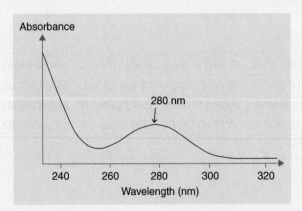

Typical UV spectrum of proteins in solution. The absorbance maximum at 280 nm is due to the aromatic amino acids.

CORRECTION FOR LIGHT SCATTERING

If the protein is not dissolved, the accuracy of the determination can be influenced by scattering of light from particulate matter (undissolved protein). This will result in an apparent increase in the absorbance of the sample. The general monograph for total protein in Ph. Eur. describes how absorbance due to light scattering at 280 nm can be determined. Filtration using a 0.2 μm filter that does not absorb the protein or centrifugation may reduce the effect of light scattering.

CALCULATIONS

1.0 mL of erythropoietin concentrated solution is diluted to 10.0 mL with 4 g/L of ammonium hydrogen carbonate R. The absorbance is measured to $A = 0.732$ after correction for light scattering, using a 1 cm cuvette. The specific absorbance of 7.43 stated in the monograph is used for the calculation. The concentration, c, of protein in the measured solution may be calculated using Beer's law.

The specific absorbance is the absorbance in a 1 cm cuvette containing 1% erythropoietin: $1\% = 1\,g/100\,mL = 10\,mg/ml$

$$A_{1\,cm}^{1\,percent} = b \times 10\,mg/mL \text{ and } A = bc$$

The concentration is isolated by the relative calculation:

$$\frac{A}{A_1^{1\%}cm} = \frac{c}{10\,mg/mL} \Rightarrow c = \frac{A \times 10\,mg/mL}{A_1^{1\%}cm}$$

$$c = \frac{0.732 \times 10\,mg/ml}{7.43} = 0.99\,mg/mL$$

The protein content in the Test solution was calculated to be 0.99 mg/mL. As the preparation was diluted 10 times prior to measuring UV absorbance the concentration of erythropoietin in erythropoietin concentrated solution was

$$10 \times 0.99\,mg/mL = 9.9\,mg/mL$$

The stated amount on the preparation was 10.0 mg/mL. The percentage content of erythropoietin is

$$\frac{9.9\,mg/mL}{10.0\,mg/mL} \times 100\% = 98\%$$

The total amount of protein was within the stated limits for the test (80–120%).

21.6.4 Glycan Analysis

Several biopharmaceuticals are glycoproteins and for these substances, impurity testing may also include an evaluation of the content of sugar groups, either by glycan profiling or by quantitation of specific monosaccharides. This is important as differing sugar contents may affect both the efficacy and the safety of the raw material.

Glycan profiling is typically performed after removal of the glycan groups from the protein backbone and is most easily performed for N-linked glycans. An enzyme, PNGase F, is available to facilitate cleavage of the N-linkage. After cleavage, the glycan groups are typically labelled with a fluorescence marker and analysed by LC with a strong anion exchange column and fluorescence detection. This procedure for deglycosylation and subsequent glycan detection is discussed in Box 21.8.

Alternatively, the content of sugar groups can be based on measurement of *sialic acid*.

The end sugar in glycans is most often sialic acid. The amount of sialic acid can therefore be related to the amount of oligosaccharides connected to the protein. Sialic acid is a collective term used for *N*- or *O*-substituted forms of the monosaccharide neuraminic acid. The amount of sialic acid is normally calculated as the amount of the most common form, *N*-acetylneuraminic acid.

21.7 Assay of Biopharmaceuticals (Active Substance)

The two most widespread approaches for assay of active substance of biopharmaceuticals are LC for quantitation and biological assays for activity/potency. Biological activity can, as mentioned above, be determined using different techniques such as cell proliferation, cell differentiation, or an enzyme assay (several assays may exist for a given protein). For antibodies, cell-based immunoassays and assays based on ligand-binding and affinity may be used. A detailed description and discussion of these tests are outside the scope of this textbook.

Assay by LC is exemplified in Box 21.12 for glucagon.

Box 21.12   Determination of Glucagon (Human) by LC According to USP

STRUCTURE

> HSQGTFTSDY SKYLDSRRAQ DFVQWLMNT

$C_{153}H_{225}N_{43}O_{49}S$	3482.75

ASSAY
PROCEDURE

Solution A	Dissolve 16.3 g of monobasic potassium phosphate in 750 mL of water, adjust with phosphoric acid to a pH of 2.7 (±0.05), add water to 800 mL, add 200 mL of acetonitrile, and degas.
Solution B	Prepare a degassed solution of acetonitrile and water (4 : 6).
Mobile phase	See the table below. (Note that the ratio of Solution A to Solution B can be adjusted to obtain a retention time of about 21 minutes for the main peak.)

Time (min)	Solution A (%)	Solution B (%)
0	61	39
25[a]	61	39
29	12	88
30	12	88
31	61	39
70	61	39

[a]The end time of the isocratic elution can be adjusted so that the gradient begins after the fourth desamido peak elutes (relative retention time about 1.4). The rest of the program is then adjusted accordingly with this offset.

System suitability solution	Reconstitute a vial of USP r-glucagon RS in 0.01 N hydrochloric acid to obtain a solution having a concentration of about 0.5 mg/mL. Let it stand at 50 °C for 48 hours. At least 7% total of all four desamido glucagons should be present in the solution.
Standard solution	Reconstitute a vial of USP r-glucagon RS in 0.01 N hydrochloric acid to obtain a solution having a concentration of about 0.5 mg/mL.
Sample solution	0.5 mg/mL of glucagon in 0.01 N hydrochloric acid

Chromatographic system

Mode	LC
Detector	UV 214 nm
Column	3 mm × 15 cm; 3 μm packing L1 (octadecyl silane)
Column temperature	45 °C
Flow rate	0.5 mL/min
Injection volume	15 μL

System suitability

Samples	System suitability solution and Standard solution

Suitability requirements

Resolution	Four peaks eluting after the glucagon peak that correspond to the desamido glucagons are clearly visible. The resolution between the main peak and the first eluting desamido peak is NLT 1.5, System suitability solution
Tailing factor	NMT 1.8 for the glucagon peak, Standard solution
Relative standard deviation	NMT 2.0%, Standard solution

Analysis

Samples	Standard solution and Sample solution

Calculate the percentage of glucagon ($C_{153}H_{225}N_{43}O_{49}S$) in the portion of glucagon taken:

$$\text{Result} = (r_U/r_S) \times (C_S/C_U) \times 100$$

r_U = peak response from the Sample solution; r_S = peak response from the Standard solution; C_S = concentration of the Standard solution (mg/mL); C_U = concentration of the Sample solution (mg/mL).

Acceptance criteria	90–105% on the anhydrous basis

DISCUSSION

The determination of glucagon is performed using a reversed phase LC system. The principle is the same as in assay of small molecule APIs. The content of glucagon is calculated based on the peak response of the sample solution and the peak response of the standard solution. When calculating the per cent amount of glucagon (related to the stated amount) the result should be corrected for the water content determined under TESTS.

21.8 Monoclonal Antibodies

For therapeutic monoclonal antibodies, no individual pharmacopoeia monographs exist. These biopharmaceuticals are regulated by general monographs in Ph. Eur. (*rDNA technology, products of* regulates all products produced by rDNA technology and *Monoclonal antibodies for human use* regulates the monoclonal antibodies), and USP (*Analytical pro-*

cedures for recombinant therapeutic monoclonal antibodies). Ph. Eur. provides a set of general tests for both production, identification, impurities, and assay while USP provides a set of general techniques that in combination will reveal the most common impurities observed for monoclonal antibodies. A brief description of therapeutic monoclonal antibodies and the analytical procedures described in the general monograph in USP can be found in Box 21.13. As can be seen from Box 21.13, the tests applied for monoclonal antibodies are similar to the tests described previously in this chapter.

Box 21.13   Monoclonal Antibodies for Therapeutic Use

Most recombinant monoclonal antibodies for therapeutic use are IgG-type antibodies. They may be of different subclasses, showing differences in amino acid sequence of the constant region and having different numbers of disulfide bonds. The molecular mass of IgG-type antibodies is approximately 150 kDa. They consist of two identical heavy and two identical light polypeptide chains (approximately 50 and 25 kDa each, respectively), which are connected by disulfide bonds. A schematic drawing of an IgG-type antibody is shown below.

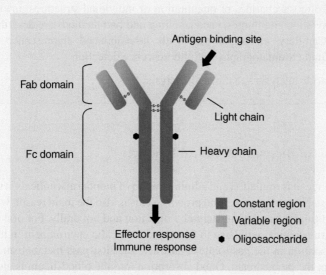

Schematic drawing of an IgG-type antibody

In contrast to naturally occurring polyclonal antibodies in blood, therapeutic monoclonal antibodies have specificity for a target and are derived from a single clone of cells. Monoclonal antibodies are glycoproteins with a glycosylation site on each of the chains in the Fc domain. In addition, some antibodies have glycosylation sites in the Fab domain.

The specificity of an antibody is dependent on its antigen binding site. This is a variable part of the antibody on the end of both Fab domains.

USP has a general chapter on analytical procedures for recombinant therapeutic proteins. The methods described in this chapter apply to monoclonal antibodies for therapeutic and prophylactic use and for use as *in vivo* diagnostics.

The monograph describes validated tests for purity assessment and oligosaccharide analysis. It also describes other methods or procedures that can be used if they provide equivalent or better results after validation.

The following tests are described for purity assessment:

– Size-exclusion chromatography – for measuring monomer and high-molecular-weight species
– Capillary SDS electrophoresis (reduced and non-reduced) – for quantitation of low-molecular-weight species such as non-glycosylated molecules, half antibodies, and fragments, and as a stability-indicating assay. Denaturation with SDS allows for analysis of the complete antibody under non-reducing conditions, and the analysis of light and heavy chains under reducing conditions
– Oligosaccharide analysis – analysis of *N*-linked oligosaccharides of monoclonal antibodies

Profiling of oligosaccharides or quantitation of individual structures. involves most often deglycosylation and fluorescence labelling and two methods are described:
Method A: Capillary electrophoresis with laser-induced fluorescence detection
Method B: Liquid chromatography with fluorescence detection

– Oligosaccharide analysis – sialic acid analysis

21.9 Analysis of Biopharmaceutical Products

The main challenge in formulation and administration of biopharmaceuticals is the physical and chemical instability of peptides and proteins. This is also the main reason why biopharmaceuticals traditionally are administered by injection and not orally. For oral administration, the biopharmaceutical will have to withstand the acidic environment in the stomach, enzymatic degradation in the gastrointestinal tract, and first-pass metabolism in the liver prior to arrival in the bloodstream. Good absorption into the bloodstream is also desirable and for proteins, the adsorption through biological membranes normally is poor.

Therefore, most biopharmaceuticals are still administered as injections and the formulations are either liquid, dried (lyophilized), or suspension formulations. From a manufacturing and user perspective liquid formulations are most convenient, while dried or suspension formulations are introduced to improve the stability of the product. Formulations intended for injection are often less complex than formulations intended for oral (tablets) or topical (creams and ointments) administration. Due to this, dilution of the formulation followed by chromatographic separation and detection is often sufficient for both identification and quantitation of the active substance in a biopharmaceutical formulation. For some products, peptide mapping is used for identification of the active ingredient. For these products, an initial isolation and/or purification step may be necessary prior to cleaving the protein into peptides. This step is included to remove excipients and stabilizers in the formulation that will interfere with the mapping procedure.

21.10   Bioanalysis of Biopharmaceuticals Using LC-MS/MS

Biopharmaceuticals are monitored in biological matrices for the same reasons as small molecule drugs, during drug development, in clinical studies, and for follow-up of the effect. Traditionally, biological assays using immunological techniques have been the gold standard for determination of peptides and proteins in biological matrices. Lately LC-MS/MS has become more and more common due to the increased sensitivity of the MS systems. By applying LC-MS/MS some of the drawbacks of immunological techniques can be addressed. This results in methods with improved selectivity, reduced interferences, and easier multiplexing. The improvement of selectivity is achieved as MS determination is based on differences in the mass-to-charge (m/z) ratio. This makes differentiation between parent drug and metabolites and between endogenous peptides/proteins and the drug–peptide or protein analogue easier when these consist of a different number and/or kind of amino acids as they will produce a different m/z ratio. The immunological response may, on the other hand, be unaltered.

One complicating factor in bioanalysis of biopharmaceuticals is the wide range of endogenous proteins present in blood. Although these peptides and proteins may have very different physicochemical properties, they are all built from the same set of amino acids and it may be difficult to isolate the desired biopharmaceutical from endogenous proteins and peptides. This is especially difficult if the biopharmaceutical also is identical in amino acid composition to an endogenous peptide or protein. This means that a drug-free matrix used for preparation of quantitation standards may consist of the endogenous counterpart of the substance. If this is the case, either a standard addition approach must be used for quantitation, the standards must be prepared in a surrogate matrix prepared to mimic the original biological matrix, or a surrogate analyte must be used for quantitation.

Another complicating factor for bioanalysis of proteins by LC-MS/MS is the level of proteins in blood, which varies from low abundant proteins leaking from tissues (pg-ng/mL level) to the high abundant serum proteins such as albumin (mg/mL level). This large dynamic range makes determination of low abundant biopharmaceuticals by mass spectrometry difficult without isolation from the high abundant species.

Which sample preparation approach and analytical procedure to choose for quantitation of biopharmaceuticals in biological matrices by LC-MS/MS depends on both the size of the biopharmaceutical and the level of the biopharmaceutical in the biological matrix. ESI is the most commonly applied ionization method in LC-MS/MS analysis of peptides and proteins. ESI produces multiple-charged ions and as the size of the biopharmaceutical increases the ion current is spread on a larger number of charge states and the analytical sensitivity is reduced. Smaller biopharmaceutical peptides such as glucagon can therefore be analysed directly with sufficient sensitivity, while for larger biopharmaceuticals a proteolytic step is normally included. In this step the protein is digested into peptides and quantitation is performed using one or more peptides specific for the protein. The peptides will have less charge states and better chromatographic separation than the protein it was derived from. Triple quadrupole mass spectrometers operated in selected reaction monitoring (SRM) mode are most frequently used for protein analysis in bioanalysis due to the high sensitivity of this instrument.

The extent of sample preparation is, in addition to the size of the analyte and its level in the biological matrix, dependent on the biological matrix to be analysed. It ranges from protein precipitation and removal of high abundant proteins through solid phase extraction to very selective extraction based on immunoaffinity (see Box 21.14).

Box 21.14   Brief Overview of Common Sample Preparation Techniques in Bioanalysis of Biopharmaceuticals

PROTEIN PRECIPITATION:

Protein precipitation is most commonly applied when the analyte is a peptide or small protein that can be kept in solution when the precipitation agent is added. Since the sample still is rather complex after precipitation, it can be used when the biopharmaceutical of interest is present in moderate to high amounts in serum or plasma.

DEPLETION OF HIGH ABUNDANT PROTEINS:

The depletion is performed by immunoaffinity removal of a set of high abundant proteins. Kits for removal of from one protein (i.e. albumin) to removal of the top 12 abundant proteins present in serum and plasma are available. By using a top 12 removal kit more than 95% of the 12 proteins that constitute of approximately 95% of all plasma proteins can be removed. This reduces the dynamic range of the sample and enables quantitation of proteins and peptides present in moderate to high amounts in serum or plasma.

SOLID-PHASE EXTRACTION (REVERSED PHASE OR MIXED MODE):

This is most commonly applied to peptides, either intact peptide drugs or to desalt and enrich proteolytic peptides after digestion of protein drug. The principle of the extraction is similar to that for small molecule drugs. The sample clean-up is more efficient than for the two above described techniques and lower detection limits are expected.

AFFINITY-BASED SAMPLE PREPARATION:

Antibodies targeting peptides and proteins can be used for selective isolation of the target drug or proteolytic peptide of the target drug. By extracting the biopharmaceutical using a selective antibody the complexity of the sample is greatly reduced. Affinity-based sample preparation enables determination of the biopharmaceutical present at low levels in biological matrices.

Index

Introduction to Pharmaceutical Analytical Chemistry, Second Edition.
Stig Pedersen-Bjergaard, Bente Gammelgaard and Trine Grønhaug Halvorsen.
© 2019 John Wiley & Sons Ltd. Published 2019 by John Wiley & Sons Ltd.

Printed and bound by CPI Group (UK) Ltd, Croydon, CR0 4YY

27/10/2024

14580303-0004